Lecture Notes in Computer Science 1166

Edited by G. Goos, J. Hartmanis and J. van Leeuwen

Advisory Board: W. Brauer   D. Gries   J. Stoer

Springer
Berlin
Heidelberg
New York
Barcelona
Budapest
Hong Kong
London
Milan
Paris
Santa Clara
Singapore
Tokyo

Mandayam Srivas   Albert Camilleri   (Eds.)

Formal Methods in Computer-Aided Design

First International Conference, FMCAD '96
Palo Alto, CA, USA, November 6-8, 1996
Proceedings

 Springer

Series Editors

Gerhard Goos, Karlsruhe University, Germany

Juris Hartmanis, Cornell University, NY, USA

Jan van Leeuwen, Utrecht University, The Netherlands

Volume Editors

Mandayam Srivas
SRI International (EL-262)
333 Ravenswood Avenue, Menlo Park, CA 94025, USA
E-mail: srivas@csl.sri.com

Albert Camilleri
Hewlett-Packard Company M/S 5596
8000 Foothills Boulevard, Roseville, CA 95747-5596, USA
E-mail: ac@hprpcd.rose.hp.com

Cataloging-in-Publication data applied for

Die Deutsche Bibliothek - CIP-Einheitsaufnahme

Formal methods in computer aided design : first international
conference ; proceedings / FMCAD '96, Palo Alto, CA,
November 6 - 8, 1996. Mandayam Srivas ; Albert Camilleri
(ed.). - Berlin ; Heidelberg ; New York ; Barcelona ; Budapest ;
Hong Kong ; London ; Milan ; Paris ; Santa Clara ; Singapore ;
Tokyo : Springer, 1996
  (Lecture notes in computer science ; Vol. 1166)
  ISBN 3-540-61937-2
NE: Srivas, Mandayam [Hrsg.]; FMCAD <1, 1996, Palo Alto, Calif.>;
  GT

CR Subject Classification (1991): B.1.2, B.1.4, B.2.2-3, B.6.2-3, B.7.2, B.7.3,
F.3.1, F.4.1, I.2.3, D.2.4, J.6

ISSN 0302-9743
ISBN 3-540-61937-2 Springer-Verlag Berlin Heidelberg New York

© Springer-Verlag Berlin Heidelberg 1996
Printed in Germany

Typesetting: Camera-ready by author
SPIN 10549200   06/3142 – 5 4 3 2 1 0   Printed on acid-free paper

Preface

This volume contains the proceedings of the International Conference on *Formal Methods in Computer-Aided Design (FMCAD'96)*, organized November 6–8, 1996, in Palo Alto, CA. This conference was a sequel to a series of IFIP WG 10.2/10.5 sponsored conferences with similar themes most recently held in 1992 and 1994 under the banner *Theorem Provers In Circuit Design (TPCD)*. The objective of this conference was to cover all relevant formal aspects of work in computer-aided system design including verification, synthesis, and testing. The title of the conference was changed from TPCD to reflect the broadening of the scope of TPCD to cover all approaches to formal methods including model checking and BDDs in addition to theorem proving. A special focus of this conference will be on the integration of complementary techniques and tools. The conference covered original research in the area, as well as case studies, technology transfer, and other practical experiments. The call for papers invited submissions of papers containing original research contributions. Of the 65 submissions received 25 were selected for presentation at the conference.

The conference included three invited lectures. The invited lectures were given by David Dill (Stanford University, USA) on *Validity Checking for Combinations of Theories with Equality*, by Kurt Keutzer (Synopsys, Inc., USA) on *The Need for Formal Methods for Integrated Circuit Design*, and J Strother Moore (Computational Logic, Inc., USA) on *ACL2 Theorems About Commercial Microprocessors*. The call for papers also invited submissions for tutorials on tools. Of the 6 tutorial proposals, 4 were chosen. The tutorials were about the following tools: VIS (Verification Interacting with Synthesis), PVS (Prototype Verification System), DRS (Digital Design Derivation), and HOL Light.

The program of FMCAD'96 was selected by a program committee consisting of D. Borrione (TIMA, France), R. Brayton (University of California, Berkeley, USA), R. Bryant (CMU, USA), A. Camilleri (Tutorials Chair, Hewlett Packard Company, USA), R. Camposano (Synopsys, Inc., USA), L. Claesen (IMEC, Belgium), E. Clarke (CMU, USA), C. Delgado Kloos (Universidad Politecnica de Madrid, Spain), M. Fujita (Fujitsu Labs, USA), S. German (IBM, Yorktown Heights, USA), M. Gordon (University of Cambridge, UK), O. Grumberg (Technion, Haifa, Israel), W. Hunt (Computational Logic, Inc., USA) S. Johnson (University of Indiana, USA), R. Kumar (FZI, Karlsruhe, Germany), M. Leeser (Northeastern University, USA), P. Loewenstein (Sun Microsystems, USA), K. McMillan (Cadence Berkeley Lab, USA), V. Nagasamy (VSIS Inc., USA), C. Seger (Intel, Oregon, USA), M.K. Srivas (Program Chair, SRI International, USA), J. Staunstrup (Technical University, Denmark), V. Stavridou (Queen Mary and Westfield College, UK), P.A. Subrahmanyam (AT&T, USA), J. Van Tassel (Texas Instruments, USA),

The following researchers helped in the evaluation of the submissions, and we are grateful for their efforts: A. Aisles, H. R. Andersen, S. Ben-David, S. Bensalem, A. Bouhoula, P.T. Breuer, C.-T. Chou, F. Corella, D. Deharbe, B. Dutertre, C. Eisner, A. Emerson, L.S. Fernandez, Y. Gil, R. Hojati, A.J. Hu,

H. Hulgaard, A. Jain, J. Jain, S. Jha, P. Kellomaki, D. Kindred, S. Kimura, F. Krohm, A. Kuehlmann, Y. Kukimoto, A.M. Lopez, N.M. Madrid, W. Marrero, A. Mets, H. Miller, C. Moon, S. Panda, L. Pierre, S. Rajan, R. Ranjan, R. Sharp, T. Shiple, J. Sifakis, V. Singhal, S. Tasiran, M. Velev, A. Wahba, and E. Wolf.

We gratefully acknowledge the help of Judith Burgess of SRI International, who is responsible for publicity, registration, and site arrangements. FMCAD'96 has received financial support from Cadence Berkeley Labs, Hewlett-Packard Company, LSI Logic Corporation, Rockwell International, SRI International, Synopsys, Inc., and Texas Instruments. We thank all sponsors for their generosity.

September 1996   Mandayam Srivas
                                                        Albert Camilleri

Table of Contents

The Need for Formal Methods for Integrated Circuit Design

Kurt Keutzer

Synopsys, Inc.
Mountain View, CA

1 Introduction

The world is increasingly dependent on electronic systems. Electronic systems have requirements on the correctness of their functioning, their speed and their reliability. Of these three, functional correctness is the most fundamental requirement because the speed and reliability of an incorrectly functioning electronic system is of no interest. Moreover, of these three requirements functional correctness is becoming the largest bottleneck in design. To function correctly circuits must be designed, implemented and manufactured correctly.

This paper aims to elucidate the various problems involved in verifying the correctness of an integrated circuit design, its implementation and manufacture and to identify areas where formal methods may be applicable.

A wide variety of approaches may be used to design integrated circuits but the majority of integrated-circuit designs are partially or entirely designed using logic synthesis. For this reason the design methodology assumed here is automated synthesis from either a behavioral or register-transfer level (rtl) model described in a hardware-design language (hdl). While this paper is not intended as a tutorial on IC design using synthesis it does attempt to supply enough details to clarify the context.

2 Taxonomy

Many different kinds of verification methods are typically used to ensure the correctness of an integrated circuit, and to understand where formal methods of verification are appropriate it will be necessary to first understand the verification problems that must be solved, whatever the method.

Given a specification of the functionality of an integrated circuit, *design verification* is the verification that the design of an IC is consistent with the specification of the IC. By the *design* of an IC we mean the description which embodies the designers intent. For example in an rtl-synthesis methodology, the design is described as an hdl model of the circuit. This hdl model, typically coded in Verilog or VHDL, can be both simulated and synthesized.

For an integrated circuit, *implementation verification* is the process of verifying that the physical layout of the IC is consistent with the initial design entry. For example, in an **rtl**-synthesis methodology this is the verification that the steps of logic synthesis, logic optimization, test synthesis and physical design have produced a physical layout that is consistent with the initial **hdl** model of the circuit.

After the production of the physical layout and masks *manufacture verification* is the verification that the manufacturing process has not introduced defects into the production of the actual chip. This is most commonly referred to as *manufacture test* or simply *test*. The quality of manufacture test is dependent on the modeling of physical defects and correlating these models with defect statistics and this topic will not be dealt with further in this paper.

The remainder of this paper will present how the tasks of design and implementation verification are currently handled and will further identify what problem areas are amenable to formal methods. First we will begin with a detailed description of the design methodology employed in a synthesis environment.

3   The Synthesis Design Methodology

In providing computer-aided design tools in general, and verification tools in particular, it is very important to understand the design methodology in which they will be used. Many potentially useful tools fail to find application because they do not integrate into any common design procedure. In this section we will go into detail into the synthesis design methodology and assume that synthesis is being used to produce a circuit that will be manufactured in a semi-custom technology [28] such as gate-arrays or standard-cells. Originally *synthesis methodology* was synonymous with *rtl-synthesis methodology* and an in-depth discussion of this methodology is given in [14]. Over the last few years the level of design entry for synthesis has risen from **rtl** to behavioral and even to the algorithmic level. The majority of designers using synthesis are currently using **rtl** synthesis with some beginning to use behavioral synthesis. For this reason this paper will primarily focus on the problems in an **rtl** synthesis methodology with some discussion of the problems in a behavioral synthesis methodology. In this section we will give a simple overview of each of these methodologies. Some of the material in this section is drawn from [21].

3.1   Developing a Specification

In any design methodology, integrated circuit design typically begins with the development of some sort of specification. If a decision has been made to implement an industry standard, such as the HDLC protocol, then the

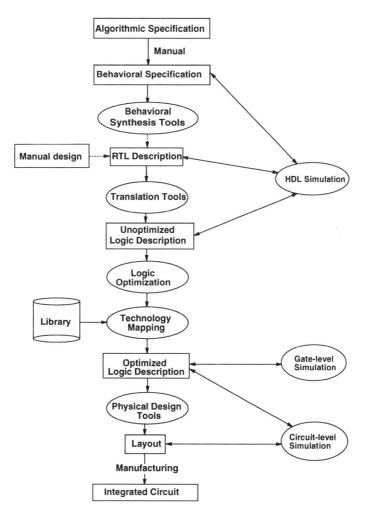

Fig. 1. A typical synthesis design flow

specification may be largely determined by the standards document. If an integrated circuit with a similar functionality already exists then the data available on that circuit, including the chip itself, may form the specification. Simulation models may already be available for the desired circuit. For behavioral or **rtl** specifications the actual design typically begins with a partitioning of the functionality into individual functional units. Control is separated, insofar as possible, from data and execution units, such as adders and multipliers. Then **rtl** models of the individual control and data units are evolved in an **hdl** such as VHDL. In **rtl** models the detailed transfers between registers are described in the **hdl** and the registers are either declared using the `hdl` syntax or are inferred. Simulation drivers, to provide input stimulus, and simulated monitors, to analyze the results, are constructed. As the design evolves it is verified through simulation within this environment. An actual implementation is synthesized to ensure that timing and area constraints are met, and finally to produce the actual circuit. We now proceed to individually describe the different levels of the design process.

3.2   Algorithmic Level

The highest level of abstraction in the design process is called the algorithmic level. At this level the designer usually describes the system using high level constructs like mathematical equations. The following simple example illustrates this approach. Assume the designer wants to design a circuit that calculates the product of two complex numbers. The following equations represent the specification of such a system at the algorithmic level.

$$RE + IMj = (A + Bj) * (C + Dj) \tag{1}$$

The real and imaginary part of the product can be calculated in the following way:

$$RE = (A * C) - (B * D) \tag{2}$$
$$IM = (A * D) + (B * C) \tag{3}$$

At this level, no other details are specified, such as how these equations are calculated and under which timing constraints. Based on this specification the designer wants to find the best design in terms of speed, area and power of the circuit. Most of the current CAD support at this level is aimed at simulation and modeling [8], but there is some initial work on providing synthesis from algorithmic-level descriptions down to the behavioral and register-transfer levels [30, 34].

3.3   Behavioral Level

Modeling a circuit's functionality at the *behavioral level* means modeling the functionality correctly, but without regard to exact clock-cycle by clock-cycle behavior. A behavioral model does not specify what hardware-units are available and how they might be used to calculate the desired output. But it does provide more information than the algorithmic level. In our example, timing constraints are added to the specification. For this example, it is also specified that the input data are available on a single input port on successive cycles. RE and IM are required on separate output ports on the same clock cycle.

Although there is not much difference between the algorithmic level and the behavioral level for this example, the differences become more explicit for more complicated examples. A digital filter is a good example. At the algorithmic level the filter is described using characteristic properties like its bandwidth. At the behavioral level, the exact operations are described that are necessary to have a filter that has this specific bandwidth.

The functional behavior can be represented in a dataflow graph as shown in Figure 2. The arcs in the graph represent the flow of data, while the vertices represent the operations performed on the data. The type of operation is indicated on the vertex. The arcs in this graph correspond to data-dependencies between operations. They indicate which operations must be executed, before another operation can be performed The top-most vertex corresponds to a multiplication that can only be performed when the data A and C are present. The vertex under that one corresponds to a multiplication that can only be performed when B and D are present. The subtraction can only be performed when the two previous multiplications are done. The boundary conditions indicating the required input/output behavior of the circuit under design are indicated in the table of Figure 2. Note that although the clock-cycle behavior of inputs and outputs is specified, it is not specified how many clock-cycles the computation itself may take. It is also not indicated which hardware components are going to be used to execute the calculations. This is the reason why this corresponds to a behavioral description and it is through eliminating these details that behavioral descriptions are able to raise the level of abstraction.

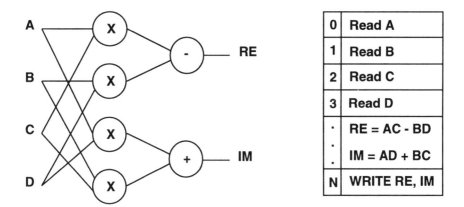

Fig. 2. Modeling a circuit's functionality at the behavioral level.

Before the behavioral model can be implemented as an IC, the behavioral model must be evolved into a *register-transfer level* model representing the *microarchitecture* of the circuit. Behavioral synthesis takes the `hdl` model of a circuit and *schedules* the cycle-to-cycle register-transfer behavior and *allocates* the functional units in the microarchitecture. Overviews of the different approaches to behavioral synthesis are given in [9] and in [13].

3.4 Register-Transfer Level

If a behavioral synthesis methodology is used then the **rtl** model is automatically produced by behavioral synthesis. If only **rtl** synthesis is used then the designer manually enters an **rtl** that reflects his understanding of the specification. A register-transfer level model is a model in which the operations of a sequential circuit are described as synchronous transfers between functional units such as arithmetic-logic units and register files. These transfer functions are under the control of an independent controller and are synchronized to a clock.

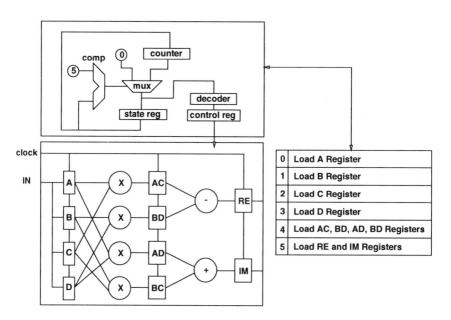

Fig. 3. The rtl-model of the behavioral description of Figure 2

One of the many possible **rtl**-models that can be derived from the behavioral description is shown in Figure 3. The boxes represent registers and the circles represent arithmetic operators. Four multipliers, an adder, a subtractor and ten registers are provided to perform the calculations. This is a refinement of the behavioral level, where it is not indicated which functional unit is used to perform the operations, but just which calculations have to be performed.

After a schedule is determined the actual order in which the different operations are performed on the different hardware units is typically controlled by a finite state machine (FSM). Its specification can be derived from the table represented in Figure 3. One way this FSM can be implemented is shown in the top part of Figure 3. By making use of an incrementer, we move from one state to another. The state register saves the state information. Once state 5 is reached, the machine is reset to state 0. A comparator is used for that purpose. A decoder makes sure that based on the state information the control signals are sent to each register, putting each register in write or read mode so that the correct operations are performed.

3.5   Logic Synthesis

The first step in logic synthesis consists of translating the rtl description of a circuit described in a hdl such as VHDL into combinational logic and registers. A state assignment is an assignment of a Boolean encoding to symbolic states in a finite-state machine. If a state assignment is already present in the description then such a translation is easy to understand. The architecture of the circuit, including the register elements, is considered to be fixed in the rtl model. To synthesize the register elements of the rtl model the registers declared in the model are simply instantiated. The combinational logic implementing the transfers between registers is then synthesized from the transfer statements in the rtl description. The simplicity of this translation process relies heavily on the fact that logic optimization will follow.

3.6   Sequential Optimization

Finite-state machines are easily described in a register-transfer style and the optimization of finite-state machines often requires the assignment of a Boolean encoding to symbolic state variables. Even when a Boolean encoding to the states has been made by the designer in the rtl model, another encoding may result in a circuit with less area. Once an encoding is arrived at, it then remains to optimize the combinational logic implementing the state-transition function, as well as the logic associated with the rest of the register transfers. This is accomplished using the techniques in the next section.

3.7   Boolean Optimization

Speed, small area, low power-dissipation and high testability are all desirable properties of an integrated circuit and hard constraints on one or more of these properties are likely to be part of the specification of an integrated circuit. The burden of realizing these constraints in an rtl synthesis system rests largely on the Boolean or logic optimization procedures. As a result, a large battery of optimization techniques for area and performance optimization have been developed.

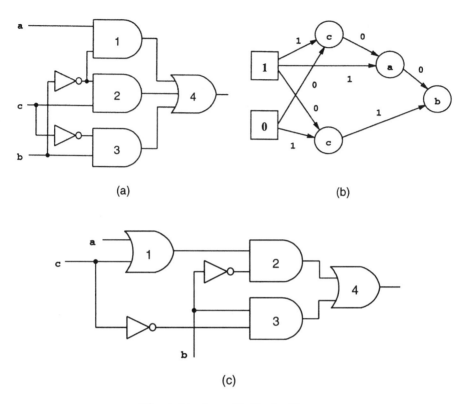

(a)

(b)

(c)

Fig. 4. Boolean Optimizations

With the aim of improving the area of a circuit, the Boolean expression $a \cdot \bar{b} + \bar{b} \cdot c + b \cdot \bar{c}$ may be re-expressed as $(a+c)\bar{b} + b \cdot \bar{c}$. These transformations are shown as they would be reflected on a circuit in the progression from Figure 4(a) to Figure 4(c). On the other hand, if the delay along the path from c to gate 1 to gate 2 to gate 4 is unacceptable in Figure 4(c) then the reverse transformation might be applied to move signal c forward in order to avoid an additional gate delay. This brings us back to the circuit in Figure 4(a). Finally, if (delay-fault) testability was the primary design requirement then Figure 4(c) would be the preferred circuit as it is hazard-free robustly-path-delay-fault testable [14].

3.8 Technology Dependent Optimizations

The Boolean optimizations described in Section 3.7 are all performed with only a general model of the area and delay. Technology mapping is the problem of implementing the combinational network and registers using exactly the cells available in the technology library of a particular vendor.

The goal is to make optimal use of all of the gates in the library to produce a circuit with critical-path delay less than a target value and minimum area. Aside from the fact that all technology dependent transformations are performed on the circuit netlist, the transformations are, from a logical standpoint, no different than those of the previous section. The result of technology dependent optimization is a netlist (graph) of cells in the library of a particular manufacturer. This netlist can then be turned over to the physical design tools for the physical design and manufacture.

3.9   Layout/Analog Level

The final level before the integrated circuit can be manufactured is the layout level. It is the most detailed level and getting a layout with minimized area and optimized timing is one of the most time-consuming steps in the design of integrated circuits. Therefore different design approaches, like gate-array, standard-cell or even general-cell approaches, have been adopted to shorten the layout time. It is the designer's responsibility to choose between different styles, to optimize the layout area, within a certain limit on the layout time.

In general the layout problem can be divided into two subproblems: placement of the different cells (or blocks) and routing of their interconnection. Placement of the different cells is very important, because the best routing may not be possible with an improper placement. If cells with interconnections between them are placed far apart on a chip, the interconnections become long. This means that they occupy more area and that they introduce more delay. Routing also needs to be automated, because connections are excessively numerous in digital systems, making routing by hand practically impossible.

The netlist of cells produced by technology mapping is placed in a two-dimensional grid. Computer-aided design tools to accomplish this have been available for some time, long enough in fact for a nice survey of the techniques to have been written on the topic [29].

3.10   Manufacture

From the two-dimensional grid resulting from physical design a series of *masks* to guide the three-dimensional processing of the integrated circuit is performed. This procedure and the economic forces behind the choice of a fabrication process are nicely reviewed by Muroga in Chapter 2 of [28].

In the following sections we indicate how the various verification tasks are performed in a synthesis design-environment.

4 Implementation Verification in a Synthesis Design Flow

Having outlined a synthesis design flow in the previous section, we now proceed to discuss the need for formal methods in this flow. We will not be in any way attempting a state-of-the-art survey of the application of formal verification research to integrated-circuit design but we will simply try to present our own assessment of the problems. We will begin with a discussion of the use of *implementation verification* in the stages between hdl entry and final layout. In Section 5 we consider the application of techniques of *design verification* to the original hdl model and the need for formal methods in developing the initial specification of the integrated circuit.

In a synthesis design flow, as illustrated in Figure 1, the steps from hdl entry to layout are highly automated and the tools employed along the way are considered to be highly reliable. Nevertheless, formal verification techniques have primarily been applied to the problem of implementation verification. There are a couple reasons for this. The first is that, although the synthesis design flow is highly automated, manual intervention can, and inevitably does, occur at many stages in the process. A second reason is simply that implementation verification has turned out to be more tractable than design verification. Each of the steps of implementation verification in a logic synthesis methodology is detailed below.

4.1 Verifying Boolean Optimization and Technology Dependent Optimization

Because Boolean equivalence checking is the workhorse for many other techniques in implementation verification, we will consider the problem of Boolean equivlance checking first. In Boolean optimization a wide variety of optimization techniques are employed and on a single combinational sub-circuit tens of thousands of individual Boolean transformations may be applied. Boolean equivalence checking can be applied to address the problem of verifying the correctness of these Boolean transformations.

Currently the most effective solution to the problem of verifying these transformations is to represent the pre-optimized circuit and the post-optimized circuit in a canonical form and compare the canonical forms of the two. A number of different canonical forms have been suggested for representing Boolean functions. A truth-table was probably the first such representation. Another early one was the representation of a function by its complete set of prime implicants. This form is also known as Blake's canonical form [4]. A very interesting canonical form is the binary decision diagram (bdd). While the bdd was known to Akers as early as 1978 [2], it was Bryant in 1986 [6] who showed that reduced-ordered bdd's are a canonical form for Boolean functions. Shannon [31] used an information theoretic argument to show that almost all Boolean functions require exponential size when represented graphically. So, in general, it is difficult to predict whether a canonical representation of a Boolean function will typically yield a manageable (polynomial) size. Moreover, in [6] and in [7] Bryant gives an example of one Boolean function, namely multiplication, that has an exponential representation as a **bdd**. Nevertheless, a number of later researchers showed that in practice **bdd**'s have polynomial representations for a very wide class of practical circuits [24, 17, 25, 5]. This research has turned out to be extremely useful. At present **bdd** based verification procedures are proving to be a very efficient mechanism for verifying Boolean optimization. For example to verify our optimization example given in Figure 4 note that the **bdd** corresponding to the circuit Figure 4(a) and the **bdd** corresponding to the circuit Figure 4(c) are equivalent as shown in Figure 4(b).

Boolean equivalence checking can be generalized to attack the more common problem of verifying the equivalence of two sequential circuits simply by applying the restriction that the registers of the two circuits must be the same. Research on techniques for showing the equivalence of combinational circuits has reached sufficient maturity that there are now a number of commercial programs which address this problem. These include: Compass' *VFormal*, Chrysalys' *Design Verifier*, Abstract Hardware Limited's *Check-off*, and Synopsys' *Formality*. These tools are just beginning to gain commercial acceptance.

4.2 Physical Design

If we have the ability to verify the equivalence of two sequential circuits at the gate level then we can easily extend this to verify the actual physical layout of the circuit. In the first step a transistor-level representation of the circuit is extracted from the final layout (see for example [33]). Combinational and sequential portions are then regenerated from the transistor level description using symbolic simulation. Because the register placement does not change after layout, the extracted circuit may be checked for equivalence against the pre-layout version of the circuit using the sequential circuit verification technique described in Section 4.1.

4.3 Sequential Optimization

In addition to Boolean optimizations, sequential optimizations may also be applied to a circuit. For example, we may wish to verify that after performing a sequential optimization on a circuit, such as re-encoding the state-encoding of a finite-state machine or retiming the registers of a sequential circuit, that the resulting optimized sequential machine is equivalent to the original one. More generally if we view the sequential machine as a finite-state machine then it is natural to cast the problem of showing that two sequential circuits are equivalent as the problem of finite-state machine equivalence. If we regard these **fsm**'s as language acceptors then the problem of verifying the equivalence of sequential machines A_1 and A_2 as the problem of checking for the equivalence of their associated languages, *i.e.* $L(A_1) \equiv L(A_2)$. The classical approach to this problem is to check if $(L(A_1) \cap \overline{L(A_2)}) \cup (L(A_2) \cap \overline{L(A_1)})$ is empty. If this set is empty then then the two automata are equivalent. Computing $(L(A_1) \cap \overline{L(A_2)})$ involves first complementing A_2. Since hardware only implements deterministic fsm's, complementing A_2 is trivial. Computing $(L(A_1) \cap \overline{L(A_2)})$ apparently requires building the product of machines A_1 and (A_2), but a procedure that gives a depth-first traversal of the product machine without explicitly building the machine is given in [16]. Despite this performance improvement that the depth-first search procedure supplies, each individual state in the product machine must be traversed and this is prohibitive for many sequential circuits.

The advantage of using bdd's for representing the next-state Boolean functions was known since the work of Supowit and Friedman [32] but a novel observation was that the reachable states could be represented by a set, and a bdd could be used to compactly represent the characteristic function of that set [11, 22, 10]. This important observation led to a dramatic increase in the size of the sequential machines that could be verified. As a result of this work, the complexity of sequential verification is not dependent on the number of reachable states in the automata, but on whether the structure of the next-state functions of the sequential machine lends itself to a compact representation as a bdd. Despite these advances not all sequential machines can be formally verified in this way. There is a real need to be able to verify sequential optimizations of sequential circuits and there is a strong desire among designers to be freed from the constraint of keeping register equivalence between two circuits two be verified. For these reasons advancing the state-of-the-art in the implementation verification of sequential circuits is an important problem. For reference, the current state-of-the-art in sequential verification is comprehensively reviewed in Section 32 of [20].

4.4 RTL Translation Tools in Logic Synthesis

This author knows of no system, either currently in use or under development, which *formally* verifies the registers and combinational logic synthesized from the hdl model against the original hdl model. Implementation verification systems typically perform a simplified logic translation of the hdl model and then verify the circuit which results from a more complex logic synthesizer against the results of the less complicated one using the formal methods described in Section 4.1.

It would be ideal to formally verify the logic synthesizer itself, following the paradigm of compiler correctness proofs ; however, although the transformations performed by the logic synthesizer are straightforward, they are numerous, particularly for languages such as VHDL. As a result, any proof of correctness of a complete logic synthesizer would be quite large and it seems unlikely that a single individual would be able to fathom it. A more promising approach is to formally verify a very simple logic synthesizer, as described above. Once verified this synthesizer could then be used to verify the results of a more sophisticated translation.

4.5 Verifying the Results of Behavioral Synthesis

A designer using behavioral synthesis is naturally concerned that the resulting rtl model of the circuit that behavioral synthesis produces is consistent with the intention in the original hdl model. This entails verifying that the scheduling of operations and the allocation of functional units is consistent with the original behavioral model. One formal approach to verifying this was given in [15]. Unfortunately this approach is not practical for behavioral models of any significant size. An alternative approach is to focus on the verification of the software implementation of the behavioral transformations themselves [26]. This may improve the designer's confidence in the software tool but it does not substitute for verifying the implementation of the design. As a result there is a real need for formal methods capable of verifying the correctness of behavioral synthesis.

4.6 Summary Regarding Implementation Verification

The problem of verifying the implementation of two gate-level sequential circuits with equivalent placement of registers has moved from research, to engineering and ultimately to commercial reality. The problem of extracting a gate-level model of a sequential circuit from its physical layout is also practically solved today. The problem of verifying that a gate-level circuit is consistent with an hdl model of the circuit is, while not well formalized, practically manageable. As a result, the implementation verification problem for an rtl-synthesis methodology is currently tractable. Probably the first advance that we will see in this arena is a greater robustness in tools for implementation verification in an rtl-synthesis flow. This means diminished runtimes, less memory consumption and the ability to handle a broader class of circuits. A significant advance for these tools would be the ability to verify the equivalence of two sequential circuits which have little correlation between their registers.

Once these issues are addressed the key remaining problem in implementation verification is verifying an rtl model for consistency against a corresponding behavioral model.

5 Design Verification in a Synthesis Design Flow

Recall that in a synthesis methodology design verification is the problem of verifying that the hdl-model is consistent with the specification of the integrated circuit. There are really two problems where formalism is needed: The first is developing a formal specification of the IC and the second is actually performing the verification of the hdl-model relative to that specification.

5.1 Formal Specifications

A wide variety of languages for specifying hardware have been offered and the *Computer Hardware Description Language* conference has been devoted to this topic for over a decade. On the informal side these include standard programming languages such as C and hdl's such as VHDL and Verilog. On the formal side these include languages such as the S/R language in the Cospan system [18].

Generally speaking, languages originating from the "formal community" have not gained acceptance in the IC design community, nor have languages widely used for design, such as Verilog, been of much interest to the formal verification community. A couple of exceptions is the work on formally defining subsets of VHDL [12] and Verilog [19]. The languages that have been most successful at bridging the gap between the formal verification community and the design community are languages which focus on formalizing a subset of the design problem such as ESTEREL [3].

Despite these efforts there is no commonly used language for describing hardware that has a well-defined formal semantics. The lack of a well-founded semantics for hdl's causes problems for designers as well as for automated tools such as simulators and synthesis systems. As a result research on techniques for specifying hardware is well motivated. Research on describing the semantics of widely used languages such as Verilog or VHDL is also be of practical value. It is hoped that with continued research in each of these areas designers will eventually have a well-founded language in which to describe the functionality of their circuits.

5.2 Design Verification

If current hardware specifications are almost never formal it naturally follows that formal methods are not often employed in design verification. Currently design verification is accomplished almost exclusively by simulation of the hdl model.

Formal verification techniques have shown some success at design verification; however, these successes have occurred in narrow application domains. In particular a *model checking* approach has proved useful in verifying cache coherency protocols for multiprocessors [27, 23, 1] and communication protocols.

Nevertheless, design verification is a problem that is getting increasingly harder. In many design groups the effort devoted to functionally verifying a design eclipses all other aspects of the design process. Equation 4 makes it easy to understand why the problem of design verification is getting increasingly difficult.

$$\frac{logic\_transistors}{chip} \times \frac{lines\_in\_design}{logic\_transistors} \times \frac{bugs}{lines\_in\_design} = \frac{bugs}{chip} \quad (4)$$

The synthesis of an hdl model might only generate an average of 10 logic transistors per line of hdl.[1] Thus if we wish to fully exploit current processing capability of manufacturing one million transistors we may need to write, and verify, 100,000 lines of hdl. According to Moore's Law, processing capabilities are approximately doubling every 18 months so if we want to ensure the correctness of the design of an IC we must reduce the number of bugs per line of hdl by half every 18 months. Simulation alone cannot result in this reduction.

5.3 Summary Regarding Design Verification

Design verification is the most urgent problem in IC design and, as a result, formal verification methods have the greatest potential for impact in this area. In order for formal approaches to design verification to gain acceptance, specification languages need to become more intuitive and easier for designers to use. To gain broader application formal verification techniques, such as model checking, need to be able to handle a broader class of circuits in a computationally efficient manner.

6 Conclusions

This paper has reviewed the typical rtl and behavioral synthesis design flows and identified areas where formal methods can be applied. In particular, the problems of design and implementation verification were identified as amenable to formal techniques and a number of well-motivated areas for research were identified. These included:

– Formalisms to discipline human reasoning during the generation of an rtl or behavioral-level design;
– Formal semantics for synthesizable subsets of existing hardware-design languages such as VHDL and Verilog;
– Formal methods for specification which are intuitive and easy to use;
– Formal means for expressing properties of integrated circuits;
– Formal methods for verifying properties of integrated circuits;

---

[1] The term "logic transistor" is used to distinguish it from transistors on a chip due to random-access memory, read-only memory and other regular components. These dense components can dramatically increase the size of a design but do not add significantly to the size of the hdl models.

- Formal methods for verifying that an rtl representation is consistent with a behavioral-level representation;
- More robust methods for verifying that two sequential circuits are equivalent.

References

1. A.Eiriksson. Integrating formal verification methods with a conventional project design flow. In *Proceedings of the 33rd ACM/IEEE Design Automation Conference*, pages 666–671. IEEE Computer Society Press, Los Alamitos, CA, June 1996.
2. S. B. Akers. Binary Decision Diagrams. In *IEEE Transactions on Computers*, volume C-27, pages 509–516, June 1978.
3. G. Berry, S. Moisan, and J-P. Rigault. ESTEREL: Towards a synchronous and semantically sound high level language for real-time applications. In *Proceedings of the Proceedings of the IEEE Real-Time Systems Symposium*, pages 30–37. IEEE, New York, December 1983.
4. Archie Blake. *Canonical expressions in Boolean algebra*. PhD thesis, University of Chicago, March 1938.
5. K. Brace, R. Bryant, and R. Rudell. Efficient implementation of a bdd package. In *Proceedings, 27th Design Automation Conference*, June 1990.
6. R. Bryant. Graph-based algorithms for boolean function manipulation. *IEEE Transactions on Computers*, C35(8), August 1986.
7. R. Bryant. On the complexity of VLSI implementations and graph representations of boolean functions, with applications to integer multiplication. *IEEE Transactions on Computers*, C40(2):205–213, February 1991.
8. J. T. Buck, S. Ha, E. A. Lee, and D. G. Messerschmitt. Ptolemy: A framework for simulating and prototyping heterogeneous systems. *International Journal in Computer Simulation*, 4(2):155–182, 1994.
9. R. Camposano and W. Wolf. *High-Level VLSI Synthesis*. Kluwer Academic Publishers, Norwell, MA, 1991.
10. E. M. Clarke, J. R. Burch, and K. L. McMillan. Sequential circuit verification using symbolic model checking. In *Proceedings, 27th Design Automation Conference*. ACM/IEEE, June 1990.
11. O. Coudert, C. Berthet, and J. C. Madre. Verification of synchronous sequential machines using symbolic execution. In *Proceedings of the International Workshop on Automatic Verification Methods for Finite State Systems, Grenoble, France*, volume 407 of *Lecture Notes in Computer Science*, pages 365–373. Springer-Verlag, New York, June 1989.
12. D. Deharbe and D. Borrione. A qualitative finite subset of VHDL and semantics. In *Correct Hardware Design and Verification Methods*, volume 987 of *Lecture Notes in Computer Science*. Springer-Verlag, New York, 1995.
13. G. DeMicheli. *Synthesis and Optimization of Digital Circuits*. McGraw-Hill, New York, NY, 1994.
14. S. Devadas, A. Ghosh, and K. Keutzer. *Logic Synthesis*. McGraw-Hill Book Company, 1994.
15. S. Devadas and K. Keutzer. An automata-theoretic approach to behavioral equivalence. *Integration, the VLSI Journal*, 12:109–129, December 1991.

16. S. Devadas, H-K. T. Ma, and A. R. Newton. On the verification of sequential machines at differing levels of abstraction. In *IEEE Transactions on Computer-Aided Design*, pages 713–722, June 1988.

17. M. Fujita, H. Fujisawa, and N. Kawato. Evaluation and improvements of Boolean comparison method based on binary decision diagrams. In *Proceedings of the Int'l Conference on Computer-Aided Design*, pages 2–5, November 1988.

18. I. Gertner and R. P. Kurshan. Logical analysis of digital circuits. In M. R. Barbacci and C. J. Koomen, editors, *Proceedings of the Eighth International Symposium on Computer Hardware Description Languages and their Applications*, pages 47–67. IFIP, North-Holland, Amsterdam, 1987.

19. Mike Gordon. The semantic challenge of Verilog HDL. In *Proceedings, Tenth Annual IEEE Symposium on Logic in Computer Science*, pages 136–145, June 1995.

20. A. Gupta. Formal hardware verification methods : A survey. In R. K. Brayton, E. M. Clarke, and P. A. Subrahmanyam, editors, *Formal Methods in System Design*, volume 1 (Nos. 2/3). Kluwer Academic Publishers, Boston, October 1992.

21. K. Keutzer and P. Vanbekbergen. Computer-Aided Design of Electronic Circuits. In D. Christiansen, editor, *Electronic Engineer's Handbook: Fourth Editon*. McGraw-Hill, 1997. (to appear).

22. J. Kukula. A technique for verifying finite state machines. Technical Report 3A, IBM Technical Disclosure Bulletin, August 1989.

23. P. Loewenstein and D. L. Dill. Verification of a multiprocessor cache protocol using simulation relations and higher-order logic. In E. M. Clarke and R. P. Kurshan, editors, *Proceedings of the Workshop on Computer-Aided Verification (CAV 90)*, volume 3 of *DIMACS Series in Discrete Mathematics and Theoretical Computer Science*. American Mathematical Society, Springer-Verlag, New York, 1991.

24. J-C. Madre and J-P. Billon. Proving circuit correctness using formal comparison between expected and extracted behavior. In *Proceedings, 25th Design Automation Conference*. ACM/IEEE, June 1988.

25. S. Malik, A. R. Wang, R. Brayton, and A. Sangiovanni-Vincentelli. Logic Verification using Binary Decision Diagrams in a Logic Synthesis Environment. In *Proceedings of the Int'l Conference on Computer-Aided Design*, pages 6–9, November 1988.

26. Michael McFarland. A practical application of verification to high-level synthesis. In *Proc. of the 1991 International Workshop on Formal Methods in VLSI Design*, January 1991.

27. K. McMillan and J. Schwalbe. Formal verfication of the gigamax cache coherency protocol. Technical report, Carnegie Mellonn University, august 1990.

28. S. Muroga. *VLSI System Design*. Wiley-Interscience, New York, 1982.

29. Bryan T. Preas and Michael J. Lorenzetti, editors. *Physical Design Automation of VLSI Systems*. Benjamin-Cummings, 1988.

30. S. Ritz, M. Pankert, V. Zivojnovic, and H. Meyr. High level software synthesis for the design of communication systems. *IEEE J. on Selected Areas in Communications*, 11:348–358, April 1993.

31. Claude E. Shannon. The synthesis of two-terminal switching circuits. *Bell System Technical Journal*, 28:59–98, January 1949.

32. K. J. Supowit and S. J. Friedman. A New Method for Verifying Sequential Circuits. In *Proceedings of the 23rd Design Automation Conference*, pages 200–207, June 1986.

33. T. Szymanski and C. Van-Wyck. GOALIE:a space-efficient system for VSLI artwork analysis. In *IEEE Transactions on Computer-Aided Design*, volume CAD-3, pages 278–280, 1984.

34. P. Zepter and T. Grotker. Generating synchronous timed descriptions of digital receivers from dynamic dataflow system level configurations. In *Proc. of European Design and Test Conference*, February 1994.

Verification of All Circuits in a Floating-Point Unit Using Word-Level Model Checking

Yirng-An Chen*, Edmund Clarke*, Pei-Hsin Ho**, Yatin Hoskote**,
Timothy Kam**, Manpreet Khaira**, John O'Leary**, Xudong Zhao**

Abstract This paper presents the formal verification of all sub-circuits in a floating-point arithmetic unit (FPU) from an Intel microprocessor using a word-level model checker. This work represents the first large-scale application of word-level model checking techniques. The FPU can perform addition, subtraction, multiplication, square root, division, remainder, and rounding operations; verifying such a broad range of functionality required coupling the model checker with a number of other techniques, such as property decomposition, property-specific model extraction, and latch removal. We will illustrate our verification techniques using the Weitek WTL3170/3171 Sparc floating point coprocessor as an example. The principal contribution of this paper is a practical verification methodology explaining what techniques to apply (and where to apply them) when verifying floating-point arithmetic circuits. We have applied our methods to the floating-point unit of a state-of-the-art Intel microprocessor, which is capable of extended precision (64-bit mantissa) computa- tion. The success of this effort demonstrates that word-level model checking, with the help of other verification techniques, can verify arithmetic circuits of the size and complexity found in industry.

1 Introduction

The floating-point division flaw [SB94, Coe95] in Intel Corp.'s Pentium underscores how hard the task of verifying a floating-point arithmetic unit is, and how high the cost of a floating-point arithmetic bug can be. About one trillion test vectors were used and none uncovered the bug. The recall and replacement of the chips in the field cost Intel $470 million. Since the Pentium processor flaw came to light, there has naturally been new interest in improved methods for functional verification of arithmetic hardware - especially in formal methods, which provide exhaustive coverage of the implementation's behavior. This paper describes the formal verification of a complete floating-point unit, described at the structural level in a hardware description language, using word-level model checking techniques and gives verification results for a recent Intel microprocessor. We have verified the correct implementation of the addition, subtraction, multiplication, square root, division, remainder, and rounding operations. This work is the first large-scale application of word-level model checking techniques.

---

* School of Computer Science, Carnegie Mellon University, Pittsburgh, PA 15213 USA
** Intel Development Labs, 5200 NE Elam Young Parkway, M/S JFT-102, Hillsboro, OR 97124-6497 USA

The principal contribution of this paper is to demonstrate how word-level model checking can be applied to practically and efficiently verify arithmetic circuits in state-of-the-art microprocessors. We will focus on the techniques we found useful for various classes of circuits, rather than the details of the model checking algorithm itself which are covered in [CKZ96]. We illustrate our techniques with respect to the design of the Weitek WTL3170/3171 floating-point coprocessors. We chose the Weitek part because substantial detail has been published about its architecture and algorithms[BSC+90], though we think it is simpler than the Intel design we actually verified. However, we emphasize that the verification methodology we propose is very general and our techniques are applicable to other floating-point and integer arithmetic circuits as well. We will report the results of verifying the FPU from an Intel microprocessor using these techniques in a separate section.

Previous work in formally verifying arithmetic hardware has either used BDD-based algorithms [Bry91, Bry95], or theorem proving techniques [VCM94], or a combination of both [KL93]. The first approach has the disadvantage that it requires extremely detailed, bit-level specifications that are difficult to formulate correctly. Moreover, the bit-level specifications of operations like multiplication can be exponentially complex. The latter two approaches are relatively laborious and require users with substantial special training to guide the proof. However, they allow specifications to be written cleanly, in terms of the usual arithmetic operations upon arbitrary-precision integers. Our work demonstrates that word-level model checking algorithms combine the best of both worlds, admitting a high degree of automation while allowing very abstract specifications.

The remainder of this paper is organized as follows. Section 2 contains a brief introduction to the word-level model checking techniques we used in verifying the FPU. Section 3 describes the functional units in the Weitek FPU and the techniques that can be used to verify them. In particular, we will discuss property decomposition, property-specific model extraction, latch removal, and verification by invariants. Section 4 presents results obtained in applying these techniques to a floating-point unit from an Intel microprocessor.

2 Word-Level Model Checking

Symbolic model checking [McM93] is a very efficient technique for verifying the correctness of sequential circuits. It is based on binary decision diagrams (BDDs) and has been very successful in verifying the control logic of industrial circuits. However, BDDs are sometimes unable to represent the data path of circuits efficiently (e.g. multipliers and shifters), preventing their wide-spread use in the verification of arithmetic circuits. Recently, data structures that allow such an efficient representation have been derived from BDDs, such as binary moment diagrams (BMD) [BC95] and multi-terminal BDDs (MTBDD) [CMZ+93]. These representations have further been combined to form hybrid decision diagrams (HDDs) [CFZ95].

2.1 Hybrid Decision Diagrams

A BDD is a directed acyclic graph with a total order on the occurrence of variables from root to leaf. Multi-terminal BDDs have a similar structure. However, BDDs have Boolean leaves, while MTBDDs have integer leaves and therefore represent functions from Booleans to integers. Functions from integers to integers can also be represented when the input is encoded in binary form. Efficient algorithms exist that compute common arithmetic operations when operands are given in this form. A BMD is another representation for functions that map Boolean vectors to integers. This representation is more compact for some useful arithmetic functions which have exponential size if represented using MTBDDs.

Both BMDs and MTBDDs have been integrated into the verification system at Intel through the use of hybrid decision diagrams (HDDs). In particular, for state variables in the circuit corresponding to data bits, this hybrid representation behaves like a BMD; while for the state variables corresponding to control signals, it behaves like a MTBDD. By using HDDs in this manner, this system is able to handle complex circuits containing both complex control logic and wide data paths. The reader is referred to [BC95, CMZ$^+$93, CFZ95] for detailed descriptions of BDDs, BMDs and HDDs.

2.2 Specifying Word-Level Properties

The HDD-based verification system allows the expression of properties involving relationships among data variables. Unlike a BDD-based system where properties can only reason about state variables, the HDD-based system allows properties involving relations between the values of data variables called words. A word is an array or bit vector of state variables. The value of the word is the value of the unsigned integer represented by that bit vector. An arithmetic expression can be constructed from words in the circuit, constants and arithmetic operations on words. In our word-level extended CTL, any Boolean combination of strict or nonstrict inequalities of integer expressions with arithmetic operations such as addition, subtraction, multiplication can be specified. For example, the property

$$\mathbf{AG}((p < 0) \to \mathbf{AX}((p = (-2 \cdot b + 3 \cdot r)) \wedge p \geq 0))$$

specifies that it is always the case that if p is negative, then in the next clock phase the value of p is equal to $-2 \cdot b + 3 \cdot r$ and p becomes non-negative.

Extended CTL allows a wide range of abstract specifications on data variables, which are not expressible in a system using a Boolean representation.

2.3 Model Checking with Word-Level Properties

Model checking is a technique that, given a state-transition graph and a temporal logic formula, determines which states satisfy the formula. In symbolic model checking systems [McM93], BDDs are used to represent the transition relations and sets of states. The model checking process is performed iteratively on these

BDDs. Symbolic model checking has dramatically increased the size of circuits that can be formally verified. However, model checking algorithms cannot be used directly for verifying arithmetic circuits. Expressions that involve variables with integer values cannot be handled in a clean and efficient manner. Word-level model checking overcomes this problem by extending the original algorithms to evaluate arithmetic expressions using hybrid decision diagrams [CKZ96].

In word-level model checking, the transition relation and formulas not involving words are implemented using BDDs as in the original algorithm. HDDs are used only to compute word-level expressions. The BDD representing the set of variable assignments that make an algebraic relation true is built using the HDD representations of the expressions within the algebraic relation. After the BDDs for atomic formulas have been computed, the BDDs for temporal formulas are computed in the same way as in standard model checking. The iterative computations are exactly the same in both cases. The power of this extended system will be evident from the verification results presented in this paper.

3 Verifying a Floating-Point Arithmetic Unit

3.1 Weitek Floating-Point Coprocessor

In this section we present a typical FPU, the Weitek WTL3170/3171 Sparc floating-point coprocessor [BSC$^+$90], and the techniques we have found useful to verify each type of circuit found in the coprocessor. The verification methodology we use for floating-point arithmetic circuits is very general and our techniques are applicable to other arithmetic circuits as well. In section 4, we shall report verification results specific to an Intel FPU design. We will focus on the mantissa computations, as they pose the most interesting verification challenges. The exponent unit., which is shared among several of the FPU's functions, is relatively simple to verify.

Figure 1 shows a block diagram of the Weitek FPU. It consists of circuits performing addition/subtraction, multiplication, square root, division and rounding. The verification methodology presented here does not follow this structural decomposition, but rather verifies one arithmetic operation at a time. The exponent operation for all units but the ALU are verified separately. The rounding operation and exponent adjustment during rounding is quite difficult to specify as part of the preceding arithmetic operation and is also verified separately. The FPU takes two floating-point operands (M_1, E_1) and (M_2, E_2) and generates a result floating-point number (M_{out}, E_{out}). In the following, we assume the mantissas to be k_M bits wide, where k_M is the internal precision of the machine. The verification of arithmetic operations is described in detail in the sequel. However, all the verification tasks follow a general methodology which is described here.

One general technique to tackle complex verification tasks is *property decomposition*, which is applicable at different levels of description. The task of verifying the correctness of the FPU is decomposed into properties asserting that individual FP sub-circuits compute the correct arithmetic functions. At a

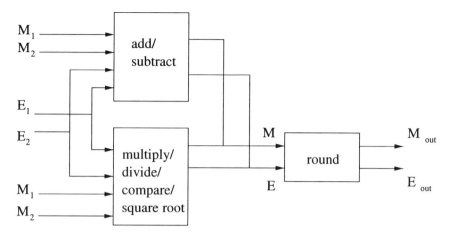

Fig. 1. Block Diagram for WEITEK Co-processor

lower level, the property of the circuit is often further decomposed into arithmetic properties relating the mantissas and other properties of the exponents.

An FPU can be divided into a number of sub-circuits, each computing a particular arithmetic function on the operands. Depending on the machine instruction or micro-operation, different sub-circuits may be activated while others irrelevant to the operation are not used. Thus the model to be verified can be simplified with respect to the property being verified. For example, to verify the division circuit, only the division and the exponent sub-circuits need to be included into the model. In addition, for each sub-circuit and for each property to be verified with respect to the sub-circuit, a *property-specific model extraction* can be performed on the sub-circuit to simplify away the parts of the sub-circuit that is irrelevant to the property. The property-specific model extraction is done by an automatic tool that was independently developed at Intel. The tool seems to be very similar to the *per-function reduction* in [BBDEL96].

Several of the arithmetic algorithms used in the FPU are inherently iterative. For example, the square root and division computations begin by generating an initial result consisting of a few high order bits, then refining the initial result in successive iterations by computing lower-order bits until the desired precision is achieved. Iterative algorithms in general can be verified by proving an invariant, a technique borrowed from software verification. An invariant is a property relating the registers used in the computation during each iteration. The proof of an invariant has two parts: a proof that the initialization phase of the circuit causes the invariant to hold, and a proof that if the invariant holds before an iteration of the algorithm, then it continues to hold after the iteration. From these two results we can conclude that the invariant always holds, subject to the precision limitations of the physical registers.

To improve performance, floating-point circuits are often heavily pipelined. Verification of sequential hardware is in general more complicated than that for combinational ones. We can verify pipelined designs in two steps: combinational verification of their functionality and sequential verification of their pipeline control. For the former, we can remove the pipeline latches and treat the whole circuit as a single combinational block. Latch removal results in simpler specifications as well as more efficient verification. When we verify the pipeline control, we can abstract away the datapath as we are interested only in the sequential behavior of the control signals. For each pipeline latch, latch removal takes away the latch and its clock signal, and then reexpresses the latch output as a logic function of its inputs and its enable and reset signals. Note that latch removal can result in incorrect verification if the arithmetic circuit is inherently sequential or iterative in nature, as are division and square root.

Dynamic variable reordering is sometimes useful in verifying arithmetic circuits, especially for iterative circuits that contain a non-trivial control part. We have generalized Rudell's dynamic variable reordering algorithm [Rud93] to work on HDDs and incorporated it into our verification system.

A significant portion of the verification effort involved the user familiarizing himself/herself with the actual circuit description so as to be able to state properties correctly in terms of appropriate circuit signals, be able to ignore irrelevant parts of the circuit and be able to formulate the properties in a manner so as to best manage verification complexity.

3.2 Square Root

This section describes the verification of the mantissa computation for the square root operation. The Weitek paper states that the square root and division operations share a common datapath, but it leaves the details of the square root operation - for example, radix, unspecified [BSC$^+$90]. For illustration, we consider here a non-restoring, radix-2 algorithm that is commonly found in the literature [BV85, OLHA95], and is implemented on a separate datapath.

The algorithm proceeds iteratively, as follows. The partial square root *proot* contains all the root mantissa bits computed thus far, and is used in each iteration to guess what the next partial root should be. The guess is always twice the partial square root plus the guess bit. The partial remainder *prem* contains the difference between the radicand mantissa and the previous guess squared. If the partial remainder is positive in a given cycle, then the square of the previous guess was less than the radicand, hence the most recently guessed bit was correctly presumed to be 1. If the partial remainder is negative, then the square of the previous guess was greater than the radicand, and so the most recently guessed bit was incorrectly assumed to be 1, and the partial root must be corrected accordingly. The guess bit b is initially $2^{2 \cdot k_M}$ and is shifted right two bits every cycle.

Because the square root algorithm is iterative in nature, we verify it by proving a loop invariant. In particular, we prove that the partially constructed root *proot*, the partial remainder *prem*, and the guess bit b have the following relationship at each iteration i:

$$prem_i < 0 \rightarrow -2 \cdot proot_i + b_i \leq prem_i$$

$$prem_i \geq 0 \rightarrow 2 \cdot proot_i + b_i > prem_i$$

We denote the conjunction of the above properties by INV_i. It can be proved mathematically that if the loop invariant is true at each iteration, when the algorithm terminates, the result is correct [OLHA95].

We prove the invariant by induction on the number k_M of iterations. According to the algorithm, in the 0'th iteration the registers are initialized as follows.

$$prem_0 = radicand$$
$$proot_0 = 0$$
$$b_0 = 2^{2 \cdot k_M}$$

The radicand is positive and less than $2^{2 \cdot k_M}$, so INV_0 should hold.

In subsequent iterations the algorithm updates the registers as follows.

$$prem_{i+1} = \begin{cases} prem_i - proot_i - b_i, & 0 \leq prem_i \\ prem_i + proot_i - b_i, & prem_i > 0 \end{cases}$$

$$proot_{i+1} = \begin{cases} (proot_i + 2 \cdot b_i)/2, & 0 \leq prem_i \\ (proot_i - 2 \cdot b_i)/2, & prem_i > 0 \end{cases}$$

$$b_{i+1} = b_i/4$$

For these later iterations we have to verify that if INV_i is true in a given clock cycle, and the registers are updated as above, then INV_{i+1} will be true in the subsequent clock cycle. That is, we use the word-level model checker to verify the following assertion.

$$INV_i \Rightarrow INV_{i+1}$$

These two results (the invariant holds initially, and is preserved by the register updates) prove that the invariant always holds.

Because the extended CTL used in the model-checking tool supports Boolean combinations of integer inequalities with subtractions, additions and multiplications, the invariant INV_i can be formulated "as is" for the model checker. To verify the properties, we must first automatically obtain a property-specific extraction of the model. In addition, to relate the values of variables in iteration i and $i + 1$, we can introduce a set of "history" variables into the abstracted model that store the previous values of *prem*, *proot*, and *b*.

3.3 Division

In this section, we discuss the verification of floating-point division. Weitek WTL3170/3171 uses a radix-4 SRT division algorithm. The similar algorithms can also be found in [Fri61, Atk68]. Since the SRT division algorithm is also iterative, the loop invariant verification technique introduced in the previous section also applies here. Several published papers [CKZ96, BC95] have also shown how to verify radix-4 SRT division circuits using model checking techniques and thus our description here is brief. Given the mantissa d (from M_1 in Figure 1) of the dividend and the mantissa b (from M_2 in Figure 1), the radix-4 SRT division algorithm iteratively computes a partial remainder r_i and a quotient digit q_i. The partial remainder r_0 is initialized to $d/4$ and the quotient digit q_0 is initialized to zero. Each iteration the algorithm gets the quotient digit from a lookup table and subtracts $q_i \cdot b$ from the partial remainder r_i that has been shifted left by 2 bits. In other words, $r_{i+1} = 4 \cdot r_i - q_i \cdot b$. The algorithm terminates when enough quotient bits have been computed. Suppose that the quotient digits are within the range $\{-n, -n+1, \ldots, -1, 0, 1, \ldots, n-1, n\}$ for some positive n. Then a radix-4 SRT division algorithm is guaranteed to be correct if both of the following properties are true in each division loop [Atk68]:

$$r_{i+1} = 4 \cdot r_i - q_i \cdot b$$

$$|r_i| \leq \frac{n \cdot b}{3}$$

The loop invariant INV_i that we want to verify with our verifier is the conjunction of the two properties above. We want to verify that the invariant INV_0 is true initially and also $INV_i \Rightarrow INV_{i+1}$. Since the quotient digits that WTL3170/3171 uses are $\{-3, -2, -1, 0, 1, 2, 3\}$, the second property simply becomes $|r_i| \leq b$.

Again, INV_i can be expressed in the extended CTL very easily. Although the extended CTL does not support the division operator, we can easily transform some inequalities with divisions to inequalities with only multiplications. For example, the second property of INV_i (for any constant n) can be specified as follows.

$$\mathbf{AG}((3 \cdot r_i \leq n \cdot b) \wedge (-3 \cdot r_i \leq n \cdot b))$$

The property-specific model extraction and variable ordering techniques discussed previously are also useful here to verify the loop invariant.

3.4 Multiplier

A floating-point multiplier consists of two parts, an integer multiplier and an exponent unit. This section describes the integer multiplier as shown in Figure 2. Depending on the precision, Weitek FP multiplier operates actually in two modes. We first describe the more common multiplier configuration for single precision. For double precision, multiplication is accomplished in two passes using the carry-save adder array. The verification of the double precision case is a straightforward extension and will not be described here.

As shown in Figure 2, the multiplier input M_1 is first encoded through the Booth encoder. The value $3 \cdot M_2$ is produced by the 3X generator. Second, the multiplier selects which multiple (1X or 3X) of the multiplicand is used for each partial product. This step is called partial product selection. Third, the carry-save adder array adds all partial products to produce two numbers. Finally, an adder produces the final product. Note that for single precision, the multiplexer

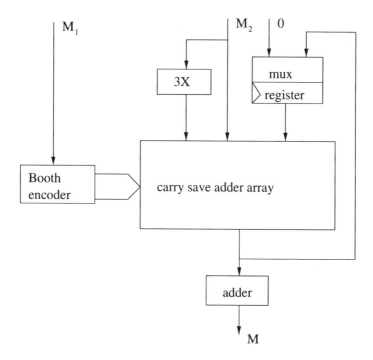

Fig. 2. Block Diagram of FP multiplier

is controlled so that the output of the CSA array is not fed back as one of its inputs.

The overall property to verify the mantissa part of the multiplier is $M = M_1 \cdot M2$. To reduce the size of the HDDs required for this word-level model checking, a few decomposition techniques must be used. The Weitek multiplier is pipelined, so the pipeline latches must first be removed so that the multiplier circuit can be considered as a combinational function.

Even with word-level model checking, one has to be careful during the construction of BMDs to make sure that no intermediate computation results in an exponential representation. A simple-minded way of constructing the output BMD function M starts from the inputs and progressively builds the intermediate Boolean functions as BDDs. Once the BDD functions for each output bit are obtained, they are collected and composed into a word BMD function. This method unfortunately will not work on wide multipliers as there is no compact BDD representation of the output bits of a multiplier. This problem can be overcome by the following approach.

Hamaguchi *et. al* in [HMY95] proposed a backward construction method for obtaining a *BMD function. A cut is first defined across the output(s) and is swept towards the inputs by iteratively moving one gate across the cut at a time. A *BMD representing the function from the cut to the output is always maintained during the backward substitution. We improve on this method by not evaluating a new BMD function for each gate in the circuit. Instead we introduce the notion of auxiliary variables to mark multiple cuts on the circuit. Small intermediate BDDs are built to represent functions separated by auxiliary variables. We obtain the BMD representation of the output word M in terms of the auxiliary variables that are the immediate inputs of M. Then we backward substitute the next intermediate function into the output BMD function, and continue until the latter is formulated in terms of primary inputs M_1 and M_2. This process is fully automatic. The simple $M = M_1 \cdot M_2$ property can be verified directly on the circuit after unlatching and specification of auxiliary variables.

3.5 Adder / Subtracter

The block diagram in Figure 3 shows the FP add/subtract circuit from the Weitek FP coprocessor. Note that it is more complicated than its integer counterpart. Given two floating-point numbers, they must first be aligned before their mantissas can be added/subtracted. This is done by comparing the relative magnitude of the two exponents and swapping (M_1, E_1) and (M_2, E_2) if $E_1 < E_2$. It then shifts the mantissa with a smaller exponent $|E_1 - E_2|$ places to the right. The larger exponent will become the exponent of the result. The result of mantissa addition will be within the bound [2, 4). If the sum is greater than two, the overflow mantissa must be shifted to the right. This is accomplished by the rounder circuit. On the other hand, after subtraction, it is possible that the resulting mantissa has leading zeros. Normalization is accomplished in such cases by multiple left shifting the mantissa. Let L be the amount of left shifting performed for normalization. The resulting exponent then must also be decreased by L. This is done in the add/subtract unit. Thus, the output of the adder/subtracter is in the range [1,4).

The FP addition and subtraction properties can be decomposed and verified by case analysis. First, four combinations of signs of the two inputs are grouped into two cases: true addition and true subtraction. True addition and subtraction are the actual operations performed by the circuit. True addition includes addition of two numbers of the same sign, and subtraction between two numbers of different signs. Similarly, true subtraction refers to subtraction of two numbers of the same sign, and addition of two numbers of different signs. Furthermore, we decomposed each case into sub-properties, which are verified, according to the difference in the exponents.

3.6 Rounding unit

The result from a floating-point operation is finally fed to the rounding unit to be rounded so that it can be represented by a floating-point number of the required precision. The Weitek paper does not give the implementation of the rounding logic. However, it is the specification methodology that is of interest here and we describe it in this section.

In simple terms, we wish to verify that the output of the rounder is within one bit of the input. Thus, one approach is to specify the rounding operation

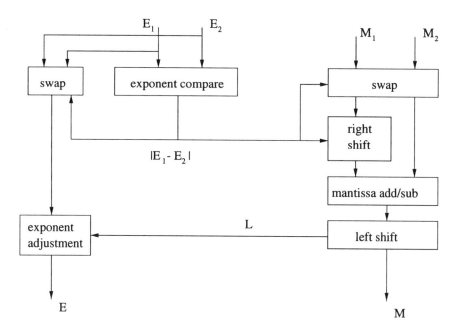

Fig. 3. Block diagram of FP adder/subtracter

as a relation between the input and output mantissa and exponents as follows (shown for single precision):

$$M - 2^{-23} < M_{out} < M + 2^{-23}$$

$$E = E_{out}$$

This specification has the advantage of being very general and independent of the specific rounding mode that is used. However, we have to split this specification into several cases to make the verification tractable. This splitting is most easily done on the basis of the rounding modes.

Most floating-point systems support four rounding modes for each precision:

- round to zero
- round to positive infinity
- round to negative infinity
- round to nearest/even.

It is required that the result of an arithmetic operation should be the same as it would be if it were computed with infinite precision and then rounded using one of the specified rounding modes. To simulate the effect of infinite precision in the implementation, the rounding unit extracts a few extra bits from its input besides the fraction bits and the leading 1 bit (L). These extra bits are called the round bit (R) and the sticky bit (S). In a normalized input, the R bit is simply the bit to the right of the least significant bit (M_0) of the mantissa and the S bit is the OR of all the bits to the right of the R bit. If the input to the rounder is not normalized, i.e., it is in the range [2,4), it is right shifted by one to bring it in the range [1,2) and the exponent is incremented before the rounding operation is carried out. In the following, we assume that the input mantissa is normalized. The final data format for normalized mantissa in single precision is shown in Figure 4.

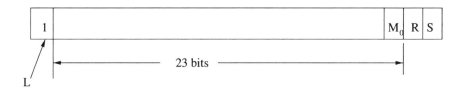

Fig. 4. Input mantissa format for single precision

To specify the relation between the input and output of the rounding unit for each rounding mode, we compute an increment bit (I) for each rounding mode from the R and S bits. The relation between the input and output mantissa is then expressed in terms of the I bit. This enables a natural decomposition of the basic specification given earlier on the basis of the rounding modes since the computation of the I bit depends on the rounding mode. The desired behavior of the rounding unit can then be specified as follows: it adds the I bit to the M_0 bit of the input mantissa. If this addition causes an overflow, the mantissa is shifted right and the exponent is incremented. The data is then chopped to the desired precision to give the final result mantissa.

The specifications for each mode then take the following form (shown for single precision), where *shift* is 1 if the addition of the I bit causes the result mantissa to be greater than or equal to 2, and 0 otherwise:

$$M_{out} = \left(M + I \cdot 2^{-23}\right) \cdot 2^{-shift}$$

$$E_{out} = E + shift$$

The computation of the I bit differs for each rounding mode and also depends on the sign of the input. Lack of space precludes a more detailed description of the specifications (see page A-24 in [HP96] for an example of I bit calculation). It is possible that the implementation also computes a similar I bit to help in the rounding. It is important that the computation of this I bit in the specification does not mimic the logic for computation of the I bit in the actual implementation. This ensures that the specification is at a higher level of abstraction than the implementation.

The above specifications apply only to *normal* numbers where the exponent value falls in the acceptable range. In case of *overflow* or *underflow*, the setting of appropriate exception flags is verified. The actual data output by the rounder in such cases depends on the implementation. In any event, the above specifications can be modified for the special cases to appropriately reflect the desired behavior of the rounding hardware.

4 Experimental Results on an Intel Microprocessor

We applied all techniques discussed above to the FPU of an Intel microprocessor. The microprocessor performs all the floating-point operations mentioned in the previous sections in 64-bit extended precision. We are able to verify the entire floating-point unit of the microprocessor using our word-level symbolic model checking system. The work shows that our techniques do apply to arithmetic circuits found in actual industrial microprocessors. The table below summarizes the figures from the experiments.

Macro-Instruction	No. Of var. in extracted model	No. of properties verifies	Memory required	BDD nodes allocated	CPU time
DIV	287	4	18.8M	756K	194s
SQRT	415	16	18.5M	445K	239s
REM	369	8	9.8M	246K	1538s
MUL	1961	2	3.9M	1923K	508s
ADD	1251	2	22.1M	838K	660s
SUB	1247	9	96.0M	3947K	38525s
RND	295	80	23.6M	692K	2034s
EXP	65	4	8.3M	157K	26s

Table 1. Verification results on an FPU from an Intel microprocessor

The experiments were done on an HP 9000 workstation with 256MB RAM. REM is a partial remainder circuit that is verified using loop invariant techniques similar to those used in the verification of the division circuit. EXP is the exponent unit which produces the exponent result for the multiply, divide and square root operations. The second column shows the number of state variables in the property-specific extractions of the original designs. The automatic property-specific extraction of the design drastically reduced the number of state variables in several models which enabled the verification to succeed. The third column shows the number of properties verified for each macro-instruction. There are eighty properties verified for the rounder because several different cases have to be considered for different modes and precisions and verification of exceptions. The fourth column shows the maximum memory required for a verification run. The experiments show that all the verifications can be done on a machine with about 100MB of memory. The fifth column shows the maximum number of

BDD nodes allocated by the symbolic model checker during the verification of all properties for the macro-instruction.

The last column shows the CPU time spent on the verification of all properties for the macro-instruction. All the experiments except the verification of the subtracter can be done in less than an hour. The subtracter takes more time in comparison to the adder primarily because there are more cases to be considered.

5 Conclusions

In this paper, we have presented a methodology for the formal verification of a complete floating-point arithmetic unit and have shown the results of this methodology applied to a recent Intel microprocessor. In particular, we have verified the correct implementation of floating-point addition, subtraction, multiplication, division, remainder, square root and rounding operations in a fairly efficient and automated fashion. To the best of our knowledge, this is the first comprehensive effort of this magnitude in the verification of complex floating-point circuits in a state-of-the-art FPU design. The results presented here are important evidence of the capability of an automated model checking system.

Our verification uses the technique of word-level model checking. The experimental results show that it was highly effective in our difficult verification tasks. Its compact representation and the efficient manipulation of arithmetic functions is made possible by the word-level HDD representation. From our experience, word-level model checking can be performed fairly automatically and the specification is at an appropriate level of abstraction.

Different verification techniques were also discussed, all of which are crucial to the successful verification of the circuits covered. In this regard, the paper contributed a practical verification methodology for efficient verification of complex arithmetic circuits.

References

[Atk68]   D. E. Atkins. Higher-radix division using estimates of the divisor and partial remainders. *IEEE Transactions on Computers*, C-17(10):925–934, October 1968.

[BBDEL96] R.E. I. Beer, S. Ben-David, C. Eisner, and Avner Landver. Rulebase: an industry-oriented formal verification tool. In *Proceedings of the 33rd Design Automation Conference*. IEEE Computer Society Press, June 1996.

[BC95]   R. E. Bryant and Y. A. Chen. Verification of arithmetic functions with binary moment diagrams. In *Proceedings of the 32nd ACM/IEEE Design Automation Conference*, pages 535–541. IEEE Computer Society Press, June 1995.

[Bry91]   R. E. Bryant. On the complexity of vlsi implementations and graph representations of boolean functions with application to integer multiplication. *IEEE Transactions on Computers*, 40(2):205–213, 1991.

[Bry95]   R.E. Bryant. Bit-level analysis of an srt divider circuit. Technical report, Carnegie Mellon University, 1995.

[BSC+90]   M. Birman, A. Samuels, G. Chu, T. Chuk, L. Hu, J. McLeod, and J. Barnes. Developing the wtl3170/3171 sparc floating-point coprocessors. *IEEE Micro*, pages 55–64, February 1990.

[BV85]   J. Bannur and A. Varma. The vlsi implementation of a square root algorithm. In *Proceedings of the 7th Symposium on Computer Arithmetic*, pages 159–165. IEEE Computer Society Press, 1985.

[CFZ95]   E. M. Clarke, M. Fujita, and X. Zhao. Hybrid decision diagrams – overcoming the limitations of mtbdds and bmds. In *Proceedings of the 1995 Proceedings of the IEEE International Conference on Computer Aided Design*, pages 159–163. IEEE Computer Society Press, November 1995.

[CKZ96]   E. M. Clarke, M. Khaira, and X. Zhao. Word level symbolic model checking – a new approach for verifying arithmetic circuits. In *Proceedings of the 33rd ACM/IEEE Design Automation Conference*. IEEE Computer Society Press, June 1996.

[CMZ+93]   E. M. Clarke, K. McMillan, X. Zhao, M. Fujita, and J. Yang. Spectral transforms for large boolean functions with applications to technology mapping. In *Proceedings of the 30th ACM/IEEE Design Automation Conference*, pages 54–60. IEEE Computer Society Press, June 1993.

[Coe95]   T. Coe. Inside the pentium fdiv bug. *Dr. Dobbs Journal*, 20(4):129–135, April 1995.

[Fri61]   C. V. Frieman. Statistical analysis of certain arithmetic binary division algorithms. *IRE Transaction*, pages 91–103, January 1961.

[HMY95]   K. Hamaguchi, A. Morita, and S. Yajima. Efficient construction of binary moment diagrams for verifying arithmetic circuits. In *Proceedings of the 1995 IEEE International Conference on Computer Aided Design*, pages 78–82, November 1995.

[HP96]   J. L. Hennessy and D. A. Patterson. *Computer Architecture: A Quantitative Approach*. Morgan Kaufmann, 1996.

[KL93]   R. P. Kurshan and L. Lamport. Verification of a multiplier: 64 bits and beyond. In C. Courcoubetis, editor, *Proceedings of the Fifth Workshop on Computer-Aided Verification*, June/July 1993.

[McM93]   K. L. McMillan. *Symbolic Model Checking*. Kluwer Academic Publishers, 1993.

[OLHA95]   J. O'Leary, M. Leeser, J. Hickey, and M. Aagaard. Non-restoring integer square root: a case study in design by principled optimization. In *Proceedings of the Theorem Provers in Circuit Design '94*, volume 901 of *Lecture Notes in Computer Science*. Springer-Verlag, 1995.

[Rud93]   R. Rudell. Dynamic variable ordering for ordered binary decision diagrams. In *Intl. Conf. on Computer Aided Design*, Santa Clara, Ca., November 1993.

[SB94]   H. P. Sharangpani and M. L. Barton. Statistical analysis of floating point flaw in the pentium processor(1994). Technical report, Intel Corporation, November 1994.

[VCM94]   D. Verkest, L. Claesen, and H. De Man. A proof of the nonrestoring division algorithm and its implementation on an alu. *Formal Methods in System Design*, 4:5–31, January 1994.

*BMDs Can Delay the Use of Theorem Proving for Verifying Arithmetic Assembly Instructions

Laurent Arditi

Université de Nice – Sophia Antipolis, France.
Laboratoire I3S, CNRS ura 1376.
email: arditi@unice.fr
WWW: http://wwwi3s.unice.fr/~arditi/lolo.html

Abstract

We address the problem of formally verifying arithmetic instructions of microprocessors implemented by microprograms that contain loops. We try to avoid theorem proving techniques using a new symbolic representation: Binary Moment Diagrams (*BMDs).

In order to use *BMDs for verifying sequential circuits as well as microprograms, we extend this representation and define several bit-vector level operators. This extension is then integrated into an automatic verification system. We illustrate the paper with examples and we discuss power and weakness of *BMDs.

1 Introduction

1.1 Motivations

One of the major problems that has to be solved when formally verifying microprocessors is to close the gap between different abstraction levels, from the electronic level to the assembly language level. An assembly language instruction is usually specified as an integer operation while it is implemented as a sequence of Boolean or bit-vector operations. Moreover, the most complex arithmetic instructions are implemented using loops of bit-vector operations and so we are faced with a combinational explosion. To handle this open problem, two methodologies arise:

1. using theorem proving techniques to perform inductive proofs, or

2. using efficient data structures that represent functions from Boolean variables to Boolean values like Binary Decision Diagrams (BDDs) [7], or to numeric values like Binary Moment Diagrams (*BMDs) [10].

We believe that these two approaches are complementary since the first one is powerful and support abstraction very well but requires more user intervention than the second. Moreover, the second methodology can easily be integrated into a verification system based on symbolic model checking or symbolic execution. These fully automatic techniques are sufficient to verify the major part of the processor instruction sets. When it is not efficient anymore – to verify data-dependent loops or when the combinational explosion cannot be avoided – theorem proving techniques are required.

1.2 Contribution of the Paper

In this paper we focus on the verification of microprograms that contain loops in order to implement complex arithmetic instructions. The first contribution of the paper is to show that *BMDs are useful to verify some arithmetic instructions like multiplication or squaring so that they delay the use of theorem proving techniques.

The second contribution is an extension of the set of *BMD operators in order to include bit-vector level operators used in microprogram descriptions like concatenation, bit field extraction and rotations. This extension is then integrated into the system SVP that is dedicated to the formal verification of microprocessors [2, 3].

1.3 Related Works

The formal verification of hardware (see [23] for a survey on this subject) is a large research topic including different verification aspects (functionality, properties, timing, power consumption, layout) and different classes of hardware (combinational or sequential circuits, synchronous or asynchronous circuits). Here we address the problem of formally verifying synchronous processor functionalities. The verification of processors was first tackled using theorem proving techniques [22, 27, 17]. Impressive proofs were performed [28, 32, 31] but in our opinion these proofs required to much user guidance from specialized experts. Some interesting results have been obtained using more automatic techniques [18, 13] but they do not verify operative parts of processors. Therefore they cannot perform functional verifications of complex arithmetic circuits. Completely automatic methods [20, 12, 5] are lacking for abstraction mechanisms and so generally focus on proving low-level descriptions.

The problem of combining automation and high-level proofs with abstraction is not limited to processor verification but also extends to several circuits and protocols. Some researchers solved this problem using two different approaches. The first one consists of combining efficient decision procedures for finite domains with theorem proving [33, 30]. The second one allows a kind of abstraction in BDD-based provers [16, 25]. They provide a support for non-Boolean data types [26] or they give symbolic representations derived from BDDs. Most of these representations are described in [8].

But, to our knowledge, no processor including microprogram loops to implement arithmetic instructions have been verified, either using theorem provers or automatic methods.

Layout of the paper: in Section 2 we describe our microprocessor verification methodology. Section 3 is a quick overview of *BMDs. Our extension of *BMDs is detailed in Section 4. In Section 5 we show the verification of a multiplication instruction and compare our results with other works. We also give results from the verifications of other arithmetic microprograms. We conclude in Section 6.

2 Microprocessor Verification

Formally verifying a processor consists of proving its instruction set with respect to its electronic implementation. This verification is usually decomposed in several proofs between two adjacent abstraction levels [28, 32].

Here we consider the specification being the instruction set architecture ($level_i$) and the implementation being the microprogram level ($level_{i+1}$). At all levels, the processor activity is modeled by state transitions. A state transition is a set of *concurrent* assignments to the processor components.

2.1 Verification of Simple Instructions

The formal verification is a technique for proving that a transition at $level_i$ ($trans_i$) is realized at $level_{i+1}$ by composing a set of transitions at $level_{i+1}$ ($trans_{i+1}^1 \circ \ldots \circ trans_{i+1}^n$). Implementation states are mapped to specification ones through abstraction (abs) mechanisms. To verify the correctness between two levels one must show that the following diagram commutes[1]:

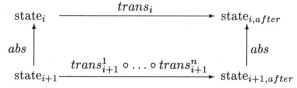

The system SVP verifies the commutation of such a diagram using symbolic execution combined with an efficient computer algebra system. It has successfully been used to verify some microprocessors with no loop in their microprograms : Tamarack [28] has been automatically proved from the electronic block level to the phase level, then to the microprogram level, and finally to the assembly instruction level [2]. The overall proof took less than 4 minutes on a Sun IPC workstation. AVM-1 [32] has been proved at the same levels in 30 minutes. The DP-32 processor has been specified from its original VHDL description [4] and then verified with SVP in 8 minutes [3]. A real processor called MTI, designed by CNET in France, has been verified from the microprogram level to the assembly level. Several errors have been found in its implementation.

---

[1][13] proposes a slightly different diagram to verify pipelined processors

2.2 Verification of Loop Instructions

The problem we specifically address in this paper is the verification of microprograms that contain loops. That is, in the above diagram, $trans_{i+1}^1 \circ \ldots \circ trans_{i+1}^n$
is not linear anymore but includes some "while" loops like in

$$trans_{i+1}^1 \circ \ldots \circ (\textbf{while } C \textbf{ do } (trans_{i+1}^p \circ \ldots \circ trans_{i+1}^q)) \circ \ldots \circ trans_{i+1}^n$$

This problem is solved using theorem proving when C is data-dependent.
When C is data-independent and the instruction is arithmetic, we prefer to use
symbolic execution with *BMDs because it is much more automatic.

Let us illustrate the notion of data-dependency on an example. Suppose
we want to describe the implementation of a multiplication instruction. We
give below two different algorithms that are functionally equivalent: the inputs
X and Y are two integers in $[0; 2^8[$, the result is R. Other variables are just
temporary buffers. After the completion of both algorithms we have $R = X * Y$.
The first algorithm, ALGO$_1$, is a multiplication by iterated additions while the
second, ALGO$_2$, is a loop of additions and shiftings.

```
ALGO₁:                        ALGO₂:
    R := 0;                       R := 0;
    Rₓ := X;                      Rₓ := X;
    while Rₓ > 0 do               C := 8 − 1;
        R := R + Y;               while C ≥ 0 do
        Rₓ := Rₓ − 1;                 if Rₓ mod 2 = 1 then R := R + Y * 2⁸;
    end while                         R := R div 2;
                                      Rₓ := Rₓ div 2;
                                      C := C − 1;
                                  end while
```

To symbolically execute these descriptions, we start by assigning *new* symbolic values to all variables: $X = \alpha$, $Y = \beta$,... The execution kernel then performs the assignments while simplifying expressions. When the loops are entered
for the first time, $R_x = \alpha$ and $C = 7$.

The algorithm ALGO$_1$ is not tractable anymore using symbolic execution
since we can not decide to enter the loop or to finish the execution because
$R_x > 0 \Leftrightarrow \alpha > 0$ is unknown. Backtracking to explore all cases is an infinite
task. In the algorithm ALGO$_2$, the condition $C \geq 0 \Leftrightarrow 7 \geq 0$ is decidable. In the
loop body, C is only decremented so its value is always a constant. Therefore
symbolic execution with a good symbolic representation is applicable on ALGO$_2$
but is not on ALGO$_1$.

2.3 Choosing the Best Verification Methodology

The three different verification methodologies used in SVP are described here.

1. The first one is symbolic execution with the help of a dedicated simplification system. This approach has many advantages: it is powerful enough

to handle integer and bit-vector level descriptions as well as abstractions but it does not hide data path functionalities. It is completely automatic and the proof times are usually very good.

2. The second methodology is also based on symbolic execution but symbolic values are represented using *BMDs. We needed to extend *BMDs in order to get exactly the same bit-vector operations provided by the previous method.

3. The last technique is theorem proving. We chose to use the Coq proof assistant [19] because it is an expressive and powerful higher-order logic system that deserves to be studied for formal hardware verification, derivation and synthesis.

SVP chooses the best methodology, among the three described above, in order to provide as much as possible automation and efficiency. Thus, the first attempt is done using the first methodology. It may fail due to a combinational explosion or a data-dependent loop. In the second case, recursive functions in the Coq formalism are automatically generated from the control graph. Then the proof is achieved with human guidance in Coq [1]. When the failure is due to a combinational explosion, an attempt is done using *BMDs. If it fails, Coq is used.

We have no place to detail our three methodologies here. We only show in the rest of the paper that *BMDs are useful to handle proof of some arithmetic instructions.

3 Binary Moment Diagrams

In this section we give a quick definition of *BMDs, their properties and their application for hardware verification. We summarize ideas found in [9, 10, 8].

3.1 Definition

*BMDs inherit from (reduced-order) BDD concepts: they represent functions over Boolean variables by acyclic graphs. Each node is labeled by a variable identifier and each variable appears at most once on a path according to a strict order. But *BMDs differ from BDDs in some aspects shown below.

Let f be a function from $\{0,1\}^n$ to $\mathbb{Z}$. BMDs are based on the Reed-Muller expansion of f with respect to a variable x. x is a Boolean variable so it has values 0 or 1 and f is decomposed in two cofactors[2]:

$$f = (1 - x) \cdot f_{\overline{x}} + x \cdot f_x$$

$f_{\overline{x}}$ is the value of f when $x = 0$, f_x is its value when $x = 1$. This can be rewritten as

$$f = f_{\overline{x}} + x \cdot f_{\dot{x}} \qquad \text{where} \quad f_{\dot{x}} = f_x - f_{\overline{x}}$$

---

[2] $+, -, \cdot$ represent addition, subtraction and multiplication for integers.

The first particularity of BMDs is that they decompose a function in two moments: $f_{\overline{x}}$ is the *constant moment*. It is the function value when $x = 0$. The other moment, $f_{\dot{x}}$, is the *linear moment*. It represents the variation of f when x changes from 0 to 1. A BMD is a tree with a variable identifier at its root and two BMDs as children that are the constant and linear moments.

BMDs have numeric values, called *weights*, on terminals (like MTBDDs [15]) but also on edges (like EVBDDs [29]). Weights are multiplied along the tree traversal, so the structure is called Multiplicative Binary Moment Diagram (*BMD). As for other compact symbolic representations, the sizes of *BMDs are reduced using sharing of common subtrees and so *BMDs are acyclic oriented and rooted graphs.

Bryant and Chen proposed synthesis functions by applying addition and multiplication on *BMDs. Moreover they gave rules in order to keep a canonical form so that a test of equality of two *BMDs is a direct comparison of weights and pointers.

Having defined *BMD, we show now their usefulness in hardware verification.

3.2 Using *BMDs in Hardware Verification

*BMDs are well-suited to represent integer values of bit-vectors. Let V be the vector of n bits $x_0, x_1, \ldots, x_{n-1}$. It has unsigned integer value $\sum_{i=0}^{i=n-1} 2^i * x_i$. Therefore it is represented with $n = 3$ by the *BMD of Figure 1 (a). Unsigned, 2's complement and sign-magnitude integers are represented by *BMDs which sizes grows linearly with the number of bits while the sizes have exponential growths with MTBDDs. Addition and multiplication representations also grow linearly with the word size as shown on Figure 1 (b) and (c).

Furthermore, *BMDs can efficiently represent Boolean functions *and, or, xor, not*: they are numeric functions with codomain $\{0; 1\}$. Therefore, *BMDs can also be used to construct representations of combinational circuits described at the gate-level [24]. But in several cases, BDDs are more efficient [8].

Notations. In the rest of the paper, we note $\langle w, Z \rangle$ a *BMD with *weight* w. Z is a *vertex*; it can be a *terminal*, noted Λ, or a decomposition of Z with respect to x, noted $\lfloor x, Z_{\overline{x}}, Z_{\dot{x}} \rfloor$. We use $+, -, *$ for addition, subtraction and multiplication on *BMDs and sometimes abbreviate notation of terminals $\langle w, \Lambda \rangle$ by just writing w.

4 The Integration of *BMDs into a Verification System

In this section we show how we extend *BMDs so that they may be integrated in a verification system working at the integer and bit-vector levels. Our goal is to use the same specification methodology, and the same symbolic execution kernel, independently of the chosen representation (symbolic bit-vectors or *BMDs).

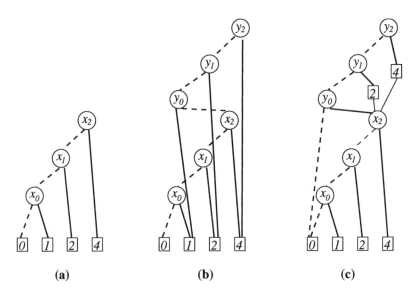

Fig. 1: *BMD representations for an unsigned word (a), addition (b) and multiplication (c). Weight equal to 1 are not shown, dashed lines are constant moments, plain lines are linear moments.

4.1 Our *BMD Representation of Bit-Vectors

We want to deal with *BMDs exactly as if they were bit-vectors, so we have to take care of the vector size. Therefore we augment the concept of *BMD by attaching additional information. We call $*BMD_{(s)}$ a *BMD representation of a bit-vector of size s. We write $Z_{(s)}$ to note that Z is a *BMD representation of the unsigned value of a s-bit vector[3]. Whereas *BMDs represent functions from $\{0,1\}^n$ to $\mathbb{Z}$, $*BMD_{(s)}$ represent functions from $\{0,1\}^n$ to integers in $[0;2^s[$. *BMD representations of Boolean functions are $*BMD_{(1)}$. We note $*BMD_{()}$ the set of all $*BMD_{(s)}$ for any $s > 0$.

4.2 Primitive Functions

Usual bit-vectors operations cannot be easily defined by a structural induction on *BMDs. They are more naturally defined using the inductive structure of a vector: it is composed of a head bit and a tail bit-vector. Therefore we first define three primitives on $*BMD_{()}$ in order to reproduce the bit-vector structure: *cons* takes as arguments a bit B and a bit-vector V. It returns a bit-vector which least significant bit is B and most significant ones are V. The *lsb* primitive returns the least significant bit, and *msbs* the most significant ones so that the following equations hold:

$$lsb(cons(B,V)) = B \quad \text{and} \quad msbs(cons(B,V)) = V$$

---

[3] s is not related to the number of variables of Z but it is the minimal number of bits required to represent all possible values of Z.

The *cons* **primitive** takes as inputs a *BMD$_{(1)}$ representing a bit and a *BMD$_{(s)}$ representing a bit-vector. The result is a *BMD$_{(s+1)}$:

$$\text{(1)} \qquad cons(B_{(1)}, V_{(s)}) \stackrel{\text{def}}{=} (B + 2 * V)_{(s+1)}$$

The *lsb* **primitive.** We only handle unsigned words so the least significant bit is 1 when the word value is odd, and 0 when it is even. Therefore, the *lsb* function is determined by the parity of the *BMD$_{(s)}$, assuming that $s > 0$:

$$\text{(2)} \qquad s > 0 \;\rightarrow\; lsb(V_{(s)}) \stackrel{\text{def}}{=} (odd(V))_{(1)}$$

The *odd* function on *BMDs is strongly related to properties of parity with respect to multiplication and addition. It returns the terminal *BMD $\langle 1, \Lambda \rangle$ when the input is odd and $\langle 0, \Lambda \rangle$ when it is even. But in most cases we cannot decide whether a function is always odd or even, so the result of *odd* is a non-terminal *BMD. We define here this function by an induction on the tree structure of a *BMD. The terminal case is when the input is a terminal. The parity is then just the parity of its weight:

$$\text{(3)} \qquad odd(\langle w, \Lambda \rangle) \stackrel{\text{def}}{=} \langle w \bmod 2, \Lambda \rangle$$

We can also return a terminal *BMD when the weight is even because the product of an even number by any number is always even:

$$\text{(4)} \qquad w \bmod 2 = 0 \;\rightarrow\; odd(\langle w, Z \rangle) \stackrel{\text{def}}{=} \langle 0, \Lambda \rangle$$

In the general case, the weight is odd and it is now useless. Let Z be a vertex. Starting from the moment decomposition of Z with respect to variable x, we expand $odd(Z)$:

$$odd(Z) \stackrel{\text{def}}{=} odd(Z_{\overline{x}} + x \cdot Z_{\dot{x}})$$

Since the parity of a sum is the exclusive disjunction of the parities we have

$$odd(Z) \stackrel{\text{def}}{=} odd(Z_{\overline{x}}) \text{ xor } odd(x \cdot Z_{\dot{x}})$$

From [9] we have $a \text{ xor } b = a + b - 2ab$ so

$$odd(Z) \stackrel{\text{def}}{=} odd(Z_{\overline{x}}) + odd(x \cdot Z_{\dot{x}}) - 2 * odd(Z_{\overline{x}}) * odd(x \cdot Z_{\dot{x}})$$

as $\forall g,\, odd(x \cdot g) = x \cdot odd(g)$ when $x \in \{0, 1\}$,

$$odd(Z) \stackrel{\text{def}}{=} odd(Z_{\overline{x}}) + x \cdot (odd(Z_{\dot{x}}) - 2 * odd(Z_{\overline{x}}) * odd(Z_{\dot{x}})) \qquad \square$$

This shows that $odd(Z)$ has constant moment $odd(Z_{\overline{x}})$ and linear moment $odd(Z_{\dot{x}}) - 2 * odd(Z_{\overline{x}}) * odd(Z_{\dot{x}})$. Thus the odd function for *BMDs is defined in the general case when w is odd as follows:

(5)

$w \bmod 2 \neq 0 \rightarrow$

$odd(\langle w, \lfloor x, Z_{\overline{x}}, Z_{\dot{x}} \rfloor \rangle) \stackrel{\text{def}}{=} \langle 1, \lfloor x, odd(Z_{\overline{x}}), odd(Z_{\dot{x}}) - 2 * odd(Z_{\overline{x}}) * odd(Z_{\dot{x}}) \rfloor \rangle$

The $msbs$ primitive is built using the same recursive schema:

(6) $\qquad\qquad s > 0 \rightarrow msbs(V_{(s)}) \stackrel{\text{def}}{=} (half(V))_{(s-1)}$

where $half$ returns the *BMD representation for the half of an input *BMD. The terminal case is[4]:

(7) $\qquad\qquad\qquad half(\langle w, \Lambda \rangle) \stackrel{\text{def}}{=} \langle w \div 2, \Lambda \rangle$

and the particular case where the weight is even is directly resolved:

(8) $\qquad\qquad w \bmod 2 = 0 \rightarrow half(\langle w, Z \rangle) \stackrel{\text{def}}{=} \langle w \div 2, Z \rangle$

The general case, when w is odd, is reduced to:

(9)

$w \bmod 2 \neq 0 \rightarrow half(\langle w, \lfloor x, Z_{\overline{x}}, Z_{\dot{x}} \rfloor \rangle) \stackrel{\text{def}}{=} \langle 1, \lfloor x, A, B \rfloor \rangle$

$\qquad\qquad$ where $A = (w \div 2) * Z_{\overline{x}} + half(Z_{\overline{x}})$

$\qquad\qquad$ and $B = (w \div 2) * Z_{\dot{x}} + half(Z_{\dot{x}}) + odd(Z_{\overline{x}}) * odd(Z_{\dot{x}})$

To efficiently implement these primitives, we used some optimizations: Equations 5 and 9 are simplified when a recursive call to $odd(Z_{\overline{x}})$ or $odd(Z_{\dot{x}})$ returns a terminal *BMD. Moreover, as odd and $half$ follow the same recursive schema, and because they are often used together, we defined a function that computes both results in only one traversal. We implemented the primitives in order to keep the canonical form and maintain a hash table of previously computed results.

Proof. The general proof of correctness of our three primitives corresponds to the verification of the following equation:

$$\forall s > 0, \ \forall X_{(s)} \in {}^*\text{BMD}_{(s)}, \quad cons(lsb(X_{(s)}), msbs(X_{(s)})) = X_{(s)}$$

We do not have enough place to completely detail the proof in this paper. It is achieved by first unfolding with Equations 1, 2 and 6. Then, we get the goal $odd(X) + 2 * half(X) = X$ which is resolved in three cases by Equations 3 and 7, 4 and 8, 5 and 9.

---

[4] $\div$ denotes integer division.

Example. Let $f(v_2, v_1, v_0) = v_2 + 3v_1 + 4v_0 + 2$. Our primitives compute $odd_f = v_2 + v_1 - 2v_2v_1$ and $half_f = v_2v_1 + v_1 + 2v_0 + 1$. One can verify that in this case $2 * half_f + odd_f = f$.

4.3 Bit-vector operators

Having defined the primitives, we can now define several bit-vector operators involved in microprograms.

The *concat* operator returns the concatenation of two *BMD$_{()}$:

$$concat(Z_{(s)}, Z'_{(s')}) \stackrel{\text{def}}{=} (Z + 2^s * Z')_{(s+s')}$$

Bit field extraction is implemented using our primitives. $field(Z_{(s)}, i, j)$ extracts bits of $Z_{(s)}$ with positions between i and j (bit at position 0 is the least significant one):

$$s > 0 \;\rightarrow\; field(Z_{(s)}, 0, 0) \stackrel{\text{def}}{=} lsb(Z_{(s)})$$

$$s > j > 0 \;\rightarrow\; field(Z_{(s)}, 0, j) \stackrel{\text{def}}{=} cons(lsb(Z_{(s)}), field(msbs(Z_{(s)}), 0, j - 1))$$

$$s > j \geq i > 0 \;\rightarrow\; field(Z_{(s)}, i, j) \stackrel{\text{def}}{=} field(msbs(Z_{(s)}), i - 1, j - 1)$$

The conditional operator, *ite*, represents a multiplexer output. The condition is a *BMD$_{(1)}$, the two possible results are *BMD$_{(s)}$:

$$ite(C_{(1)}, Z_{(s)}, Z'_{(s)}) \stackrel{\text{def}}{=} (C * Z + (1 - C) * Z')_{(s)}$$

Other operators are easily defined using the three primitives and the previous operators. For example we also provide bitwise logical operations, rotations and shifts, or the two outputs (*add* and *carry*) of an adder described at the bit-vector level. For example :

$$add(Z_{(s)}, Z'_{(s)}) \stackrel{\text{def}}{=} field((Z + Z')_{(s+1)}, 0, s - 1)$$

$$carry(Z_{(s)}, Z'_{(s)}) \stackrel{\text{def}}{=} field((Z + Z')_{(s+1)}, s, s)$$

where $+$ is the *BMD addition as described is [9].

5 Examples and Experimental Results

We illustrate here the use of *BMD$_{()}$ to verify arithmetic instructions. We compare our results with similar works and discuss the usefulness of *BMDs.

5.1 Verification of a Multiplication Microprogram

We focus here on a multiplication microprogram so that all other processor details are irrelevant and their verifications are orthogonal. We consider four n-bit registers: X, Y which hold the two inputs and R_l, R_h in which the result is built (they are the least and most significant parts of the result). The instruction semantics is as follows:

$$multi: \quad \begin{aligned} R_l &\leftarrow & field(X * Y, 0, n - 1) \\ R_h &\leftarrow & field(X * Y, n, 2n - 1) \end{aligned}$$

The microprogram of this instruction implements the algorithm ALGO2 presented in 2.2 but here all variables are bit-vectors and the register R is split in R_l and R_h. Moreover, R_x is now a part of R_l. The "div 2" operations of ALGO2 are translated into shiftings of bit-vectors that are in fact combinations of bit extractions and concatenations. The microprogram follows:

$$\mu_1: \quad \begin{aligned} R_l &\leftarrow & X \\ R_h &\leftarrow & \langle 0, \Lambda \rangle_{(n)} \\ C &\leftarrow & \langle n - 1, \Lambda \rangle_{(\log_2 n)} \end{aligned}$$

while $C \geq \langle 0, \Lambda \rangle_{(\log_2 n)}$ **do**

$$\mu_2: \quad \begin{aligned} R_l &\leftarrow & concat(msbs(R_l), ite(lsb(R_l), lsb(add(R_h, Y)), lsb(R_h))) \\ R_h &\leftarrow & concat(msbs(ite(lsb(R_l), add(R_h, Y), R_h)), \\ & & ite(lsb(R_l), carry(R_h, Y), \langle 0, \Lambda \rangle_{(1)})) \\ C &\leftarrow & C - \langle 1, \Lambda \rangle_{(\log_2 n)} \end{aligned}$$

The verification starts from symbolic states: each component initial value is a *BMD$_{()}$ representation of a symbolic bit-vector like the one of Figure 1 **(a)**. At the abstract level it executes the *multi* semantics. At the microinstruction level it executes the microprogram to build the final *BMD$_{(n)}$ values of R_l and R_h using our extended algebra. Finally the values of R_l and R_h at both levels are compared. This verification is completely automatic using symbolic execution. The user has not to worry about the concrete representation of bit-vectors.

5.2 Results and Discussion

The verification of this multiplication instruction is fast. Table 1 shows the final *BMD size (in number of nodes) of $concat(R_l, R_h)$ for different word sizes (n). The third column shows the size of the largest intermediate *BMD that is generated during the verification. This gives an idea of the memory requirements. In the fourth column we show running time needed to perform the complete verification including the symbolic execution, the synthesis of *BMDs and garbage collections. The last column gives garbage collection time. The experiment was performed on a Sun SparcStation 10 running a Scheme interpreter.

One can see that the final size grows linearly in n ($2n+1$), that the maximum size is polynomial ($1/2n^2 + 3/2n + 1$) and running time has a cubic growth.

n	final size	max size	time (s)	GC time (s)
8	17	45	1.150	0.000
16	33	153	4.433	0.000
32	65	561	38.333	6.433
64	129	2145	359.900	73.133
128	257	8385	3203.166	746.500

Table 1: Verification results of multiplication microprogram.

We tried different variable orderings: bits of X first or last, bits of X and Y interlaced, most significant bits first or last, etc. Our experiments have shown that the variable ordering is not crucial: final *BMD sizes are $2n+1$ in the best cases and $4n-1$ in the worst ones. Sizes of largest *BMDs are exactly the same for all orderings and running times are very similar.

Similar verifications have been done using theorem proving but required much human intervention especially to find a correct invariant.

Several multipliers[5] were verified using BDDs and related data structures. The best results have been obtained using special optimizations [11, 6]. Our results are more promising because BDD is not a good representation for multiplication: it has an exponential growth [7] while *BMD has a linear one [10].

Bryant and Chen verified multipliers with *BMDs [10] using the hierarchical verification methodology proposed in [29]. They first proved that each component is correct with respect to its word-level specification. This is performed using BDDs, a conversion from BDD to *BMD and an encoding. Then they proved that the composition of components is correct with respect to the word-level specification of the multiplier. Their running times are better than ours but should not be compared: they adopted a hierarchical methodology. We have integrated *BMDs in our verification system so that this is completely transparent from a user point of view. Therefore we had to follow a non-hierarchical methodology that is less efficient but more automatic.

Hamaguchi *et al.* proposed another automatic verification technique with *BMDs[24]. It is based on the substitution operation: the substitution of a variable in a *BMD by another *BMD. They construct a *BMD representation of a circuit starting from its output and iteratively replacing variables up to the inputs. Since a Boolean variable can only be substituted by a *BMD representation of a Boolean function, this technique does not apply to the abstraction levels considered in this paper but only to the gate level.

---

[5]Most of them are combinational. Actually, very few integer multiplication instructions are now microcoded.

5.3 Other Results

We applied our methodology to verify microprograms implementing other arithmetic instructions: squaring, exponentiation and division.

For the squaring instruction, the sizes of the final and of the largest *BMDs have quadratic growths. For the exponentiation (X^Y), the sizes are linear if X is a constant. If Y is a constant and X is symbolic, there is an exponential blowup due to the complexity of the largest intermediate *BMD. And, not surprisingly, when X and Y are both symbolic, the size of largest and final *BMDs grow exponentially (about $O(4^n)$).

The division is implemented by a microprogram very similar to the one of the multiplication. Our experiments have shown that the sizes of *BMDs representing quotient and remainder grow exponentially (about $O(3^n)$). This seems to be a consequence of the fact that the microprogram contains a comparative test of two *BMDs: $f - g < 0$. But finding solutions for this problem is difficult because *BMD, unlike BDD, is not an "easily invertible representation" (see [9] for further explanations).

6 Conclusion

In order to formally verify complex arithmetic assembly instructions of microprocessors, we experimented the use of *BMDs as a symbolic representation of bit-vectors. We defined *BMD$_{()}$ which is an extension of *BMD and a *BMD$_{()}$ algebra that includes almost all operators needed to write bit-vector level descriptions. This extension has been integrated in a larger verification framework.

Our experimental results have shown that the limitations of our *BMD$_{()}$ algebra are the ones of *BMDs. Actually, *BMD is not a panacea to verify arithmetic instructions or circuits. A lot of problems are not tractable using this representation only. It seems that operations that are easily performed with BDDs are difficult with *BMDs and vice versa. Thus a combination of different representations should be a good way to deal with such difficult verifications. Recently, decision diagrams mixing different decompositions have been proposed [14, 21]. An algebra similar to the one we described here for *BMD$_{()}$ should easily be defined for these representations.

Our *BMD$_{()}$ algebra could also be used to treat a more actual – but much more difficult – problem: the verification of floating point microprograms. This verification is hard, even when using theorem proving, because of the complexity of the data and the need for approximations.

Acknowledgments. The author would like to thank H. Collavizza for fruitful discussions, kind advices, and constructive comments on early versions of this paper.

References

1. L. Arditi. Formal verification of microprocessors: a first experiment with the Coq proof assistant. Technical Report RR-96-31, Université de Nice – Sophia Antipolis. Laboratoire I3S, May 1996.

2. L. Arditi and H. Collavizza. An object-oriented framework for the formal verification of processors. In *European Conference on Object-Oriented Programming*, volume 952 of *LNCS*, 1995.

3. L. Arditi and H. Collavizza. Towards verifying VHDL descriptions of processors. In *European Design Automation Conference with EURO-VHDL*, 1995.

4. P. J. Ashenden. *The VHDL Cookbook*. Public Domain, Dept. Computer Science, University of Adelaide, South Australia, first edition, July 1990.

5. D. L. Beatty and R. E. Bryant. Formally verifying a microprocessor using a simulation methodology. In 31st *ACM/IEEE Design Automation Conference*, 1994.

6. J. Bern, C. Meinel, and A. Slobodová. Efficient OBDD-based boolean manipulation in CAD beyond current limits. In 32nd *ACM/IEEE Design Automation Conference*, 1995.

7. R. E. Bryant. Graph-based algorithms for boolean function manipulation. *IEEE Transactions on Computers*, C-35(8):677–691, Aug. 1986.

8. R. E. Bryant. Binary decision diagrams and beyond: Enabling technologies for formal verification. In *IEEE Internationl Conference on Computer-Aided Design*, 1995.

9. R. E. Bryant and Y.-A. Chen. Verification of arithmetic functions with binary moment diagrams. Technical Report CMU-CS-94-160, School of Computer Science, Carnegie Mellon University, June 1994.

10. R. E. Bryant and Y.-A. Chen. Verification of arithmetic circuits with binary moment diagrams. In 32nd *ACM/IEEE Design Automation Conference*, 1995.

11. J. Burch. Using BDDs to verify multipliers. In 28th *ACM/IEEE Design Automation Conference*, 1991.

12. J. Burch, E. Clarke, K. McMillan, and D. Dill. Sequential circuit verification using symbolic model checking. In 27th *ACM/IEEE Design Automation Conference*, 1990.

13. J. Burch and D. Dill. Automatic verification of pipelined microprocessor control. In *Computer-Aided Verification*, volume 818 of *LNCS*, 1994.

14. E. Clarke, M. Fujita, and X. Zhao. Hybrid decision diagrams. overcoming the limitations of MTBDDs and BMDs. In *IEEE Internationl Conference on Computer-Aided Design*, 1995.

15. E. Clarke, K. McMillan, X. Zhao, M. Fujita, and H.-Y. Yang. Spectral transforms for large Boolean functions with application to technology mapping. In 30th *ACM/IEEE Design Automation Conference*, 1993.

16. E. C. Clarke, O. Grumberg, and D. E. Long. Model checking and abstraction. In *ACM Symposium on Principles of Programming Languages*, 1992.

17. A. Cohn. A proof of correctness of the Viper microprocessor: the first level. In *VLSI Specification, Verification and Synthesis*. Kluwer Acad. Publ., Jan. 1987.

18. F. Corella. Automatic high-level verification against clocked algorithmic specifications. In 11th *International Conference on Computer Hardware Description Languages and their Applications*. Elsevier Science Publishers B.V., 1993.

19. Cornes, Courant, Filliâtre, Huet, Manoury, Mũnoz, Murthy, Parent, Paulin-Mohring, Saïbi, and Werner. The Coq Proof Assistant. Reference Manual. Version 5.10. Technical report, INRIA Rocquencourt – CNRS-ENS Lyon, 1996.

20. O. Coudert, C. Berthet, and J.-C. Madre. Verification of synchronous sequential machines based on symbolic execution. In *Automatic Verification Methods for Finite State Systems*, volume 407 of *LNCS*, 1989.

21. R. Drechsler, B. Becker, and S. Ruppertz. K*BMDs: a new data structure for verification. In *European Design and Test Conference*, 1996.

22. M. Gordon. Proving a computer correct using LCF–LSM hardware verification system. Technical Report 42, Computer Laboratory, The University of Cambridge, Sept. 1983.

23. A. Gupta. Formal hardware verification methods: a survey. *Formal Methods in System Design*, 1(2/3):151–238, Oct. 1992.

24. K. Hamaguchi, A. Morita, and S. Yajima. Efficient construction of binary moment diagrams for verifying arithmetic circuits. In *IEEE Internationl Conference on Computer-Aided Design*, 1995.

25. R. Hojati and R. K. Brayton. Automatic datapath abstraction in hardware systems. In *Computer-Aided Verification*, volume 939 of *LNCS*, 1995.

26. A. D. Hu, D. L. Dill, A. J. Drexler, and H. C. Yang. Higher-level specification and verification with BDDs. In *Computer-Aided Verification: Fourth International Workshop*, volume 663 of *LNCS*, 1992.

27. W. Hunt Jr. *FM8501: a Verified Microprocessor*. PhD thesis, Institute for Computing Science, University of Texas at Austin, 1986.

28. J. Joyce. *Multi Level Verification of Microprocessor-Based Systems*. PhD thesis, University of Cambridge, Computer Laboratory, 1989.

29. Y.-T. Lai and S. Sastry. Edge-valued binary decision diagrams for multi-level hierarchical verification. In 29th *ACM/IEEE Design Automation Conference*, 1992.

30. S. Rajan, N. Shankar, and M. Srivas. An integration of model checking with automated proof checking. In *Computer-Aided Verification*, volume 939 of *LNCS*, 1995.

31. M. K. Srivas and S. P. Miller. Applying formal verification to a commercial microprocessor. In *International Conference on Computer Hardware Description Languages*, 1995.

32. P. J. Windley. Formal modeling and verification of microprocessors. *IEEE Transactions on Computers*, 44(1), Jan. 1995.

33. Z. Zhu, J. Joyce, and C. Seger. Verification of the Tamarack-3 microprocessor in a hybrid verification environment. In *Higher Order Logic Theorem Proving and Its Applications*. 6th *International Workshop*, volume 780 of *LNCS*, 1993.

Modular Verification of Multipliers*

Kavita Ravi   Abelardo Pardo   Gary D. Hachtel   Fabio Somenzi

Dept. of Electrical and Computer Engineering
University of Colorado at Boulder

Abstract

We present a new method for the efficient verification of multipliers and other arithmetic circuits. It is based on modular arithmetic like Kimura's approach, and on composition, like Hamaguchi's approach. It differs from both in several important respects, which make it more robust. The technique builds the residue Algebraic Decision Diagram (ADD) of as many variables as the number of outputs in the multiplier and composes the implementation circuit from the outputs to the inputs into the residue. Finally, the residue ADD is checked against the specification.

1   Introduction

The verification of arithmetic circuits has been the subject of much recent work [8, 9, 14, 20, 6, 17, 13].

Multipliers have been recognized earlier on as difficult cases for verification methods based on decision diagrams. Bryant [8] proved that the BDDs for array multipliers grow exponentially in size regardless of the variable order. Experiments show that the growth is proportional to $2^{3n/2}$, and that it is impractical to build BDDs for multipliers with $n > 12$. Henceforth, several approaches have been proposed based on some relaxation of the variable ordering requirement. For instance, Burch proposed splitting one input variable into several ones, so as to eliminate the reconvergent fanout at the root of the BDD explosion [9]. Jain later extended this idea by allowing a variable to appear multiple times along a path in the BDD, each time with a different index [14]. Plessier, on the other hand, devised a technique based on a semi-canonical representation, that is, one that mixed pure BDDs with a structural representation of the multiplier [20]. A similar approach was presented in [1]. All the approaches were able to deal with multipliers with 16 bits at most.

A major step forward in the verification of multipliers was made possible by the introduction of Multiplicative Binary Moment Diagrams (*BMDs) by Bryant [6]. *BMDs afford very compact, word-level representations of many arithmetic functions, including multipliers, but not dividers. Word-level representations are functions mapping bit vectors to integers. Two verification methods for multipliers based on *BMDs have been devised. Bryant has proposed a hierarchical

---

*This work was supported in part by NSF/DARPA grant MIP-94-22268 and SRC contract 95-DJ-560.

approach in which the building blocks of the multiplier being verified—for instance a carry-save adder—are checked against their word-level specifications. Once the individual sub-blocks have been verified, they are composed to yield the function of the entire block.

This approach had been originally proposed by Lai and Sastry [18]. However, with Edge-Valued BDDs instead of *BMDs, the word level specification of multipliers had remained intractable. Bryant's hierarchical verification of multipliers is very fast and can deal with very large word lengths. The word-level specification in terms of *BMDs is a natural and concise form of behavioral specification for arithmetic circuits. However, the hierarchical approach assumes that the specification has the same hierarchy as the implementation, and that the corresponding blocks in the specification and the implementation are equivalent. Both conditions may be invalidated by optimization techniques applied to the implementation. In particular, logic optimization may make changes whose validity depends on the *don't care* conditions at the periphery of a block, consequently destroying structural and hierarchical correspondence between the two. For instance the *don't care* conditions may be inputs that are never be presented to the block, or distinct outputs produced by the block that elicit the same response at the primary outputs. Several optimization techniques commonly in use rely on this *don't care* information to reduce area, delay, and power, and hence make hierarchical verification inapplicable.

Hamaguchi proposed a method to overcome this limitation, which is a non-hierarchical variant of Bryant scheme [13]. Assuming that the multiplier is to compute $X \cdot Y = Z$, the procedure is given three inputs: A *BMD for the specification $X \cdot Y$, a *BMD for the result Z, and a gate-level circuit for the implementation. Regarding the gate-level circuit as a DAG, with vertices corresponding to gates and an arc (g_1, g_2) if gate g_1 feeds gate g_2, the gates of the implementation are composed into the *BMD of Z one by one in some reverse topological order. If the implementation is correct, the *BMD obtained at the end of the process is identical to the *BMD of the specification. This method has been successfully applied to multipliers with up to 64 bits. It is much slower that the hierarchical approach, but more general. One serious limitation of Hamaguchi's method, however, is in dealing with incorrect implementations. When the implementation of the multiplier contains errors, the *BMDs obtained by composition normally become very large. This can be intuitively justified by considering that the compactness of *BMDs for multipliers relies on a very high degree of recombination, which is easily destroyed by even small changes in the function.

So far we have considered only methods that rely primarily on decision diagrams for the verification of multipliers. Other methods are not hindered by the possibly exponential size of the canonical representations, and hence have been successful at verifying moderate sized multipliers. The more advanced of these other methods rely on finding corresponding points in the two circuits being compared [4, 21, 15]. These methods have been successfully applied to 16-bit multipliers, and can probably deal with larger instances as well. However, they are restricted to comparing specifications and implementations that have quite similar structures. Unfortunately, even related variants of the same type of multiplier—for instance a two's complement array multiplier based on carry-save adders—tend to have few internal corresponding points.

A forerunner to this paper is Kimura's method to verify multipliers using modular arithmetic. Kimura uses the Chinese Remainder Theorem and the fact that residues of multiplication depend only on the residues of the operands, consequently keeping under check the growth of the size of representation (Binary Decision Diagrams). However, his method makes an assumption that is valid only for a particular type of Binary Decision Diagrams (BDDs) for which the sizes can be much larger than the most efficient representation of BDDs. A detailed discussion of Kimura's approach is deferred to Section 2.2.

The method presented in this paper is also based on modular arithmetic and uses the Chinese Remainder Theorem. A variation of BDDs, Algebraic Decision Diagrams (ADDs), are used for the representation of the residues. The gate-level representation of the implementation is composed, from the outputs to the inputs, into the residue ADD. The final ADD is checked against the specification ADD for correctness. The main virtue of this method is the reduced size of verification as compared to BDDs and the reduced size of verifying incorrect circuits as compared to *BMDs.

This paper is organized as follows: In Section 2, we introduce decision diagrams and present Kimura's method. Section 3 contains our techniques applied to residue verification. Section 4 presents experimental results and Section 5 presents the conclusions.

2 Preliminaries

2.1 Decision Diagrams

Binary Decision Diagrams (BDDs) [7, 3] are an efficient data structure for the representation of logic functions. A BDD representing a set of functions $\{f_1, \ldots, f_n\}$ is a directed acyclic graph (DAG) with n roots—one for each function—and two leaves. One leaf represents the constant 1 and the other represents the constant 0. An internal node N of the DAG is labeled with a variable v and has two children, T and E. The function represented by N, denoted by f_N is given by:

$$f_N = v \cdot f_T + v' \cdot f_E.$$

It is customary to impose the restriction that the variables be ordered along all paths in the DAG and that no isomorphic subgraphs exist. Under these restrictions, BDDs provide a canonical representation of logic functions.

The success of BDDs in solving seemingly intractable problems has motivated researchers to consider variants of the basic data structure that support a wider variety of applications [2, 11, 18, 6] or that are very efficient for some classes of problems [19, 16]. Among these variants, Algebraic Decision Diagrams (ADDs) extend BDDs by allowing an arbitrary number of leaves. The leaves may store values drawn from a given set—for instance, the integers. ADDs are formally

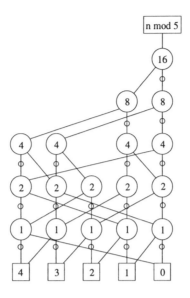

Figure 1: ADD for the residue modulo 5 of a five-digit integer.

defined as boolean functions over a boolean algebra whose carrier is larger than the number of the leaves. Thanks to this definition, all theorems of boolean algebra apply to ADDs [2]. Figure 1 shows the ADD for the function that associates to each number between 0 and 31 its residue modulo 5. In the figure, each variable corresponds to a bit of the binary representation of the argument, and each internal node is labeled with the weight of the corresponding variable. Hence, the most significant bit of the input is at the top. The arcs with a small circle point to the E children. Input 24, which translates to 11000 in binary, corresponds to the leftmost path in the diagram. The path leads to the constant 4, which is indeed the residue of 24 modulo 5.

ADDs are unsuited to the word-level representation of arithmetic circuits: The number of leaves grows exponentially with the number of variables for many common functions like $f(x) = \sum_{i=0}^{n-1} 2^i x_i$. However, as we shall see, this is not a drawback when dealing with modular arithmetic. Furthermore, ADDs require less space per node than other forms of decision diagrams like Edge-Valued BDDs or *BMDs. The manipulation algorithms are also simpler and usually faster. Therefore, we have chosen ADDs to represent residues.

2.2 Modular Arithmetic

Let Z_m be the additive group modulo m, and let $m = m_1 \cdots m_p$, where the m_i are pairwise relatively prime. The Chinese remainder theorem states that, for $x \in Z_m$ and $x_i = x \bmod m_i$, the correspondence $x \leftrightarrow (x_1, \ldots, x_p)$ is a bijection.[1] It follows that for $(x, y) \in Z_m^2$, $x = y \Leftrightarrow (x_1, \ldots, x_p) = (y_1, \ldots, y_p)$. To check for

---

[1] A proof of this result, in the context of an excellent introduction to modular arithmetic, can be found in [12].

identity of two functions $f, g : D \mapsto Z_m$, it is sufficient to check for identity of f mod m_i and g mod m_i, for $i = 1, \ldots, p$.

The properties of modular arithmetic have been used for a long time to design error correcting circuits, and fast arithmetic circuits. Recently, Clarke *et al.* [10] suggested the use of the Chinese remainder theorem as an abstraction mechanism in formal verification, but gave no details and presented no examples of application. The first reported results of the application of modular arithmetic to formal verification are due to Kimura [17].

He observed that the BDDs of a function that depends only on the residue modulo k of its input have maximum width equal to k. Since the residues of multiplication depend only on the residues of the operands, their BDDs grow polynomially with the large word length of the multiplier. The problem, however, is that one cannot build the BDDs of the multiplier and then obtain the BDDs for the residues, because the first step cannot be carried out for large word lengths. To circumvent this difficulty, Kimura introduces an operation, called rsd_bdd, which converts a BDD into another one with width less than or equal to k.

Kimura's procedure uses rsd_bdd while building the BDDs for the multiplier, to prevent the intermediate results from growing too large. The correctness of the procedure relies on a lemma (Lemma 4 in [17]) that states that rsd_bdd commutes with the boolean operations. Let $[f]_{\mathrm{mod}k}$ be the result of applying rsd_bdd to f with a maximum prescribed width of k, and let ∘ be a boolean operation. Then, Lemma 4 of [17] states that, if $f = g \circ h$, then $[f]_{\mathrm{mod}k} = [g]_{\mathrm{mod}k} \circ [h]_{\mathrm{mod}k}$.

However, the proof of the lemma is incorrect, as shown by the counterexample of Figure 2. In the upper part two BDDs, g and h, are shown along with the BDD for their intersection f. The index of each variable indicates its weight. Since the intersection has at most one node for each variable, $[f]_{\mathrm{mod}k}$ is the same as f. The lower part of Figure 2 illustrates the application of rsd_bdd to g for a maximum prescribed width of 2. For both nodes labeled 2, the residue is 0. Hence, the second node encountered is suppressed. The application of rsd_bdd to h produces the same result, which is also, therefore, $[g]_{\mathrm{mod}2} \wedge [h]_{\mathrm{mod}2} \neq [f]_{\mathrm{mod}2}$.

The proof of Lemma 4 is valid for what Kimura calls *levelized* BDDs, that is, BDDs in which nodes with identical children are not suppressed. (Others call these diagrams *quasi-reduced*.) Unfortunately, levelized BDDs are larger than BDDs, and require specialized manipulation algorithms.

The application of residue arithmetic to verification is not restricted to multipliers. Indeed, every pair of combinational circuits can be compared by interpreting their outputs as integers, and comparing a suitable number of residues. However, for functions that do not depend on the residue of the input, the computation of residues may not afford any simplification. For instance, although the BDDs for the residues of a multiplier are polynomial, the BDDs for each individual output of the multiplier are unaffected by the application of the residue function and therefore are exponential.

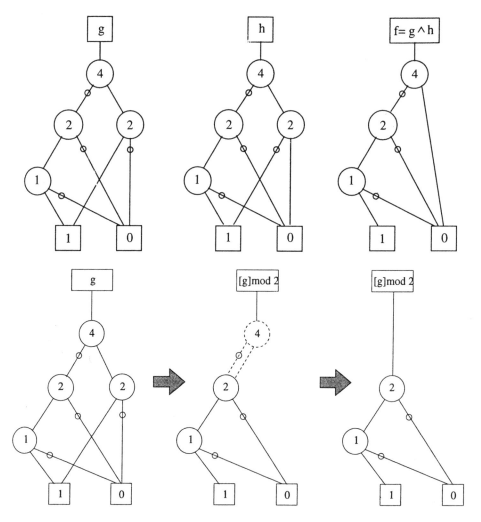

Figure 2: Counterexample to Lemma 4 of [17].

3 Efficient Computation of Residues

3.1 Verification Scheme

In what follows, the circuits to be verified are assumed to have $2n$ inputs and q outputs. The choice of $2n$ is merely for convenience of exposition.

Verification proceeds as follows. Two circuits, $\mathcal{S}$ (specification) and $\mathcal{I}$ (implementation) are given, described by gate-level circuits. A bijection between the inputs as well as the outputs of the two circuits is given. (Typically, corresponding inputs/outputs have the same name in the two circuits.) The specification may be incomplete: A predicate D over the input variables tells for what input values the output is unspecified. From the number of primary outputs, q, a list of moduli is derived, in a way described in Section 3.2. For each modulus, m_i, an ADD, M, is built, which represents the residue of a q-bit binary number modulo m_i. (The case of $q = m_i = 5$ was shown in Figure 1.) The input variables of M are the output variables of $\mathcal{S}$. The gates of $\mathcal{S}$ are then composed into M, starting from the gates driving the primary outputs and proceeding towards the primary inputs. The composition process is described in detail in Section 3.3. Call the result of composition R_S. The process is then repeated for the implementation, obtaining R_I. If $D = 0$, then R_S and R_I must be identical, which can be checked in constant time. If $D \neq 0$, a procedure is called that traverses R_S, R_I, and D simultaneously and tells whether there exist assignments to the inputs for which the output is specified and the implementation disagrees with the specification.

3.2 Choice of Moduli

The minimum requirement for the choice of the moduli is that their product should meet of exceed the maximum value of the circuit output. A common approach [17] is to take the first p primes as moduli, where p is large enough to satisfy the above criterion. This choice, however, is not the most efficient. Consider the verification of a 16-bit multiplier. The product of the moduli should exceed 2^{32}. If the moduli are the first p primes, then $p = 10$ and the largest modulus is 29. However, if 2 is replaced by 16384, and 3 is replaced by 9, the resulting list is still composed of relatively prime numbers, and it is now possible to drop the last three (19, 23, and 29). The largest modulus is now 16384, which is larger than 29. However, computation of residues modulo 2^k amounts to considering only the k least significant outputs. In multipliers, these least significant outputs have relatively small BDDs, regardless of the word length, which can be built quite easily for $k \leq 14$. In particular, 16384 is an easier modulus than 29, 23, or 19. Therefore, the revised list of moduli is shorter and eliminates the three most expensive computations.

Once the k least significant bits of the output have been verified by using 2^k as modulus, there is no need to consider them any further. This corresponds to applying all the remaining moduli only to the $q - k$ most significant outputs. Alternatively, one can multiply $m_2, \ldots, m_p$ by 2^k. The resulting moduli are no longer relatively prime. Hence, their scaling by 2^k does not increase the size of the additive group beyond $m = \prod_{i=1}^{p} m_i$. However, the correspondence between $x \in Z_m$ and its residues modulo $m_i \cdot 2^k$ is still a bijection, because $x \equiv y$ (mod $m_i \cdot m_1$) implies $x \equiv y$ (mod m_i). Furthermore, for $k \leq q$,

$$\sum_{j=0}^{q-1} x_j 2^j \bmod (m_i \cdot 2^k) = 2^k \cdot \left(\sum_{j=k}^{q-1} x_j 2^{j-k} \bmod m_i \right) + \sum_{j=0}^{k-1} x_j 2^j.$$

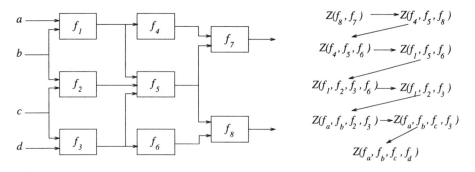

Figure 3: Example of function composition.

The term $\sum_{j=0}^{k-1} x_j 2^j$ is the residue modulo 2^k. Once this term is known to be the same for all corresponding output values, it can be ignored. Likewise, the constant factor 2^k can be dropped. Hence, multiplying all residues but the first by 2^k corresponds to ignoring the k least significant outputs. Although the direct verification of k outputs will reduce the number of outputs, consequently the size of the circuit, during residue verification, our method does not benefit from this reduction because the reduction in the number of outputs does not translate in practice into smaller diagrams. Hence, we do not normally use this optimization.

3.3 Function Composition

The procedure of function composition can be formulated as follows: Given a function g and another function f that depends on a variable x, obtain the function such that the dependency on x is replaced by the dependency on g.

Composition of BDDs and ADDs is based on Boole's expansion theorem:

$$f|_{x=g} = g \cdot f_x + g' \cdot f_{x'}.$$

In BDDs or ADDs, this operation is executed by splitting on the operands until x becomes the top variable and then calling ITE.

Typically, a circuit is described as a multiple output function, $Z(f_m, \cdots, f_n)$ composed of several interconnected combinational blocks (may be single gates or several gates in an implemented circuit). Each block is represented by the boolean function of its output as a function of its input variables. Both the function and the variable associated with a block are denoted by f_i. Consider an example where the given function is $Z(f_7, f_8)$ (Figure 3). Functions f_7 and

funct *Compose_Residues(Circuit)* ≡
  Create_Functions_For_Nodes(Circuit);
  ListOfPrimes = Create_List_Of_Primes(Circuit);
  foreach *(Prime ∈ ListOfPrimes)* **do**
    Residue = Create_Residue_ADD(Prime);
    while *({v|v ∈ Support(Residue) ∧ v ∉ PrimaryInputs} ≠ ∅)* **do**
        v = Select_Variable_To_Compose(Residue);
        f = Function_Associated_With_Variable(v);
        NewResidue = Compose_Variable(Residue, v, f);
        Residue = NewResidue;
    od
    Store_Prime_Info_In_Circuit(Circuit, Prime);
  od
end

Figure 4: Residue composition procedure.

f_8 are specified respectively as $f_7(f_4, f_5)$ and $f_8(f_5, f_6)$ respectively. Applying the two compositions $Z(f_7, f_8)|_{\{f_7=f_7(f_4,f_5)\}\{f_8=f_8(f_5,f_6)\}}$ results in the function $Z(f_4, f_5, f_6)$. With repeated application of compositions of the intermediate nodes in the circuit, the final function obtained is in terms of the circuit inputs.

Since the composition of the circuit into the residue ADD is repeated for every prime for both the specification and the implementation (see Section 3.1), composition is a critical operation. We describe techniques in the following subsections to make composition more efficient. Figure 4 illustrates a first-approach pseudo-code for the verification method based on the above described composition.

3.3.1 Scheduling of Function Compositions

The order in which the variables are composed with the residue may affect the number of function compositions performed. To avoid repetition we adopt the following rule (henceforth referred to as *One-Time Rule*): A function is composed into the residue ADD only after all the functions associated with its fanout blocks have been composed into the residue. This rule will ensure that each block is composed only once into the residue in spite of reconvergence in the circuit.

In Figure 3 if function f_7 is composed first and function f_5 is composed second, the same function f_5 will reappear in the residue when the function f_8 is composed and f_5 will require a second composition. But the *One-Time Rule* would dictate that function f_5 be composed only after both f_7 and f_8 have been composed.

However observance of this rule still leaves different ways to schedule the compositions of blocks in the residue, the two extremes being: A block is composed as soon as all its fanouts are part of the residue as against the composition of a block is deferred until we cannot proceed any further without it.

The first criterion, which we will denote as "as soon as possible" (ASAP), orders the blocks in the graph by means of a topological sort from the outputs to the inputs of the circuit. The nature of this sort guarantees that a block is considered for composition only when all its fanout variables are already part of the residue. Empirical data indicated this approach to be inefficient, in particular with multipliers. Due to large sizes of the supports of blocks close to the outputs of the circuit and the fact that every variable is brought into the residue ADD as soon as possible, the composition procedure suffered from having to manipulate ADDs with a large number of variables to do each composition for a major part of the verification process. This proved expensive in terms of time and memory occupied by the ADD package.

The second criterion, which we will denote as "as late as possible" (ALAP), brings a variable into the support of the residue only when no further compositions can be done without it. The order of the blocks, in this case, is obtained by a topological sort of the blocks starting from the inputs to the outputs of the circuit and then composing the blocks in the reverse order. This nature of sort conforms to the *One-Time Rule* that all the fanouts of a block are present in the residue when it is composed. Experimental results show that the evolution of the peak support size (largest support) during the composition process is better for the ALAP heuristic.

3.3.2 Grouping Function Compositions

When multiple compositions are performed on the same function, the compositions may be ordered in different ways, as long as they obey the above mentioned criterion. However, it may be possible to perform several compositions simultaneously. Simultaneous composition proves to be more efficient than composing one function at a time in cases where the support of the functions being composed completely overlap.

The topological sort of the blocks, in order to compute their order of composition, will label blocks with numbers in increasing order. In choosing the ALAP heuristic, the labeling would indicate the farthest distance of each block from the primary inputs. Since the ALAP heuristic obeys the *One-Time Rule*, two blocks with the same labeling do not appear in the support of each other; hence may be composed in any order. All nodes with the same label are grouped into the same composition "layer." A characteristic of this layer is that no further compositions are possible until some/all of these blocks are composed into the residue. An advantage of composing the entire layer before proceeding to the next layer is to reduce the lifetime of variables present in the ADD. It is more efficient to simultaneously compose variables with complete overlap in support. Such blocks frequently occur in multiplier circuits. This reduces the number of ITE [3] calls and the lifetime of each variable over the various compositions. This may eventually result in smaller intermediate results in composition and reduce the peak (maximum) size of the composed ADD.

3.3.3 Variable Re-usage

The composition procedure is very sensitive to the order of the variables that are being composed. In principle every block of the circuit needs a variable assigned to it, but not all the variables are active in the verification process at any instant i.e., they do not all appear in the residue ADD. Since the number of active variables at any time has a direct effect on the memory requirement, there is a strong motivation to reduce the number of variables being used at any point in time. When a layer is composed in, the variables labeling the nodes of this layer are no longer required and may be used again. After the composition of each layer, these variables just released are repositioned to be reused by the new nodes in the subsequent layer.

Another important fact is that compositions are efficient when the variables in the support of the nodes being composed are lower in order of BDD/ADD variables than the variables of the nodes themselves. Hence with each layer composition, the variables of the nodes in the support of each layer are positioned below the nodes being composed by sifting the variable across the ADD. They are also placed immediately below the variables of their fanout nodes in order to decrease the number of ITE calls.

3.3.4 Storing Intermediate Results

Due to the size of the residue in the middle of the composition process, the storage of intermediate results in a *computed table* is key to avoid recomputation and achieve high performance. The number of such distinct results increases with the size of the residue. However, these results are needed only as many times as there are pointers to this node in the ADD. Hence these results can be freed after the node has been looked up as many times. This results in big memory savings for the composition procedure with respect to a lossless computed table; and it is faster than an ordinary cache-based (lossy) computed table [3].

The final pseudo-code for the residue verification procedure with the above described optimizations is presented in Figure 5.

4   Experimental Results

We implemented the procedure of Figure 5 in the VIS [5] system. In this section, we report preliminary results. We ran our experiments on a Sun Ultra 1(167MHz) with 192MB of Main Memory. Two different implementations of a 16-bit multiplier were considered. One was a simple array implementation of a multiplier based on ripple carry adders. The other implementation was a Braun multiplier. The experiments involved the verification of correct and faulty multipliers.

The results of successful verification are shown in Table 1. Here, two correct implementations of the above described multipliers were compared. The 14 least significant outputs were verified by building the BDDs directly. This direct verification is equivalent to checking the circuits modulo a number that is an power of 2, i.e., 16384. The remaining moduli that were required for verification

```
funct Verify_with_Residues(Specification, Implementation)  ≡
  ListOfPrimes = Create_List_Of_Primes(Specification);
  ListOfLayers = Partition_Circuit_Into_Layers(Specification, Implementation);
  foreach (Prime ∈ ListOfPrimes) do
    Residue = Create_Residue_ADD(Prime);
    foreach (Circuit ∈ (Specification, Implementation)) do
      foreach (Layer ∈ ListOfLayers) do
        Reorder_and_Reuse_Variables();
        Determine_New_Variables_Required();
        Assign_New_And_Recyled_Variables_To_Support_Of_Layer(Circuit, Layer);
        Nodes_To_Be_Composed = Create_Functions_For_Nodes(Circuit, Layer);
        NewResidue = Compose_Layer_Into_Residue(Residue, Nodes_To_Be_Composed);
        Residue = NewResidue;
      od
    od
    Compare_Specification_and_Implementation_Residues;
    if (Fail)Break; fi
  od
end
```

Figure 5: Final residue composition procedure.

are listed in the table, along with the times taken to compose the two circuits into the respective residue ADDs. Both heuristics for ordering the compositions—the ASAP and ALAP methods—were tried. The ALAP performed significantly better. All the other optimizations explained in Section 3.3 were used. These optimizations had a significant impact on the verification process. The last row in Table 1 gives the total time taken to verify the two 16-bit multipliers.

In order to illustrate the effectiveness of this method, we compared several faulty implementations to the correct ones. The faults in the multiplier descriptions were injected in 2 different ways. One set contained random bugs in both circuits where the functions of the nodes were incorrect. The second set of bugs were chosen so that the 14 least significant outputs were correct. These circuits revealed the bugs in the early residues.

Table 2 shows the verification of some faulty multipliers that were generated from the original two implementations. The *16x16.fa* multiplier is the simple-minded implementation with the full-adders. The *16x16.br* is the Braun implementation. The directly verified outputs contribute to a big gain in time required for bug-detection in the lower order output bits. Other bugs are detected early by the first prime used.

Modulus/Direct	Time (sec)
Direct (16384)	217.49
5	62.22
7	120.38
9	211.33
11	347.59
13	616.93
17	1519.79
Reusable Variable Assignment	67.13
Total	3206.67

Table 1: Verification of correct 16-bit multiplier.

Bug Type	Circuits	Memory	Detection	Time (secs)
Random	16x16.fa vs. 16x16.bug1	43MB	Direct Verification	293.33
	16x16.br vs. 16x16.bug2	44MB	Direct Verification	199.15
	16x16.fa vs. 16x16.bug3	49MB	Direct Verification	266.6
Biased Random	16x16.fa vs. 16x16.bug4	68MB	Prime 5	839.55
	16x16.br vs. 16x16.bug5	183MB	Prime 5	1931.58

Table 2: Verification of Faulty Multipliers.

5 Conclusions

We have presented a technique for the verification of multipliers based on modular arithmetic. Initial experiments show that our technique is currently applicable to the verification of medium size multipliers (16x16 bits) and that it is robust with respect to errors. Our method can deal with smaller multipliers than Hamaguchi's, but it performs well when the two multipliers being compared are not equivalent. One reason for the difference is that the *BMD representation of multipliers is more compact than the residue-based representation. However, the argument applies only to correct multipliers. For incorrect circuits both representations grow in size, but the *BMD incurs a more severe blow-up, because the conditions that guarantee the high recombination of the subgraphs cease to hold.

The modular arithmetic approach is less affected by the reduced recombination, because its effectiveness stems from abstraction: Each residue ADD does not carry all the information in the original multiplier description. By contrast, the *BMD representation preserves all the information, and relies exclusively on the ability to recombine identical subgraphs to achieve compactness.

Since our technique works on a netlist of gates, no knowledge of the multiplier structure is required. The technique is also applicable to circuits different to multipliers.

We propose to investigate further the composition algorithms variable ordering, and composition scheduling . (For instance, we want to study the use of implication techniques in determining the scheduling order.) We also plan to include in our experiments a wider variety of multiplier designs (both correct and incorrect), and to evaluate the effectiveness of modular techniques on other arithmetic circuits, including dividers.

References

[1] P. Ashar, A. Ghosh, and S. Devadas. Boolean satisfiability and equivalence checking using general binary decision diagrams. In *Proceedings of the International Conference on Computer Design*, pages 259–264, Cambridge, MA, October 1991.

[2] R. I. Bahar, E. A. Frohm, C. M. Gaona, G. D. Hachtel, E. Macii, A. Pardo, and F. Somenzi. Algebraic decision diagrams and their applications. In *Proceedings of the International Conference on Computer-Aided Design*, pages 188–191, Santa Clara, CA, November 1993.

[3] K. S. Brace, R. L. Rudell, and R. E. Bryant. Efficient implementation of a BDD package. In *Proceedings of the 27th Design Automation Conference*, pages 40–45, Orlando, FL, June 1990.

[4] D. Brand. Verification of large synthesized designs. In *Proceedings of the International Conference on Computer-Aided Design*, pages 534–537, Santa Clara, CA, November 1993.

[5] R. K. Brayton et al. VIS: A system for verification and synthesis. Technical Report UCB/ERL M95/104, Electronics Research Lab, Univ. of California, December 1995.

[6] R. Bryant and Y.-A. Chen. Verification of arithmetic circuits with binary moment diagrams. In *Proceedings of the Design Automation Conference*, pages 535–541, San Francisco, CA, June 1995.

[7] R. E. Bryant. Graph-based algorithms for boolean function manipulation. *IEEE Transactions on Computers*, C-35(8):677–691, August 1986.

[8] R. E. Bryant. On the complexity of VLSI implementations and graph representations of boolean functions with application to integer multiplication. *IEEE Transactions on Computers*, 40(2):205–213, February 1991.

[9] J. R. Burch. Using BDDs to verify multipliers. In *Proceedings of the Design Automation Conference*, pages 408–412, San Francisco, CA, June 1991.

[10] E. M. Clarke, O. Grumberg, and D. E. Long. Model checking and abstraction. In *Proceedings of the 19th ACM Symposium on Principles of Programming Languages*, January 1992.

[11] E. M. Clarke, K. L. McMillan, X. Zhao, M. Fujita, and J. C.-Y. Yang. Spectral transforms for large boolean functions with applications to technology mapping. In *Proceedings of the Design Automation Conference*, pages 54–60, Dallas, TX, June 1993.

[12] T. H. Cormen, C. E. Leiserson, and R. L. Rivest. *An Introduction to Algorithms*. McGraw-Hill, New York, 1990.

[13] K. Hamaguchi, A. Morita, and S. Yajima. Efficient construction of binary moment diagrams for verifying arithmetic circuits. In *Proceedings of the International Conference on Computer-Aided Design*, pages 78–82, San Jose, CA, November 1995.

[14] J. Jain, M. Abadir, J. Bitner, D. S. Fussell, and J. A. Abraham. IBDDs: An efficient functional representation for digital circuits. In *Proceedings of the European Conference on Design Automation*, pages 440–446, Brussels, Belgium, March 1992.

[15] J. Jain, R. Mukherjee, and M. Fujita. Advanced verification techniques based on learning. In *Proceedings of the Design Automation Conference*, pages 420–426, San Francisco, CA, June 1995.

[16] U. Kebshull, E. Schubert, and W. Rosenstiel. Multilevel logic synthesis based on functional decision diagrams. In *Proceedings of the European Conference on Design Automation*, pages 43–47, Brussels, March 1992.

[17] S. Kimura. Residue BDD and its application to the verification of arithmetic circuits. In *Proceedings of the Design Automation Conference*, pages 542–545, San Francisco, CA, June 1995.

[18] Y.-T. Lai and S. Sastry. Edge-valued binary decision diagrams for multilevel hierarchical verification. In *Proceedings of the Design Automation Conference*, pages 608–613, Anaheim, CA, June 1992.

[19] S.-I. Minato. Zero-suppressed BDDs for set manipulation in combinatorial problems. In *Proceedings of the Design Automation Conference*, pages 272–277, Dallas, TX, June 1993.

[20] B. Plessier, G. Hachtel, and F. Somenzi. Extended BDDs: Trading off canonicity for structure in verification algorithms. *Journal of Formal Methods in System Design*, 4(2):167–185, February 1994.

[21] S. M. Reddy, W. Kunz, and D. K. Pradhan. Novel verification framework combining structural and OBDD methods in a synthesis environment. In *Proceedings of the Design Automation Conference*, pages 414–419, San Francisco, CA, June 1995.

Verification of IEEE Compliant Subtractive Division Algorithms

Paul S. Miner[1] and James F. Leathrum, Jr.[2]

[1] NASA Langley Research Center, Hampton VA 23681-0001 USA,
E-mail: p.s.miner@larc.nasa.gov
[2] Old Dominion University, Department of Electrical and Computer Engineering,
Norfolk VA 23529 USA, E-mail: leathrum@ee.odu.edu

Abstract. A parameterized definition of subtractive floating point division algorithms is presented and verified using PVS. The general algorithm is proven to satisfy a formal definition of an IEEE standard for floating point arithmetic. The utility of the general specification is illustrated using a number of different instances of the general algorithm.

1 Introduction

As computing systems become more complex, it becomes increasingly difficult to ensure that testing fully exercises the design. This was made abundantly clear by the infamous bug in the floating point unit of the Intel Pentium(TM) microprocessor. The bug consists of five missing entries in a lookup table. Pratt [21] provides a thorough analysis of this error. He provides compelling arguments that a thorough manual analysis of a design may still allow errors to evade detection. This is particularly true if the flaw is in a region of the design that is thought to be unreachable. Machine assisted reasoning is crucial to prevent such errors. Two recent verifications of Taylor's SRT divider [20] illustrate how theorem proving techniques can be used to prevent omissions in lookup tables similar to that employed by the Pentium [5, 18]. These verifications describe a relationship from a verified algorithm to a hardware design. In order to complete the verification, it is necessary to relate the algorithm to a specification of the floating point operation.

Our work provides a link between the IEEE floating point standards [7, 8] and a class of verified division algorithms. A strength of theorem prover based verification is that it allows verification of classes of algorithms. Once a class is verified with respect to the standard, it can be used routinely in the development of verified hardware. Taylor's SRT divider is an instance of the class we have verified. Other instances include the classical restoring and non-restoring division algorithms. Thus, our general theory provides a standard specification for IEEE compliant subtractive division.

Section 3 presents a verification of the class of subtractive division algorithms. We illustrate the utility of the general verification by exhibiting a number of instances. Section 4 extends these algorithms to provide verified IEEE compliant rounding. This verification proceeds in two stages. Section 4.1 presents the verification of a standard algorithm to round floating point results in accordance with the standard. Section 4.2 composes the rounding algorithm with the general division algorithm to provide a verified IEEE compliant division algorithm.

2 Related work

Much of the previous work in theorem prover based verification of floating point algorithms has focused on verifying core algorithms and a corresponding hardware design. The first efforts targeted verified implementations of binary non-restoring algorithms. Leeser, O'Leary, et al. present a verification (using *Nuprl*) of a binary non-restoring square root algorithm and its implementation [10, 13]. Verkest, et al. present a similar verification (using *nqthm*) of a binary non-restoring division algorithm [22].

In response to the flaw in the Pentium [21], several researchers investigated theorem prover based verifications of SRT division hardware. Clarke, German, and Zhao used the ANALYTICA theorem prover to verify Taylor's [20] radix-4 SRT division circuit [5]. Their verification includes an abstract representation of the lookup table and a proof that it defines all necessary values for the quotient selection logic. Rueß, Srivas, and Shankar [18] generalize this work using the Prototype Verification System (PVS [17]). They present a general verification of arbitrary radix SRT division algorithms, instantiate their theory with Taylor's radix-4 SRT division circuit, and verify a description of the hardware. Included in their work is a technique to verify a concrete representation of the lookup table. The PVS type system automatically generates proof obligations to ensure that blank entries in the table are inaccessible.

In all of the efforts above, the verifications address functional correctness of core algorithms and associated hardware designs. They do not address the issue of relating the algorithms to a formal definition of floating point operations. Harrison has presented a verification of two floating point algorithms; square root, and a CORDIC [23] natural logarithm algorithm [9]. For both algorithms, Harrison relates the proofs to a floating point interpretation. Although he does not present hardware descriptions, he does address some of the preliminary error analysis necessary to provide correct rounding. The IEEE standards for floating point arithmetic unambiguously state that each operation

> *shall be performed as if it first produced an intermediate result correct to infinite precision and with unbounded range, and then that result rounded according to one of the modes ... [7, 8]*

Barrett manually verified a general rounding algorithm with respect to a Z formalization of IEEE 754 [2, 3]. When Barrett performed his verification, there was no machine assisted reasoning for the Z specification language. Some tools for machine assisted application of Z have recently been developed [14]. Thus far, these have not been applied to floating point verification.

Recently, the microcode for the floating point division and square root algorithms of the $AMD5_K86^{TM}$ microprocessor has been mechanically verified using the $ACL2$ theorem prover [16, 19]. Both algorithms assume correct hardware for floating point multiplication, addition, and subtraction. Both verifications include detailed analysis of rounding and proof that the delivered result is rounded in accordance with the IEEE standard. In addition, the verifications guarantee that all intermediate results of the algorithms fit the datapath of the existing floating point hardware. Aagaard and Seger employ a combination of theorem proving and model-checking techniques to verify a floating point multiplier against a formal definition of IEEE multiplication [1].

Our work is a generalization of the Rueß, Srivas, and Shankar verification. In addition, to the SRT algorithms, our verification encompasses most of the algorithms presented in [6]. In addition, our verification includes a formal path relating the algorithm to the IEEE standard.

2.1 Brief introduction to PVS

PVS [17] is a verification system that provides support for general purpose theorem proving. The specification language is a higher order logic augmented with dependent types. Theories can be parameterized, and the dependent type mechanism allows for stating arbitrary constraints on theory parameters. The type system of PVS includes predicate subtypes and is therefore undecidable. PVS frequently generates proof obligations to ensure that expressions are well typed. PVS has powerful decision procedures, so many proofs involving simple arithmetic expressions can be discharged automatically. In addition, PVS provides a collection of pre-proven results in the **prelude**. Also included with PVS are libraries providing support for bit-vectors and finite sets. In PVS, the real numbers are a base type, and other numeric types are defined as subtypes of the reals. This allows specifications to freely mix operations on numeric types.

3 General verification of subtractive division algorithms

There are two principle classes of floating point division algorithms. The subtractive algorithms use shifting followed by addition/subtraction to generate quotient digits in time linear with respect to the size of the operands. Multiplicative algorithms, such as Goldschmidt's Algorithm or Newton-Raphson iterations provide fewer iterations, but the operations in each iteration grow increasingly complex. Ercegovac and Lang present a detailed study of subtractive algorithms for both division and square root [6]. The general division algorithm is presented in PVS providing a parameterized class of verified subtractive division algorithms.

3.1 General algorithm definition

Subtractive algorithms generate one quotient digit per iteration. They are designed so that in iteration i the remainder is no larger than r^{-i} for a radix-r

algorithm, assuming certain constraints on the dividend and divisor. Ercegovac and Lang present a series of interrelated factors which differentiate subtractive division algorithms: the radix (r), the quotient-digit set $(\{-a, -a +
1, ..., -1, 0, 1, ..., a - 1, a\})$, the range of the divisor (b), and the quotient-digit
select function (qs) [6]. These factors are all parameters to the general division
theory. The formal parameters for theory *general_division* are

```
r  : {i : posint | i > 1},
a  : {i : posint | ceiling((r-1)/2) <= i & i < r},
b  : {i : posint | 1 < i & i <= r},
(IMPORTING divide_types[r, a, b])
qs : qs_type
```

The first three parameters are defined using predicate subtypes of the positive
integers. In addition, the types for a and b are constrained by the value of r.
The type signature for function qs depends on all of the previous parameters.
The declaration for type *qs_type* is imported from theory *divide_types*. Theory
divide_types declares constant $\rho = \frac{a}{r-1}$ and the following types:

```
D_type     : TYPE = {d : real    | 1 <= d & d < b}
dividend   : TYPE = {x : posreal | 1/r <= x & x < rho}
p_type(D)  : TYPE = {p : real    | abs(p) <= rho*D}
qs_type    : TYPE =
    [D : D_type, p : p_type(D) ->
     {q : subrange(-a,a) | abs(r*p - q*D) <= rho*D}]
```

Constant ρ denotes the redundancy factor of the quotient-digit set. It represents a trade-off in design complexity between the quotient selection function
and generation of divisor multiples. The type of the divisor is constrained by
D_type, which is defined to include the numeric range of the significand of a
normalized floating point number. The type of the dividend is constrained to
ensure that it satisfies constraints imposed on the partial remainder. Parameterized type *p_type(D)* encodes an invariant, dependent on divisor D, that the
partial remainder must satisfy during execution of the algorithm. Finally, the
type of the quotient selection function, *qs_type*, is restricted to functions that,
given a divisor D and a partial remainder p such that $|p| \leq \rho \cdot D$, return a digit
q such that $-a \leq q \leq a$ and $|r \cdot p - q \cdot D| \leq \rho \cdot D$.

The subtractive division algorithms, given divisor D and dividend X, are
characterized by the following recurrence equations for the quotient q and the
partial remainder p:

$$p_i = r \cdot p_{i-1} - qd_i \cdot D$$
$$q_i = r \cdot q_{i-1} + qd_i$$

where $q_0 = 0$, $p_0 = X$, and qd_i is the quotient digit selected for iteration i. A
PVS specification of these equations is

```
divide(X,D)(n) : RECURSIVE [# p : p_type(D), q: integer #] =
    IF n=0 THEN (# p := X, q := 0 #)
          ELSE (# p := r*p(divide(X,D)(n-1)) - qd_n*D,
                 q := r*q(divide(X,D)(n-1)) + qd_n #)
                WHERE qd_n = qs(D,p(divide(X,D)(n-1))) ENDIF
    MEASURE n
```

By using record types for the range of the function, the definition is a direct transliteration of the recurrence equations. In addition, by declaring the partial remainder to be of type *p_type(D)*, PVS automatically generates a proof obligation to ensure that the invariant is satisfied. This obligation is proven using the type constraints on the quotient selection function.

3.2 Verification of the general algorithm

To simplify later definitions we define the abbreviations:

Definition 1.

$$p(X, D)(n) \mathrel{\hat=} p(divide(X, D)(n))$$
$$q(X, D)(n) \mathrel{\hat=} q(divide(X, D)(n))$$

PVS strategy (induct-and-simplify) proves

Lemma 2. $p(X, D)(n) \cdot r^{-n} = X - q(X, D)(n) \cdot r^{-n} \cdot D$

The invariant property of the partial remainder guarantees

Lemma 3. $\left| \dfrac{p(X, D)(n) \cdot r^{-n}}{D} \right| \le r^{-n}$

To strengthen the convergence property, the algorithm includes a corrective step after the final iteration.

Definition 4.

$$P(X, D)(n) \mathrel{\hat=} \begin{cases} (p(X, D)(n) + D) \cdot r^{-n} & \text{if } p(X, D)(n) < 0 \\ (p(X, D)(n) - D) \cdot r^{-n} & \text{if } p(X, D)(n) = D \\ p(X, D)(n) \cdot r^{-n} & \text{otherwise} \end{cases}$$

$$Q(X, D)(n) \mathrel{\hat=} \begin{cases} (q(X, D)(n) - 1) \cdot r^{-n} & \text{if } p(X, D)(n) < 0 \\ (q(X, D)(n) + 1) \cdot r^{-n} & \text{if } p(X, D)(n) = D \\ q(X, D)(n) \cdot r^{-n} & \text{otherwise} \end{cases}$$

Expanding the definitions of P and Q, and using lemma 2, we prove

Theorem 5 (correctness). $\dfrac{X}{D} = Q(X, D)(n) + \dfrac{P(X, D)(n)}{D}$

Similarly expanding the definition of P and using lemma 3, we prove

Theorem 6 (convergence). $\dfrac{P(X, D)(n)}{D} < r^{-n}$

Thus, $Q(X, D)(n)$ contains n radix-r digits of the quotient $\dfrac{X}{D}$.

3.3 Example instantiations

A number of quotient selection functions have been developed for use with the general subtractive division algorithm. These include functions for three radix-2 algorithms and five radix-4 algorithms. The radix-2 algorithms are restoring, non-restoring, and SRT. The radix-4 algorithms include two using fixed multiples of the divisor for comparison with p (using $a = 3$ and $a = 2$), and three distinct radix-4 SRT lookup tables. The tables developed are for

1. A quotient selection function that uses an approximation of p that is either exact, or underestimated. Examples include designs that compute an estimate of the next remainder as in Taylor's circuit [20] (verified in [18]), or use a carry-save adder for the computation of the partial remainder [6]
2. A quotient selection function that uses an over approximation of p. This is the case if the partial remainder is computed using a signed-digit adder [6]
3. A selection function where the approximation error is towards zero. In this case, the lookup table is folded in half.

The verification amounts to proving that each quotient digit select function satisfies the type constraints of qs_type. For the radix-2 functions and the two radix-4 functions based on comparison multiples, PVS strategy (grind) was sufficient to discharge the proof obligation. The verification of the lookup tables requires a little more effort. The process of verifying lookup tables in PVS was first illustrated by Rueß, Srivas, and Shankar [18] and repeated here with respect to the general division algorithm. However, the development of two new lookup tables proved insightful in demonstrating the usefulness of PVS in the design process. Not only were omissions from the table detected, but PVS allowed the faster development of new tables. Each table was developed starting from the base table in [18] and then modified to meet the new requirements. PVS was used to indicate which table entries were incorrect under the new specifications, allowing the designer to easily modify the table. Each new table was fully designed and verified in less than three hours elapsed time.

4 Verification with respect to the IEEE standard

This section illustrates how to extend the above general algorithm to provide a verified IEEE compliant implementation of division. This does not constitute a full proof of compliance, it just illustrates how non-exceptional cases of division can be realized. This verification step is performed with respect to a formal specification of IEEE 854 defined using PVS [15]. The verification is performed in two stages. First, a generalization of the basic guard, round, and sticky bit rounding algorithm is shown to satisfy the requirements of the standard. Then, the general subtractive algorithm is shown to provide sufficient information to utilize this rounding scheme. The verified rounding scheme is applicable for all floating-point algorithms. In addition, the theory mapping the standard to the general subtractive algorithm includes a number of intermediate results that apply to all floating point division algorithms.

4.1 Rounding scheme

The IEEE Standards for floating-point arithmetic [7, 8] require support for four rounding modes. The default mode is *round to nearest even*, and requires that the returned value be the floating point number nearest to the exact result. If the exact result is halfway between two floating point numbers, the standards require that it be rounded to the one with an even least significant digit. The other three modes round the result towards positive infinity, negative infinity, or zero. The discussion in this section uses the following fact about real numbers:

Lemma 7. *For any integer $b > 1$, a nonzero real number z can be uniquely decomposed into three parts: a sign, $sgn(z) \in \{-1, 1\}$; an integer exponent $e(z)$; and a significand $1 \leq sig(z) < b$ such that*

$$z = sgn(z) \times sig(z) \times b^{e(z)}$$

The following definition is from the PVS prelude:

Definition 8. The fractional part of a real number z is defined by

$$\{z\} \,\hat{=}\, z - \lfloor z \rfloor$$

The next two definitions are from Miner's PVS formalization of IEEE 854 [15]. The function, *round*, converting real number z to an integer for each of the four rounding modes required by the IEEE standard is defined by:

Definition 9.

$$round(z, to\_pos) \,\hat{=}\, \lceil z \rceil$$
$$round(z, to\_neg) \,\hat{=}\, \lfloor z \rfloor$$
$$round(z, to\_zero) \,\hat{=}\, sgn(z) \times \lfloor |z| \rfloor$$
$$round(z, to\_near) \,\hat{=}\, \begin{cases} \lfloor z \rfloor & \text{if } \{z\} < 1 - \{z\} \\ \lceil z \rceil & \text{if } 1 - \{z\} < \{z\} \\ 2 \times \lfloor \frac{\lceil z \rceil}{2} \rfloor & \text{otherwise} \end{cases}$$

This definition is extended to round a real number z to a real number with at most p base-b digits of precision using the following function:

Definition 10.

$$round\_scaled(z, mode) \,\hat{=}\, round(z \cdot b^{p-1-e(z)}, mode) \cdot b^{e(z)-(p-1)}$$

A common algorithm for implementing IEEE compliant rounding to p digits of precision uses two extra digits and a sticky-flag [12]. The first extra digit is called the guard digit and ensures that the computed result can be normalized while preserving p digits of precision. This is necessary for multiplication and division algorithms. The second extra digit is called the round digit, and is used to control rounding for every mode except *to_zero*. Finally, a sticky flag is required to distinguish the case when the infinitely precise result lies halfway between two representable values for mode *to_near*. The PVS theory implementing the Guard, Round, Sticky (GRS) scheme has been generalized to allow an arbitrary even radix. Thus it works for both base-2 and base-10 instances of IEEE 854. The principle function realizing the GRS rounding scheme is

Definition 11. For $s \in \{-1, 1\}$, $n \in \mathbf{N}$, base-b digit d, and *sticky* $\in \mathbf{B}$

$$grs(s, n, d, sticky, to\_pos) \; \hat{=} \; \begin{cases} n+1 \text{ if } s \geq 0 \wedge ((d > 0) \vee sticky) \\ n \quad \text{ otherwise} \end{cases}$$

$$grs(s, n, d, sticky, to\_neg) \; \hat{=} \; \begin{cases} n+1 \text{ if } s < 0 \wedge ((d > 0) \vee sticky) \\ n \quad \text{ otherwise} \end{cases}$$

$$grs(s, n, d, sticky, to\_zero) \; \hat{=} \; n$$

$$grs(s, n, d, sticky, to\_near) \; \hat{=} \; \begin{cases} n+1 \text{ if } (d > \frac{b}{2}) \vee (d = \frac{b}{2} \wedge (sticky \vee odd?(n))) \\ n \quad \text{ otherwise} \end{cases}$$

To complete the definition of rounding using the GRS scheme, we use the following functions to scale the result and extract the relevant fields:

Definition 12.

$$\begin{aligned} scaled(z) &\; \hat{=} \; \lfloor |z| \times b^{p-1-e(z)} \rfloor \\ round\_digit(z) &\; \hat{=} \; \lfloor |z| \times b^{p-e(z)} \rfloor \bmod b \\ sticky\_flag(z) &\; \hat{=} \; \{|z| \times b^{p-e(z)}\} \neq 0 \end{aligned}$$

Function *scaled* extracts the p most significant base b digits of z scaled to a natural number. Function *round_digit* extracts the $(p+1)$th digit, and *sticky_flag* is true if and only if there are significant digits beyond the $(p+1)$th. For a result in a sign and magnitude representation, functions *scaled* and *round_digit* can be realized in hardware by extracting the appropriate digits from the given result. Implementing the *sticky_flag* requires some additional logic; the actual realization will vary depending upon the algorithm employed to compute the result. For comparison to *round_scaled*, we define

Definition 13.

$$\begin{aligned} round\_grs(z, mode) \\ \hat{=} \; sgn(z) \times b^{e(z)-(p-1)} \times \\ grs(sgn(z), scaled(z), round\_digit(z), sticky\_flag(z), mode) \end{aligned}$$

The principle result of this section is

Theorem 14. $round\_grs(z, mode) = round\_scaled(z, mode)$

The PVS proof of this result consists of a fairly simple case analysis, except for mode *to_near*.

The initial PVS proof for mode *to_near* included a complicated case analysis, where it was difficult to exploit symmetry. IEEE floating point numbers are defined using a sign and magnitude representation. However, the definition of *round_scaled* does not take advantage of this representation. Thus, the first proof for mode *to_near* included an unnatural case split on the sign of the argument. Since the GRS rounding scheme is defined in terms of a sign and magnitude representation, the cases for negative arguments do not align in the same manner as for the positive arguments. Thus, there was little opportunity to reuse proofs from the corresponding positive cases. The PVS proof has been simplified using the following lemma (which was proven using the PVS strategy (grind)):

Lemma 15. *round(z, to_near) = sgn(z) × round(|z|, to_near)*

Even without the case split due to the sign, the proof for mode *to_near* still involves a difficult case analysis. This case analysis consists of relating the values of the round digit and sticky flag to the corresponding cases from the specification for rounding mode *to_near*.

4.2 Relating the general algorithm to the standard

The verified rounding scheme asserts that in order to achieve IEEE compliant rounding to p significant digits, it is sufficient to compute $(p + 1)$ truncated significant digits and determine if there are any remaining non-zero digits. The general subtractive division algorithm ensures at least $n - 1$ digits of precision after n iterations. Furthermore, if the computed remainder is nonzero, then there are additional digits in the infinitely precise result. Since it is possible for the radix of the division algorithm to be different from that for the representation of floating point numbers, the PVS theory has to relate the potential division radices to those allowed by the standards.

There are some simple results that describe the range of possible values for floating point division. Let x denote a base-b floating point number with p significant digits. A finite floating point number x is represented using three fields: a sign $\sigma_x \in \{0, 1\}$, an integer exponent E_x, and a significand d_x, where d_x is a function from $\{0, \ldots, (p-1)\}$ to $\{0, \ldots, (b-1)\}$. The numerical value of d_x is given by

$$v_d(x) = \sum_{i=0}^{p-1} d_x(i) \cdot b^{-i}$$

The numerical value of floating point number x is given by

$$v(x) = (-1)^{\sigma_x} \cdot b^{E_x} \cdot v_d(x)$$

A floating point number x such that $1 \leq v_d(x) < b$ is normalized. Furthermore, if its exponent also falls within the range allowed by the standard, it is a *normal* floating point number.

The IEEE standard requires that all operations be performed as if to infinite precision and then rounded. The infinitely precise quotient of two floating point numbers x and y is $v(x)/v(y)$. A number of general facts about floating point division have been proven using PVS. These include the following:

Lemma 16. $\dfrac{v(x)}{v(y)} = (-1)^{\sigma_x - \sigma_y} \cdot b^{E_x - E_y} \cdot \dfrac{v_d(x)}{v_d(y)}$

This lemma states that a floating point division algorithm can be decomposed into simple operations on the sign and exponent fields combined with an algorithm to compute the quotient of the significands.

Lemma 17. *For normalized floating point numbers x and y,*

$$b^{-1} + b^{-(p+1)} \leq \frac{v_d(x)}{v_d(y)} \leq b - b^{(1-p)}$$

Lemma 18. *For normalized floating point numbers x and y, if $v_d(x) < v_d(y)$ then*

$$\frac{v_d(x)}{v_d(y)} \leq 1 - b^{(1-p)}$$

Lemmas 17 and 18 assert that the result of a floating point division algorithm requires at most one left shift to normalize the result, and furthermore, the necessity of post-divide normalization can be determined by comparing the magnitude of the significands. Therefore, rounding cannot affect the exponent field for a division operation. These results are true for any floating point division algorithm; they do not depend on a particular instantiation.

These results allow us to define the following exponent adjustment function for the quotient of normalized floating point numbers x and y:

Definition 19. $Adj(x,y) = \begin{cases} 1 \text{ if } v_d(x) < v_d(y) \\ 0 \text{ otherwise} \end{cases}$

The above results allow us to define a template for a floating point division algorithm:

Definition 20. For normal floating point numbers x and y such that $E_{\min} \leq E_x - E_y - Adj(x,y) \leq E_{\max}$, let $z = \frac{v_d(x)}{v_d(y)}$ in

$$\begin{aligned}
&div\_algorithm(x,y,mode) \\
&\hat{=}\ finite(\ XOR(\sigma_x, \sigma_y), \\
&\qquad\qquad E_x - E_y - Adj(x,y), \\
&\qquad\qquad nat2d(grs(\ (-1)^{\sigma_x - \sigma_y}, scaled(z), \\
&\qquad\qquad\qquad round\_digit(z), sticky\_flag(z), mode)))
\end{aligned}$$

Constructor *finite* generates a finite floating point number from a sign, an exponent, and a digit vector. Function *nat2d* converts a natural number n, where $n < b^p$, to a digit vector. Definition 24 will complete the algorithm by providing *scaled*, *round_digit*, and *sticky_flag* for the given quotient.

Theory *ieee_divide*, relating the general algorithm to the IEEE standard, uses the same formal parameters as theory *general_division*. It also uses additional parameters needed to import the IEEE theories. The base for the floating point representation is b, the radix for the division algorithm, r, is selected so that

$r = b^i$ for some positive integer i. This ensures shift operations can be effectively used in the computation of the partial remainder. The divisor, $v_d(y)$, already satisfies the type constraint D_type. The dividend needs to be shifted right either one or two base-b digits to ensure that it satisfies the initial type constraints for the partial remainder. Finally, the algorithm must be be iterated enough times to ensure $p + 1$ significant digits. The following definitions invoke the general algorithm using arguments selected to ensure the algorithm computes enough information for IEEE compliant rounding.

Definition 21. Let $pre = 2 - \lfloor \rho \rfloor$, $i = \log_b(r)$, and $steps = \lceil \frac{p+2+pre}{i} \rceil$ in

$$\text{Quotient}(x, y) \;\hat{=}\; b^{pre} \cdot Q(v_d(x) \cdot b^{-pre}, v_d(y))(steps)$$
$$\text{Remainder}(x, y) \;\hat{=}\; b^{pre} \cdot P(v_d(x) \cdot b^{-pre}, v_d(y))(steps)$$

Using theorems 5 and 6, respectively, we establish the next two results:

Lemma 22. $\dfrac{v_d(x)}{v_d(y)} = \text{Quotient}(x, y) + \dfrac{\text{Remainder}(x, y)}{v_d(y)}$

Lemma 23. $\dfrac{\text{Remainder}(x, y)}{v_d(y)} < b^{(pre - i \cdot steps)} \leq b^{-(p+2)}$

This result combined with lemma 17 ensures that the computed quotient has at least $p + 1$ significant digits for correct rounding. The arguments to complete definition 20 are

Definition 24.

$$scaled\_div(x, y) = \lfloor \text{Quotient}(x, y) \cdot b^{(p-1)+A\hat{d}j(x,y)} \rfloor$$
$$round\_digit\_div(x, y) = \lfloor \text{Quotient}(x, y) \cdot b^{p+A\hat{d}j(x,y)} \rfloor \bmod b$$
$$sticky\_flag\_div(x, y) = (\{\text{Quotient}(x, y) \cdot b^{p+A\hat{d}j(x,y)}\} + \text{Remainder}(x, y)) > 0$$

All that remains is to prove the following:

Theorem 25 (IEEE_compliant). *For normal floating point numbers x and y such that $E_{\min} \leq E_x - E_y - Adj(x, y) \leq E_{\max}$*

$$fp\_div(x, y, mode) = div\_algorithm(x, y, mode)$$

Function *fp_div* defines the floating point division operator required by the IEEE standard [15]. The first part of the proof consists of showing that the restrictions on the exponents of x and y ensure that the result of *fp_div* returns a normal floating point number. The second part of the proof uses the following identities:

Lemma 26.

$$scaled\_div(x, y) = scaled\left(\frac{v(x)}{v(y)}\right)$$

$$round\_digit\_div(x, y) = round\_digit\left(\frac{v(x)}{v(y)}\right)$$

$$sticky\_flag\_div(x, y) = sticky\_flag\left(\frac{v(x)}{v(y)}\right)$$

The PVS proofs of these three identities involve algebraic manipulation of expressions composed of exponents, absolute values, and the integer floor function. The proofs are not conceptually difficult, but neither do they yield to the brute-force proof strategies of PVS.

4.3 IEEE compliant division

The PVS theories are structured so that a designer can generate an instance of a verified algorithm with a minimum of effort. First, the designer selects a quotient selection function based on the constraints of the development effort. One such choice might be to use a radix-4 SRT lookup table based on using a signed-digit adder for computing the partial remainder. Suppose further, that the implementation is for single-precision IEEE 754 arithmetic. The following sequence of PVS declarations provides a verified algorithm for the design:

```
IMPORTING
  divide_types[4,2,2],
  signed_digit_lookup

qs: qs_type[4,2,2] = q_sel

IMPORTING
  ieee_divide[4,2,2,qs,24,192,127,-126]
```

Theory signed_digit_lookup contains the definition of an SRT4 lookup table suitable for use with a signed-digit adder. It contains the proof that the quotient selection function *q_sel* satisfies type *qs_type[4,2,2]*. Theory ieee_divide defines function *div_algorithm* and includes the proofs from the previous section. With the instantiation of the quotient selection function, the function *div_algorithm* completely describes a verified IEEE compliant division algorithm composed of operations that have simple hardware realizations. The designer must verify that a hardware design correctly implements this algorithm, but need not be concerned with the functional correctness of the algorithm. Given a collection of verified quotient selection functions, a designer can select one appropriate to the constraints of a particular development effort.

5 Suggestions for future work and lessons learned

5.1 Suggestions for future work

There are a number of ways to build on the work presented here. The basic style of the specification is common for a large number of floating point algorithms. The most obvious class of algorithms to consider is subtractive square root algorithms. Another good candidate for exploration is whether a similar general schema can be developed for division through multiplication algorithms.

Another example to be considered is the generalized CORDIC algorithm [23]. The algorithm has already been defined in PVS, and the solutions to the defining

CORDIC equations have been verified. These proofs required the addition of some axioms describing properties of trigonometric and hyperbolic functions. The limited support (in PVS) for reasoning about these functions made analysis of accumulated error and convergence difficult, thus remaining as future work.

From the above mentioned work, it should be possible to build up a library of general floating point algorithms verified with respect to the IEEE standard. Such a library would present a developer of floating point hardware with a variety of options, provided there was a good link between the verified library and hardware development tools.

5.2 Lessons learned

The PVS language features allow direct definition of recurrence equations. The proof strategies are effective for establishing the functional correctness of recurrence based algorithms. However, issues such as IEEE compliant rounding required difficult and time consuming proofs. There were two primary sources of difficulty. The first is that these details often involve complex case analysis where each case has a different structure. The difficulty is compounded by the fact that most of these cases do not succumb to the automatic proof strategies. The second major source of difficulty was that most of the verification effort was in a domain where there is limited support from the prover. During this verification effort, most proof steps consisted of algebraic manipulation of expressions involving exponentiation. In contrast to the basic arithmetic functions, there are a limited number of pre-proven properties about exponentiation. Although the prelude includes some facts about exponentiation, these are not organized to enable effective automatic rewriting.

The PVS type system can be used to effectively restrict types in definitions. However, this may lead to extra effort proving irrelevant type correctness conditions. Finding the right balance can be difficult. During the verification relating the algorithm to the IEEE standard, much of the effort consisted of repeatedly discharging the same collection of type correctness conditions.

6 Concluding remarks

Formal verification is an enabling technology. This general verification will allow designers to focus their efforts on more advanced optimizations of hardware implementations secure in the knowledge that the routine aspects of the design have been addressed. However, a great deal of work is still needed to make formal verification a useful technology. In particular, a designer should not be required to generate all of the supporting theories for well known algorithms and hardware structures. A large set of libraries should be available from which to select the pieces required to complete and verify a design.

The ultimate goal of the work is to assist in the development of verified hardware. With that in mind, the general algorithm was presented in a standard form that can be easily transformed to an equivalent tail-recursive description [11].

This provides a top-level specification for deriving a hardware realization using a transformational system such as DRS [4]. This process has been tested with the general subtractive division algorithms presented in this paper.

The work presented here is a major step toward establishing an environment conducive to development of formally verified floating point hardware. The primary benefit of the work is not the fact that subtractive division algorithms are shown to be compliant with IEEE standards. Instead, this work demonstrates that with proper foresight in developing and verifying generalized solutions, it is then much easier for future designers to verify particular instantiations of those general solutions. As a complete library of verified floating point algorithms emerges, future floating point implementations should have a much higher confidence of correctness.

Acknowledgments

We are grateful for the many helpful comments and suggestions made by Steve Johnson, Mandayam Srivas, and the anonymous reviewers.

Electronic Sources

All of the results presented in this paper have been verified mechanically using PVS 2.0 Alpha Plus (patch level 2.394). The PVS theories and proof files for the work presented in this paper are available from

`ftp://air16.larc.nasa.gov/pub/fm/larc/fp_div/`

References

1. Mark D. Aagaard and Carl-Johan H. Seger. The formal verification of a pipelined double-precision IEEE floating-point multiplier. In *Proceedings 1995 IEEE/ACM International Conference on Computer-Aided Design*, pages 7–10, San Jose, CA, November 1995.
2. Geoff Barrett. A formal approach to rounding. In *Proceedings 8th Symposium on Computer Arithmetic*, pages 247–254, 1987.
3. Geoff Barrett. Formal methods applied to a floating-point number system. *IEEE Transactions on Software Engineering*, 15(5):611–621, May 1989.
4. Bhaskar Bose. DRS - derivational reasoning system: A digital design derivation system for hardware synthesis. In T. Hilburn, G.Suski, and J. Zalewski, editors, *Safety and Reliability in Emerging Control Technologies*. Elsevier Science Publishers, Oxford, UK, 1996. ISBN 0 08 042610 7, to be published.
5. E. M. Clarke, S. M. German, and X. Zhao. Verifying the SRT division algorithm using theorem proving techniques. In Rajeev Alur and Thomas A. Henzinger, editors, *Computer-Aided Verification, CAV '96*, number 1102 in Lecture Notes in Computer Science, New Brunswick, NJ, July/August 1996. Springer-Verlag.
6. Miloš D. Ercegovac and Tomás Lang. *Division and Square Root: Digit-Recurrence Algorithms and Implementations*. Kluwer Academic Publishers, 1994.

7. IEEE-754. *Standard for Binary Floating-Point Arithmetic*, 1985. ANSI/IEEE Std 754-1985.

8. IEEE-854. *Standard for Radix-Independent Floating-Point Arithmetic*, 1987. ANSI/IEEE Std 854-1987.

9. J. Harrison. Floating point verification in HOL. In E.T. Schubert, P.J. Windley, and J. Alves-Foss, editors, *Proceedings 8th International Workshop on Higher Order Logic Theorem Proving and its Applications*, volume 971 of *Lecture Notes in Computer Science*, pages 186–199, Aspen Grove, UT, USA, September 1995. Springer Verlag.

10. J. O'Leary, M. Leeser, J. Hickey, and M. Aagaard. Non-restoring integer square root: A case study in design by principled optimization. In T. Kropf and R. Kumar, editors, *Proc. 2nd International Conference on Theorem Provers in Circuit Design (TPCD94)*, volume 901 of *Lecture Notes in Computer Science*, pages 52–71, Bad Herrenalb, Germany, September 1994. Springer Verlag. published 1995.

11. Steven D. Johnson. *Synthesis of Digital Design from Recursion Equations*. The MIT Press, Cambridge, MA, 1984. ACM Distinguished Dissertation 1984.

12. Israel Koren. *Computer Arithmetic Algorithms*. Prentice Hall, 1993.

13. M. Leeser and J. O'Leary. Verification of a subtractive radix-2 square root algorithm and implementation. In *Proceedings International Conference on Computer Design 1995 (ICCD '95)*, pages 526–531, October 1995.

14. Irwin Meisels and Mark Saaltink. The Z/EVES reference manual (draft). Technical Report TR-95-5493-03, ORA Canada, December 1995.

15. Paul S. Miner. Defining the IEEE-854 floating-point standard in PVS. Technical Memorandum 110167, NASA, Langley Research Center, Hampton, VA, June 1995.

16. J Strother Moore, Tom Lynch, and Matt Kaufmann. A mechanically checked proof of the correctness of the kernel of the $AMD5_k86$ floating point division algorithm. (Submitted), URL:http://devil.ece.utexas.edu:80/~lynch/divide/divide.html, March 1996.

17. Sam Owre, John Rushby, Natarajan Shankar, and Friedrich von Henke. Formal verification for fault-tolerant architectures: Prolegomena to the design of PVS. *IEEE Transactions on Software Engineering*, 21(2):107–125, February 1995.

18. H. Rueß, N. Shankar, and M. K. Srivas. Modular verification of SRT division. In Rajeev Alur and Thomas A. Henzinger, editors, *Computer-Aided Verification, CAV '96*, number 1102 in Lecture Notes in Computer Science, pages 123–134, New Brunswick, NJ, July/August 1996. Springer-Verlag.

19. David M. Russinoff. A mechanically checked proof of correctness of the AMD K5 floating-point square root algorithm. unpublished manuscript, July 1996.

20. George S. Taylor. Compatible hardware for division and square root. In *Proceedings 5th Symposium on Computer Arithmetic*, pages 127–134, 1981.

21. V. Pratt. Anatomy of the Pentium bug. In P.D. Mosses, M. Nielsen, and M.I. Schwartzbach, editors, *TAPSOFT '95: Theory and Practice of Software Development*, volume 915 of *Lecture Notes in Computer Science*, pages 97–107, Aarhus, Denmark, May 1995. Springer Verlag.

22. D. Verkest, L. Claesen, and H. De Man. A proof of the nonrestoring division algorithm and its implementation on an ALU. *Formal Methods in System Design*, 4:5–31, January 1994.

23. J. S. Walther. A unified algorithm for elementary functions. In *Proceedings of Spring Joint Computer Conference*, pages 379–385, 1971.

Hierarchical Verification of Two-Dimensional High-Speed Multiplication in PVS: A Case Study

Harald Rueß

Universität Ulm, Abtl. Künstliche Intelligenz
D-89069 Ulm, Germany
ruess@informatik.uni-ulm.de

Abstract. It is shown how to use the PVS specification language and proof checker to present a hierarchical formalization of a two-dimensional, high-speed integer multiplier on the gate level. We first give an informal description of iterative array multiplier circuits together with a natural refinement into vertical and horizontal stages, and then show how the various features of PVS can be used to obtain a readable, high-level specification. The verification exploits the tight integration between rewriting, arithmetic decision procedures, and equality that is present in PVS. Altogether, this case study demonstrates that the resources of an expressive specification language and of a general-purpose theorem prover permit highly automated verification in this domain, and can contribute to clarity, generality, and reuse.

1   Introduction

Verifying functional correctness results about arithmetic circuits poses some serious challenges to current hardware verification techniques. Almost all automated approaches to hardware-verification, such as model-checking, rely on *binary decision diagrams* (BDDs) or some variant of these as the underlying representation form. BDD representations of arithmetic circuits such as multiplication or division, however, have been shown to grow exponentially with the word size [Bry94]. Consequently, it is not feasible to directly apply BDD-based methods to prove the overall correctness of many interesting arithmetic circuits with wide data paths. Moreover, such methods suffer from a lack of expressiveness for describing high-level specifications. Theorem-proving based approaches, on the other hand, exploit the regular structure of arithmetic circuits and use inductive reasoning to establish correctness results for arbitrary word sizes. A major complaint against the use of mechanical theorem provers for hardware verification, however, has been that they often require significant human guidance.

The overall aim in this paper is to explore the use of techniques based on high-level specification and highly automated formal proof construction in PVS to systematically verify the functional correctness of arithmetic circuits with fixed, but arbitrary sized operands. The paper is based upon a case study: a specification for a two-dimensional, high-speed integer multiplier, its realization on the gate-level, and its formal verification. The basic strategy for constructing this correctness proof relies on a natural decomposition of the two-dimensional multiplier circuit into vertical and horizontal stages.

This paper is organized as follows: Section 2 discusses the main features of the PVS specification language and the proof-checker. Section 3 informally introduces (unsigned) *iterative array multiplication* and informally describes the decomposition of the verification problem into a hierarchy of verification tasks. The steps in the hierarchy are independent from each other, but the description of these verification steps proceeds strictly bottom-up. Gate-level implementations of low-level building blocks such as *carry-propagate* and *carry-save* addition rows and their automatic verification are developed in Section 4, and the formalization of the vertical and horizontal stages of the *iterative array multiplier* are described in Section 5. It abstracts this multiplier in terms of a series of carry-save addition steps followed by a final carry-propagate step. Next, it is formally verified in Section 6 that the overall functional correctness of the iterative array multiplier follows from the functional correctness of the lower-level building blocks. These higher-level proof steps are independent of details of the gate-level implementations of the lower-level building blocks. The PVS language support for formalizing *generic hardware components* is discussed in Section 7. Finally, Sections 8 and 9 end with a comparison with related work and some conclusions.

Altogether, this case study demonstrates that the resources of an expressive specification language and a general-purpose theorem prover permit highly-automated verification in the domain of arithmetic circuits, and can contribute to clarity, generality, and reuse.

2   An Overview of PVS

The PVS system combines an expressive specification language with a productive, interactive proof checker that has a reasonable amount of theorem proving capabilities, and has been used for reasoning in domains as diverse as microprocessor verification, protocol verification, and algorithm and architectures concerning fault-tolerance [ORSvH95].

The PVS specification language builds on classical typed higher-order logic with the usual base types bool, nat, rational, real, ... and the function type constructor [A -> B]. The type system of PVS is augmented with *dependent types* and *abstract data types*. A distinctive feature of the PVS specification language are *predicate subtypes* {x:A | P(x)}. These subtypes consist of exactly those elements a of type A satisfying predicate P(a). Predicate subtypes are used to explicitly constrain the domain and ranges of operations in a specification and to define partial functions. Predicate subtypes are used, for example, to define the integer ranges below[n] ($[0, \ldots, n)$) or subrange[-m, n] ($[-m, \ldots, n]$). In general, type-checking with predicate subtypes is undecidable, and the type-checker generates *type correctness conditions* (TCCs) corresponding to predicate subtypes. A large number of TCCs are discharged by specialized proof strategies, and a PVS expression is not considered to be fully type-checked unless all generated TCCs have been proven correct.

PVS specifications are packaged as *theories* that can be parametric in types and constants. A built-in *prelude* and loadable *libraries* provide standard specifications and proved facts for a large number of theories; for example, fixed-sized bit-vectors.

An N-bit bit-vector is represented as an array bvec[N], i.e. a function from the type below[N] of natural numbers less than N to bit. The parameter N is constrained to be a positive natural number posnat since bit-vectors of length 0 are not permitted. Bits are represented as elements of type upto[1] which consists of the numbers 0 and 1.

The function bv2nat (see [1]) from the bit-vector library defines the unsigned interpretation of bit-vectors of length N as natural numbers.

```
                                                                          1
bv_nat[N: posnat]: THEORY
 BEGIN
  IMPORTING bitvectors@bv[N]

  bv2nat_rec(n: upto[N], bv:bvec[N]): RECURSIVE nat =
  IF n = 0 THEN 0 ELSE exp2(n-1) * bv(n-1) + bv2nat_rec(n-1, bv) ENDIF
  MEASURE n

  bv2nat(bv:bvec[N]): below[exp2(N)] = bv2nat_rec(N, bv)
  CONVERSION bv2nat
 END bv_nat
```

It is defined in the theory bv_nat that is parameterized with respect to the length N of bit-vectors. bv2nat_rec is an auxiliary function used only to define bv2nat, and interprets the lower portion of a bit vector as a natural number. Recursive function definitions in PVS must have an associated MEASURE function to ensure termination. The type-checker automatically generates type correctness proof obligations to show that the argument to every recursive call is decreasing according to the given measure. Moreover, the CONVERSION directive causes the type-checker to implicitly inject the conversion bv2nat whenever it finds a bit-vector instead of the required number. In this way it is possible to *overload* bit-vectors xv with their unsigned interpretations bv2nat(xv). This kind of overloading is commonly used in computer arithmetic textbooks to express concise and readable circuit descriptions.

Many more basic facts about bit-vectors are predefined in the bit-vector library of PVS. Concatenation of bit-vectors xv and yv, for example, is denoted by xv o yv, and extraction of (i+j+1) many bits i through j from bit-vector xv of type bvec[N] is denoted by xv^(i,j); xv(i) is shorthand for xv^(i,i). Obviously, extraction is restricted, using predicate subtyping, to the cases (i < N) and (i >= j).

Proofs in PVS are presented in a sequent calculus. The atomic commands of the PVS prover component include induction, quantifier instantiation, automatic conditional rewriting, simplification using arithmetic and equality decision procedures and type information, and propositional simplification using binary decision diagrams. The **skolem*** command, for example, repeatedly introduces Skolem constants for universal-strength quantifiers, and **assert** combines rewriting with decision procedures.

Finally, PVS has an LCF-like strategy language for combining inference steps into more complicated proof strategies. The strategy **then**, for example, sequentially applies a sequence of sub-strategies, and the defined rule **grind** combines rewriting with propositional simplification using BDDs and decision procedures. This strategy, in combination with case analysis, has been very effective for proving many typical hardware theorems.

3   Iterative Array Multipliers

Many multiplier circuits are based on the grade school principle of multiplying any two given **N** bit numbers: computing the partial products, and adding the partial products to obtain the required result. The two basic operations of multiplication, the generation of partial products and their summation, may be merged. This merging speeds up the multiplication, since the overhead that is due to the separate controls is avoided. Such multipliers, which consist of identical cells, each capable of forming a new partial product and adding it to the previously accumulated partial product, are called *iterative array multipliers* [Kor93].

To illustrate the operation of an iterative array multiplier for unsigned numbers, examine the 4×5 parallelogram shown in Figure 1. This circuit multiplies two **N** = 5 wide operands, say **av** and **xv**. It adds the first two partial products (i.e. **av * xv(0)** and **av * xv(1)**) in row one after proper alignment. The results of the first row are then added to **av * xv(2)** in the second row, and so on. The basic cell for such an array multiplier is a full adder **FA** accepting one bit of the new partial product, and a carry-in bit. In the first four rows there is no horizontal carry propagation. In other words, a *carry-save* type addition is performed in these rows, and the accumulated partial product consists of intermediate sum and carry bits. A horizontal carry propagation is allowed only in the last row. The last row of cells in this figure is a *ripple-carry adder*. Consequently, an iterative array multiplier circuit with operands of size **N** is abstracted in terms of a series of (**N** - 1) carry-save steps followed by a carry-propagate step. Block diagrams of these stages of the *vertical decomposition* or vertical refinement are depicted in Figure 2 together with word-level equations between appropriately shifted input and output vectors. The word-level condition **sv'** = 2^N *** cv +** **sv** of the carry-propagate step, for example, states that the (unsigned) interpretation of the output sum vector **sv'** equals the sum of the input sum vector **sv** and the carry-in vector **cv** shifted by **N** bits.

Moreover, looking at the circuit in Figure 1, every vertical stage can naturally be decomposed into different functional units. The *horizontal decompositions* of **csa_step(i)** and **cpa_step** are depicted in Figure 3. For every vertical stage,

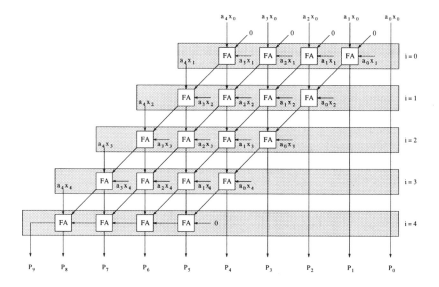

Fig. 1. Iterative Array Multipliers

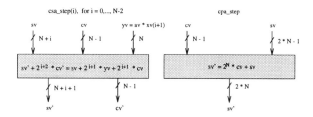

Fig. 2. Vertical Refinement

the decomposition consists of three functional units. The left-most unit simply computes the upmost bit of the next partial sum `av(N - 1) * xv(i + 1)`, the right-most unit shifts through the lower `(i + 1)` bits of the current partial sum `sv`, and the central unit consists of a carry-save adder row of length `(N - 1)`. The final step of the multiplier is decomposed in a similar way (see Figure 3) with the central unit being a `(N - 1)`-bit carry-propagate adder row.

Informally, the correctness of the horizontal refinement of the carry-save step is obtained by the following equality reasoning:

$$sv' + 2^{i+2}cv'$$
$$= \{ \text{ decompositions } \}$$
$$2^{N+i}sv'(N+i) + 2^{i+1}sv'(N+i-1,i+1) + sv'(i,0) + 2^{i+2}cv'$$
$$= \{ \text{ algebra } \}$$
$$2^{N+i}sv'(N+i) + 2^{i+1}(sv'(N+i-1,i+1) + 2cv') + sv'(i,0)$$
$$= \{ \text{ functional behavior } \}$$
$$2^{N+i}yv(N-1) + 2^{i+1}(sv(N+i-1,i+1) + cv + yv(N-2,0)) + sv(i,0)$$
$$= \{ \text{ algebra } \}$$
$$2^{N-1}2^{i+1}yv(N-1) + 2^{i+1}yv(N-2,0) + 2^{i+1}cv$$
$$+2^{i+1}sv(N+i-1,i+1) + sv(i,0)$$
$$= \{ \text{ compositions } \}$$
$$2^{i+1}yv + s^{i+1}cv + sv$$

The composition and decomposition steps above are based on the following arithmetic equations over unsigned interpretations of bit-vectors:

$$xv(i-1,0) + 2^i xv(n-1,i) = xv \qquad \text{(composition)}$$
$$xv \; o \; yv = 2^m xv + yv \quad \text{(decomposition)}$$

Here, xv and yv are bit-vectors of length n and m, respectively, and o denotes concatenation of bit-vectors. The correctness proof for the horizontal refinement of the carry-propagate step works analogously. Finally, it is straightforward to prove, by induction, the overall correctness of iterative array multiplication (see also Section 6) from the word-level specifications of the vertical stages.

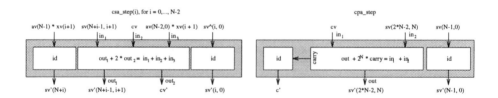

Fig. 3. Horizontal Refinement

Altogether, this two-dimensional decomposition of the circuit leads to the following hierarchical strategy for proving the overall correctness of this multiplier:

- prove the correctness of gate-level implementations of carry-save and carry-propagate adders;
- assuming correct adder rows, prove the correctness of the horizontal decompositions;
- assuming the correctness of the horizontal decomposition prove the correctness of the vertical decomposition, and, consequently, the overall correctness of the circuit.

4 Implementation and Verification of Adders

This section describes gate-level implementations of iteratively generated N-bit carry-propagate and carry-save adder rows. Their arithmetic behavior is stated in terms of equations over word-level interpretations of the input and output signals. The proofs of these characteristic theorems are carried out automatically in PVS.

The theory cpa in [2] describes the implementation and the corresponding correctness statement for the carry-propagate adder. This theory is parameterized with respect to the size N of the bit-vector inputs and output. It imports a theory full_adder[1] which contains a gate-level implementation and the arithmetic specification of a full adder with output carry bit and sum bit, and the theory bv which specificies the bit-vector type (bvec[N]) and related bit-vector functions. The carry bit that ripples through the full adders is specified recursively by means of the function nth_cin. The function cpa defines the carry output to be the N-th carry, and the n-th sum output of the carry propagate adder is defined to be the sum output of the n-th full adder applied to the n-th bits of the input bit-vectors and the n-th carry.

```
                                                                        2
cpa[N: posnat]: THEORY
  BEGIN
   IMPORTING full_adder, bitvectors@bv[N], bitvectors@bv_nat

   xv, yv: VAR bvec[N]; n: VAR below[N]; bj: VAR upto[N]

   nth_cin(j, xv, yv):RECURSIVE bit =
    IF j = 0 THEN 0
    ELSE carry(FA(xv(j - 1), yv(j -1), nth_cin(j - 1, xv, yv))) ENDIF
   MEASURE j

   cpa(xv, yv): [# carry: bit, sum: bvec[N] #] =
    (# carry := nth_cin(N, xv, yv),
       sum   := (LAMBDA n: sum(FA(xv(n), yv(n), nth_cin(n, xv, yv))))) #)

   cpa_inv: LEMMA
    bv2nat_rec(j, sum(cpa(xv, yv)))
    = bv2nat_rec(j,xv) + bv2nat_rec(j,yv) - exp2(j) * nth_cin(j, xv, yv)

   cpa_char: THEOREM
    sum(cpa(xv, yv)) = xv + yv - exp2(N) * carry(cpa(xv, yv))
  END cpa
```

Having described an implementation, we now characterize its functionality. The theorem cpa_char in [2] expresses the conventional correctness statement

---

[1] The theories full_adder, bit_mult, and csa are not included in this paper for lack of space.

of an adder row on the *word-level*, and relates the gate-level implementation of
cpa with the functional level. This characteristic theorem of cpa is an immediate consequence of the invariant cpa_inv in ⟨2⟩, and the proof of this invariant
proceeds by induction on the variable j using an induction scheme for the type
upto[N].

```
(induct-and-simplify "j" :name "upto_induction[N]"
                        :exclude "FA"
                        :theories  "full_adder")
```

In this case, the use of the parameters **exclude** and **theories** yields a certain
economy of proof construction, since they direct the induction strategy to apply the arithmetic characterization of the full adder as specified in the theory
full_adder instead of blindly unfolding the definition of the function FA. It takes
about 15 seconds to complete the proof on a Sun Sparcstation 10.

Correctness proofs for N-bit carry-propagate adders have been presented for
many different logics and proving systems. Cyrluk et al. [CRSS94], for example, define a specialized proof strategy for proving a slight variant of the above
carry-propagate adder and other hardware theorems. An alternative approach
of synthesizing word-level properties of iteratively generated circuits from properties of the functional units has been proposed in [HDL90]. As shown above,
however, the improved general-purpose induction strategies of the PVS prover
already suffice to automatically prove the correctness of inductively generated
carry-propagate adders.

In a similar way, one defines a carry-save addition row as a function csa such
that the theorem csa_char holds.

```
                                                                    3
    csa(xv, yv, cv: bvec[N]): [# sum, carry: bvec[N] #] = ...

    csa_char: THEOREM
       sum(csa(xv, yv, cv)) = xv + yv + cv - 2 * carry(csa(xv, yv, cv))
```

The PVS proof script of this theorem is identical with the proof strategy used
above for the correctness result of carry-propagate addition.

5   Formalization of Horizontal and Vertical Composition

Formalization of the horizontal and the vertical refinements of the iterative array
multiplier, as discussed in Section 3, are developed in this section.

The formalization of the carry-propagate step as depicted in Figure 3 is
straightforward. The function cpa_step in ⟨4⟩ calls the carry-propagate adder
on the carry vector cv and the upper half of the sum vector sv, and the resulting output sum vector is created by concatenating the carry output with the
sum output and the lower half of the input sum.[2]

---

[2] Here, a conversion from bit to bvec[1] is assumed.

```
                                                                    4
cpa_step((cv: bvec[N - 1]), (sv: bvec[2 * N - 1])): bvec[2 * N] =
   LET cpa = cpa(sv^(2 * N - 2, N), cv) IN
      carry(cpa) o sum(cpa) o sv^(N - 1, 0)
```

The formalization of carry-save steps csa_step(i) is more complicated and requires the concept of *dependent types*, since the types of both the input parameters and the result depends on the value of the index i.

```
                                                                    5
csa_step(i:below[N - 1])
        ((sv: bvec[N+i]), (cv: bvec[N-1]), (yv:bvec[N]))
  : [# sum: bvec[N + i + 1], carry: bvec[N - 1] #] =
   LET csa = csa(sv^(N + i - 1, i + 1), yv^(N - 2, 0), cv) IN
      (# sum   := yv(N - 1) o sum(csa) o sv^(i, 0),
         carry := carry(csa)                          #)
```

The function csa_step uses PVS's dependent typing features and the result of this function is a tuple consisting of output sum and carry-out vector. It is a straightforward transliteration of the block diagram of the carry-save step in Figure 3. Note also that the role of dependent types for hardware specification (and verification) has already been noted in [HDL90].

Finally, the carry-save and the carry-propagate stages are combined to form an iterative array multiplier in exactly the same way as described in Section 3.[3]

```
                                                                    6
connect_csa((i: upto[N - 2]), (av, xv: bvec[N])):
  RECURSIVE [# carry: bvec[N - 1], sum: bvec[N + i + 1] #] =
  (IF i=0 THEN
      csa_step(0)(bit_mult(av, xv(0)), bvec0, bit_mult(av, xv(1)))
   ELSE
      csa_step(i)(sum(connect_csa(i - 1, av, xv)),
                  carry(connect_csa(i - 1, av, xv)),
                  bit_mult(av, xv(i + 1)))
   ENDIF)
  MEASURE i

iam(av, xv: bvec[N]): bvec[2 * N] =
  cpa_step(carry(connect_csa(N-2)(av,xv)),sum(connect_csa(N-2)(av,xv)))
```

The iterative array multiplier iam in $\boxed{6}$ consists of (N - 2) stages of appropriately wired carry-save steps csa_step(i) followed by a final carry-propagate step. This finishes the formalization of this multiplier.

---

[3] bvec0 denotes a constant 0 bit-vector of appropriate length.

6 Correctness of the High-Level Design

The mechanized correctness proof of the iterative array multiplier is described in this section. This correctness proof is parameterized[4] with respect to the lower-level building blocks in Section 4 such as carry-save and carry-propagate adder rows and formation of partial products. It does not, however, refer to the details of the gate-level implementations and only uses the characteristic behavioral theorems such as cpa_char in ⎡2⎤ and csa_char in ⎡3⎤.

The correctness proofs of the horizontal refinements need the following decomposition and composition lemmas as has been demonstrated in Section 3. Let n,m be positive natural numbers, and xv, yv be bit-vectors of sizes n and m, respecively, and i of type subrange[1,n-1]; then:[5]

```
                                                                    7

    decompose: LEMMA xv o yv = xv * exp2(m) + yv
    compose  : LEMMA xv^(i - 1, 0) + xv^(n - 1, i) * exp2(i) = xv
```

First, the correctness of the horizontal refinement of the carry-save step is expressed by the following lemma:

```
    csa_step_char: LEMMA                                            8

      FORALL ((i:below[N-1]),(sv: bvec[N+i]),(cv: bvec[N-1]),(xv: bvec[N]))
         sum(csa_step(i, sv, cv, xv))
         = sv + exp2(i + 1) * xv + exp2(i + 1) * cv
           - exp2(i + 2) * carry(csa_step(i, sv, cv, xv))
```

Unfortunately, the PVS prover needs some guidance to apply the **compose** lemma above in an appropriate way, since the rewrite engine of PVS does not apply *commutative-associative* matching; the appropriate matching terms, however, are typically found in independent branches of a nested +-structure. Altogether 15, mostly trivial, interactive steps are necessary to complete this proof.

Second, the behavioral specification given in Figure 2 for the carry-propagate step

```
                                                                    9

    cpa_step_char: LEMMA
      FORALL (cv: bvec[N - 1], sv: bvec[2 * N - 1]):
         cpa_step(cv, sv) = sv + cv * exp2(N)
```

is proved automatically. Skolemization followed by unfolding the definition cpa_step and repeated calls to the decision procedures discharge this proof

```
    (then (skosimp*) (expand "cpa_step") (repeat (assert)))
```

---

[4] In a way described more clearly in Section 7.

[5] Slight variants of these lemmas are stated and proved in the PVS bit-vector library.

after introducing lemmas `compose`, `decompose`, and the behavioral constraint `cpa_char` (see [2]) as rewrites to the prover.

Third and finally, the lemmas `csa_step_char` and `cpa_step_char` permit an easy induction proof of the behavioral correctness `iam_char` of the multiplier.

```
                                                                        10
  av, xv: VAR bvec[N]; i: VAR upto[N - 2]

  iam_inv: LEMMA
   sum(connect_csa(i, av, xv))
   = av * bv2nat_rec(i+2, xv) - exp2(i+2) * carry(connect_csa(i, av, xv))

  iam_char: THEOREM iam(av, xv) = av * xv
```

The induction scheme for proving `iam_inv` in [10] is given by

```
(induct "i" :name "upto_induction[N - 2]")
```

Both the base case and the induction step are proved with a combination of calls to the decision procedures and rewriting.

7   The use of Generic Hardware Components

Behavioral specifications of many hardware components are sufficient to specify and reason about higher-level building blocks; other details are often irrelevant. Consequently, developments on higher levels abstract from specific gate-level implementations of the basic building blocks. In this way, *generic hardware components* allow verification of circuits to be done in a hierarchical fashion. It is demonstrated in this section how features of the PVS specification language such as parameterization of theories and predicate subtypes support this use of generic hardware components.

In our example, formalizations of higher-level designs related to the horizontal and vertical decomposition (see Section 3) can be developed in the theory `iam` in [11]. This theory is parameterized with respect to three *generic* hardware components; namely, a carry-save adder `csa`, a "carry-propagate" adder `cpa`, and the function `bit_mult` which computes a partial product.

Behavioral constraints associated with generic hardware components are specificed in PVS in the `ASSUMPTIONS` part, and any instantiation of the formal parameters incurs the obligation to prove all of the `ASSUMPTIONS`. Obviously, the last two generic hardware components in [11] can be realized by the gate-level adder implementations in [3] and [2], respectively. A useful variation of the circuit in Figure 1, however, can be realized by replacing the final *ripple-carry adder* with a fast two-operand carry-lookahead adder, if a shorter overall execution time is desired. In this case, all developments within theory `iam` are "inherited" for this new circuit as long as the assumption `cpa_char` in [11] can be shown to hold.

```
                                                                    11
iam[ (IMPORTING bitvectors@bv, ...)
 N        : upfrom(3),
 bit_mult: [bvec[N], bit -> bvec[N]],
 csa: [bvec[N - 1], bvec[N-1], bvec[N-1] -> [# carry, sum: bvec[N-1] #]],
 cpa: [bvec[N - 1], bvec[N - 1] -> [# carry: bit, sum: bvec[N - 1]#]]
]: THEORY
 BEGIN
  ASSUMING
   xv, yv, cv: VAR bvec[N - 1]; av: VAR bvec[N]; x: VAR bit

   bit_mult_char: ASSUMPTION bit_mult(av, x) =av * x
   csa_char     : ASSUMPTION
     sum(csa(xv, yv, cv)) = xv + yv + cv - 2 * carry(csa(xv, yv, cv))
   cpa_char     : ASSUMPTION
     sum(cpa(xv, yv)) = xv + yv - exp2(N-1) * carry(cpa(xv, yv))
  ENDASSUMING

   (... see Sections 5 and 6 ...)
 END iam
```

Instead of using explicit **ASSUMPTIONS** constraints on the generic hardware
components as shown above, one could also define predicate subtypes for each of
these components and parameterize the theory **iam** in 11 with respect to these
subtypes. The type **BIT_MULT** in 12, for example, characterizes the set of valid
bit_mult functions.

```
                                                                    12
BIT_MULT[N: posnat]: THEORY
  BEGIN
   BIT_MULT: TYPE =
    { bit_mult: [bvec[N], bit -> bvec] |
        FORALL (xv: bvec[N], y: bit):  bit_mult(xv, y) = xv * y }
  END BIT_MULT
```

In PVS, theory parameterization must be used to define this family of types
indexed over **N**.

```
                                                                    13
iam[ (IMPORTING BIT_MULT, ...)
 N: upfrom[3], bit_mult: BIT_MULT[N], csa: CSA[N], cpa: CPA[N]
]: THEORY
 BEGIN ... END iam
```

Using the type definition in 12 and analogously defined types **CSA[N]** and
CPA[N] the parameterized theory **iam** in 11 is rewritten in a very concise way
as shown in 13.

8 Related Work

The well-publized FDIV error of initial releases of the Pentium floating-point divider sparked renewed interest in the verification of arithmetic circuits [Bry95, CB95, KS96, RSS96, ML96, CGZ96, LO95, AS95, KS96]. For the large number of recent publications in this area, we restrict ourselves to a comparison to work that we think is most closely related to this paper.

Chen and Bryant [BC94, CB95] use *Binary Moment Diagrams* (BMDs) to verify iterative multipliers for word sizes up to 256 in a hierarchical fashion. At the lower levels, a set of component modules are described at both the bit-level (as combinational circuits) and at the word-level (as algebraic expressions). They verify that the bit-level implementation of each block implements its word-level specification. At the higher levels, the system is described as an interconnection of components having word-level representations, and the specification is also given at the word-level. The verification task makes sure that the composition of the block functions corresponds to the system specification. The verification is, after the manual decomposition of the problem and a suitable choice for the variable ordering, fully automatic. The array multiplier they are verifying is similar to the one in Figure 1, and their hierarchical decomposition of this multiplier is analogous to the one presented in this paper. The verification approach based on BMDs, however, only supports the verification of fixed-sized circuits, while our approach is based on inductive theorem proving and proves, once and for all, the correctness of circuits of arbitrary size. Moreover, circuit descriptions in their specification language ACV are still rather low-level, since this language does not allow to express parameterized designs and sequencing features.

Kapur and Subrahmaniam [KS96] mechanically verify multiplier circuits in RLL by developing a common top-level specification that abstracts multiplication in terms of a component that computes the partial sums and another that adds these partial sums to compute the final product; a similar decomposition of multipliers that mimic the usual process of longhand multiplication also been been proposed by [Chi92]. The top-level specification in terms of conditional equations is parameterized over circuit-specific details such as the number of partial sums computed in each stage and the adder interconnection. Kapur and Subrahmaniam "instantiate" this top-level scheme with various multipliers such as Wallace tree and an array multiplier. Bit-vectors are encoded in [KS96] as linear lists. One consequence of this encoding are side conditions about the length of bit-vector representation in their specifications. This complication is avoided in our approach by using bit-vectors of fixed-sized length, since, in this case, illicit expressions are already detected by the type-checker. Moreover, the specifications in this paper reflect the structure of the underlying circuit and the sizes of the data-path more closely than the ones given in [KS96]; one reason for this is the represention of right-shifts in [KS96] by appending trailing zeros to the bit-vectors. Also, the functional representation of the array multiplier in this paper allows one to generate verified gate-level implementations.

The construction of the correctness proofs in [KS96] requires a minimum of user interactions (such as manual orientation of some equations) and are very fast; this is mainly due to efficient rewriting and the discovery of appropriate induction schemes by cover set induction. With our functional approach of generating the circuits by means of recursive functions, however, selection of appropriate induction schemes was always straightforward.

9 Concluding Remarks

A two-dimensional high-speed multiplier circuit has been specified and formally verified in a hierarchical way using the PVS system. It took about 2 weeks for an — back then — inexperienced PVS user to complete this case study. Altogether, the specifications of this multiplier consist of 5 separate theories with about 220 lines of code.

This verification exercise demonstrates the value of an expressive specification language in mechanized verification. In particular, the use of dependent types together with predicate subtypes and theory parameterization allows the writing of specifications that are clear, concise, and generic. Such high-level descriptions not only minimize the pitfalls of introducing errors in initial design specification but also open the door to using these specifications as design documents. Moreover, the modular specification approach and the use of generic hardware components supports modular proof development and has the advantage that some variations of a particular circuit design can be verified by just redoing one part of the proof. It is, for example, a simple matter to substitute the carry-propagate adder used in this paper with a (faster) carry-lookahead adder.

The construction of the proof requires about 50 user interactions, most of which are as trivial as introducing rewrites to the prover or explicitly calling the decision procedures. All the lemmas about unsigned interpretations of bit-vectors that were required to finish the proof could already be found in the bit-vector library; the main lemmas in [7], however, had to be restated in a different form to make the PVS correctness proof "more automatic".

Altogether, the combination and tight integration of term rewriting with efficient decision procedures in PVS proved to be a useful workhorse, since most theorems could be proved fairly automatically. Only the proofs at the higher-levels needed some guidance. It is conjectured, however, that a PVS system extended with standard theorem proving techniques such as associative-commutative term rewriting (and perhaps a slight variation of the current induction strategy) could prove all the theorems in this paper fully automatically.

Another promising approach of automating this and other hardware-related proofs is to extend PVS with efficient decision procedures for theories about bit-vectors of fixed, but arbitrary size N. Moreover, further work is needed to demonstrate the general applicability of the hierarchical decomposition of two-dimensionally generated circuits into vertical and horizontal stages as proposed in this paper.

References

[AS95]   M.D. Aagaard and C.J.H. Seger. The Formal Verification of a Pipelined Double-Precision IEEE Floating-Point Multiplier. In *Proc. of ICCAD'95*, pages 7–10. IEEE Computer Science Press, 1995.

[BC94]   R.E. Bryant and Y.A. Chen. Verification of Arithmetic Circuits with Binary Moment Diagrams. Technical Report CMU-CS-94-160, School of Computer Science, Carnegie Mellon University, 1994.

[Bry94]   R.E. Bryant. Verification of Arithmetic Functions with Binary Moment Diagrams. Technical Report CMU-CS-94-160, School of Computer Science, Carnegie Mellon University, Pittsburgh, PA 15213, 1994.

[Bry95]   R.E. Bryant. Bit-Level Analysis of an SRT Divider Circuit. Technical Report CMU-CS-95-140, School of Computer Science, Carnegie Mellon University, Pittsburgh, PA 15213, April 1995.

[CB95]   Y.A. Chen and R.E. Bryant. ACV: An Arithmetic Circuit Verifier. 1995.

[CGZ96]   E.M. Clarke, S.M. German, and X. Zhao. Verifying the SRT Division Algorithm using Theorem Proving Techniques. In R. Alur and T.A. Henzinger, editors, *CAV'96*, number 1102 in Lecture Notes in Computer Science, pages 111–122. Springer-Verlag, 1996.

[Chi92]   S.K. Chin. Verified Functions for Generating Signed-Binary Arithmetic Hardware. *IEEE Transactions on Computer-Aided Design*, 11(2):1529–1558, December 1992.

[CRSS94]   D. Cyrluk, S. Rajan, N. Shankar, and M. Srivas. Effective Theorem Proving for Hardware Verification. In R. Kumar and Th. Kropf, editors, *Theorem Provers in Circuit Design*, number 901 in Lecture Notes in Computer Science, 1994.

[HDL90]   F.K. Hanna, N. Daeche, and M. Longley. Specification and Verification Using Dependent Types. *IEEE Transactions on Software Engineering*, 16(9):949–964, September 1990.

[Kor93]   I. Koren. *Computer Arithmetic Algorithms*. Prentice-Hall, 1993.

[KS96]   D. Kapur and M. Subramaniam. Mechanically Verifying a Family of Multiplier Circuits. In R. Alur and T.A. Henzinger, editors, *CAV'96*, number 1102 in LNCS, pages 135–146. Springer Verlag, 1996.

[LO95]   M. Leeser and J. O'Leary. Verification of a Subtractive Radix-2 Square Root Algorithm and Implementation. In *Proc. of ICCD'95*, pages 526–531. IEEE Computer Society Press, 1995.

[ML96]   P.S. Miner and J.F. Leathrum. Verification of IEEE Compliant Subtractive Division Algorithms. 1996. FMCAD'96, This Volume.

[ORSvH95] S. Owre, J. Rushby, N. Shankar, and F. von Henke. Formal Verification for Fault-Tolerant Architectures: Prolegomena to the Design of PVS. *IEEE Transactions on Software Engineering*, 21(2):107–125, February 1995.

[RSS96]   H. Rueß, M. Srivas, and N. Shankar. Modular Verification of SRT Division. In R. Alur and T.A. Henzinger, editors, *CAV'96*, number 1102 in Lecture Notes in Computer Science, pages 123–134. Springer Verlag, 1996.

Acknowledgments This case study has been conducted while visiting the formal verification group at SRI International, Menlo Park. I would like to thank S. Owre, J. Rushby, N. Shankar, and M. Srivas for all the support and useful feedback, and H. Pfeifer for proof-reading.

Experiments in Automating Hardware Verification Using Inductive Proof Planning

Francisco J. Cantu⋆   Alan Bundy⋆⋆   Alan Smaill⋆⋆   David Basin⋆⋆⋆

Abstract. We present a new approach to automating the verification of hardware designs based on planning techniques. A database of methods is developed that combines tactics, which construct proofs, using specifications of their behaviour. Given a verification problem, a planner uses the method database to build automatically a specialised tactic to solve the given problem. User interaction is limited to specifying circuits and their properties and, in some cases, suggesting lemmas. We have implemented our work in an extension of the *Clam* proof planning system. We report on this and its application to verifying a variety of combinational and synchronous sequential circuits including a parameterised multiplier design and a simple computer microprocessor.

1 Introduction

Confidence in the correctness of hardware designs may be increased by formal verification of the designs against specifications of their desired behaviour. Although this is common knowledge, formal verification is almost completely neglected by industry; what interest there is centres on 'push-button' systems based on model checking, where weak decidable logics are used for problem specification. But such systems are necessarily limited and cannot be applied to many important classes of problems. For example, parameterised designs and their specifications can almost never be expressed. However, systems based on more expressive logics are often resisted because proof construction is no longer fully automated and most circuit designers have neither the time, the patience, nor the expertise for semi-interactive theorem proving.

To increase the acceptance of formal methods in domains with undecidable verification problems, we must automate proof construction as much as is practically possible. The problem to overcome is the large search space for proofs. Even when semi-decision procedures like resolution are available, completely automated theorem proving is not viable because such techniques are too general and cannot exploit structure in the problem domain to restrict search. Heuristics are called for, perhaps augmented with user interaction to overcome incompleteness. One example of this is the system *Nqthm*, which uses a fixed set of heuristics

⋆ Center for Artificial Intelligence, ITESM, Mexico. Supported by CONACYT grant 500100-5-3533A

⋆⋆ Department of Artificial Intelligence, University of Edinburgh, UK. Supported by EPSRC grant GR/J/80702

⋆⋆⋆ Max-Planck-Institut für Informatik, Saarbrücken, Germany to automate the construction of proofs by induction. Proof construction is automated but sometimes the user must interrupt the prover and suggest lemmas to stop it from exploring an unsuccessful branch. A second example is embodied by tactic-based proof development systems like *HOL*, *Lambda*, *PVS*, and *Nuprl*, where users themselves raise the level of automation by writing tactics (programs that build proofs using primitive inference rules) for particular problem domains. Here incompleteness is addressed by interactive proof construction: rather than writing a 'super-tactic' which works in all cases, users interactively combine tactics to solve the problem at hand and directly provide heuristics.

We present a new approach to hardware verification that falls in between the two approaches above and offers automation comparable to systems like *Nqthm* but with increased flexibility since one can write new domain specific proof procedures as in the tactic approach. We have designed a set of tactics for constructing hardware proofs by mathematical induction. These tactics include tactics developed previously at Edinburgh for automating induction proofs [7] augmented with tactics specifically designed for hardware verification. Rather than having the user apply these tactics interactively, we use ideas from planning to automate proof construction. Each tactic is paired with a *method* which is a declarative specification of its behaviour. Given a verification problem, a planner uses the method database to build automatically a specialised tactic to solve the given problem. User interaction is mainly limited to specifying circuits and their properties. As in the case of *Nqthm*, additional lemmas are occasionally needed to overcome incompleteness, but this is rare; see section 4 for statistics on this and section 5 for a comparison to *Nqthm*.

We have implemented our approach to planning in an extension of the *Clam* proof planning system [8]. We report on this and its application to a variety of combinational and synchronous sequential circuits including a parameterised multiplier design and a simple computer microprocessor. Although we haven't attempted the verification of commercial hardware systems yet, we provide evidence that this approach is a viable way of significantly automating hardware verification using tactic-based theorem provers, that it scales well, and that it compares favourably with both systems based on powerful sets of hardwired heuristics like *Nqthm* and tactic-based approaches. We believe that it offers the best of both worlds: one can write tactics that express strategies for building proofs for families of hardware designs and the planner integrates them in a general proof search procedure. This provides an 'open architecture' in which one may introduce heuristics to reduce search in a principled way. This also suggests a possible methodology for designing provers for verification domains like hardware where experts design domain specific tactics (and methods) and systems such as ours help less skilled users construct correctness proofs.

The remainder of this paper is organised as follows. In section 2 we describe proof planning and extensions required for hardware verification. In section 3 we present our methodology for hardware verification and illustrate it with two examples: the verification of a parallel array multiplier and a simple computer design proposed by Mike Gordon (and hence called the *Gordon computer*). Results of experiments in applying proof planning to verify other combinational and synchronous sequential circuits are reported in section 4. In section 5 we compare some of our results with approaches used in systems such as *Nqthm, PVS* and *HOL*. Finally, in section 6, we summarise advantages and limitations of our approach and discuss future work.

2 Verification Based on Proof Planning

We begin by reviewing proof planning and methods for induction. These two parts contain only a summary of material that has been published in detail elsewhere: see [4, 7] for more on proof planning and [6, 1] for details on rippling. After, we discuss extensions required to apply *Clam* to hardware verification.

2.1 Proof Plans

Proof planning is a meta-level reasoning technique for the global control of search in automatic theorem proving. A proof plan captures the common patterns of reasoning in a family of similar proofs and is used to guide the search for new proofs in the family. Proof planning combines two standard approaches to automated reasoning: the use of tactics and the use of meta-level control. The meta-level control is used to build large complex tactics from simpler ones and also abstracts the proof, highlighting its structure and the key steps.

The main component of proof planning is a collection of methods. A *method* is a specification of a tactic and consists of an input formula (a sequent), preconditions, output formulae, postconditions, and a tactic. A method is *applicable* if the goal to be proved matches the input formula of the method and the method's preconditions hold. The preconditions, formulated in a meta-logic, specify syntactic properties of the input formula. Using the input formula and preconditions of a method, proof planning can predict if a particular tactic will be applicable without actually running it. The output formulae (there may be none) determine the new subgoals generated by the method and give a schematic description of the formulae resulting from the tactic application. The postconditions specify further syntactic properties of these formulae. For each method there is a corresponding general-purpose tactic associated to that method. Methods can be combined at the meta-level in the same way tactics are combined using tacticals.

The process of reasoning about and combining methods is called *proof planning*. When planning is successful it produces a tree of methods, called a *proof plan*. A proof plan yields a composite tactic, which is built from the tactics associated with each method, custom designed to prove the current conjecture. Proof plan construction involves search. However, the planning search space is typically many orders of magnitude smaller than the object-level search space. One reason is that the heuristics represented in the preconditions of the methods ensure that backtracking during planning is rare. Another reason is that the particular methods used have preconditions which strongly restrict search, though in certain domains they are very successful in constructing proofs. There

Methods:

1. symbolic_evaluation	Simplifies symbolic expression
a. elementary	Tautology checker
b. eval_definition	Unfold definition
c. reduction	Apply reduction rules
d. term_cancellation	Cancel-out additive terms in equation
2. induction_strategy	Applies induction strategy
a. induction	Selects induction scheme and induction variables
b. base_case	Applies symbolic_evaluation to the base case
c. step_case	Applies ripple and fertilise to the step case
i. ripple	Applies rippling heuristic
- wave	Applies wave-rules
ii. fertilise	Simplifies induction conclusion with induction hypothesis
3. generalise	Generalises common term in both sides of equation
4. bool_cases	Does a case split on Boolean variable
5. difference_match	Matches and annotates differences on two expressions
6. apply_lemma	Applies existing lemma to current subgoal

Fig. 1. Method Database

is of course a price to pay: the planning system is incomplete. However, this has not proved a serious limitation of the proof planning approach in general [7] nor in our work where proof plans were found for all experiments we tried.

The plan formation system upon which our work is built is called *Clam*. Methods in *Clam* specify tactics which build proofs for a theorem proving system called *Oyster*, which implements a type theory similar to *Nuprl's*.

Methods in *Clam* are ordered and stored in a database, c.f. figure 1. Each of the names in the left-hand side is a method and the indentation of names represents the nesting of methods (a method can call subroutines called sub-methods, which include methods themselves), i.e. an inner method is a sub-method of the one immediately outside it. Given a goal, *Clam* tries the methods in order. If an applicable method is found then the output yields new subgoals and this process is applied recursively. If none is applicable, then *Clam* backtracks to a previous goal or reports failure at the top level. A number of search strategies exist for forming plans. These include depth-first, breadth-first, iterative deepening, and best-first search. We have conducted our experiments in hardware verification using the depth-first planner.

2.2 Induction

A number of methods have been developed in *Clam* for inductive theorem proving and we used these extensively to prove theorems about parameterised hardware designs. Induction is particularly difficult to automate as there are a number of search control problems including selection of an induction rule, control of simplification, possible generalisation and lemma speculation, etc. It turns out though that many induction proofs have a similar shape and a few tactics can collectively prove a large number of the standard inductive theorems.

The induction strategy is a method for applying induction and handling subsequent cases. After the application of induction, the proof is split into one or more base and step cases. The *base case* method attempts to solve the goal using simplification and propositional reasoning (with the sub-method *symbolic_evaluation*). If necessary, another induction may be applied. The *step case* method consists of two parts: rippling and fertilisation. The first part is implemented by the *rippling* method. Rippling is a kind of annotated rewriting where annotations are used to mark differences between the induction hypothesis and conclusion. Rippling applies annotated rewrite rules (called *wave-rules* which are applied with the *wave* method) which minimise these differences. Rippling is goal directed and manipulates just the differences between the induction conclusion and hypothesis while leaving their common structure preserved; this is in contrast to rewriting based on normalisation, which is used in other inductive theorem provers such as *Nqthm* [3]. Rippling also involves little search, since annotations severely restrict rewriting. The second part of the step case, fertilisation, can apply when rippling has succeeded (e.g. when the annotated differences are removed or moved 'out of the way', for example, to the root of the term). The *fertilise* method then uses the induction hypothesis to simplify the conclusion.

2.3 Extensions

The above completes our review of *Clam* and some of the methods for induction. We were able to apply *Clam* to hardware verification with fairly minor modifications and extensions.

First there were modifications to *Clam* methods which resulted, in part, from the fact that the experiments reported here are the largest that have been carried out with *Clam* and revealed some inefficiencies which required improvement. For example: the method *symbolic_evaluation* was extended with a *memoisation* procedure for efficiently computing recursive functions in verifying the Gordon computer, and with an efficient representation of the meta-logic predicate *exp_at* that finds a sub-expression in an expression; the *fertilisation* method was extended to deal with a post-rippling situation that arose in verifying a parameterised version of the *n*-bit adder and which has appeared in many other combinational circuits; the *generalisation* method was extended with type information to generalise over terms of type bool; the *equal* sub-method was extended to apply a more general form of rewrite equations in the hypotheses list of the sequent to simplify the current goal. We also needed to extend *Clam*'s database of induction schemas. To do this we formalised in *Oyster* new data-types appropriate for hardware such as a data-type of words, which are lists of booleans. Then we derived new induction schemas based on these which we added to *Clam*'s collection, for example induction over the length of words (i.e., a special case of list induction), simultaneous induction over two words of the same length, and induction on words where step cases are generated by increment and addition of words, and the like.

Finally, we developed a number of new methods. For example, the method *bool_cases* was added to solve goals by exhaustive case-analysis over booleans. This suffices often to automate base cases of induction proofs about parameterised designs. The method *difference_match* was added to reason about sequential circuits modelled as finite-state machines; Also, the method *term_cancellation* was added to strengthen arithmetic reasoning in *Clam*.

3 Verification Methodology

We now describe how proof planning can be used for hardware verification and afterwards provide two examples.

The user begins by giving *Clam* definitions, in the form of equations, which define the implementation of the hardware and its behavioural specification. Then, the user provides *Clam* with the conjecture to be proven. For combinational hardware this is typically an equation stating that the specification is equal to an abstract form of the implementation:

$$\forall x_1, \ldots, x_n \; spec(x_1, \ldots, x_n) = abs(imp(x_1, \ldots, x_n))$$

For synchronous sequential hardware this is typically an implication stating that if the specification is equal to an abstract form of the implementation at some time t, then the specification must be equal to an abstract form of the implementation at all time t' greater that t. If the specification and the implementation involve different time scales, then we must provide a mapping f that converts times from one scale to the other:

$$\forall t, t' : time \; \forall x_1, \ldots, x_n$$

$$spec(t, x_1, \ldots, x_n) = abs(imp(f(t), x_1, \ldots, x_n)) \wedge t \leq t'$$

$$\longrightarrow$$

$$spec(t', x_1, \ldots, x_n) = abs(imp(f(t'), x_1, \ldots, x_n))$$

To create a plan, the user instructs *Clam* to find a proof plan for the conjecture by selecting one of the built-in planners, e.g., to find a plan using depth-first search. At this point *Clam* has been loaded not only with definitions and the conjecture but also the method database and wave-rules. If *Clam* finds a plan, the user may execute the tactic associated with it to construct an actual proof. Otherwise, the user can correct a bug in the theorem statement or in the definitions or suggest a lemma (which will provide new wave-rules) and try again. After *Clam* completes a proof plan, the tactic produced is executed to build an actual proof. Failure at this stage occurs rarely and happens only when a method improperly describes a tactic; in such cases, we improve the method or the tactic and plan again.

Example: A parallel array multiplier

We now describe the verification of an nm-bit parallel array multiplier. The external behaviour of the multiplier is expressed by the formula:

$$word2nat(x) \times word2nat(y) \,.$$

We represent words by lists of bits, and if x and y are words of length n and m respectively, then $word2nat$ returns the natural numbers which they represent and the above specifies their product. Multiplication of binary words can be implemented by a simple parallel array multiplier using binary additions. Consider, for example, multiplying a 3-bit word by a 2-bit word. This is represented by:

$$
\begin{array}{ccccc}
 & & x_2 & x_1 & x_0 \\
 & & \times & y_1 & y_0 \\
\hline
 & & x_2 y_0 & x_1 y_0 & x_0 y_0 \\
 & x_2 y_1 & x_1 y_1 & x_0 y_1 & \\
\hline
z_4 & z_3 & z_2 & z_1 & z_0 \\
\end{array}
$$

To multiply an n bit word by an m bit word the array multiplier uses $n \times m$ AND gates to compute each of these intermediate terms in parallel, and then, m binary additions are used to sum together the rows. This requires a total of $n \times m$ one-bit adders. The following equations formalise the above as a recursive description of a (parameterised) implementation:

$$mult(x, nil) = zeroes(length(x))$$

$$mult(x, h :: y) = cadd(mult\_one(x, h) <> zeroes(length(y)), mult(x, y), 0)$$

Here $cadd$ is the definition of an n-bit adder with arguments of the same length, $mult\_one$ multiplies a word times a bit, $<>$ appends two lists (built from nil and 'consed' together using $::$), $zeroes(n)$ yields n zeroes. To show the equivalence of the specification and the implementation, we give $Clam$ the conjecture

$$\forall x, y : word \; word2nat(x) \times word2nat(y) = word2nat(mult(x, y))$$

This theorem has been proof planned by $Clam$ using the depth-first planner in about 1 minute. Figure 2 displays the structure of the proof plan. The plan requires 15 levels of planning (number on the left) and 32 plan steps (numbers on the nodes of the tree). Each node in the tree indicates the application of a method. In particular, steps 1, 3, 12, 14, 20, 22 and 29 correspond to the application of the induction method. We briefly explain steps 1, 2 and 7 corresponding to the first induction. In step 1, the induction method analyses the conjecture and available definitions and uses heuristics (similar to those used in $Nqthm$) to suggest an induction on the word y. In step 2, the $base\ case$,

$$word2nat(x) \times word2nat(nil) = word2nat(mult(x, nil))$$

is simplified by symbolic evaluation using the base equation of $mult$, $word2nat$, and multiplication by zero. This yields

$$0 = word2nat(zeroes(len(x)))$$

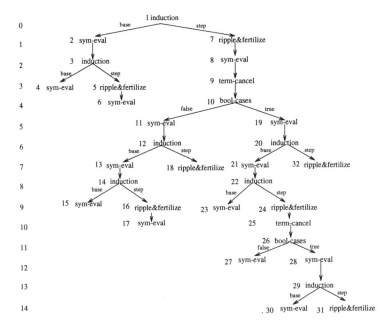

Fig. 2. Proof plan for verifying an nm-bit multiplier

which is solved by another induction, symbolic evaluation, rippling, fertilise and another application of symbolic evaluation. In step 7, for the *step case*, the induction conclusion:

$$word2nat(x) \times word2nat(v0 :: v1) = word2nat(mult(x, v0 :: v1))$$

is simplified by rippling with the recursive equations of *mult*, the wave-rule obtained from the verification of the n-bit adder, *word2nat*, distributivity of times over plus, and fertilisation, to yield:

$$word2nat(x) \times (bitval(v0) \times 2^{length(v1)}) + word2nat(x) \times word2nat(v1)$$

$$=$$

$$word2nat(mult\_one(x, v0) <> zeroes(length(v1))) +$$

$$(word2nat(x) \times word2nat(v1) + bitval(false))$$

This equation is solved by the methods indicated in steps 8-32 in figure 2. In order to generate the proof plan, we provided wave-rules which correspond to the verification of the n-bit adder, distributivity of times over plus, associativity of times, and a non-definitional wave-rule of plus on its second argument. These wave-rules come from lemmas which we previously verified using proof planning. Hence, we use proof planning to develop, hierarchically, theories about hardware.

Finally, we ask *Oyster* to execute our proof plan. This consists of executing a tactic for each of the methods indicated in the proof plan, following the structure displayed in figure 2 in a depth-first manner. The development of the proof plan took about 40 hours distributed over a two weeks period. The most time-consuming part was identifying the required lemmas.

Example: the Gordon computer

This is a 16-bit microprocessor, with 8 programming instructions, no interrupts, and a synchronous communication interface with memory, designed by Mike Gordon and his group at Cambridge University and verified interactively using the HOL system [12]. The specification is given in terms of the semantics of the 8 programming instructions. Each instruction consists of the set of operations that determines a new computer state, where a state is determined by the contents of the memory, the program counter, the accumulator and the idle/running status of the computer. The execution of an instruction defines a transition from a state into a new state and this transition determines the time-unit of an instruction-level time-scale. Thus, for each instruction we must specify the way in which each of the four components of a computer state are calculated. The implementation is at the register-transfer level. It consists of a data-path and a microprogrammed control unit. A computer state at the register-transfer level is determined by the contents of 11 components: the memory, the program counter, the accumulator, the idle/running flag, the memory address register, the instruction register, the argument register, the buffer register, the bus, the microcode program counter and the ready flag. The control unit generates the necessary control flags to update the computer registers. Communication between the bus and the registers is regulated by a set of gates. The implementation uses a microinstruction time-scale. The number of microinstructions required to compute a given instruction is calculated automatically by using the ready flag in the microcode, and the associated time at the microinstruction-level time-scale mapped onto the respective time at the instruction-level time-scale. A translation from the relational description used by Gordon into a functional description required by *Clam* was done by hand. This translation can be automated. For instance, *PVS* provides assistance in producing the functional representation from the relational one [15]. The correctness theorem asserts that the state of the computer at the specification level is equal to an abstract state of the implementation level each time an instruction is executed. After doing the extensions explained in the next section, the verification proceeded without user intervention. See [9] for more details.

4 Experiments

We have applied the methodology just described to a variety of combinational and sequential circuits: some circuits are from the *IFIP WG10.2 Hardware Verification Benchmark Circuit Set* (n-bit adder, parallel multiplier, Gordon computer) and from other sources [14]. Table 1 displays some statistics. A detailed analysis of these proofs can be found in [9].

Circuit	time			lemmas
	planning (min:sec)	plan-development (hours)	Clam-development (hours)	
COMBINATIONAL				
n-bit adder				
parameterised	4:50	16	100	0
word interpretation	15:50	8	16	0
little endian	5:30	8	16	2
big endian	0:57	4	0	1
look-ahead carry	2:40	8	0	1
n-bit alu				
parameterised	4:40	56	0	0
little endian	8:30	16	0	1
big endian	5:35	8	0	0
nat. number interpretation	4:50	8	0	0
n-bit shifter				
parameterised	4:30	32	30	0
big endian	3:20	16	0	0
n-bit processor unit				
big endian	5:00	8	2	0
parallel array				
nm-bit multiplier	1:03	40	0	6
n-bit incrementer				
little endian	0:58	4	0	1
big endian	0:06	2	0	1
word arithmetic				
addition	0:08	4	8	1
subtraction	0:10	4	0	2
multiplication	0:15	4	0	2
quotient	0:35	8	4	3
remainder	0:30	8	0	5
exponentiation	0:26	4	0	2
factorial	0:40	8	0	3
SEQUENTIAL				
n-bit counter	2:04	20	0	1
Gordon computer	45:00	200	360	0

Table 1. Some Circuits Verified using Proof Planning

We shall explain here what these numbers measure. The first column lists the circuits. A parameterised representation means that there is an explicit parameter n corresponding to the length of a word. *Big-endian* means that the most significant bit is at the end of the list; *little-endian* means that the least significant bit is at the end of the list. Reasoning about big-endian representations when using lists to represent words is easier than the little-endian counterparts because in the big-endian case there is no need to traverse the list to access the least significant bits. In a word interpretation, the specification has type word, as the implementation does, so that the verification theorem establishes the equality of the specification and the implementation on the type word. The word arithmetic operations establish a relationship between the set of words and the set of the natural numbers.

The next column lists timings, broken down three ways: first, planning time, is the time *Clam* took to generate a plan [4]. The time to execute the proof plan is not included here. This time is in general much higher, mainly because *Oyster* the object-level theorem prover uses a complex type theory formalism with type checking proof obligations which are time consuming. For instance, generating the proof plan for the incrementer takes just 6 seconds, but its execution in *Oyster* took 5:40 minutes. Second, plan-development time, which is the time spent developing the proof plan, from understanding the problem, finding the right representation for the specification and the implementation, debugging them, and finding and justifying missing lemmas, until a proof plan was obtained. This time doesn't include the time spent extending, tuning, and debugging *Clam*, which is the Clam-development time displayed in the third column. Some of these times may appear excessive, but, to give an accurate picture, we are accounting for everything, including many one-time costs.

The Clam-development time column includes time spent writing new methods, extending existing methods, experimenting with method orderings, writing new predicates and modifying existing ones for the meta-language, finding new induction schemes, and debugging *Clam* itself. The time to obtain a proof tended to be high for the first circuit of a certain kind, but dramatically decreased for subsequent circuits. For instance, the n-bit adder was the first circuit we verified and took about 116 man-hours, of which approximately 100 are of work extending previous methods such as *fertilise* and *generalise*, writing new methods such as *bool_cases* and *term_cancellation*, experimenting with method ordering, and writing new predicates such as *find_type* to provide the proof planner with boolean type information. The rest of the circuits in the table utilised these extensions and their proofs were obtained in shorter times because these extensions were already there. The word-interpretation version, which took 16 hours, required 16 hours of extensions to the methods *generalise* and *normalise*. The little-endian version, required 16 hours to derive in *Oyster* a new induction scheme (double induction on two words of the same length) and this is used in many of the other verification proofs. The big-endian representation, which took just 4 hours, used all the previous extensions. When we tried the parameterised version of the n-bit alu, it turned out that all the extensions required were already done for the parameterised version of the n-bit adder including method ordering; the 56 hours reported were mainly spent debugging the specification. Thus, the time for the little-endian and big-endian versions became shorter (16 and 8 hours, respectively). For the word arithmetic the extensions required included deriving new induction schemes such as induction with increment of a word and addition of two words, which made the proofs very easy to find. The

---

[4] Experiments were done in a Solbourne 6/702 dual processor with 128Mb of memory, which is equivalent to a two-processor SparcStation 20

multiplier didn't require any extensions; most of the time was spent in finding the lemmas required by the proof.

The Gordon computer required a huge effort to scale-up *Clam* capabilities. The very scale of the specification required that we make a number of extensions to *Clam*, such as memoisation, so that the system would more gracefully handle large terms. Also significant is that the theorem involves two different time-scales with automatic calculation of the number of cycles for each instruction; methods like *difference_match* were designed to handle this and to automate time abstraction. Again, all these extensions are required for the verification of similar circuits, so the 360 hours of development time (9 man-weeks) is a one-time effort and the verification effort should significantly decrease for new circuits of the same kind. The 200 hours of plan-development time (5 man-weeks) include learning about microprocessor architecture, translating the original relational specification of the computer into a functional one, finding appropriate abstraction mechanisms, formalising recursive definitions for the implementation devices (memory, program counter, accumulator, microcode program counter, etc), and debugging the specification and the implementation.

The final column in our table indicates the number of lemmas used in the proof planning. The Gordon computer didn't require any lemmas, the multiplier requires lemmas about the n-bit adder, distributivity of $+$ over $\times$, and associativity of times. In general, proof planning utilises very few lemmas compared with most other systems.

5 Comparisons

We compare *Clam/Oyster* with *Nqthm*, *PVS* and *HOL*. *Clam/Oyster* is a *fully-expansive* system, so it provides the user with the security characteristic of these systems, as opposed to systems like *Nqthm* and *PVS* which don't necessarily incorporate this feature [2]. In a tactic-based fully expansive system, the execution of a tactic results in the execution of the inference rules that support that tactic. Theorem provers like *HOL* and *Nuprl* are also fully-expansive systems.

5.1 Nqthm

Most of the circuits in the table have been verified elsewhere using *Nqthm*. For instance, the n-bit adder was verified (big-endian) in half a man-day, where the discovery of the required lemmas was the most difficult part[17]. A combinational processing unit (alu and shifter) was verified by Warren Hunt as part of the verification of the FM8501 microprocessor. This processing unit is verified in 3 theorems corresponding to the word, natural number, and two's complement interpretations. It took about 2 months effort, runs in a few seconds[5], and used about 53 lemmas [11]. Although the processor unit reported here is less complicated than FM8501's because we don't include the two's complement interpretation, we use just 2 lemmas in its proof planning.

---

[5] Personal communication

As an experiment, some of these circuits were re-implemented in *Nqthm* by a newcomer to formal verification [16]. The following table summarises the results:

Circuit	time		
	run (min:sec)	proof-development (hours)	lemmas
COMBINATIONAL			
n-bit adder (big endian)	3:40	32	6
n-bit adder look-ahead carry	3:20	12	6
n-bit alu (big endian)	13:30	40	11
n-bit shifter (big endian)	2:30	24	2
n-bit processor unit (big endian)	0:37	6	13
n-bit incrementer (big endian)	0:35	20	3

Most of the proof-development time was spent determining the lemmas required by the proof. For instance the proof of the *n*-bit adder uses the following lemmas: commutativity of plus, commutativity of times, and addition, and multiplication by recursion on the second argument. Of these, *Clam* requires only the lemma addition by recursion on the second argument. In general, *Clam* required fewer lemmas than *Nqthm* to verify these circuits. On the other hand, *Nqthm* provides a very stable implementation, and was much easier to use. There is of course a danger in such quantitative comparisons as metrics can oversimplify and even mislead. For example, the development times of novice users can be orders of magnitude higher than experts and moreover experts have better insights on how to structure problems to avoid lemmas. Still, we have tried to compare like with like: both the *Clam* and *Nqthm* proofs were carried out by relative beginners to both automated theorem proving and formal reasoning about hardware.

5.2 PVS

The *n*-bit adder, the *n*-bit alu, and the Tamarack microprocessor [6] have been implemented in *PVS* [10, 15]. The run time for verifying each of these circuits was 2:07, 1:27 and 9:05 minutes respectively in a Sun SparcStation 10. These low run-times are explained by the built-in decision procedures available to *PVS*. In these verifications the user must provide the induction parameters and use a predefined proof strategy. There are two ways in which *Clam* could enhance *PVS* automation facilities: (1) The decisions for selecting an induction scheme and induction variables are done by the user. These decisions could be automated by interfacing *Clam* and *PVS*. (2) There are proof strategies packaged as sets of *PVS* commands for a certain kind of circuits. The adder and the alu use the same proof strategy, the Tamarack and the pipelined Saxe microprocessor share another proof strategy. Our set of methods and the planning mechanism comprise these proof strategies, and can in principle create new ones by customising a composite tactic for a new conjecture using the same methods.

---

[6] A refined implementation of the Gordon computer. Its verification in HOL and PVS is also more abstract, as tri-state values and gates to access the bus, and the input of manual information through the switches and the knob, are not considered. However, Tamarack-3 includes an option for asynchronous communication with memory.

5.3 HOL

The *Gordon computer* was originally designed and verified using *HOL* by Mike Gordon and his group [12] and later implemented and verified as the Tamarack microprocessor by Jeffrey Joyce [13]. The verification took about 5 weeks of proof-development effort and required the derivation of at least 200 lemmas including general lemmas for arithmetic reasoning and temporal logic operators which are now built into *HOL*. It didn't require to tune *HOL* and runs in about 30 minutes in a modern machine[7]. Although the architecture of the *Gordon computer* verified here is less complicated than the architecture of Tamarack-3 verified by Joyce because we assume a synchronous communication with memory, we don't use lemmas in its proof planning. There are also two ways in which *Clam* could enhance *HOL* automation facilities: (1) the selection of induction parameters (scheme and variables), and (2) the generation of proof strategies for families of circuits. A project to interface *Clam* and *HOL* is about to start [5].

6 Conclusions

We have described how a hardware verification methodology based on proof planning can be applied to guide automatically a tactic-based theorem prover in verifying hardware designs. We have applied this technique to verify a variety of combinational and synchronous sequential circuits. Our experience shows that the *Clam* system and the proof planning idea carry over well to this new domain, although a number of extensions in the details (as opposed to the spirit) of *Clam* and the development of domain specific tactics and methods were required.

Overall, our experience was quite positive. We investigated several kinds of parameterised circuits and were able to develop methods which captured heuristics suitable for reasoning about families of such designs. We reported on our development time for both proving particular theorems and doing extensions to *Clam*. Although the times are sometimes high for initially verifying new kinds of circuits, subsequent development times were respectable and the majority of time was spent on simply entering and debugging the specification. This provides some support to our belief that a system like *Clam* might be usable by hardware engineers, provided that there is a 'proof engineer' in the background who has worked on the design of a set of methods appropriate for the domain.

There are several directions for further work. As noted, much of our time was spent entering and debugging specifications. Part of this is due to our use of a somewhat complicated type-theory as a specification language. However, our planning approach should work with any tactic based theorem prover; one need only port the tactics associated with the current methods from one system to another so that they produce the same effects. Interfacing *Clam* with a prover like HOL could significantly reduce specification and tuning times. It would also provide us immediate access to the large collection of tactics and theorems already available for this system. An interface to a standard hardware

---

[7] Personal communication

description language would also ease specification and make it possible to integrate better proof planning into the hardware design cycle. Another area for further work concerns improving the power or efficiency of some of our methods. For example, reasoning about boolean circuits by case evaluation, could be replaced by a more efficient routine based on BDDs. We also will investigate the development of a temporal reasoning methods for reasoning about circuits with asynchronous interfaces. Finally, we plan to apply our approach to the verification of commercial circuits including microprocessors systems.

References

1. David Basin and Toby Walsh. *Annotated Rewriting in Inductive Theorem Proving*. Journal of Automated Reasoning, 16:147–180, 1996.
2. Richard J. Boulton. *Efficiency in a fully-expansive theorem prover*. Technical Report 337, University of Cambridge Computer Laboratory, 1994.
3. R.S. Boyer and J.S. Moore. *A Computational Logic*. Academic Press, 1979.
4. A. Bundy. *The use of explicit plans to guide inductive proofs*. In proc. of the 9th Conference on Automated Deduction, pp 111–120. Springer-Verlag, 1988.
5. A. Bundy and M. Gordon. Automatic Guidance of Mechanically Generated Proofs. Research proposal, Edinburgh-Cambridge, 1995.
6. A. Bundy, A. Stevens, F. van Harmelen, A. Ireland, and A. Smaill. *Rippling: A Heuristic for Guiding Inductive Proofs. Artificial Intelligence*, 62:185–253, 1993.
7. A. Bundy, F. van Harmelen, J. Hesketh, and A. Smaill. *Experiments with Proof Plans for Induction*. Journal of Automated Reasoning, 7:303–324, 1991.
8. A. Bundy, F. van Harmelen, C. Horn, and A. Smaill. *The Oyster-Clam system*. In M.E. Stickel, editor, *10th International Conference on Automated Deduction*, pages 647–648. Springer-Verlag, 1990. Lecture Notes in Artificial Intelligence 449.
9. Francisco J. Cantu. *Inductive Proof Planning for Automating Hardware Verification*. PhD thesis, University of Edinburgh, 1996. Forthcoming.
10. D. Cyrluk, N. Rajan, N. Shankar, and M.K. Srivas. *Effective Theorem Proving for Hardware Verification*. In 2nd TPCD Conference, Springer-Verlag, 1994.
11. Warren Hunt. *FM8501: A Verified Microprocessor*. Technical report 47, Institute for Computing Science, University of Texas at Austin, 1986.
12. Jeff Joyce, G. Graham Birtwistle, and M. Gordon. *Proving a Computer Correct in Higher-order Logic*. Tech. Report 100, U. of Cambridge Computer Lab., 1986.
13. Jeffrey J. Joyce. *Multi-level Verification of Microprocessor-based Systems*. Technical Report 195, University of Cambridge Computer Laboratory, 1990.
14. M. Morris Mano. *Digital Logic and Computer Design*. Prentice Hall, Inc, 1979.
15. S. Owre, J.M. Rushby, N. Shankar, and M.K. Srivas. *A Tutorial on Using pvs for Hardware Verification*. In 2nd TPCD Conference, Springer-Verlag, 1994.
16. Victor Rangel. *Metodos Formales para Verificacion de Hardware: Un Estudio Comparativo*. Master's thesis, Instituto Tecnologico de Monterrey, Mexico, 1996.
17. V. Stavridou, H. Barringer, and D.A. Edwards. *Formal specification and verification of hardware: A comparative case study*. In Proceedings of the 25th ACM/IEEE Design Automation Conference, pages 89-96. IEEE, 1988.

Verifying Nondeterministic Implementations of Deterministic Systems[1]

Alok Jain
Department of ECE
Carnegie Mellon University
Pittsburgh, PA 15213
email: alok.jain@ece.cmu.edu

Kyle Nelson
IBM Corporation
AS/400 Division
Rochester, MN 55901
email:kln@vnet.ibm.com

Randal E. Bryant
School of Computer Science
Carnegie Mellon University
Pittsburgh, PA 15213
email:randy.brant@cs.cmu.edu

Abstract. Some modern systems with a simple deterministic high-level specification have implementations that exhibit highly nondeterministic behavior. Such systems maintain a simple operation semantics at the high-level. However their underlying implementations exploit parallelism to enhance performance leading to interaction among operations and contention for resources. The deviation from the sequential execution model not only leads to nondeterminism in the implementation but creates the potential for serious design errors. This paper presents a methodology for formal verification of such systems. An abstract specification describes the high-level behavior as a set of operations. A mapping relates the sequential semantics of these operations to the underlying nondeterminism in the implementation. Symbolic Trajectory Evaluation, a modified form of symbolic simulation, is used to perform the actual verification. The methodology is currently being used to verify portions of a superscalar processor which implements the PowerPC architecture. Our initial work on the fixed point unit indicates that this is a promising approach for verification of processors.

1. Introduction

Some modern circuit designs with a simple deterministic high-level specification have implementations that exhibit highly nondeterministic behaviors. Systems that operate on an externally visible stored state such as memories, data paths and processors often exhibit these behaviors. The implementation of these systems frequently overlap the execution of tasks in an effort to enhance performance while maintaining the appearance of sequential execution.

A large class of systems that exhibit such behavior are processors. At the high-level, the sequencing model inherent in processors is the sequential execution model. However, the underlying implementation of processors use pipelines, multiple instruction issue, and nondeterministic protocols to interact with other subsystems. The resulting interaction among instructions and contention for resources leads to nondeterminism in the implementation. Such implementations contain many subtle features with the potential for serious design errors. The methodology outlined in this paper is able to bridge the wide gap between the abstract specification and the implementation's often radical deviation from the sequential execution model.

A methodology for formal verification must ensure that such a system functions correctly under all possible execution sequences. Since there is an infinite number of execution sequences, we verify each operation individually and then reason about stitching arbitrary operations together to form execution sequences.

The goal is to develop a methodology with which a designer can show that an imple-

---

1. This work partially funded by Semiconductor Research Corporation # 95-DC-068.

mentation correctly fulfills an abstract specification of the desired system behavior. The abstract specification describes the high-level behavior of the system independent of any timing or implementation details. As an example, the natural specification of a processor is the instruction set architecture. The specification is a set of *abstract assertions* defining the effect of each operation on the user-visible state. The verification process must bridge a wide gap between the detailed circuit and the abstract specification. In spanning this gap, the verifier must account for issues such as system clocking, pipelines and interfaces with other subsystems. To bridge this gap between the abstract behavior and the circuit, the verification process requires some additional mapping information. The mapping defines how state visible at the abstract level is realized in the detailed circuit. The mapping has both spatial and temporal information. Designers are typically aware of the mapping but do not have a rigorous way of recording or documenting the information.

Our specification is thus divided into two components: the *abstract specification* and the *mapping information*. The distinction serves several purposes. Several feasible circuit designs can be verified against a single abstract specification. An abstract specification can be used to verify both an unpipelined and a pipelined version of a processor. The abstract specification describes the instruction set of the processor independent of any pipeline details. The task of the mapping is to relate the abstract specification to the complex temporal behavior and nondeterministic interactions of the pipelined processor. As an example, an instruction might stall in a pipeline stage waiting to obtain the necessary resources. The order and timing in which these resources are obtained often varies leading to nondeterministic behavior. Our verification methodology will verify the circuit under all possible orders and timing.

The distinction between the abstract specification and mapping enables hierarchical verification. This paper concentrates on a single level of mapping that maps the abstract specification into a specific implementation. In the future, one can envision an entire series of implementation mappings. Each level in the mapping serves to make the assertion more concrete. A verification task could be performed at each level. A series of mappings could also be used to perform modular simulation. Simulation models could be developed at each abstraction level and models at different levels of abstractions could be intermixed using the mapping information.

Once the abstract assertions have been individually verified, the methodology must be able to stitch operations together in order to reason about execution sequences. The mapping has information about how to stitch instructions together. Since the abstract assertions use a sequential execution model, stitching operations at the abstract specification level requires a simple sequencing of the instructions. However, the underlying nondeterministic implementation requires operations to interact and overlap in time. The mapping ensures that the operations can interact and overlap to create arbitrary execution sequences.

The abstract specification and the mapping are used to generate the *trajectory specification*. The trajectory specification consists of a set of *trajectory assertions*. Each abstract assertion gets mapped into a trajectory assertion. The verification task is to verify the set of trajectory assertions on the circuit. A modified form of symbolic simulation called *Symbolic Trajectory Evaluation*[1] is used to perform the verification task. We use the term trajectory specification and trajectory assertions partly for historical reasons. Our trajectory assertions are a generalization of the trajectory assertions introduced by Seger[4]. The justification is that the assertions define a set of trajectories in the simulator.

The formal verification methodology presented in this paper is currently being used to verify a superscalar processor which implements the PowerPC architecture[12]. The processor has several complex features such as pipeline interlocks, multiple instruction issue, and branch prediction to advance the state of the art in verification.

1.1. Related Work

Beatty[2][3] laid down the foundation for our methodology for formal verification of processors. The instruction set was specified as a set of declarative abstract assertions. A downward implementation mapping was used to map discrete transitions in the abstract specification into overlapping intervals for the circuit. The overlapping was specified by a *nextmarker* which defined the nominal end of the current instruction and the start of the next instruction. However this work had one basic limitation. The verification methodology could handle only bounded single behavior sequences. The mapping language was formulated in terms of a formalism called *marked strings*[2]. Marked strings could not express divergent or unbounded behavior.

We have extended the verification methodology so as to handle a greater level of non-determinism behavior. As a motivating example, consider a fixed point unit of a processor performing a bitwise-or operation that fetches two source operands (A and B) from a dual-ported memory subsystem. Assume that the operands might not be immediately available. The fixed point unit might have to wait for an arbitrary number of cycles before either of the operands are available. Furthermore the operands might be received in different orders: Operand A might arrive before B, operand B might arrive before A, or both might arrive simultaneously. The verification methodology should be able to verify the correctness of the circuit under any number of wait cycles and all possible arrival orders. Marked strings cannot express such an operation. Our formulation is in terms of state diagrams. State diagrams allow users to define unbounded and divergent behavior.

Symbolic Trajectory Evaluation (STE) has been used earlier to verify trajectory assertions. Beatty[3] mapped each abstract assertion into a set of symbolic patterns. STE was used to verify the set of symbolic patterns on the circuit. The set of symbolic patterns corresponded to a single sequence of states in a state diagram. Seger[4] extended STE to perform fixed point computations to verify a single sequence of states augmented with a limited set of loops. In our work, trajectory assertions are general state diagrams. We have extended STE to deal with generalized trajectory assertions.

Our work has some resemblance to the capabilities provided by the Symbolic Model Verifier (SMV)[5][6]. SMV requires a closed system. The environment is modeled as a set of machines. In some sense the state diagrams in our mapping correspond to creating an environment around the system. The state diagrams corresponding to the inputs can be viewed as generators that generate low-level signals required for the operation of the processor. State diagrams corresponding to outputs can be viewed as acceptors that recognize low-level signals on the outputs of the processor. However, there is one essential difference. Though SMV does provide the capability of describing the environment, it does not provide a methodology for rigorously defining these machines and stitching them together to reason about infinite execution sequences. The other difference is that the model-checking algorithm in SMV requires the complete next-state relation. It would be impossible to obtain the entire next-state relation for a complex processor. On the other hand, we use STE to evaluate the next-state function on-the-fly and only for that part of the processor that is required by the specification.

Kurshan[7][8] has used the concept of mappings to perform hierarchical verification. The specification (called a *task*) is represented by a deterministic automaton. The implementation is modeled as a state machine. Verification is cast in terms of testing whether the formal language of the implementation is contained in the formal language of the task. The user can provide reduction transformations as a basis of complexity management and hierarchical verification. Reductions are homomorphic transformations which correspond to abstraction mappings. Inverse reductions (called *refinements*) correspond to our implementation mappings. However, Kurshan does not have the concept of stitching tasks together to verify an infinite sequence of tasks. Also, Kurshan uses language containment as opposed to Symbolic Trajectory Evaluation as the verification task.

Burch[13] has proposed a verification methodology that avoids the need to explicitly define the mapping. Instead, an abstraction mapping is implicitly defined by flushing the pipeline and then using a projection function to hide some of the internal state. The methodology has been used to verify a pipelined version of the DLX architecture. However, the abstraction mappings are limited to internal states in the processor. In addition to mapping internal states, our implementation mappings allow users to define nondeterministic input/output protocols.

In the area of processor verification theorem proving has been used in the past to verify processors[9][11]. However, in these cases either the processor was very simple or a large amount of manual intervention was required. STE was used to verify the Hector processor[3]. However, Hector is a very simple processor similar to the PDP-11[10]. We are using the verification methodology to verify the PowerPC architecture. Recently, we verified the pipeline interlocks in the decode stage of the fixed point unit.

Details of how to specify the system behavior in our methodology are discussed in Section 2. The system behavior is specified in terms of a set of abstract assertions and the implementation mapping. An abstract assertion and the corresponding mapping for a small example are shown in Section 3. The abstract assertions are mapped into a set of trajectory assertions. Details of how to verify the set of trajectory assertions on a circuit are discussed in Section 4. Some of the ongoing work on using the formal verification methodology to verify a superscalar processor is presented in Section 5.

2. Specification

The specification consists of the abstract specification and the mapping. The mapping is a nondeterministic mapping formulated in terms of a variation of state diagrams called *control graphs*.

2.1. Abstract Specification

Hardware Description Languages seem to be the obvious choice for expressing abstract specifications. However, Hardware Description Languages tend to overspecify systems since they tend to describe a particular implementation rather than an abstract specification. It seems that these languages are appropriately named. They "describe" rather than "specify" hardware. We have defined a *Hardware Specification Language* to specify hardware. Here we are presenting the basic form of the Hardware Specification Language. We will later in this section extend the language to the symbolic and vector domains. The specification is an abstract machine. The abstract machine is associated with a set of single bit *abstract state elements* (S_v). Transitions in the machine are described as a set of *abstract assertions*. Each abstract assertion is an implication of the form: $P \Rightarrow Q$, where P is the precondition and Q is the postcondition. P and Q are abstract formulas of the form:

- **Simple abstract formula**: (s_i **is** 0) and (s_i **is** 1) are abstract formulas if $s_i \in S_v$.
- **Conjunction**: (F_1 **and** F_2) is an abstract formula if F_1 and F_2 are abstract formulas.
- **Domain restriction**: ($0 \rightarrow F$) and ($1 \rightarrow F$) are abstract formulas if F is an abstract formula.

Mathematically an abstract assertion can be defined as follows: Assume S is the set of assignments to the abstract state elements. Therefore $S = \{0, 1\}^n$, where $n = |S_v|$. Define a satisfying set for an abstract formula as the set of abstract state assignments that satisfy the abstract formula. The satisfying set of an abstract formula F, written $Sat(F)$, is defined recursively:

- $Sat(s_i \text{ is } 0)$ is a subset of S with 2^{n-1} abstract state assignments with the ith position in each assignment being 0. Similarly $Sat(s_i \text{ is } 1)$ is a subset of S with 2^{n-1} abstract state assignments with the ith position in each assignment being 1.
- $Sat(F_1 \text{ and } F_2) = Sat(F_1) \cap Sat(F_2)$.
- $Sat(0 \rightarrow F) = S$, and $Sat(1 \rightarrow F) = Sat(F)$.

Each assertion defines a set of transitions in an abstract Moore machine. For an abstract assertion $A = P \Rightarrow Q$, the sets $Sat(P)$ and $Sat(Q)$ are the sets of abstract state assignments that satisfy the precondition and postcondition respectively. The transitions associated with the assertion can now be defined as: $Trans(A) = (Sat(P) \times Sat(Q)) \cup ((S - Sat(P)) \times S)$. If the abstract machine starts in a set of states that satisfy the precondition, then the machine should transition into a set of states that satisfy the postcondition. The set $Sat(P) \times Sat(Q)$ represents the set of transitions when the precondition is satisfied. If the machine starts in a set of states that do not satisfy the precondition, then the assertion does not place any restrictions on the set of transitions. The set $(S - Sat(P)) \times S$ represents the set of transitions when the precondition is not satisfied.

Let i index the set of abstract assertions. The intersection $\bigcap_i Trans(A_i)$ defines a non-deterministic Moore machine corresponding to the abstract specification.

Consider the example of a bitwise-complement operation. A bitwise-complement operation requires two abstract state elements, SA and ST, where SA is the source operand and ST is the target operand. Assuming SA and ST are single bit operands, the bitwise-complement operation can be specified by using 2 abstract assertions.

(SA **is** 0) ➡ (ST **is** 1)   // Assertion A_1

(SA **is** 1) ➡ (ST **is** 0)   // Assertion A_2

For this example, $n = 2$ and the set S has 4 state assignments, $S = \{00, 01, 10, 11\}$, where the left bit is the logic value associated with SA and the right bit is the logic value associated with ST. The transitions corresponding to these assertions are shown in Figure 1. Let us consider assertion A_1. Now $Sat(\text{SA is } 0) = \{00, 01\}$ and $Sat(\text{ST is } 1) = \{01, 11\}$. The solid edges represent transitions when the precondition is satisfied, i.e. the set $Sat(P) \times Sat(Q)$. The shaded edges represent the transitions when the precondition is not satisfied i.e. the set $(S - Sat(P)) \times S$. A similar case can be made for assertion A_2. The intersection of the transitions associated with both assertions gives the Moore model for the specification. In this particular model, nondeterminism is used only to represent the choice of the next input. In other cases, nondeterminism can be used to represent unspecified or don't care behavior such as the output of the multiplier while performing an add operation in a processor.

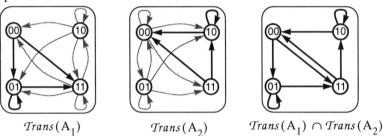

$Trans(A_1)$ $\qquad$ $Trans(A_2)$ $\qquad$ $Trans(A_1) \cap Trans(A_2)$

Figure 1. Nondeterministic Moore machine model for assertions.

We have presented the basic form of the abstract assertions. They can be extended to the symbolic and vector domain. In the symbolic domain, the domain restriction in abstract formulas is extended to be of the form: $(e \rightarrow F)$ where e is a Boolean expression over a set of symbolic variables. We introduce the notation $(s_i \text{ is } e)$ as a shorthand for $(\bar{e} \rightarrow (s_i \text{ is } 0))$ **and** $(e \rightarrow (s_i \text{ is } 1))$. In the vector domain, the abstract state elements are extended to represent different word sizes. A symbolic abstract assertion defines a set of scalar abstract assertions. A bitwise-complement operation for k-bit word size can be specified with a single symbolic abstract assertion. Assume that a is a symbolic k-bit variable that denotes the current value of the source operand. The symbolic abstract assertion for the bitwise-complement operation is:

(SA **is** a) ➡ (ST **is** ã)

The ~ operator is the bitwise-complement operator. The symbolic abstract assertion represents 2^k scalar assertions corresponding to all possible assignments to the variable a.

2.2. Control Graphs

Control graphs are state diagrams with the capability of synchronization at specific time points. Mathematically a control graph can be represented as a tuple: $G = \langle V, U, E, s, t \rangle$, where

- V is the set of *state vertices*.
- U is the set of *event vertices*.
- E is the set of directed edges.
- s is the source, $s \in U$.
- t is the sink, $t \in U$.

There are two types of vertices in a control graph: 1) Event vertices representing instantaneous time points and 2) State vertices representing some non-zero duration of time. Event vertices are used for synchronization. A control graph has a unique event vertex with no incoming edges called the source vertex. Also the graph has a unique event vertex with no outgoing edges called the sink vertex. Nondeterminism is modelled as multiple outgoing edges from a vertex.

2.3. Implementation Mapping

The implementation mapping provides a spatial and temporal mapping for the abstract machine. The mapping has three main components:

- Set of *node elements*.
- Single *main machine*.
- Set of *map machines*.

The mapping defines a set of node elements (N_v) that represent inputs, outputs and internal states in the circuit.

The main machine defines the flow of control for individual system operation. However we need to reason about sequence of operations. Therefore, the main machine also defines how individual operations can be stitched together to create valid execution sequences. The main machine is a tuple of the form: $M = \langle V_m, U_m, E_m, s_m, t_m, Y_m \rangle$, where

- $\langle V_m, U_m, E_m, s_m, t_m \rangle$ is a control graph.
- Y_m is a vertex cut of the control graph, $Y_m \subseteq U_m$.

The source s_m denotes the start of an operation and the sink t_m denotes the end of the operation. The vertex cut Y_m denotes the *nominal end* of the operation. The nominal end of an operation is defined to be the timepoint where the next operation can be started. The vertex cut divides the set of vertices into 3 sets, X_m, Y_m and Z_m where $X_m \cup Y_m \cup Z_m = V_m \cup U_m$. The set X_m represents the set of vertices to the left of the vertex cut and the set Z_m represents the set of vertices to the right of the vertex cut. The vertex cut places certain restrictions on the set of edges in the graph. All paths from vertices in X_m to vertices in Z_m have to pass through an event vertex in Y_m. In addition there cannot be any paths from vertices in Z_m to vertices in X_m. The restriction ensures that the pipeline is committed to an instruction once it has started.

For an unpipelined circuit, $Y_m = \{t_m\}$, since the current operation must end before the next operation can start. However, for pipelined circuits, Y_m is a cutset of event vertices between the source and sink. The cutset defines the time point when the next operation can be started, thus overlapping the current and next operations. As an example, consider the main machine for a simple pipelined processor shown in Figure 2. The vertices F, D, and E are state vertices representing the fetch, decode and execute stages respectively. The vertices s, p, q, and t are event vertices. Note that the nominal end cutset indicates that the next operation can be fetched when the current operation has entered the decode stage.

Nominal End Cutset

Figure 2. Main machine for a simple pipelined processor.

Individual instructions can be stitched together to create execution sequences as shown in Figure 3. The machines M^1, M^2, M^3 are multiple copies of the main machine with the superscripts indexing successive instructions. The machine M^* corresponds to the infinite execution sequence obtained from stitching all instructions. Stitching is performed by aligning the source vertex of an instance of the main machine with the nominal end cutset of the previous main machine and taking the cross-product of all the aligned machines.

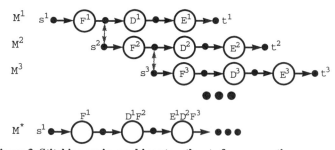

Figure 3. Stitching main machines together to form execution sequences.

In addition to the main machine, each simple abstract formula in the abstract specification is mapped into a map machine. The map machine provides a spatial and temporal mapping for the simple abstract formula. A control graph defines the temporal component of the mapping. Each state vertex in the control graph is associated with a node formula. The node formulas define the spatial component of the mapping. Here we will present the basic form of the node formulas. Later in the section we will extend them to the symbolic and vector domain. Node formulas are of the form:

- **Simple node formula**: $(n_i \text{ is } 0)$ and $(n_i \text{ is } 1)$ are node formulas if $n_i \in N_v$.
- **Conjunction**: $(F_1 \text{ and } F_2)$ is a node formula if F_1 and F_2 are node formulas.
- **Domain restriction**: $(0 \rightarrow F)$ and $(1 \rightarrow F)$ are node formulas if F is a node formula.

The map machines are not completely independent. A synchronization function maps event vertices in the map control graph to event vertices in the main control graph. The synchronization function relates a simple abstract formula with the flow of control of the entire operation. Mathematically a map machine for a simple abstract formula s can be represented as a tuple of the form: $Map(s) = \langle V_s, U_s, E_s, s_s, t_s, \sigma_s, \Upsilon_s \rangle$, where

- $\langle V_s, U_s, E_s, s_s, t_s \rangle$ is a control graph.
- σ_s is a labelling function that labels state vertices with node formulas.
- Υ_s is the synchronization function, $\Upsilon_s : U_s \rightarrow U_m$.

We have presented the basic form of the node formulas and map machines. The node formulas can be extended to the symbolic and vector domain. In the symbolic domain, the domain restriction is extended to be of the form: $(e \rightarrow F)$ where e is a Boolean expression over a set of symbolic variables. We introduce the notation $(n_i \text{ is } e)$ as a shorthand for $(\bar{e} \rightarrow (n_i \text{ is } 0))$ **and** $(e \rightarrow (n_i \text{ is } 1))$. In the vector domain, node elements are extended to represent a vector of nets in the circuit. Similarly, map machines can be extended to the symbolic and vector domains. A symbolic map machine defines a set of scalar map machines. The mapping for a k-bit operand (ST) can be specified by a single symbolic machine, $Map(\text{ST is } v)$, where v is a k-bit symbolic variable. The symbolic machine captures 2^k scalar map machines corresponding to all possible assignments to the variable v.

The trajectory formula corresponding to an abstract formula is the set of trajectories defined by the abstract formula for a particular implementation. The user needs to define the map machine for simple abstract formulas. Given the map machines, the trajectory formula for an abstract formula AF, written $Traj(AF)$, can be automatically computed as follows:

- $Traj(s_i \text{ is } 0) = Map(s_i \text{ is } 0)$ and $Traj(s_i \text{ is } 1) = Map(s_i \text{ is } 1)$.
- $Traj(AF_1 \text{ and } AF_2) = Traj(AF_1) \parallel Traj(AF_2)$, where $\parallel$ is the parallel composition operator. Parallel composition amounts to taking the cross product of the two control graphs under restrictions specified by the synchronization function. Assume that v_1 and v_2 are state vertices in the trajectory formulas for AF_1 and AF_2 respectively. Further assume that NF_1 and NF_2 are the node formulas associated with vertices v_1 and v_2 respectively. Then the node formula associated with the cross-produce vertex $(v_1 \times v_2)$ is $(NF_1 \text{ and } NF_2)$.
- $Traj(e \rightarrow AF)$ is essentially the same as $Traj(AF)$, except that the node formulas are modified. Assume that v is a vertex in the trajectory formula for AF with node formula NF. The node formula associated with the vertex v for the trajectory formula $(e \rightarrow AF)$ is $(e \rightarrow NF)$.

Note that we can verify each operation individual and then reason about stitching operations together since $Traj(AF)$ captures the effect of any preceding or following operations.

For an abstract assertion $A = P \Rightarrow Q$, $Traj(P)$ and $Traj(Q)$ are the trajectory formulas associated with the precondition and postcondition respectively. The trajectory assertion corresponding to A can be automatically computed as $Traj(P) \parallel Traj(Q)$, where $\parallel$ is the shift-and-compose operator. The shift-and-compose operator shifts the start of the machine $Traj(Q)$ to the nominal end cutset in the main machine and then performs the composition. The node formulas in $Traj(P)$ are treated as *antecedent node formulas.* The antecedent node formulas define the stimuli and current state for the circuit. The trajectory formula corresponding to the precondition is being used as a generator since it generates low-level signals for the circuit. On the other hand, node formulas in $Traj(Q)$ are treated as *consequent node formulas.* The consequent node formulas define the desired response and state transitions. The trajectory formula corresponding to the postcondition is being used as an acceptor since it defines the set of acceptable responses for the circuit.

Incidentally the shift and compose operator was also used to stitch operations together to form execution sequences in Figure 3, with the slight variation that the main machines were not associated with any node formulas.

3. Example

Assume that we want to verify the bitwise-or operation in an ALU. The abstract specification would define abstract state elements, SA, SB and ST. SA and SB serve as the source operands and ST serves as the target operand for the bitwise-or operation. Assume that a and b are symbolic variables that denote the current values of SA and SB. The abstract assertion for the bitwise-or operation is:

(SA **is** a) **and** (SB **is** b) ➡ (ST **is** a|b)

The | operator is the bitwise-or operator. Note that for a k-bit word size, this symbolic abstract assertion represents 4^k scalar assertions corresponding to all possible combinations of assignments to variables a and b.

The specification is kept abstract. It does not specify any timing or implementation specific details. Now let's assume a specific implementation of the ALU where the source operands have to be fetched from a register file and may not be immediately available. Assume that both source operands are associated with a valid signal which specifies when the operand is available. The valid signals have a default value of logic 0 and are asserted by the register file subsystem to a logic 1 when the operand is available. The ALU computes the bitwise-or and makes it available for storage back into the register file one cycle after both operands have been received. The block diagram for such an ALU is shown in Figure 4. This is a simple example that serves to illustrate some of the issues that we have encountered in our effort to verify the PowerPC architecture.

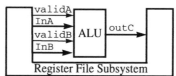

Figure 4. An implementation with a valid signal for each operand.

There are a few interesting points to note about this implementation. First, the ALU might have to wait for an arbitrary number of cycles before either of the source operands is available. Secondly, the source operands might arrive in different orders. Figure 5 shows part of an execution sequence for the implementation. The sequence is divided into various segments each of which represents a bitwise-or operation. A segment is associated with a nominal end marker. The execution sequence can be obtained by overlapping the start of the successive segment with the nominal end marker of the current segment. Note that segments can be of different widths. Segment 1 exhibits a case where the A operand arrived before the B operand. Segments 2 and 3 exhibit cases where both operands arrived simultaneously. Segment 3 represents the maximally pipelined case where the operands were immediately available. And segment 4 exhibits a case where the B operand arrived before the A operand. The goal of verification is to ensure that the circuit correctly performs a bitwise-or operation under any number of wait cycles and under all possible arrival orders.

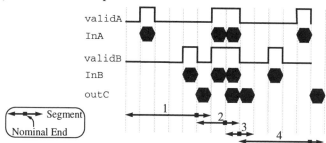

Figure 5. Part of an execution sequence for the bitwise-or operation.

The main machine for such an implementation is shown in Figure 6. Each state vertex represents a clock cycle. One or more cycles might be required to fetch the operands. This is represented by the state vertex `fetch`. Multiple cycles are required when the operands are not immediately available. After obtaining the operands, the result of the bitwise-or operation is available in the next cycle represented by state vertex `execute`. The control graph is modeling nondeterministic behavior. The machine can remain in the `fetch` state for an arbitrary number of cycles and then transition into the `execute` state. The main machine captures all possible segments in the execution sequence. The event vertex `fetched` represents the nominal end of the segment.

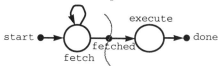

Figure 6. Main machine for the bitwise-or operation.

The map machine for the abstract formula (`SA` **is** `v`) is a nondeterministic machine shown in Figure 7. The symbolic variable `v` is serving as a formal argument. Later on the formal arguments will get replaced by the actual arguments. The labelling inside the state vertices represents the value of the valid signal for the A operand. The actual node formulas associated with the state vertices are shown in shadowed boxes in the figure. Since the operand might not be immediately available, the implementation might have to wait for an arbitrary number of cycles. This is represented by state vertex `wait`. The operand is received in state vertex `fetch`. After obtaining the A operand, the implementation might have to wait again for an arbitrary number of cycles for the B operand. This is represented by state vertex `fill`. The vertices `M.start` and `M.fetched` represent the event vertices `start` and `fetched` in the main machine. So the start of the map machine is synchronized to the start of the main machine. The end of this machine is synchronized to the event vertex fetched in the main machine where M.fetched represents the time point where both operands have been fetched.

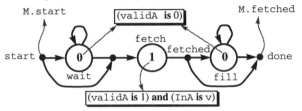

Figure 7. Map Machine for (SA is v).

The map machine for the abstract formula (SB is v) is the same as for state element SA except that the node formulas refer to the B operand.

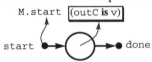

Figure 8. Map machine for (ST is v).

Finally the map machine for (ST is v) is shown in Figure 8. The start of this machine is synchronized with the start of the main machine. This indicates that the result of the computation for the previous operation should be available at the start of the current operation.

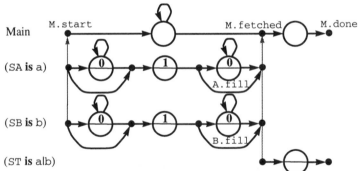

Figure 9. Alignment of machines for the bitwise-or operation.

The map machines get aligned with the main machine as shown in Figure 9. The abstract formulas (SA is a) and (SB is b) appear in the precondition of the abstract assertion. The formal argument in the corresponding map machine gets replaced by the actual arguments a and b. The node formulas in the map machines are treated as antecedent node formulas. Antecedent node formulas are shown in the upper half of the state vertex. The event vertices are synchronized as specified by the synchronization function. The abstract formula (ST is a|b) appears in the postcondition. The formal argument v is replaced by the expression a|b. The node formulas are treated as consequent node formulas. Consequent node formulas are shown in the lower half of the state vertex. The synchronization point is shifted to the nominal end cutset in the main machine.

The trajectory assertion for the bitwise-or operation corresponds to the composition of these machines. For this example, composition amounts to performing the cross product construction of the main, SA and SB machines between event vertices M.start and M.fetched, followed by the cross-product construction of the main and ST machines between event vertices M.fetched and M.done. However, the implementation mapping requires one additional piece of information. Notice that the state vertices A.fill and B.fill represent the fact that one of the operands has been received and the implementation is waiting for the other. When both operands have been received, the implementation will go ahead and compute the bitwise-or. So in the cross-product construction, we want to invalidate the cross product of vertices A.fill and B.fill. With this additional information, the composition results in the control graph shown in Figure 10. The resultant control graph captures all possible orderings and arbitrary number of wait cycles. In the top path, the A operand was received before the B operand. In the bottom path, the B operand was received before the A operand. In the middle path, both source operands were received simultaneously. The result of the bitwise-or operation is available in the final state vertex.

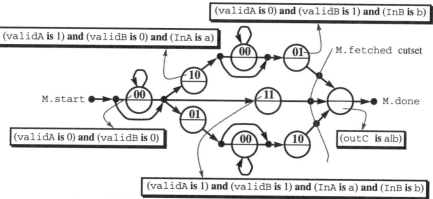

Figure 10. Trajectory Assertion for bitwise-or operation.

The resultant control graph serves as the trajectory assertion. In general the trajectory specification consists of a set of trajectory assertions. Each trajectory assertion can be represented as a tuple: $T = \langle V, U, E, s, t, \sigma_a, \sigma_c \rangle$, where

- $\langle V, U, E, s, t \rangle$ is a control graph.
- σ_a labels state vertices with antecedent node formulas.
- σ_c labels state vertices with consequent node formulas.

The trajectory assertions express the system behavior relative to a path in a control graph. The antecedent labelling along a path places restrictions on the inputs and internal states of the circuit. The consequent labelling specifies the set of allowed circuit responses along the path.

Trajectory assertions can be classified into three categories in increasing order of generalization: 1) Single Sequence 2) Acyclic 3) Generalized. The control graph associated with a single sequence trajectory assertion has a linear sequence of vertices. An acyclic trajectory assertion has a directed acyclic control graph. And the generalized trajectory assertion has any arbitrary control graph.

4. Verification

The main concern in the verification task is that representing the next-state function or relation for large systems is not computationally feasible. Therefore, we use Symbolic Trajectory Evaluation (STE) which evaluates the next-state function on-the-fly and only for that part of the circuit required by the trajectory assertion. STE uses a partially ordered system model. In the past, STE has been used to verify only single sequence trajectory assertions or single sequences augmented with simple loops. We extended STE to deal with acyclic and generalized trajectory assertions.

4.1. Partially Ordered System Model

The logic set is extended to $\{0, 1, X\}$. The logic value X denotes an unknown and possibly indeterminate logic value. The logic values are assumed to have a partial order with logic X lower in the partial order than both logic 0 and 1 as shown in Figure 11.

Figure 11. Partial Order for logic values 0,1,X.

Assume that $\tilde{N}$ is the set of node assignments in the partially ordered system. Therefore $\tilde{N} = \{0, 1, X\}^m \cup \top$ where $m = |N_v|$. The elements of the set $\tilde{N}$ define a complete lattice $[\tilde{N}, \sqsubseteq]$ with X^m as the bottom element and $\top$ as an artificial top element. The partial order in the lattice is a pointwise extension of the partial order shown in Figure 11. The symbols $\sqcup$ and $\sqcap$ denote the least upper bound and greatest lower bound operations respectively.

In the partial order system, node formulas are represented by an element of the lattice. The restricted form of the node formula ensures that it can be represented by a lattice element. The lattice element corresponding to a node formula F, written $Lat(F)$, is defined recursively:

- $Lat(n_i \text{ is } 0)$ is an element of the lattice with i^{th} position in the element being 0 and the rest of the positions being X's. $Lat(n_i \text{ is } 1)$ is an element of the lattice with the ith position being 1 and the rest of the positions being X's.
- $Lat(F_1 \text{ and } F_2) = Lat(F_1) \sqcup Lat(F_2)$.
- $Lat(0 \rightarrow F) = X^m$ and $Lat(1 \rightarrow F) = Lat(F)$.

4.2. Extensions for acyclic trajectory assertions.

An acyclic trajectory assertion can be verified by enumerating all the paths from source to sink and using STE separately on each path. However, there can be an exponential number of paths in a graph. We use path variables to encode the graph and verify all paths in one single run of the simulator. STE has been extended with a conditional simulation capability. Conditional simulation allows STE to advance to the next-time instant under some specified Boolean condition. Under the condition, the simulator advances to the next-time instant. Under the complement of the condition, the simulator maintains the previous state of the circuit. The trajectory assertions are symbolically encoded and STE conditionally evaluates the next-state function for each vertex in topological order. As an example, assume an acyclic trajectory assertion of the form shown in Figure 12. The path variables shown in the figure are introduced to encode the graph. For every vertex w with an outdegree of $d(w)$, $\lceil log(d(w)) \rceil$ number of variables are introduced to encode the edges. The conditions are accumulated at the vertices as shown. An expression at a vertex specifies all possible conditions under which the vertex can be reached from the source. The simulator can now compute the vertices in topological order with the vertex expressions being used as constraints for computing the excitation function. The number of additional path variables introduced is $\sum_{w \in W} \lceil log(d(w)) \rceil$, where $W = V \cup U$.

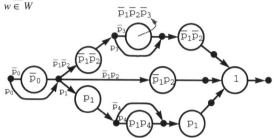

Figure 12. Symbolic encoding of acyclic trajectory assertions.

4.3. Extensions for generalized trajectory assertions.

Cycles in trajectory assertions can be dealt by performing a greatest fixed point computation[4]. Therefore, we can deal with generalized trajectory assertions. We identify the strongly connected components, obtain the corresponding acyclic component graph, and encode the acyclic component graph. The strongly connected components require a fixed point computation and are dealt with in a recursive manner. As an example, the graph in Figure 12 is the acyclic component graph of the trajectory assertion in Figure 10. Note that an exact greatest fixed point computation in a cycle requires introducing new variables for each iteration of the cycle until a fixed point is reached. The number of path variables introduced is no longer defined solely by the trajectory assertion. Therefore, for each iteration of the cycle we quantify out the path variables introduced for the cycle and reuse them in the next iteration. Quantification requires a greatest lower bound operation on the lattice. The greatest lower bound computation can lead to a pessimistic verification. A pessimistic verification can generate false negatives (correct circuit is reported to be incorrect) but will never generate a false positive (incorrect circuit is reported to be correct). In effect STE computes a conservative approximation of the reachability set.

5. Verification of a Superscalar Processor

We are testing our methodology on a superscalar processor. The processor is an implementation of the PowerPC architecture. The particular design is a fully functional prototype of the processor found in IBM's AS/400 Advanced 36 computer[12].

Presently, work has concentrated on verifying arithmetic and logical instructions with two source register operands and one target register operand in the fixed point unit

124

(FXU). The FXU has three pipeline stages, dispatch, decode and execute. Most of our attention has focussed on verifying the decode stage because this is where the bulk of the control logic resides. The decode stage is responsible for obtaining the source operands from the register file and reserving the target operand register. An instruction might stall in the decode stage because the operands are not immediately available or because the execute stage is busy processing another instruction. If the execute stage has a valid instruction, then the FXU can bypass the register file and forward the target operand data in the execute stage to the decode stage. This occurs when the source register address of the instruction in the decode stage is the same as the target register address of the instruction in the execute stage.

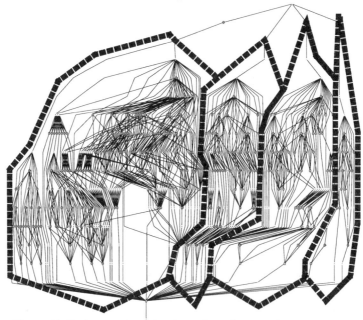

Figure 13. Trajectory Assertion for a class of instructions in the FXU.

We specified the mapping to capture these implementation details. The main machine defined the dispatch, decode and execute stages. A total of 14 map machines described the mapping for simple abstract formulas. Our tool took the abstract specification and mapping and generated the resultant trajectory assertion shown in Figure 13. The trajectory assertion corresponds to all possible cases that can arise due to interactions between the map machines. The control graph can be divided into 4 vertical slices as shown by the thick, dotted lines in the figure. We will number these slices 1 to 4 from left to right. Slice 1 is the most complex of all slices and corresponds to the case where neither of the source operands are bypassed. This slice has to deal with all possible arrival orders of the source operands from the register file. Slices 2 and 3 correspond to the case where one of the source operands is being bypassed. Slice 4 is the simplest of all slices and corresponds to the case where both source operands are bypassed.

The trajectory assertion has 261 state vertices, 57 cycles that require a fixed point computation and 4210 paths in the corresponding acyclic component graph. It would be computationally infeasible to enumerate all paths and use STE separately on each path. Instead our tool encoded the paths using 410 path variables. STE verified the entire trajectory assertion in a single verification run. We took the OR instruction as a representative of this class of instructions. The verification of the OR instruction took 3 hours of CPU time and 70 Meg of memory on an IBM Power Series 850, 133MHz 604.

6. Conclusions

We have presented a methodology for verification of systems that have a simple high-level specification but have implementations that exhibit nondeterministic behavior. The verification task verifies each individual operation. The theoretical foundation behind the methodology ensures that the operations can be stitched together to create arbitrary execution sequences. STE has been enhanced to deal with greater amounts of nondeterminism. Our work on the fixed point unit in the PowerPC architecture indicates that this is a promising approach for the verification of processors.

References

[1] R. E. Bryant, D. L. Beatty and C. J. H. Seger, "Formal Hardware Verification by Symbolic Ternary Trajectory Evaluation," *28th Design Automation Conference*, pp. 397-402, June 1991.

[2] D. L. Beatty, "A Methodology for Formal Hardware Verification with Application to Microprocessors," *PhD Thesis, published as technical report CMU-CS-93-190, School of Computer Science, Carnegie Mellon University*, August 1993.

[3] D. L. Beatty and R. E. Bryant, "Formally Verifying a Microprocessor Using a Simulation Methodology," *31st Design Automation Conference*, pp. 596-602, June 1994.

[4] C. J. H. Seger and R. E. Bryant, "Formal Verification by Symbolic Evaluation of Partially-Ordered Trajectories," *Formal Methods in System Design 6*, pp. 147-189, 1995.

[5] J. R. Burch, E. M. Clarke, K. L. McMillan and D. L. Dill, "Sequential Circuit Verification Using Symbolic Model Checking," *27th Design Automation Conference*, pp. 46-51, June 1990.

[6] K. L. McMillan, "Symbolic Model Checking," Kluwer Academic Publishers, 1993.

[7] R. P. Kurshan, "Analysis of Discrete Event Coordination," *Lecture Notes in Computer Science 430*, pp. 414-453, 1990.

[8] R. P. Kurshan, "Computer-Aided Verification of Coordinating Processes: The Automata-Theoretic Approach," *Princeton University Press*, 1994.

[9] W. A. Hunt, "FM8501: A Verified Microprocessor," *Lecture Notes in Artificial Intelligence 795*, 1994.

[10] T. K. Miller III, B. L. Bhuva, R. L. Barnes, J.-C. Duh, H.-B. Lin and D. E. Van den Bout, "The Hector Microprocessor," *International Conference on Computer Design*, pp. 406-411, 1986.

[11] M. Srivas and M. Bickford, "Formal Verification of a Pipelined Microprocessor," *IEEE software 7(5)*, pp. 52-64, September 1990.

[12] C. May, E. Silha, R. Simpson and H. Warren, "The PowerPC Architecture: A Specification for a New Family of RISC Processors," *Morgan Kaufmann Publishers*, 1994.

[13] J. R. Burch and D. L. Dill, "Automatic Verification of Pipelined Microprocessor Control," *Lecture Notes in Computer Science, Computer Aided Verification, 6th International Conference, CAV 94*, pp. 68-80, 1994.

A Methodology for Processor Implementation Verification

Daniel Lewin

Dept. of Electrical Engineering
Technion, Haifa, Israel

Dean Lorenz

Dept. of Electrical Engineering
Technion, Haifa, Israel

Shmuel Ur

IBM Israel Science and Tech.
Haifa, Israel,

Abstract

We address the problem of verification of implementations of complex processors using architectural level automatic test program generators. A number of automatic test program generators exist, and are widely used for verification of the compliance of complex processors with their architectures. We define a four stage verification process: (1) describing the processor implementation control as a Finite State Machine (2) deriving transition coverage on the FSM using methods from formal verification (3) translation of the covering tours to constraints on test programs (4) generation of test programs for each set of constraints. This process combines a high quality and well defined theoretical method along with tools used in industrial practice. There are a number of advantages of our Method: (a) The last three stages are automated (b) Implementing the FSM model involves relatively little expert designers time (c) The method is feasible for modern superscalar processors and was studied on an enhanced PowerPC processor. We describe a formal framework for the new process, identify the obstacles that are encountered in the modeling phase, and show how to overcome them.

1 Introduction

Functional verification comprises a large portion of the effort in designing a processor [1]. The investment in experts time, and computer resources is huge, and so is the cost of delivering faulty products [2]. In current industrial practice, random architectural test program generators and simulation methods are used to implement large portions of the verification plans [1][3][4][5][6]. Massive amounts of test programs are generated and run through an architecture simulation model and through the design simulation model, and the results are compared. The test space is enormous, thus even a large number of tests represent only a small section of the test space. In practice the actual coverage of the global test space is unknown, and there is a lack of adequate feedback from the verification process as to the quality of the tests that are simulated.

In order to solve some of these downfalls, a new verification methodology was introduced by R. C. Ho, C. Han Yang, M. A. Horowitz and D. L. Dill in [17]. This methodology contains three basic phases: (1) describing the processor implementation control as a Finite State Machine (2) deriving transition coverage on the FSM using methods from formal verification (3) automatic translation of the covering tours to test vectors. This method is attractive because it is formal, mostly automatic, and produces high quality tests. Another advantage of the method is that there is efficient use of expert designers time in the verification process: designers focus on the control in the section that they are responsible for, and the interactions between the sections are derived automatically by the process. Unfortunately, this methodology is not applicable to modern, real-life, superscalar processors. This is due to the size of the FSM needed to describe such processors. We show how this new methodology can be integrated into the existing industrial practice for verification of modern, complex processor implementations.

Existing industrial practice for verification is based on the following four stages:

• *Implementation Evaluation* - Identification of the sections of the implementation to be verified. These sections include modules and interactions between them that are considered to be bug prone.

- *Verification Task Definition* - For each design section identified, a list of verification tasks is defined. A verification task describes event sequences that exercise the relevant modules and interactions. Definition of the event sequences may be easy (e.g. an add instruction) or very involved (e.g. a miss predicted branch with an interrupt happening before its resolution).
- *Test Template Production* - Each verification task is translated into a Test Description Language (TDL). This is done by translating event sequences into restrictions on a sequence of instructions. The TDL can be read by an architectural test program generator.
- *Test Code Generation and Validation* - The previously produced templates (in TDL form) are fed to an architectural test program generator such as [3]. The test program generator automatically produces a set of architectural test programs for each template. These tests are run through the design simulation model, and the results are verified.

This practice has a number of problems: (1) it is expensive in terms of expert person years (2) the first three stages are not automated (3) it lacks formality, and it is difficult to asses the quality of the process.

We show how to integrate the new methodology described in [17] into current verification practice by using it to generate verification tasks automatically and applying existing industrial tools (e.g. [3]) to generate architectural tests from these tasks. The third stage in the process of [17] (automatic translation of the covering tours to test vectors) is replaced with translation of tours on the model into verification tasks. A fourth stage, as in the existing practice, is added to generate architectural test programs from these tasks. This synergy creates a new methodology that combines the advantages of the method of [17], with the applicability of the existing process. We study the application of our methodology to a superscalar PowerPC implementation in [24]. Table 1 compares this new combined methodology with current industrial practice.

Stages	Current Method	Our Method
Implementation Evaluation	Each designer evaluates his module and it's behaviors. Interaction between modules is described by lead designers using a lot of high quality (and costly) human resources.	A state machine describing the "Heart" of the implementation is created. Each designer describes her module in FSM form. Interactions are automatically generated.
Verification Task Definition	Designers define sequences of events that exercise each section. Informal description. Complex tasks may require great human effort.	Automatic. Tasks are derived from the product machine defined in the previous stage. Formal description as tours on the FSM. Requires a modest amount of computing resources.
Test Template Production	The test engineer translates event sequences into constraints for a test program generator. This stage requires a huge amount of human resources.	Automatic. Each verification task corresponds to a tour on the FSM. These tours are translated into test templates. Requires negligeable computer resources.
Test Generation and Validation	Test generation by automatic test program generator. Validation is usually done by **visually** scanning the test traces.	Test generation by automatic test program generator. Validation is done by **automatic** inspection of the traces.

Table 1. Our method vs. Current Methods

We introduce a formal framework for the new process. This framework describes the method in general, the character of the FSM model, and the test generation procedure. Within this framework we can identify more formally the limits of the industrial tools used in our method - particularly, what types of tests are hard for test generators to produce. The insight that the framework provides leads to new techniques in the crucial stage of our method - building the FSM model. Specifically we expect it to allow the modeling of modern superscalar processors while keeping the size of the model manageable.

The main contribution of this paper is the formal framework. The framework allows us to describe the crucial elements of the method, and to analyze possible solutions to problems that arise. The crux of our method is how to change the target of [17] from test vectors to verification tasks. We show how this change not only makes it possible to integrate industrial tools in the process, but makes the whole method feasible for modern processors.

The rest of this paper is organized as follows: In section 2 the last three stages of the method are described within a formal framework. Section 3 describes the first stage and the character of the FSM model. Sections 4 contains some examples of modeling.

1.1 Terminology and Background

This section describes some terms that are used in the rest of this paper. We elaborate on partition testing which is the framework in which we will analyze the quality of our process.

An *Architectural Test Program* is a test program with expected results that could be run on any actual hardware implementing the architecture. A test is not architectural if it depends on the design simulation environment to execute. An example of non-architectural tests are those produced in [17]. The *Global Test Domain* for a processor is the set of all Architectural Test Programs for the processor.

Partition testing for software [8], is a testing methodology where the global test domain is divided into subdomains and a single test from each subdomain is generated and simulated. The goal of such a partitioning is to make the division in such a way so that when tests are selected from each subdomain and there is a bug, the resulting test set will uncover it.

In [9] a heuristic method for defining such a partition is presented. The global test domain is divided into two partitions, the specification partition, and the implementation partition. The specification partition divides the domain into test sets whose elements *should* be treated in the same way by the specification. The implementation partition divides the domain into sets that *are* treated in the same way by the implementation. The final partitioning is defined by intersecting these two partitions. Verification is considered complete if a test program from each subdomain has been simulated on the design. In [8] it is shown that partition testing is most effective when the implementation partition is based on models of bugs that are likely to occur in the design. Based on the processor bug classification in [10], it is clear that processor bugs are in many cases revealed in situations where a number of conditions are present at once. Thus, each domain should represent a combination of "interesting conditions".

In the domain of protocol testing, transition coverage of Finite State Machines is commonly used to produce concrete tests for the protocol [11]. In many cases an FSM of reasonable size may be constructed from the specification of the protocol that is actually an implementation, possibly a very simplified one, of the whole protocol. Thus, inputs to the model are easily translatable into inputs for a real implementation. The major difference between protocol verification and processor verification is the size of the implementation. In [7] a CFSM model is used to drive test generation to check a PowerPC system controller and a PowerPC multi-processor exclusive access mechanism (semaphores). Building an FSM that models a complete processor, as is done for protocols, is infeasible due to the state explosion problem. Therefore, the model is built as an abstraction of the processor. The fact that there is abstraction makes the translation of tours on the model to concrete tests extremely difficult. We show how to build the model so that the translation is feasible.

2 Method

This section describes in detail the last three stages of the process within a formal framework. The reason for this out-of-order explanation is that the tools and technology used in the automatic process are major factors that shape the stages of abstraction and modeling. In section 2.1 we define the relationship between the FSM model and architectural tests. The coverage model on the FSM is described, and the process is discussed in terms of partition testing. Test generation is formally defined. In section 2.2 the difficulties of test program generation are discussed. A measure of "generate-ability" is given to verification tasks. In section 2.3 the industrial tool CCTG used to generate covering tours on the FSM model is described. In section 2.4 we explain how the Genesys expert system is used to generate test programs.

2.1 Formal Framework

We begin our description assuming that an FSM has been defined using knowledge of the implementation such that there is a mapping from the global test domain into sequences of inputs of the FSM. Using this mapping the transitions of the FSM induce a subdomain covering of the global test domain in the following manner:

Let ι be a test in the global test domain G. Let M be the mapping from the global test domain into sequences of inputs to the FSM. $M(\iota)$ is the input sequence for the FSM corresponding to ι. Input sequences corresponding to tours in the FSM are described by ordered sets of edges. For each edge $e \in E$ of the FSM we

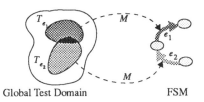

Figure 1. The Mapping M from the Global Test Domain into Tours on the FSM

define the following subdomain of G : $T_e = \{\iota \in G \mid e \in M(\iota)\}$. T_e is the set of all tests whose corresponding input sequence traverses the edge e (figure 1).

Along the line of partition testing, once we have defined the covering $\{T_e \mid e \in E\}$ of G, we need to generate a set of tests $T \subseteq G$ so that for each edge e of the FSM $T \cap T_e \neq \varnothing$. In other words, to generate at least one test from each subdomain T_e. In theory we would like to be able to specify an edge e of the FSM and have the generator output tests in T_e. However, test program generators generate tests at the architectural level, and need to have their input at that level. Our method for task generation from the FSM will solve this problem by generating for each edge a tour through the FSM that contains that edge, thereby specifying the sequence of events leading to a state from which the edge may be traversed. The tour generation stage produces *transition coverage* of an FSM by a set of tours π such that for each edge e there is a tour $\pi_e \in \Pi$ passing through it. Each tour is translatable into constraints on a *sequence* of instructions which is used as a guideline for the test program generator. Thus, existing test program generators may be used to generate the set T. More formally, the tour generation stage produces for each edge $e \in E$ a tour $\pi_e = \left(e_{i_1}, e_{i_2}, e_{i_3}, \dots \right)$ (each tour is denoted by the ordered set of edges traversed in the tour) such that $e \in \pi_e$.

A test program generator is a function $TPG : E^N \times N \to G$ such that for all $\pi \in E^N$ and all $n \in N$, $M(TPG(\pi, n)) = \pi$.[1] In other words TPG is a sort of inverse function to M. Given a sequence of inputs to the FSM π, and a number n, $TPG(\pi, n)$ is a test that when "run" on the FSM follows the tour π. The test program generator may use the number n to vary the test which is produced along parameters that do not change the corresponding tour through the FSM. Using such a test program generator the set of tests T may be generated using the set of tours $\Pi = \{\pi_e | e \in E\}$ produced from the FSM. The completeness condition on the set of tours and the condition on TPG guarantee that the resulting set of tests contains a test from each subdomain T_e.

The cover $I \equiv \{T_e | e \in E\}$ of G is an implementation partition as the edges of the FSM represent transitions in the control of the processor (see section 3). Along the same lines of the heuristic for creating an effective partition presented in [8] and [9] for software testing, we would like to intersect I with a partition based on architectural considerations. In the case of test programs, this architectural partition is very difficult to describe as each implementation subdomain induces its own architectural dimension. In addition, the true architectural partition is huge. For example an implementation partition may not distinguish between groups of instructions, but an architectural partition would. Nevertheless, in each subdomain we would like to explore the range of architectural alternatives. In order to do this we exploit the power the Genesys expert system [3][4] that contains expert knowledge on how to generate test programs with "interesting" architectural variations. More formally, we use Genesys as the function TPG, and for each $e \in E$ we generate the set of tests: $\tau_e = \{TPG(\pi_e, 1), TPG(\pi_e, 2) \dots, TPG(\pi_e, K)\} \subseteq T_e$ where K is a constant determined by the size of T_e, and the time and money available. The quality of the Genesys system states that this set will be a good representation of the architectural dimension of the subdomain T_e, or that this set will be "intelligently scattered" in T_e. With truly random program generators all that we know is that as $K \to \infty$ the test set represents more and more of the domain T_e. Systems such as Genesys attempt to control the convergence of the test set as to obtain the "best" set in least time. So, we do not intersect the cover I with an architectural cover, but effectively sample the architectural dimension of each T_e. The set T is therefore $\bigcup_{e \in E} \tau_e$.

2.2 Implementing TPG with Existing Tools

In the above defined process, the most technically challenging point is implementing the function TPG. Obviously the character of the model influences how difficult it is to implement TPG (high level abstraction means difficult TPG implementation). Since we want to use existing tools, we must design the model according to the limits of existing test program generators. Therefore it is important to understand how difficult it is for a system such as Genesys to satisfy a given test program constraint. In this section a measure of "difficulty" called *Mazeness* is introduced that gives us a concise way of talking about the difficulty of a test program constraint.

---

1. We ignore sequences that are not valid tours through the FSM. Formally, we may define M, and TPG to be constant null on these sequences.

Let Ψ be all possible instances of a single instruction.

A k *dimensional instruction constraint* c_k is a function: $c_k : \underbrace{\Psi \times \Psi \ldots \times \Psi}_{} \to \{True, False\}$.
Intuitively, the function c_k describes the sequences of length k that fit a requirement. For example a two dimensional instruction constraint could be "Both the first and second instructions are floating point instructions".

A *sequence constraint* is an ordered set of instruction constraints $C = (c_1, c_2 \ldots, c_n)$ such that for all $1 \leq i \leq n$, c_i is an i dimensional instruction constraint. Intuitively, each c_i introduces the constraints on the next instruction in the sequence, but these constraints may depend on the previous instructions in the sequence. For example a sequence constraint on two instructions could be defined as follows:

- c_1 : The first instruction is an arithmetic instruction with a target register.

- c_2 : The second instruction is a load with the same target register as the first instruction.

A sequence of instructions $(I_1, I_2, \ldots, I_n) \in \Psi^n$ satisfies a sequence constraint C, written $(I_1, I_2, \ldots, I_n) \propto C$, if for each $1 \leq i \leq n$, $c_i(I_1, I_2, \ldots, I_i) = True$.

The set $SAT(C)$ is defined as follows: $SAT(C) = \{ (I_1, I_2, \ldots, I_n) \mid (I_1, I_2, \ldots, I_n) \propto C \}$ meaning all of the instruction sequences that satisfy the sequence constraint C. The next definition describes the type of sequence constraints that most test program generators are best at satisfying. A sequence constraint $C = (c_1, c_2 \ldots, c_n)$ is called *sequentially consistent* if for all $1 \leq k \leq n-1$: $\forall ((I_1, I_2, \ldots, I_k) \in SAT(c_1, c_2 \ldots, c_k))$, $\exists I_{k+1}$ such that $c_{k+1}(I_1, I_2, \ldots, I_k, I_{k+1}) = True$ [1]

This intuitively means that the constraint set C has the property that *any* conforming prefix of instructions of length k may be continued to a conforming sequence of length $k+1$. In terms of a test program generator this means that if instructions are generated sequentially then no decision taken will block generation of a later instruction. Genesys does generate each instruction sequentially, satisfying the constraints on each instruction in the context of the current machine state defined by the instructions previously generated. Therefore, constraints that are sequentially consistent are much easier for Genesys to satisfy. If sequentially inconsistent constraints are presented to Genesys, then backtracking may be needed when generation of a previous instruction blocks the continuation of the sequence.

We now attempt to understand how hard it will be for the Genesys to generate a sequence fitting a given sequence constraint. The set of possible extensions of a given valid prefix of length k is:
$Next(C, (I_1, I_2, \ldots, I_k)) = (I_1, I_2, \ldots, I_k, *) \cap SAT(c_1, c_2 \ldots, c_{k+1})$. $Next(C, \varnothing)$ is defined as $SAT(c_1)$. If the generator chooses at each stage from a uniform distribution over the possible extensions and each choice is independent of the previous choices then the probability of reaching a given sub-sequence $I_1, I_2, \ldots, I_k$ is [2]:

$$P_C((I_1, I_2, \ldots, I_k)) = \left(\prod_{j=0}^{k-1} |Next(C, (I_1, I_2, \ldots, I_j))| \right)^{-1}$$

Where $(I_1, I_2, \ldots, I_k) \in SAT(c_1, c_2 \ldots, c_k)$ and zero otherwise. P_C defines a probability measure over the set $T(C)$ of terminal instruction sequences:

$$T(C) = \left\{ (I_1, I_2, \ldots, I_k) \in \bigcup_{j=1}^{N} \Psi^j \mid Next(C, (I_1, I_2, \ldots, I_k)) = \varnothing \wedge (I_1, I_2, \ldots, I_k) \in SAT(c_1, c_2 \ldots, c_k) \right\}$$

The probability that the generator will find an instruction sequence fitting a sequence

---

1. This implies $(I_1, I_2, \ldots, I_k, I_{k+1}) \in SAT(c_1, c_2 \ldots, c_k, c_{k+1})$

2. The problem of finding the possible extensions is left to the constraint satisfaction algorithms of the generator e.g [15][14].

constraint without having to backtrack is then $P_C(SAT(C))$. Since the generator in reality does backtrack we need to understand when there is a good chance of finding a solution without backtracking too much. Let:

$$Viable(C) = \left\{ (I_1, I_2, ..., I_k) \in \bigcup_{j=1}^{N} \psi^j \mid \exists I_{k+1}, I_{k+2}, ..., I_n \text{ so } (I_1, I_2, ..., I_n) \propto C \right\}$$

be the set of all instruction sequences that can be continued to a full sequence that satisfies C. For a sequence of instructions Q let: $Continuations(Q) = \{Q\} \cup \bigcup_{R \in Next(Q)} Continuations(R)$

be the set of all reachable sub-sequences with the prefix sequence Q. We define the *Mazeness* of the sequence constraint C as follows:

$$Mazeness(C) = \sum_{Q \in Viable(C)} \sum_{\substack{R \in Next(Q) \\ R \notin Viable(C)}} P_C(R) Cost(Continuations(R))$$

where the function $Cost$ is a function that identifies *continuations* that the generator will have difficulties backing out of. $Cost$ returns a number indicating the price (e.g. time) payed for searching a given continuation that has no viable paths in it. If we assume that the generator looks for a solution by a randomly ordered DFS search on sub-sequences, then the Mazeness describes the average cost of the generators bad decisions. The formula describes the average price payed for searching a bad continuation while traversing a viable path. (See figure 2) Note that if C is sequentially consistent, then $Mazeness(C) = 0$ because the second sum in the definition is empty.

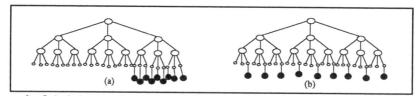

Figure 2. In both these search trees the probability of reaching a complete solution (reaching a black leaf) is $P_C(SAT(C)) = 1/3$. In spite of this, a backtracking generator will be much more efficient searching tree (b) than tree (a). If we choose the $Cost$ function to be $Cost(A) = |A|$ then $Mazeness(C)$ will be the average amount of backtracking needed to find a solution. As expected $Mazeness(C_a) = 26/3$ is larger than $Mazeness(C_b) = 2/3$.

A sequence constraint C with a larger Mazeness has a worse chance of being generated by a test program generator like Genesys that solves the constraints sequentially using a backtracking mechanism. Mazeness is a measure of the quality of a sequence constraint in terms of how efficient Genesys can be in finding a test satisfying the constraints. Low quality constraints may lead to inefficient test generation, or in practice to failure of the test generation process.

In order to use Genesys as the function TPG we must translate the tour π into a sequence constraint. Practically, this is done by translating the tour into a constraint description language that Genesys understands. Abstractly we may assume that there is a function ∇ that maps tours on the FSM into sequence constraints. Using such a mapping we have $TPG(\pi, n) = Genesys(\nabla(\pi), n)$, where n is used as the random seed for the random number generator Genesys uses to make decisions. From the discussion above it is clear that if tours derived from the FSM are mapped by ∇ onto sequence constraints that are not sequentially consistent enough (i.e. have high Mazeness) then Genesys will not be able to generate an instruction sequence fitting the constraint. The function ∇ is actually the crux of our method. As will be shown below, much of the construction of the model is motivated by the need for a realizable ∇ that not only produces constraints in a form that the generator understands, but produces constraints that are sequentially consistent enough (have low Mazeness). The model is built in such a way that a tour π is directly translatable into a sequentially consistent enough constraint. Thus, the problem of implementing ∇ is relayed to the problem of creating a model that generates "good" tours.

Now that we have established a theoretical framework for the last stages, we present some practical details involved.

2.3 Tour Generator - CCTG

The CCTG tool (D. Geist, M. Farkas, A. Landver, Y. Lichtenstein, S. Ur, Y. Wolfsthal [7]) automatically generates the set of covering tours of a CFSM. In practice, the number of transitions in the model may be large and likewise the size of Π may be huge. In CCTG a hierarchy on the state variables of the FSM is used to define flexible coverage criteria which are derivatives of transition coverage. The hierarchy can be used to direct coverage away from transitions that are not a part of the basic model.

CCTG uses *model checking* to generate tours that cover a given transition. Model checking [12] is a formal verification technique which checks whether a model satisfies an assertion specified in temporal logic. An assertion about the model serves as a query to the model checker. When the assertion is false, the model checker responds to the query by presenting a counter-example that contradicts the assertion. To generate a test that covers a specific transition, the model checker is queried with an assertion that claims that no such transition exists in the model. The counter example presented by the model checker will be a valid execution path on the model which contains the transition. The inputs that drive the model along this path, taken from the counter-example, is actually a tour that covers the transition. By using a model checker such as SMV [13] and writing the appropriate assertions, it is possible to generate a tour that covers a specific transition. For example if x and y are states in the model, then the assertion $(state(x) \rightarrow next(\neg state(y)))$ declares that "if the model is in state x than the next state cannot be y". In many cases there are additional conditions on the tour that covers a transition. Examples of such conditions are that the tour always end in a specific state, or that the tour must pass through a specific transition before passing through the transition to be covered. Such additional conditions may be asserted in temporal logic, and added to the basic assertion that is supplied to the model checker. For example the assertion $(state(x) \rightarrow next(state(y))) \rightarrow \neg \exists eventually(state(INIT)))$ asserts that after a transition from x to y, the model cannot eventually reach the state $INIT$. A counter example generated from this assertion will be a tour that contains the required transition and returns to an initial state. This power will be very important in our process, and will allow us to shrink the size of our model in many cases.

2.4 Task and Test Generation

After obtaining the set of tours from CCTG, the tours must be translated into constraints for the Genesys system. Constraints for Genesys are defined using a Test Description Language [14] that allows description of constraints on a sequence of instructions. Since Genesys is an architectural test program generator, constraints on an instruction sequence must be stated in terms of processor facilities that are visible on the architectural level, e.g. opcodes, registers, addresses, translation paths [15][16]. Expressing events in the processor implementation on the processor architecture level is usually a very difficult task (e.g. define which instruction sequences cause a specific signal to behave in a given manner). If a tour generated from CCTG represented a sequence of implementation level events (e.g. signal a was high then signal b was low etc.) then translation from tours to constraints would be almost impossible. Therefore, the model is designed so that a tour through the model represents a sequence of architectural events (some events such as cache-hit may be events that have a relatively simple translation to the architectural level). Specifically, the inputs that drive the model along a given tour represent constraints on instructions in an instruction sequence. The translation is almost syntactical, although there may be a refinement process that uses heuristics to reduce the Mazeness of the constraints produced. The test template derived from a tour π_e describes the set $M^{-1}(\pi_e) \subseteq T_e$.

The next step is to compute the constant K which is the size of the set τ_e (the set of tests that are generated in the partition T_e). Since it is impractical to generate all of T_e we use the test program generator to sample T_e. Heuristics may be applied to a test template to decide how many tests should be generated to obtain an acceptable sampling of T_e. One example of such a heuristic is that if T_e is large (i.e. the template has few constraints) then K should be larger. If the test generator TPG is high quality, meaning that τ_e converges quickly to a set that is a "good" representation of $M^{-1}(\pi_e)$, then the value of K may be smaller. In any case, a heuristic may be applied to the constraint set to obtain a value for K that takes into account both quality, and available resources. The Genesys system is then used to generate, from the Test Description Language, K test programs. These programs are then simulated on the design, and expected results are compared.

In order to apply the above process it is important to understand the concept of "a good representation of $M^{-1}(\pi_e)$", and how a high quality test program generator TPG can be implemented. This is important because we rely on Genesys to explore effectively the architectural dimension of each partition. Understanding exactly how the generator works is also an important factor in deciding on the K to use. These issues will be detailed in later research.

3 The Model

This section describes the first stage of our method in light of the framework described in the previous section. First, we compare our model with other formal models used in similar methodologies [17][18][19]. Next, general guidelines are defined for building the FSM.

The FSM that is used to derive tasks should model the behavior of the processor, and must contain enough of the behaviors of the processor so that the mappings ∇ and M are realizable. For example, if the model contained *only* the behavior of the cache subsystem, then it would be very hard to ascertain what sequences of instructions cause a particular transition in the model to be traversed! The model must contain enough of the basic behavior of the processor, such as flow of instructions in the pipe, so that instruction sequences are more apparent. This is critical to the methodology because the descriptions of instruction sequences must be understandable by test program generators in order to generate test programs in each partition.

The choice of the pipeline control as the base for the model is not arbitrary. In addition to the fact that many design bugs involve mechanisms around the pipe [10], using such a model induces a natural mapping M from the global test space onto the inputs of the resulting FSM. An important characteristic of this mapping is that the inputs to the FSM at each clock are directly translatable into constraints on the instructions entering the pipe. This allows simple translation of tours through the FSM into constraints on instruction sequences. This is one example of how the character of the automatic stages of our process influence the structure of the model.

In [17] an FSM model is used to direct processor verification, and is built directly from the design description (HDL). The model is relatively accurate as it is an abstraction based directly on the implementation. The tests produced from this model are described in terms of internal signals, and the simulation is done by 'taking control' of signals in the simulation environment. The use of an HDL based model forces this kind of simulation because there are no tools to translate the test descriptions into architectural test programs. The path through the FSM corresponds to sequences of internal processor signal values, and usually the correlation between internal signals, and test programs is difficult to ascertain. 'Controlled' simulation limits the exploration of the architectural dimension, as in each test the events are forced into a single sequence by the control signals. For example, if a test contains a floating point exception then the actual data on the datapath is ignored during the test. The exception is forced by taking control of the signal that specifies that an exception has occurred. Bugs related to the actual data that caused the exception will not be uncovered. Thus, the architectural dimension of each implementation scenario is not covered. Furthermore, the model must be very detailed because an inaccurate model would produce tests with inaccurate or meaningless timing diagrams that cannot be simulated. The size of models based on a design grows linearly with the size of the design. Thus, complicated designs are difficult (or impossible) to handle because of the size of the resulting model.

We use our model not as an accurate description of the implementation, but as a guide towards interesting design sections. Transition coverage of this model does not imply a formal correctness criteria for the design, but does guarantee that all the aspects of the design contained in the model have been tested. We use formal methods, but we are not limited to an accurate description, and thus the model does not need to include many aspects of the implementation. For instance, even without incorporating "hazard detection" mechanisms, we can describe the different behaviors of the design when encountering various hazards and have the test generator produce command sequences which cover all the hazards. The actual simulation model has the full design and does not need any information except the test program. On the other hand we must include all the "natural" flow of commands through the pipe, because this will make ∇ realizable, and keep the Mazeness of the derived constraints down.

3.1 Model structure

The model is described as a set of communicating finite state machines. There are two main advantages of describing the model with CFSMs. First, processor control around the pipeline lends itself well to modeling by CFSMs. The second advantage is that when each processor section is modeled as a separate state machine, only the *local* or *first order* interactions of the section being modeled must be foreseen. The complex interactions of all the elements of the processor control will exist as transitions in the product machine. For example, a pipe stage that requests, and waits for data from the memory subsystem only has to interact directly with the cache, and this interaction is very simple - request/ answer. More complex interactions such as a request to the cache which is a miss, and causes a page fault do not have to be specified. These complex interactions are obtained *automatically* from the composition of the sections of the model together.

The advantages cited above are not only technical; they also impact the amount of expensive expert time that must be invested to use our method. Since each section may be described separately and only first order interactions need to be explicitly specified, different designers responsible for different design sections may separately add their sections to the model. The demand for people familiar with the complete design is reduced because their expertise is usually needed to specify the complex higher order interactions between units which in our case are automatically generated.

3.2 The Size Problem

The main problem that we will encounter when modeling processor control is the size of the model. There are two aspects to the size problem. First, the tools that we use to manipulate the model have size restrictions notoriously familiar to those with experience in formal verification. Secondly, the model may

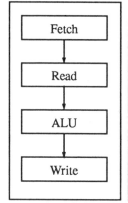

Figure 3. Pipe of an Imaginary Processor.

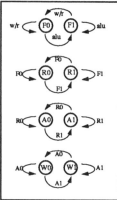

Figure 4. Naive Modeling of the Pipe.

be such that a huge number of tours are required to obtain complete transition coverage on it. If there are too many tours, we will not have enough time during processor verification to generate test instances for each one. The first of these problems is the most acute as it turns out that common pipeline structure induces a huge number of states in a model.

We begin our discussion of construction of the model with a simplified example that highlights some of the problems that are encountered, and some of the solutions.

Figure 3 describes a simple pipeline of an imaginary processor that has two instructions: *Read-Write*, and *Alu*. The pipeline has four stages: Fetch, Read, Alu, and Write. Read-Write instructions read data in the read stage, perform a computation in the Alu stage, and then write data in the Write stage. Alu instructions pass through the Read stage without doing anything, perform a computation in the Alu stage, and then pass through the Write stage without doing anything.

Figure 4 shows our first attempt at modeling the pipeline. Each stage is a separate machine, and the two states represent the two kinds of instructions that may be in the stage. The communication between the machines is used to model the flow of instructions through the pipeline. This modeling captures the whole description of the pipe. Notice that

there will be 16 reachable states in the product machine of this model (2^4)

Figure 5 shows another model of the same pipeline with only one state in the fetch stage. This new model is of course less descriptive than the first, but on the other hand, both types of instructions are *treated exactly the same way* in the actual implementation of the Fetch stage. From the standpoint of testing this specific pipeline, we do not care which instruction

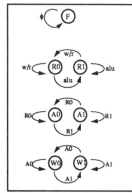

Figure 5. Late Bifurcation applied to the Fetch Stage.

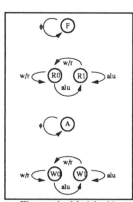

Figure 6. Model with Inconsistency

type is currently in the Fetch stage. The instruction type only becomes interesting in the Read stage where there are two distinct behaviors that we are interested in testing - and as we can see the Read stage's state is dependant on the instruction type. This model postpones the decision about the type of instruction until the READ stage. The number of states in the product FSM is reduced by a factor of 2; from 16 to 8. This technique for saving size is called *late bifurcation*. Late bifurcation means that stages should begin to differentiate between instructions flowing down the pipe only when their behavior depends on the differentiation. The question of whether the differentiation made in a certain stage must be passed down the pipe, will be addressed in the next example.

Figure 6 shows yet another model of the same pipeline. In this model we have discerned that both instruction types are treated in the same way by the Alu stage, and therefore there is only one state in this stage. We are left with an intriguing problem - notice that if there is a Read-Write instruction in the Read stage then after two clocks there *must* be a Read-Write instruction in the Write stage. Since the Alu stage does not contain information on the instruction type we are left with a number of possibilities. We can change the Write stage to be dependant on the state of the Read stage two clocks ago, although this would amount in practice to adding the states that we have already taken out of the Alu stage to the Write stage. Or, as in the model shown, a new input to the Write stage is used to determine what instruction type is currently in this stage. Taking out the differentiation in the ALU stage is different from late bifurcation because now there are input sequences that do not relate to any sequence of Read-Write, Alu instructions going down the pipe! This is an example of the most ubiquitous reason for state explosion in our pipe models - information that *must* be passed down the pipe in order to correctly model the flow of instructions through the pipe. This type of information is called *non-persistent*, and when we lose such information by using new inputs such as in the example we say that we have lost *consistency* of the inputs. In the example we have obtained a model that has inconsistency of inputs, but on the other hand notice that there are only 4 reachable states in the product machine. Changing our imaginary processor to have four types of instructions: Read, Write, Read-Write, and Alu, makes figure 6 an actual model for this processor where the late bifurcation technique has been correctly used. The naive implementation for this processor (4 states in each stage) would have 256 reachable states in the product machine in comparison to only 4 (!!) reachable states in the product machine of figure 6.

3.3 Modeling Guidelines

The example above shows how carefully the modeling of the pipe must be done, and demonstrates two important concepts: *late bifurcation*, and *consistency*. In a real pipe, instructions enter the pipe from the top as a set of bits read from memory, or in terms of an FSM model, through inputs in the first stage in the pipe. Complete information about a single instruction is propagated down the pipe as the instruction advances. Modeling this flow of information requires a huge number of states. Late bifurcation breaks up the information on a single instruction into sections that are supplied to the model in different stages. Differentiation is only made in stages whose behavior depends on knowing something new about the instruction. Translation from tours to constraints needs to unify this distributed information into instruction constraints. The use of late bifurcation does not lessen the quality of tasks derived from the model, it only reduces the number of meaningless states in terms of the interesting behaviors of the stages of the pipe.

Consistency is required to assure that tours through the FSM represent realizable architectural tests, or in other words, that the set of constraints on each instruction do not contain contradictions. Since modeling all of the possible consistency requirements would make the model huge, some consistency must be sacrificed. When consistency is relaxed there are two possible outcomes: either the constraints generated from a tour contradict and there is no test that fits the constraints, or the constraints are not sequentially consistent enough (i.e. have high Mazeness (section 2.2)). Deciding where to sacrifice consistency is the most delicate part of the modeling process. In each case when consistency is to be relaxed, the impact must be carefully analyzed. The questions to ask are: Will the lack of consistency cause all tours through the model to contain contradictions, or only a small part of them? For each transition in the model is there at least one consistent tour that passes through the transition? Will the lack of consistency cause the derived constraints to have high mazeness? The impact of lost consistency may be softened by using a number of techniques that aim at avoiding inconsistent tours. For example when an inconsistent tour is encountered by the test generator (e.g. the generator is not successful in generating a test after a number of attempts), an assertion in temporal logic may be added to the basic assertion to CCTG requiring coverage of the same transition by a tour *different* from the current one. Another useful tool is called *generalized tours*. The difference between a regular tour and a generalized tour is that for each state transition, a generalized tour lists *all* possible inputs that cause the transition while a regular tour gives only *one* of the possible inputs. Thus, a generalized tour can be used to describe a large number of equivalent regular tours, some of which may be inconsistent. The selection of a consistent set of inputs is then left to the test program generator. Generalized tours can also be used to implement different types of coverage such as hazard enumeration by requiring that the test generator produce a test for each possible input combination. Another important tool for modeling consistency is the use of auxiliary statements in temporal logic that constrain the tours produced by CCTG. For example if a generalized tour on the model contains many inconsistent input combinations, an additional assertion to CCTG can require that only legal tours are produced.

In general, consistency requires propagation of differentiation down the pipe, adding to the model aspects that we are not interested in testing. For example, the states in the ALU stage in figure 5 that differentiate between the instruction types are necessary (to enforce consistency), but do not embody interesting behaviors of the ALU stage that we are interested in testing. The ability of CCTG to generate sets of tours that conform to different coverage criteria may be used to remove behaviors added by consistency requirements from the coverage criteria. The state variables that are used to propagate non-persistent information through intermediate stages are marked so that CCTG will not specifically generate tours to cover transitions between states that differ only on these variables. This solution reduces the number of transitions that are required to be covered.

4 Examples of Modeling

We have examined the implementation of an enhanced version of a PowerPC 601 processor, and have provided solutions for modeling most of the basic elements of this superscalar processor in [24]. Here we present two examples that highlight the type of solutions presented. The first example describes modeling of the hazard detection, and data bypass mechanisms. Hazards are detected in the decode stage of each pipe. Either the pipe is stalled until the hazard does not exist any more, or forwarding mechanisms are invoked to remove the hazard. An important observation is that constraints induced by demanding that a hazard occur are of the type "in the current instruction use the same resource as some *previous* instruction in the instruction sequence". This observation is important because hazard constraints in this format can not imply impossible constraints if the instruction stream is being generated instruction-by-instruction. There are a number of methods for enumerating all the hazards in a given pipeline [20][21]. Including logic for identification of hazards in the model is very complicated, and would make the model very large. Instead, an enumeration is used while building the model to partition hazards into groups on which the behavior of the control of the processor is significantly different. This

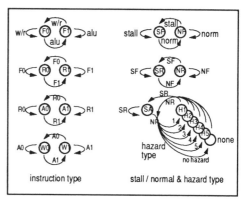

Figure 7. Propagating Instruction Type and Hazard Condition

information can be used to implement a late bifurcation model of deciding on hazards for an instruction going down the pipe. For example in the decode stage decide if there is a hazard that causes a pipe stall or not and in the execute stage decide if a certain forwarding mechanism is invoked. After an instruction has left the pipe all the decisions taken on the hazards for the instruction will define a subset of the hazards. This set is translated into a hazard constraint for the generator (such a constraint may specify using a certain register, memory location or instruction).

In order to verify that all the necessary interlock mechanisms have been correctly implemented it is desirable to exercise each hazard type at least once. Since two hazards may use the same forwarding mechanism, in order to guarantee that each hazard has been exercised there must be a transition in the model for each one. Such transitions may be added to differentiate between all types of hazards. In figure 7 late bifurcation and multiple states are used to identify hazards. In the fetch stage an instruction either stalls the pipe or not. In the ALU stage specific hazards are given to the instruction depending on whether the instruction stalled the pipe or not. Another approach to covering all the hazards is to require the test program generator to generate instances for each possible hazard that conforms to a hazard constraint in the task. This can most easily be done by duplicating the task generated from the model and replacing the hazard constraint with each possible hazard that fits the constraint.

The second example deals with modeling the mechanism for implementing precise interrupts in a superscalar processor. When an interrupt occurs the affects of the execution of some instructions might need to be reversed in order to implement precise interrupts. Because instructions may be dispatched out of order (i.e. in an order other than the program order), and may also modify the processor state out of order the reverse

mechanism is one of the more complicated, and bug prone aspects of a superscalar processor. In PPC 601 the different behaviors of the reverse mechanism depend mostly on the number of instructions that need to be undone and which pipe these instructions were executed in. Instructions in each pipe in PPC 601 have to be in program order. Therefore, the number of instructions that would need to be reversed on an interrupt depends on the relative position in the program order of the instructions that are being executed in each pipe. Keeping track of the exact program order of instructions being executed in parallel pipes requires a huge number of state variables and would make the model size unfeasible

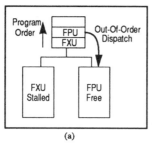
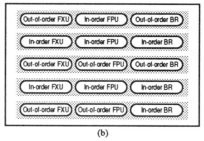

(a)   (b)

Figure 8. (a) Out of Order Dispatch (b) Example of a History Frame of Out-of-Order Dispatches

(i.e. adding a 4 bit counter to each stage in each pipe etc.). In order to save state variables, we exploit the fact that the modeling of the reverse mechanism only exists to guide the test generation towards probing the actual implementation of the mechanism, and does not have to behave exactly as the real implementation. We show a solution for which coverage of the model *implies* coverage of the implementation, but the model is much simpler than the actual implementation. A necessary condition for one pipe to run ahead of another in program order is that instructions are executed out of order, or instruction are dispatched out of order. Out of order dispatch occurs when there is an instruction in the instruction buffer waiting to be dispatched to a stalled pipe, and another instruction farther back in the program order is dispatched to another pipe which is free to accept new instructions. The above condition is easy to check by looking at the state of the model of the instruction buffer and of the stall machine of each pipeline. Instead of modeling the reverse mechanism directly, we model the history of the out of order dispatches in a window of time. If the window is large enough, coverage of all possible combinations of out of order dispatch histories implies coverage of the various states of the reverse mechanism. The out of order dispatch history is an approximation of the real state of the pipelines. For example we do not keep track of how long each instruction actually stays in the pipe, or in what order they modify the processor state. A processor model containing an out of order history module described above would produce tours and instruction sequences that cause the stalls needed to produce each history snapshot. This examples shows the power of coupling constraint generation via FSM and random architectural test generators.

5 Conclusions

The replacement of the target of the process of [17] from test vectors to verification tasks is a central issue that affects the method in two areas:

1. *The character of the model changes.* The model is derived from considerations on the implementation, but is not necessarily an abstraction of the real implementation. The basic demand on the model is that there should be a mapping between architectural tests and tours on the model. The theory shows that the crux of the model is consistency which derives low Mazeness constraints from tours on the FSM. We have identified the important concepts of persistency and late bifurcation and elaborated on their influence on the modeling process. We have identified a number of techniques for reducing model size while retaining enough consistency to allow low Mazeness constraints:

   - Do not include in the coverage criteria transitions that are only in the model to reflect propagation of information.
   - Direct CCTG away from the inconsistent tours.
   - Model enough consistency to make test generation possible with high probability.

- Force consistency on the FSM using temporal logic statements.
- Generalized tours.
- Allow inconsistencies.

2. *The character of the tests changes.* The tests produced by the method are true architectural test programs. No simulation environment control is required to run the tests. (This is an important point when using modern emulation methods) Architectural test program generators are used to explore the architectural dimension of each verification task. The quality of this process is justified by an analog to partition testing which has been thoroughly researched in the domain of software testing.

The model construction method for processors suggested in [17] is abstraction of the implementation. Such a model is a simplified version of the implementation for which there is a one-to-one mapping to the real implementation. This is the strength of this method because there is a simple correspondence between the behavior of the model and the behavior of the implementation. The main fault of this method is the size of resulting models. The simple demand that there be a realizable inverse mapping from the model to architectural tests is a much weaker constraint than what is required of an abstraction of the implementation. As a result, the above size reduction techniques can be used while abstracting from the high level design. The resulting model is therefore much smaller.

We have shown the viability of the claims in Table 1, "Our method vs. Current Methods," . Our method automates a large portion of the current methods. Complex interactions that are likely to be overlooked are generated automatically as in [17]. Thus, the method raises the probability of uncovering the so called "integration bugs" that are notoriously hard to find. The applicability of the method is limited for specific unit verification since the whole processor must be modeled to obtain acceptable constraints. The internal of each unit are abstracted and the model concentrates on the interaction of the units.

Existing methods suffer from a lack of formalism. The completeness of a set of tasks which is manually defined may only be qualitatively assessed. Our method provides a framework in which coverage can be used for assessing the completeness of a set of verification tasks. In [24], we show how these techniques apply to a real, superscalar processor with a split L1 cache which implements a number of advanced features such as branch prediction and precise interrupts.

Given the comparisons above, we believe that this new methodology should be combined with the existing methodology for implementation verification. This combination is a win-win situation: Automation of a relatively large portion of the existing methodology frees designer time for development of complex, deeply probing test cases. In addition, the new methodology covers previously overlooked corner cases. This combination will provide better verification at lower cost.

We are currently working on maturing the methodology and applying it within a complete verification of a real processor. Further research may focus on how the model may be more easily built. Existing or new microarchitecture description languages may be used during the design phase to describe a high level abstraction of each processor element. The guidelines such as late bifurcation and consistency requirements may be programmed into a translator from a micro-architecture description language into a CFSM model for use in the process. Design methodologies may be changed to incorporate description of the model in such a language during the actual design process.

Acknowledgments

We would like to thank Yossi Malka for initiating this work, Aharon Aharon for his valuable advise and Danny Geist for his helpful comments on this paper.

References

1. A. Aharon, A. Bar-David, B. Dorfman, E. Gofman, M. Leibowitz, V. Shwartzbund "Verification of the IBM RISC System/6000 by a Dynamic Biased Pseudo-Random Test Program Generator", In *IBM Systems Journal*, April 1991

2. B. Beizer, "The Pentium Bug, an Industry Watershed", *Testing Techniques Newsletter On-Line Edition*, September 1995

3. A. Aharon, D. Goodman, M. Levinger, Y Lichtenstein, Y. Malka, C. Metzger, M. Molco, G. Shurek "Test Program Generation for Functional Verification of PowerPC Processors in IBM", In proceeding of *ACM/ IEEE Design Automation Conference* 1995

4. Y. Lichtenstein, Y. Malka, A. Aharon "Model Based Test Generation for Processor Design Verification", *In Innovative Applications of Artificial Intelligence (IAAI)* AAAI Press 1994

5. Ahi A. M., Burroughs G.D., Gore A.B., LaMar S.W., Lin C.R., Wieman A.L "Design Verification of the HP9000 Series 7000 pa-risc Workstations", *Hewlett-Packard-Journal* num. 8 vol. 14 August 1992

6. A. Chandra, V. Iyengar, D. Jameson, R. Jawalker, I. Nair, B. Rosen, M. Mullen, J. Yoor, R. Armoni, D. Geist, Y. Wolfstal "AVPGEN - A Test Case Generator for Architecture Verification", *IEEE Transactions on VLSI Systems* 6(6) June 1995

7. D. Geist, M. Farkas, A. Landver, Y. Lichtenstein, S. Ur, Y. Wolfsthal "Coverage Directed Generation Using Symbolic Techniques", *FMCAD 96*

8. E. J. Weyuker, B. Jeng "Analyzing Partition Testing Strategies", *IEEE Transactions on Software Engineering* vol. 17 no. 7 July 1991

9. E. J. Weyuker, T.J. Ostrand "Theories of Program Testing and the Application of Revealing Subdomains" *IEEE Transactions on Software Engineering* vol. 6 no 3 May 1980

10. Y. Abarbanel, Y. Lichtenstein, Y. Malka, S. Ur "Coverage Driven Processor Bug Classification" Submitted to *ACM/IEEE Design Automation Conference* 1996

11. G. J. Holtzman, *"Design and Validation of Computer Protocols"*, Prentice Hall, Englewood Cliffs, NJ 1991

12. K.L McMillan *"Symbolic Model Checking"* Kluwer Academic Press, Norwell MA 1993

13. K.L McMillan *"The SMV System DRAFT"*, Carnegie Mellon University, Pittsburgh PA 1992

14. A.K. Chandra, V.S. Iyengar, R.V. Jawalekar, M.P. Mullen, I. Nair, B.K. Rosen "Architectural Verification of Processors Using Symbolic Instruction Graphs", In *Proceedings of the International Conference on Computer Design*, October 1994

15. D. Lewin, L. Fournier, M. Levinger, E. Roytman, G. Shurek "Constraint Satisfaction for Test Program Generation", *IEEE International Phoenix Conference on Communication and Computers*, 1995

16. A.K. Chandra, V.S. Iyengar "Constraint Solving for Test Case Generation", In *Proceedings of ICCD-92*, Cambridge Mass, 1992

17. R. C. Ho, C. Han Yang, M. A. Horowitz, D. L. Dill "Architecture Validation for Processors" In *ACM ISCA* 1995

18. H. Iwashita, S. Kowatari, T. Nakata, F. Hirose "Automatic Test Program Generation for Pipelined Processors", In *Proceedings of the International Conference on Computer Aided Design*, November 1994

19. D. L. Beatty, R. E. Bryant "Formally Verifying a Microprocessor Using a Simulation Methodology", In *Proceedings of the ACM/IEEE Design Automation Conference* 1994

20. T. A. Diep, J. P. Shen "Systematic Validation of Pipeline Interlock for Superscalar Microarchitectures" In *Proceedings of the 25'th Annual International Symposium on Fault Tolerance*, June 1995

21. H. Iwashita, T. Nakata, F. Hirose "Integrated Design and Test Assistance for Pipeline Controllers", IEICE Transactions Information Systems (Japan) Vol.E76-D, No. 7, July 1993

22. C. May, E. Silha, R. Simpson, H. Warren editors *"The PowerPC Architecture"*, Morgan Kaufmann, 1994

23. S. Weiss, J. E. Smith *"POWER and PowerPC"*, Morgan Kaufmann, 1994

24. D. Lewin, D. Lorenz, S. Ur "A Processor Implementation Verification Methodology", IBM Unpublished Document

Coverage-Directed Test Generation Using Symbolic Techniques

Daniel Geist, Monica Farkas, Avner Landver, Yossi Lichtenstein,
Shmuel Ur and Yaron Wolfsthal

IBM Science and Technology, Haifa Research Lab, Haifa Israel

Abstract. In this paper, we present a verification methodology that integrates formal verification techniques with verification by simulation, thereby providing means for generating simulation test suites that ensure coverage. We derive the test suites by means of BDD-based symbolic techniques for describing and traversing the implementation state space. In our approach, we provide a high-level of control over the generated test suites; a powerful abstraction mechanism directs the generation procedure to specific areas, that are the focus for verification, thereby withstanding the state explosion problem. The abstraction is achieved by partitioning the implementation state variables into categories of interest. We also depart from the traditional graph-algorithmic model for conformance testing; instead, using temporal logic assertions, we can generate a test suite where the set of state sequences (paths) satisfies some temporal properties as well as guaranteeing transition coverage. Our methodology has been successfully applied to the generation of test suites for IBM PowerPC and AS/400 systems.

1. Introduction

The goal of hardware design verification is to ensure the equivalence of a hardware design (*implementation*) to its architectural specification. Most commonly, verification methodologies pursue verification by simulation. Under this paradigm, the design is simulated against a set of tests in an environment that models the actual hardware system. Simulation results are then validated by comparison to expected results [1,11].

In practice, verification by simulation is carried out by running a relatively small subset of selected tests, since the test space is typically very large. One of the critical aspects of verification by simulation is the definition of appropriate measurable criteria for the quality of the test set in terms of *coverage* of the design. Coverage is a measure of completeness of sets of tests. Coarse coverage measures that are used in practice, with limited success, include specification based coverage [5], program based coverage [3], measures based on statistical quality control, and others.

In recent years, the progress made in formal verification technology [4] has made this technology a viable means for complementing verification by simulation, and specifically for ensuring a high level of coverage. Using formal verification methods, e.g. symbolic model checking [21], one can verify that a hardware unit will function correctly in all possible execution sequences, with all combinations of outputs. A major limitation of formal verification, however, is the size of the hardware it can verify, due to the complexity of the state space that is searched. In spite of this limitation, formal verification is today a part of the verification methodology of many organizations, where it is applied to safety-critical portions of many designs projects.

In order to apply the benefits of formal verification to bigger designs a hybrid approach is required. A possible such approach is presented in this paper; a methodology that integrates formal verification techniques with simulation, thereby providing means for generating simulation test suites that ensure coverage. The methodology draws on techniques from protocol conformance testing [16]. As shown by [15], conformance testing techniques, specifically *transition coverage* (the coverage of all state machine transitions), are applicable to hardware design verification. In our approach, as opposed to traditional conformance testing and its application to design verification, we provide a high-level of control over the generated test suites, in two key ways that are facilitated by virtue of integrating the formal verification techniques into our framework. First, we provide a powerful abstraction mechanism by which the generation procedure can be applied to specific parts of the implementation that are the focus for verification, thereby withstanding the state explosion problem inherent in generating conformance testing for the full model. The abstraction is achieved by partitioning the implementation state variables into categories of interest. This abstraction mechanism is generic and model independent, as opposed to the method of [17] where model specific knowledge was applied to limit the test suite to tests that exercise certain pipeline hazards. Second, we depart from the traditional graph-algorithmic model for conformance testing [10, 7, 9, 15]; instead, using temporal logic assertions, we can generate a test suite where the set of state sequences (paths) satisfies some temporal properties as well as guaranteeing transition coverage. Furthermore, we derive the conformance test suites by means of BDD-based symbolic techniques [8, 21] for describing the design model and traversing the design state space. This allows the handling of large models.

Our experience with this methodology has shown this approach to be of high value in the verification of complex systems. Specifically, this methodology has been successfully applied to the generation of test suites for PowerPC systems and AS/400 systems in IBM, where the targets for coverage were the multi-processor exclusive access mechanisms (semaphores) and cache hierarchies.

The rest of this paper is organized as follows. Section 2 defines terms used throughout the paper. In Section 3, we describe the methodology, which is then demonstrated by means of a simple example (sample PowerPC system) in Section 4. Section 5 contains results from our experience. A summary and conclusions are presented in Section 6.

2. Background and Definitions

First we define *coverage,* which is a measure of completeness of sets of tests. Given a set S, a *coverage model* of S is a set C of subsets of S. Each element of C is then called a *coverage task*. Given a coverage task $CT \in C$, we say that every element in CT *satisfies* that task. Given a set $F \subseteq S$, we say that F *covers* S with respect to the coverage model C if for every non-empty element $CT \in C$, $(CT \cap F) \neq \emptyset$. We also say that F provides *coverage* for S.

For example, a common software coverage model is *statement coverage*. The set to cover is all possible program executions. All executions that pass through a specific program statement comprise one coverage task. The coverage model in conformance testing is the transition coverage of the formal specification (i.e. all executions that pass through a specific transition comprise one task).

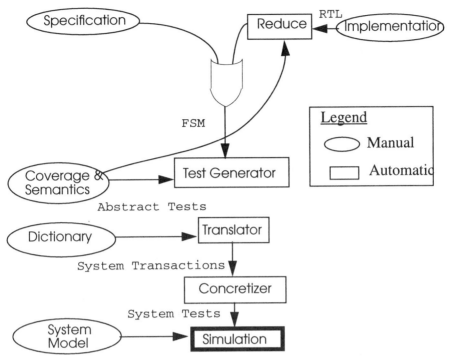

Fig. 1. The test generation process

A *Finite State Machine* (FSM) is a Kripke structure [21]. The FSM states in this paper are represented via binary state encoding (i.e. $V = \{ (v_1, ..., v_n) \mid \forall i \; v_i \in \{0, 1\} \}$ for some integer n). Each v_i is a *state variable*.

A state *transition* is a an FSM state pair (v, v') . *Transition difference* is defined via the usual vector inequality. That is, two transitions (v, v') and (w, w') are said to be *different* iff $\exists i$ such that $v_i \neq w_i$ or $v_i' \neq w_i'$. A transition is *valid* for a given FSM, if the FSM has an execution path that includes it.

Covering an FSM means covering the set of all tests for some coverage model. A *test* is just an execution path. The coverage model targeted in this paper is *transition coverage*. That is, one test is generated for each transition (v, v') that can occur in the FSM.

3. Methodology

The FSM-based system verification methodology, presented here, consists of the following steps (see Figure 1).

Building the FSM Model. The user creates an FSM representation of the architecture specification, or alternatively the implementation is translated into an FSM representation. In the first case (specification modeling), the user identifies the design functions to test, and defines a specification FSM that abstractly performs those functions or alternatively isolates the parts of the architecture that control those functions. In the second case (implementation modeling), the translation may also reduce the imple-

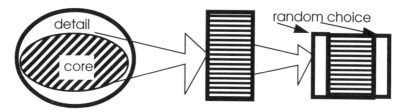

Fig. 2. Test concretizing

mentation to remove any details not relevant to the desired functional properties (e.g., parts of the implementation that deal with other functions that are not targeted). The reduction is performed by manually isolating the core design, and removing parts of the design that are not relevant.

Defining a Coverage Model The state variables underlying the system are ranked so as to specify their desired respective importance within the system under verification. The resulting core of the system retains the main functions, while the details are left out. Essentially, the ranking of state variable lets the user shape the coverage model; this is an essential quality and strength of our verification methodology.

Semantic Control path constraints on the generated tests can be defined by a temporal logic assertion. The coverage test generator described in Section 3.2, will produce a set of execution sequences that provides transition coverage while obeying the assertion. Typically, any assertion defined and used under this scheme will specify not only the transition of interest but also the initial state. Further, using assertions driven by the semantics of the FSM model, it is possible to generate execution sequences that provide coverage for some system architecture model. As an example, in an FSM-based PCI system [22], it would be possible to not only generate a transition coverage, but also to generate sequences with various PCI transaction types.

Generating Tests. A test generated by coverage test generator is *abstract* since the FSM description it was generated from contains only partial information of the Implementation Under Test (IUT). In order to convert abstract tests into concrete ones for simulation, it is necessary to add information for inputs that were abstracted out from the model (see Figure 2). This procedure is performed in two steps.

1. First, a translator is used, which, using a pre-defined dictionary, transforms the set of abstract tests into a set of lower-level tests which represent (in a higher level of detail) the exercising of the functional operations in the large design as implied by the abstract tests (Section 3.3). This processes leaves many operation parameters undefined.

2. A concretizer, driven by a full system model, transforms the test suite output by the translator into a test suite that is acceptable by the simulation environment of the IUT. In the process, the concretizer randomly generates any missing information (such as specific data values) that was not in the abstract model of the IUT. This is depicted in Figure 2. The resulting test suite has the advantage of being exhaustive on the main functions identified when building the FSM model.

3.1 Defining a Coverage Model

In practice, one typically seeks to define coverage models coarser than transition coverage. For example, most reactive systems [4] operate under the assumption of some predefined environment. It is then necessary to model the environment as one or more FSMs that provide legal input to our model. This environment becomes part of the input to the coverage test generator but we do not wish our coverage models to be dependent on the environment FSMs, because they only represent the legal external stimulus to our model. Therefore we need to be able to direct the test suite generator to ignore the environment FSMs, since covering them will result in an unnecessary large amount of tests.

Test generation is controlled by a partitioning of the FSM variables into three sets:

- **Coverage Model set** - Any two transitions that differ on the value of one of the variables belonging to the Coverage Model set are considered distinct and require separate tests. Hereafter, we refer to this set as the Coverage set. If all variable are marked as coverage then our coverage model becomes transition coverage.
- **Ignore set** - Any two transitions that differ only on the values of variables from this set are considered equivalent and do not require separate tests. Marking all variables as ignore will generate a test suite containing one test (all tests are equivalent).
- **Care set** - Care set variables are variables that we wish to consider in some but not all cases. Two transitions that only differ on the Care set variables will not necessarily require different tests. The decision whether to generate different tests should really be taken by analyzing the logic driving the FSM and identifying the states where Care set variables influence the change in state.

Intuitively, the Care set variables represent different reasons for executing a transition in the coverage set. We want to generate a test for each reason. For example, if an exception can happen because of two different reasons (e.g. divide error and storage error) we may want to generate a test for each reason.

Formally, the above partitioning separates the indices $\{1, ..., n\}$ into 3 sets. Assume, without loss of generality, that the Coverage set, Care set and Ignore set are $\{1, ..., k\}$, $\{k+1, ..., l\}$, $\{l+1, ..., n\}$ respectively, (i. e. the variables are split in order).

A state can be viewed as (X, Y, Z) where:

$$(X) = \left(v_1, ..., v_k \right), \ (Y) = \left(v_{k+1}, ..., v_l \right) \text{ and } (Z) = \left(v_{l+1}, ..., v_n \right).$$

Given two transitions $t_1 = ((X_1, Y_1, Z_1), (X_2, Y_2, Z_2))$ and $t_2 = ((X_3, Y_3, Z_3), (X_4, Y_4, Z_4))$ then the following three rules direct the coverage test suite generation:

1. If $X_1 \neq X_3$ or $X_2 \neq X_4$, then t_1 and t_2 require two different tests.
2. If $X_1 = X_3$ and $X_2 = X_4$ and $Y_1 = Y_3$ and $Y_2 = Y_4$, then t_1 and t_2 do not require different tests.

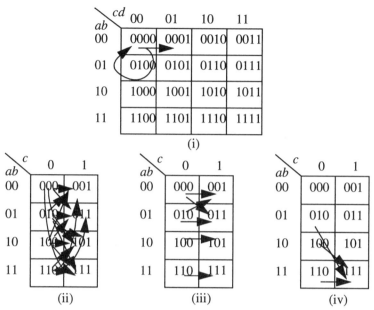

Fig. 3. A care set implementation

3. If $X_1 = X_3$ and $X_2 = X_4$, but $Y_1 \neq Y_3$ or $Y_2 \neq Y_4$, then t_1 and t_2 may require different tests depending on specific implementation of the Care set.

A Heuristic for Processing Care Set Variables

It is not always simple to determine which state transitions are specifically influenced by the Care set variables, and therefore require full coverage. We have developed a simple heuristic, presented below, that determines some of the cases where there is no influence. In other cases, Care set variables are treated like Coverage set variables.

Let $t_1 = ((X_1, Y_1, Z_1), (X_2, Y_2, Z_2))$ and $t_2 = ((X_1, Y_3, Z_3), (X_2, Y_4, Z_4))$ be two transitions that have the same coverage set values. Let $t_1' = ((X_1, Y_1), (X_2, Y_2))$ and $t_2' = ((X_1, Y_3), (X_2, Y_4))$ be the respective projections of t_1 and t_2 on the Coverage and Care variables. For an m-dimensional boolean space, B^m, a *cube* is a subspace $C_1 \times C_2 \times \ldots \times C_m$ in B^m where $C_i \subseteq \{0, 1\}$. In our heuristic, we only need to generate one test from the two transitions, t_1 and t_2, if there exists a cube C in the space of projected state transitions, such that $t_1', t_2' \in C$ and $\forall t' \in C$, t' is a projection of a valid transition. Intuitively, if any variable values on which the two transitions differ are possible, then those variables do not influence the set of transitions in the cube.

Figure 3 illustrates this heuristic. In Figure 3 there are four state variables a,b,c,d. Variables a,b belong to the Care set, c belongs to the Coverage set and d is in the Ignore set. Figure 3 (i) explains our notation: Each cell in the table is a possible state in the state machines. An arrow going from one cell to another, specifies that a valid transition exists in between the respective states. In (i) there are two arrows depicted for transition, (0000,0000) and (0000,0001). However, since d is an Ignore variable, these transitions are equivalent in the projected state.

Figure 3 (ii)-(iv) depict three example FSMs on variables a, b, c, d (with d removed). For clarity of the exposition, we only have transitions where c changes from 0 to 1. In the first example FSM (ii) we have a full set of transition from the left column to the right one. The cube (**0, **1) is all valid and any transition in this cube will satisfy coverage of the rest. In the second example FSM (iii) the minimal coverage which is also the minimum coverage requires 3 transitions: (100,101), (110, 111) and any one from the cube (0*0, 0*1). In the last example FSM (iv), we need a candidate from the cube (1*0, 111), and also a candidate from the cube (*10,111). Notice that is sufficient to choose (110, 111) in order to cover both cubes.

3.2 Test Generation

Given a temporal assertion, the test generator produces a test suite which provides coverage while ensuring that each test obeys that assertion.

Semantic Control

Model checking [21] is a formal verification technique which verifies that the model of the design satisfies an assertion (property) specified in temporal logic. An assertion serves as a query to the model checker. When the assertion is false, the model checker will answer the query by presenting a counter-example that contradicts the assertion. A counter example is a sequence of state transitions, including all changes of inputs, of a valid execution path in the model that contradicts the assertion. We use this counter example mechanism for test generation.

For simplicity, we first describe how a model checker can be utilized to generate a test that covers a specific transition. The model checker is queried with an assertion that claims that no such transition exists in the model. The inputs that drive the model along this execution path, taken from the counter example, are actually a test that covers the transition.

Specifically, let v and v' be two states in the model, then the assertion

$$(state\,(v) \rightarrow next\,(\neg state\,(v')))$$ (1)

declares that "if the model is in state v then the next state cannot be v' ". If a transition from v to v' exists then a counter-example (test) containing this transition will be generated.

In some cases, a requirement for a test will be that the final state will be in some predetermined set where it is possible to check results. Usually this is one of the initial states. Let the set of Initial states be $INIT$. The previous assertion can be changed to:

$$(state\,(v) \rightarrow next\,(state\,(v') \rightarrow \neg \exists eventually\,(state\,(INIT))))$$ (2)

This means that, after a transition from v to v' the model cannot eventually reach $INIT$. A test generated from this assertion will be one that contains the required transition and returns to an initial state.

The two assertions (1) and (2) are just examples of how a user can direct the path of a test that covers the transition. In general, It is possible to use temporal logic for specifying very complex conditions on the generated test. One can specify that the test must pass in specific transitions before or after the passing through the covered transition.

Given an FSM, it is possible to pair all possible states (all 2^n binary values where n is the number of state variables). However the number of states is exponential in n and the number of pairs is a square of the states. Only a small fraction of those will be valid model transitions and the time it will take to compute which ones are valid will make this method useless (even if checking wether a transition is valid could be done in $O(1)$). Therefore it is necessary to direct the model checker only to handle transitions which are known to be valid. A method for doing so is described below. In practice the model checking phase can be skipped since the answer is known in advance. Instead the assertion can be directly submitted to the counter example generation mechanism.

The Test Generator

The test generator keeps an internal representation of the model FSM as a transition relation. The transition relation, T is simply the set of transitions, (v, v') that are valid in the model FSM state space. The algorithm uses this representation as follows:

1. Create the transition relation T.
2. $T' = T$.
3. While ($T' \neq \emptyset$)

   A. choose a transition $(v, v') \in T'$.

   B. Generate an assertion claiming that $(v, v') \notin T$.

   C. Apply model checking to generate a counter-example for the claim, that includes transition (v, v').

   D. $T' = T' - \{ (v, v') \}$.

Note that in practice, since the assertion in step 3.B is false, we can skip model checking in step 3.C and use the counter example mechanism directly.

To support the coverage models described in Section 3.1, T' is created and maintained using the following procedure:

1. All the legal transition of the FSM are stored in a BDD.
2. All Ignore set variables are removed from the BDD. Wherever there is a branch on an Ignore set variable, its two descendents are joined with an or.
3. While there are still transitions not covered
   - A transition is chosen
   - A test that covers this transition is generated.
   - All transitions that are covered by the chosen transition (using the variable hierarchy) are removed from the BDD.

Step 3 does not necessarily produce a minimum number of tests. Finding the minimal transition coverage is *NP-Complete*. The easiest reduction is to *subset-sum* [14] which can be done by putting all the variables into the Care category.

Test Generation Efficiency

Since the counter example mechanism [12] is heavily used during test generation, we have improved it in order to make test generation faster. The counter example generation works in such a way that assertions containing "fixed point" temporal operators (*until, eventually, forever*) have subformulas evaluated many times. Each fixed point operator adds one more evaluation repetition of the rest of the assertion subformula that follows it. Thus the counter example generation may have square complexity in the size of the assertion.

To reduce the complexity we added a hash table to preserve computations of sub-formulas. However, we really benefited from subformula preservation by saving the subformula $eventually\,(state\,(INIT))$ in Equation (2) which was in all the assertions we generated. This subformula was rather expensive to compute, and by preserving it we reduced test generation time from hours to a few minutes (see Section 5).

3.3 Converting Abstract Tests to Concrete Tests

A counter-example (test) gives a sequence of states and variable values. By giving these transition meaningful names, the counter-example becomes a scenario for the system behavior, which is an abstract test. The test is abstract since the FSM description it was generated from contains only partial information of the IUT. In order to convert this abstract test to a concrete test for simulation, it is necessary to add information for inputs that were abstracted out from the model.

Translation

Conversion to a concrete test requires, as a first step, *translation* of abstract test cases into ones written in the language supported by the target simulation environment. As most contemporary simulation environments include test loaders that accept tests in a language containing high level directives, test translation should support the full richness of the target test language. For example, simulation environments for CPUs will accept as tests programs written in assembler or machine language code. Simulation environments for ASICs may have test cases written in terms of bus transactions.

The translation process views a high level operation as a long sequence of low level state transitions that drive the IUT along the high level test execution path. This is done by means of pattern matching, as in [6]. Essentially, a *parser* follows the state transitions and generates a high level operation (e.g. transaction) whenever it recognizes a pattern that indicates that such a transaction is taking place. The sequence of high level operations thus recognized is a skeleton for concrete test.

More specifically, translation is implemented by means of a parser that is defined as follows:

1. A parser state: *Parserstate*
2. The parser transitions table in the form:

$$((Parserstate = S1) \wedge (CFSMstate = X1)) \Rightarrow next\,(Parserstate) \leftarrow S2 \qquad (3)$$

3. Values for the output variables

$$((Parserstate = S1) \wedge (CFSMstate = X1)) \Rightarrow OUTPUT\,(transaction) \qquad (4)$$

In Section 4 we give an example of such a translation.

Removing redundancy. We note that during the process of translating the tests, some of the tests may become equivalent to others or redundant for the following reasons:

1. Some of the transitions are clearly not interesting. For example transitions where the FSM performs wait states or transitions immediately after the initial state.
2. The translation may decide to ignore wait states.
3. Some transitions are part of the model but are not targeted for testing.

In order to remove redundant tests, the translator can check for interesting events (whatever the user defines) in the test and discard any tests that do not contain such events. Alternately, the translator can have a post-process that discards duplicate tests that become equivalent to other tests (after translation).

Concretization

A high level test is presented to a random biased test generator [1,11,13] which "fills in the blanks" by selecting values such as addresses, data values, transaction lengths, etc. The resulting concrete test can be run on hardware simulation and checked for results. We call this application a *Concretizer*.

Biasing. Most of the work in generating biased random tests for large systems is dedicated to creating smart biasing [1]. The problem is to produce biasing which will make it likely that the tests will exercise large parts of the system. Totally random biasing is unlikely to achieve this goal with any reasonable resources. For instance if a register can get any value between 0 to 2^{32} it is very unlikely that in any of the tests it will get the value zero. It is important that this specific value be given in a high percentage of the tests as the behavior of the system usually treats it as a special value.

A random test bias generator can be created, in the following way:

- Abstract tests are created for the system
- Tests are chosen randomly from the abstract test and then are passed through the concretizer which creates concrete tests.

If the same abstract test is passed twice through the concretizer two different concrete tests will be created. This random test bias generator is used to produce massive amount of tests. It is possible that we would like to rank the abstract test so as to indicate the number of concrete tests to create from each. A ranking could be based, for instance, on the number of unique transitions, i.e. transitions that exist in no other test. All abstract tests chosen have at least one (as otherwise they will be redundant). A test with two unique transition might appear twice as many times as a test with one unique transition. The result of such biasing will produce a more even spread in the concrete tests of the transitions that exist in the FSM.

4. Example

Consider the PowerPC personal system with a PCI I/O bus system in Figure 4. For system of this size, conformance testing is impractical as the state space of the full system is very large - implying an enormous number of tests in the conformance test suite.

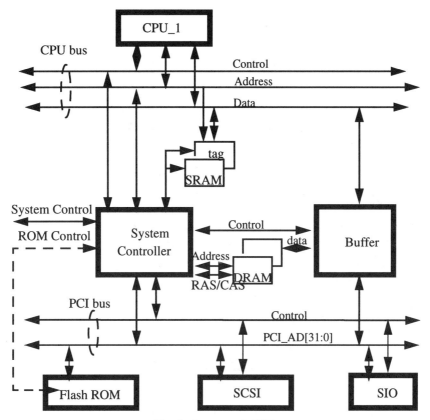

Fig. 4. An example IUT

Our solution is to create a FSM model which abstracts some parts and leaves these that are of interest, and then to generate a conformance test suite for the reduced FSM.

As a first step, a set of state variables that describe the main functions of the system is identified. Assume the goal is to verify the interaction of units this system is composed of, ignoring all operations that are internal to the units themselves. A good approach to covering all the interactions is to model the system controller. This controller determines what unit interactions are possible on this system. The controller contains the information on what interaction types are legal on the system (whatever is not supported by the controller is irrelevant) and also on the kind of concurrency that is allowed. Covering the controller, therefore, covers all system interactions.

The controller for the system in Figure 4 has the following core FSMs that govern the its behavior (Figure 5):

- when acting as a slave on the CPU bus (FSMs 1 and 2).
- when acting as a master on the CPU bus (FSM 5).
- when acting as a slave on the I/O bus (FSM 4).
- when acting as a master on the I/O bus (FSM 3).

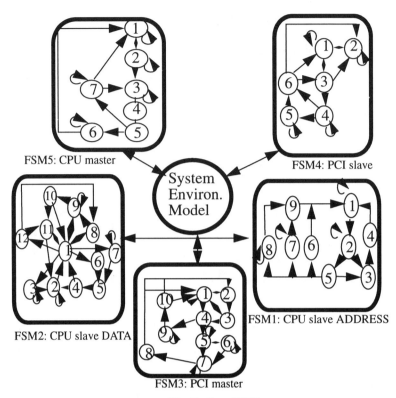

Fig. 5. Core FSMs

We identify the state variables that control the 5 FSMs as the Coverage set, and we insert the rest of the variables into the Ignore set. After generating the test suite, a test on the composite of these FSMs translates to a system level test. One or more of these FSMs has to be involved, whenever an interaction takes place between units on the system. The resulting test suite is exhaustive with respect to the core, assuring that all main functions (unit interactions) are tested. These abstract tests are then translated into concrete tests by "filling in the blanks" heuristically. The resulting test suite has the advantage of being exhaustive on the main functions.

The AWK language [2] is used to write a translator. Following is sample code:

```
$3=="pci_wait"
{print "CPU to PCI address tenure starts";var_sa=1;}
($3=="idle")&&($1=="CPU_sa_state")
 {if(var_sa==1) print "CPU to PCI address tenure ends";}
```

Here, **var_sa** is a parser state variable and **pci_wait** and **idle** describe internal states in the FSM number 4 from Figure 5.

A resulting translated test (at the transaction level) for the described system would then be for example:

```
CPU to memory address tenure
CPU to memory data tenure starts
CPU to PCI address tenure
CPU to PCI data tenure starts immediately after CPU to
memory data tenure
CPU to PCI data tenure ends
PCI to PCI starts
PCI to PCI ends
```

5. Experience

The motivation of this work came from experience in formal verification and simulation based verification. The symbolic simulation engine used was the one implemented in SMV [20]. We enhanced the SMV user control as follows:

1. Two input files that contained the list of Coverage variables and the list of Ignore variables were added.
2. Three keywords, TEST, TRANSITION and INIT_ST, were added to the SMV language in order to allow test generation assertion. For example, assertion (1) is specified as

$$TEST\,(TRANSITION) \tag{5}$$

and assertion (2) is specified as

$$TEST\,(TRANSITION\,(INIT)) \tag{6}$$

The system builds a transition relation of a model and then uses the algorithm procedures described in Section 3.2 to generate a test suite. An existential quantifier operator [21] is used to remove dependencies of the Ignore variables from the transition relation. The assertions are generated automatically for each transition and submitted to the counter-example generation procedure.

We have applied the methodology to the following complex system-level examples:

Example 1 (PowerPC 604): We used the methodology for the actual system described in Section 4. The FSM was derived manually from the implementation of the system controller. Parts of code of the controller (the core FSMs in Figure 5) were copied verbatim from the actual implementation. The communication between those FSMs was abstracted manually. The resulting FSM was relatively small. In all, 157 tests were generated. At the time we did not have a test concretizer for the system so they were never later simulated. A bug discovered almost a year later occurred in a situation which was in one of these abstract generated tests. Additionally, the process of generating tests in itself discovered a deadlock in the design: assertion (6) which requires that after going through a transition the test has to return to the initial state was used, but could not be satisfied due to the deadlock. Interestingly, this bug was not encountered earlier in spite of thorough simulation.

Example 2 (AS/400): The methodology was applied to a system containing two PowerPC processors that are using shared memory. The verification target was the PowerPC semaphore implementation. A semaphore is implemented in PowerPC by two machine instructions LWARX and STWCX [19]. A processor requests a reservation by executing a LWARX (Load and Reserve), which is a load reserved, he confirms the reservation by executing STWCX (store conditional) to the same address. However, the processor which performs the STWCX first wins, that is all other reservations are cancelled. The other processors STWCX fails (does not happen) and he needs to reissue a LWARX. Note that a successful STWCX usually stores a value that indicates to the other processors that they cannot perform a STWCX until that value is reset by the successful processor.

The goal was to cover all LWARX-STWCX sequences. Also, both processors were employing a writeback MESI [19] cache coherency protocol. A processor can keep a cache line in its cache in four states: Modified, Exclusive, Shared or Invalid. Depending on its cache state, the processor accesses main memory or uses the data in its cache. The processor also needs to answer queries (snoops) by other processors on its cache state. If a particular line is reserved then it becomes a difficult task to track reservation cancellations and the resulting state machine becomes complex. The coverage variables in the FSM model were therefore the processor reservation bit and the MESI state bits. This model plus coverage models resulted in about 400 tests. The model was not copied from any implementation but, was written according to the PowerPC specification. The added benefit of using SMV as platform for writing the model is that we could also formally verify it for cache coherency and correct reservation procedure. We used a test generator called SysGen [13] as a concretizer. An AWK script was written to translate the test cases to a form that SysGen could accept. SysGen accepted the test cases from the model as scenarios and randomly chose real addresses and data for them to build a full test. These tests were later run in hardware simulation.

Table 5 contains the results from both examples. The table indicates that while a conformance test suite for the entire FSM model would be huge, the definition of proper coverage models allows us to handle very big models while focusing the test case generation and reducing the tests required significantly. While the examples focus on system-level verification, the methodology and tool are in fact applicable to any verification domain where the IUT can be specified as a FSM.

Table 5:

Name	No. of state vars	No. of Coverage vars	Number of Ignore vars	Reachable space size	No. of Tests	Run Time
Ex. 1	26	10	16	188416	173	153s
Ex. 2	30	5	21	4.05504e+06	412	51s

6. Conclusion and Future Work

The limitations of current coverage estimation tools drives verification by simulation to rely on massive simulations without an inherent way to drive the test generation process by coverage considerations. In this paper, we presented a verification methodology that integrates formal verification techniques into verification by simulation, thereby providing means for generating simulation test suites that ensure coverage.

The methodology is based on symbolic techniques, and supports semantic control over the generation process unlike earlier graph-algorithmic approaches to conformance testing (e.g. [16, 15, 9]). Symbolic methods are slower for exploring single execution paths (such as in [15]) but they enable handling of models with much larger complexity. In practice we have not found the speed of symbolic exploration to be a bottleneck. Additionally, the methodology involves a scheme for coverage model specification so as to withstand state-space explosion problems and focus on the most important targets for verification. To support the methodology, a transition coverage generator was implemented, based on SMV [20]. The tool was used on two different real systems.

Currently we are working on extending on the application of this work in two directions: First, we are investigating application of this methodology to verification of processor pipelines [18]. Additionally, we are working on applying this methodology for generating unit level tests. At the unit level it is easier to capture more details of the implementation, since the simulation environment controls at this level are finer. The unit we are working on is an I/O controller of an AS/400 PowerPC system. The model is bigger than the examples described above and test generation takes hours, but it is not a bottleneck when comparing to running these tests in simulation.

Another improvement we are considering is the concatenation of tests to one bigger test in order to reduce the simulation time caused by the setup conditions. There are also advantages to short tests since they are easier to debug and the traces of large tests (especially in system simulation) require huge amounts of disk space. The solution may be to allow the user to specify the number of transitions to cover per test.

In order to successfully embed this methodology in an industrial verification environment, we now work on infrastructure, robustness and usability aspects in several areas. Semantic control over generation is an active area of development. We are also working on providing a mechanism by which test concretizers would support complex synchronization directives. Efficient user interfaces are also required, since (as is depicted in Figure 1) the user has to provide complex inputs to the process in several points.

Another possible area for further research is to determine whether it is possible to take a real design and automatically (driven by the coverage models) reduce it to one acceptable by the transition coverage generator.

Acknowledgments. We would like to thank our partners George Gorman, Bob Kohn in IBM Rochester and Katherine Dunning in IBM Austin for their cooperation on this work. We would like to thank Raanan Gewirtzman and Tom Walker for initiating the activity. We would like to thank Aharon Aharon and Danny Lewin on their helpful reviews. Cindy Eisner suggested preservation of subformulas in a hash table.

REFERENCES

[1] A. Aharon, D. Goodman, M. Levinger, Y. Lichtenstein, Y. Malka, C. Metzger, M. Molcho, and G. Shurek. Test program generation for functional verification of powerpc processors in ibm. In *32nd Design Automation Conference, DAC 95*, pages 279–285, 1995.

[2] A. Aho, B. Kerningham, and P. Weinberger. *The AWK Programming Language*. Addison-Wesley, 1988.

[3] I. Beer, M. Dvir, Kozitza B., Y. Lictenstein, S. Mack, W.J. Nee, Rappaport E., Schmierer Q., and Y. Zandman. VHDL Test Coverage in a BDLS/AUSSIM Environment. Technical Report 88.342, IBM Science and Technology, Haifa, Israel, 1993.

[4] I. Beer, M. Yoeli, S. Ben-David, , R. Gewirtzman, and D. Geist. Methodology and System for Practical Formal Verification of Reactive Hardware. In *Computer Aided Verification*, pages 182–193, 1994.

[5] B. Beizer. *Software Testing Techniques*. Van Nostrand Reinhold, New York, 1990.

[6] A. Benoit and D. Luckham. Validating Discrete Event Simulations Using Event Pattern Mappings. In *ACM/IEEE Design Automation Conference*, pages 414–419, 1992.

[7] E. Brinksma. A theory for the derivation of tests. In S. Aggarwal and K Sabanni, editors, *Protocol Specification, Testing, and Verification, IIIV*, pages 119–131. IFIP, North Holland, 1988.

[8] R. E. Bryant. Graph based algorithms for boolean function manipulation. *IEEE Trans. on Computers*, C-35, 1986.

[9] S.P. van de Burgt, J. Kroon, E. Kwast, and H.J. Wilts. the RNL Conformance Kit. In J. de Meer, L. Mackert, and W. Effelsberg, editors, *Proc. of the 2nd International Workshop on Protocol Test Systems*, pages 279–294. North-Holland, October 1989.

[10] W. Y.L. Chan, S. T. Vuong, and M.R. Ito. On test sequence generation for protocols. In E. Brinksma, G Scollo, and C.A. Vissers, editors, *Protocol Specification, Testing, and Verification, IX*, pages 119–131. IFIP, North Holland, 1990.

[11] A. Chandra, V. Iyengar, D. Jameson, R. Jawalkelar, I. Nair, B. Rosen, M. Mullen, J. Yoon, R. Armoni, D. Geist, and Y. Wolfsthal. AVPGEN - A Test Case Generator for Architecture Verification. *IEEE Transactions on VLSI Systems*, 6(6), June 1995.

[12] E. Clarke, O. Grumberg, K.Mcmillan, and X. Zhao. Efficient generation of counter examples and witnesses in symbolic model checking. *32nd ACM/IEEE Design Automation Conference*, pages 427–432, 1995.

[13] M. Farkas, D. Geist, and K. Holtz. *SysGen Architecture Verification Program Generator User's Guide*. IBM Science and Technology, Haifa, Israel, first edition, 1994.

[14] M. S. Garey and D. S. Johnson. *Computers and Intractability*. W. H. Freeman and Co., New York, 1979.

[15] R. C. Ho, C. H. Yang, M. A. Horowitz, and D. L. Dill. Architecture validation for processors. In *International Symposium of Computer Architecture 1995*, pages 404–413, 1995.

[16] G. J. Holzmann. *Design and Validation of Computer Protocols*. Prentice Hall, 1991.

[17] H. Iwashita, S. Kowatari, T. Nakata, and F. Hirose. Automatic test program generation for pipelined processors. In *International Conference on Computer Aided Design*, November 1994.

[18] D. Levin, D. Lorenz, and S. Ur, A methodology for processor implementation verification. In *FMCAD 96: Int. Conf. on Formal Methods in Computer-Aided Design*, NOV 1996. to appear.

[19] C. May, E. Silha, R. Simpson, H. Warren, eds. *The PowerPC Architecture*. Morgan Kaufmann, 1994.

[20] K. L. McMillan. *The SMV System DRAFT*. Carnegie Mellon University, Pittsburgh, PA, 1992.

[21] K. L. McMillan. *Symbolic Model Checking*. Kluwer Academic Press, Norwell, MA, 1993.

[22] PCI Special Interests Group, Portland, OR. *PCI Local Bus Specification*, 1995.

Self-Consistency Checking

Robert B. Jones[1,2], Carl-Johan H. Seger[1], and David L. Dill[2]

[1] Intel Development Labs, JFT-102, 2111 NE 25th Avenue,
Hillsboro, OR 97124, USA
[2] Computer Systems Laboratory, Stanford University,
Stanford, CA 94305, USA

Abstract. We introduce the notion of *self-consistency checking*, a new methodology for applying formal methods in the debugging process. Intuitively, self-consistency checks functional consistency of a circuit in two different modes or environments. Self-consistency can (1) simplify property verification and (2) enable the use of symbolic simulation in the absence of a concrete specification. We present a correctness model for property verification with self-consistency checking and the formal framework for the technique. Finally, we provide illustrative examples and partial verification results using self-consistency checking on circuits with many thousands of latches.

1 Motivation

As hardware designs become more complex, the cost of validation is becoming a larger fraction of the total design cost. Validation currently consumes a large fraction of design teams and requires massive simulation efforts. In fact, one industry expert predicts that there will soon be two or three validation engineers for every design engineer on major microprocessor design projects [Wil95].

Formal verification has been suggested as a remedy for the increasing difficulty of design validation. For a broad survey of this field, the reader is referred to Gupta [Gup92]. Formal verification methods always check the consistency of two descriptions of behavior. For example, one description may be a state graph and the other a set of temporal logic properties. Most work in formal verification rests on the assumption that one description represents the implementation of the design, while the other is a *requirements specification*, which captures the important correctness properties of the design (while usually disregarding efficiency and performance issues).

Unfortunately, many interesting properties are not verifiable in practice. There are two main problems. The first is that the system specification is often incomplete or absent. In fact, our experience indicates that specifications are rarely written; when they are written, they are rarely precise; and when they are precise, they rapidly become obsolete as the design implementation evolves. The primary contribution of this work is to provide a way to perform substantial partial verification in the absence of a functional specification. The second problem is that the verification task is far too complex for current tools and techniques. A large variety of research attempts to address this problem

[BD94b, BD94a, Cyr93, Hun85, SGGH91, Win95]. Our approach complements other methods by easing the difficulty of creating the specification itself and reducing the complexity of the ensuing property verification.

We report results applying our technique to industrial circuits with many thousands of latches—significantly larger circuits than can be handled with other techniques. Indeed, the self-consistency technique has proven useful for real-world verification tasks because it enables *partial verification in the absence of a specification*.

We motivate our methodology with a simple example: a pipelined DLX processor [HP90] as shown in Figure 1. This processor has a *data hazard* which arises when an instruction in the execution stage (EX) is dependent upon the results of a previous instruction before those results are visible in the pipeline. To remove this hazard without stalling the pipeline, the ALU result is *bypassed* back to the ALU input latches. Consider the program in Figure 2(a). The SUB is dependent on the result of the ADD, and requires bypassing of R3 (from MEM to EX) to avoid a stall. In Figure 2(b), NOPs have been added to prevent the SUB and ADD from being in the pipeline at the same time.

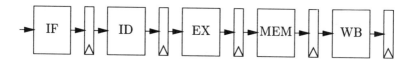

Fig. 1. DLX pipeline structure.

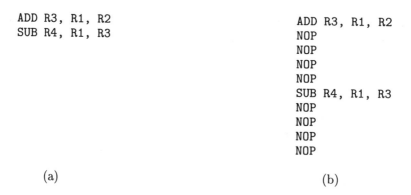

```
ADD R3, R1, R2                    ADD R3, R1, R2
SUB R4, R1, R3                    NOP
                                  NOP
                                  NOP
                                  NOP
                                  SUB R4, R1, R3
                                  NOP
                                  NOP
                                  NOP
                                  NOP

        (a)                              (b)
```

Fig. 2. Code sequences with data hazard.

Intuitively, we expect that the results (contained in the register file) of running both programs should be the same. This is because NOPs should not change register values, and the only change we have made to the program is the addition of NOPs. In fact, we should be able to add an arbitrary number of NOPs at any location in the input program without affecting the final results. The kernel of the idea is this: *the design should produce self-consistent behavior for certain classes of inputs.* If we extend the self-consistency check from pairs of instructions to five (the depth of the pipeline) instructions, then we can conclude that pipelining (including bypassing) has been implemented correctly. We will revisit this example in detail in Section 3.

Self-consistency can also simplify the creation of the specification. For example, describing the behavior of DLX when the input sequence is restricted (with a sufficient number of NOPs between consecutive instructions) is significantly simpler than describing the behavior when pipelining has to be considered. If the specification is stated in terms of current- and next-state transition relations, the pipeline registers will be exposed. However, if we restrict the input to contain sufficient NOPs between instructions, we can in fact ignore the pipe register contents in the specification. Thus, the values in the register file will depend only on the current values in the register file and the current instruction.

Much of the complexity in modern processor design arises from implementation performance enhancements like pipelining, parallelism, speculation, and out-of-order execution. Many of the bugs that arise during the design process are a direct result of the sophisticated control logic required for these performance enhancements.

Self-consistency checking can be used to verify consistency between an unenhanced design and a performance enhanced design. Since an unenhanced design is usually not available, we use the performance enhanced design in a simpler or debug mode. (We'll discuss later how this is accomplished in practice.) Our work provides the attractive option of being able to verify the correctness of the complex control logic in the absence of a specification. Verifying that a design is self-consistent can also considerably decrease the complexity of property verification. We will first develop a formal model for self-consistency, then discuss how we have applied it in practice.

2 Formalization of Self-Consistency Checking

In this section we formalize the concept of self-consistency by giving a circuit model, defining self-consistency, and providing definitions and theorems about self-consistency.

2.1 Circuit Model

To keep the definitions general we use a fairly broad definition of circuits. A circuit is modeled as a deterministic finite transition system. More precisely, a circuit, $\mathcal{C}$, is a quadruple $(\Sigma, \mathcal{Q}, q_0, \delta)$, where

$$\Sigma = \{0,1\}^k \text{ is the input alphabet } (k \geq 1),$$
$$\mathcal{Q} = \{0,1\}^n \text{ is the state set } (n \geq 0),$$
$$q_0 \in \mathcal{Q} \text{ is the initial state, and}$$
$$\delta : \mathcal{Q} \times \Sigma \to \mathcal{Q} \text{ is the transition function.}$$

We will usually deal with the behavior of $\mathcal{C}$ over sequences of inputs. Thus, let ϵ denote an input sequence of length 0 and extend the transition function to a mapping from $\mathcal{Q} \times \Sigma^*$ to $\mathcal{Q}$ as described below; we use the same symbol δ for the extended function:

$$\delta(q, \epsilon) = q$$
$$\delta(q, w\sigma) = \delta(\delta(q, w), \sigma)$$

for all $q \in \mathcal{Q}$, $w \in \Sigma^*$ and $\sigma \in \Sigma$.

The main reason for not including input and output functions is that we are primarily interested in the final state of the circuit after it has consumed some (finite) input sequence. Furthermore, to avoid unnecessary notation, we assume that there is a single initial state. Extending the definitions and results in this section to the case with a set of initial states is straightforward.

2.2   Self-consistency

Intuitively, we want a circuit to be self-consistent if equivalent input sequences lead to equivalent states. The primary difficulty here is to define precisely what "equivalent" means. In particular, there is a correlation between what we ultimately want to prove about the circuit, i.e., that the circuit satisfies some specification, and the notion of equivalence. For now, we will postpone the precise definition of these equivalences, and instead introduce a more general notion of self-consistency.

Definition 1. Given a circuit $\mathcal{C} = (\Sigma, \mathcal{Q}, q_0, \delta)$ and two relations, $\overset{\Sigma^*}{\sim} \subseteq \Sigma^* \times \Sigma^*$ and $\overset{\mathcal{Q}}{\sim} \subseteq \mathcal{Q} \times \mathcal{Q}$, we say that $\mathcal{C}$ is *self-consistent with respect to* $\overset{\Sigma^*}{\sim}$ and $\overset{\mathcal{Q}}{\sim}$ if and only if

$$\forall w_1, w_2 \in \Sigma^*. (w_1 \overset{\Sigma^*}{\sim} w_2) \implies (\delta(q_0, w_1) \overset{\mathcal{Q}}{\sim} \delta(q_0, w_2)).$$

In other words, $\overset{\Sigma^*}{\sim}$ relates input sequences that should take the circuit into states that are $\overset{\mathcal{Q}}{\sim}$ related.

Note that if appropriate input and state relations are used, self-consistency does not depend on any specification. Thus, we can determine whether a circuit is self-consistent *without* having a specification for the circuit. (One can, in some sense, view the relations as the specification. However, we have found that for non-trivial circuits, the relations are considerably easier to define than the specification.)

2.3 Use of Self-Consistency

One important use of self-consistency is in simplifying the correctness proof of a complex system. This is accomplished by performing property verification on a subset of inputs which are more amenable to verification than arbitrary inputs. Recall the simple pipeline example from Section 1: it is simpler to perform property verification if only one instruction is allowed in the pipeline at a time. If we can show that the circuit is self-consistent over the remaining inputs, we can generalize the property verification result.

In our context we are primarily interested in the final state of the circuit after it has consumed some (finite) input sequence. More formally, we assume that a specification is a predicate, $P(w, q)$, mapping $\Sigma^* \times Q$ to *true* or *false*. Intuitively, the predicate determines whether the state q is an acceptable state to reach after consuming input w. Note that, similar to [BD94b], we often use relatively complex predicates that first flush the machine before checking for some values in particular registers. We will return to this in Section 3.

We now link the relations used in self-consistency with the specification. The following definition captures the essence of this relationship.

Definition 2. Given a specification predicate $P(w, q) : \Sigma^* \times Q \rightarrow \{true, false\}$ and relations $\overset{\Sigma^*}{\sim} \subseteq \Sigma^* \times \Sigma^*$ and $\overset{Q}{\sim} \subseteq Q \times Q$, we say that $\overset{\Sigma^*}{\sim}$ and $\overset{Q}{\sim}$ are *property-preserving* if

$$\forall w_1, w_2 \in \Sigma^*. \forall q_1, q_2 \in Q. \ ((w_1 \overset{\Sigma^*}{\sim} w_2) \wedge (q_1 \overset{Q}{\sim} q_2)) \implies (P(w_1, q_1) = P(w_2, q_2)).$$

Intuitively, a pair of property-preserving input and state relations relate input sequences and initial states that should yield the "same" result.

The following theorem formalizes the use of self-consistency to simplify property verification.

Theorem 3. Let $C = (\Sigma, Q, q_0, \delta)$ be a circuit, $P(w, q)$ a specification predicate, and $\overset{\Sigma^*}{\sim} \subseteq \Sigma^* \times \Sigma^*$, $\overset{Q}{\sim} \subseteq Q \times Q$ property-preserving relations. Let $U \subseteq \Sigma^*$ be the set of "simpler" input sequences. Then, if

1. for every $w \in \Sigma^*$ there exists $u \in U$ such that $w \overset{\Sigma^*}{\sim} u$,
2. C is self-consistent with respect to $\overset{\Sigma^*}{\sim}$ and $\overset{Q}{\sim}$, and
3. for all $u \in U$, $P(u, \delta(q_0, u))$ holds,

it follows that $P(w, \delta(q_0, w))$ for every $w \in \Sigma^*$.

Proof. Consider an arbitrary word $w \in \Sigma^*$. There are two cases. If $w \in U$ the claim follows trivially from assumption 3. Otherwise, $w \in \Sigma^* - U$. By assumption 1 there exists $u \in U$ such that $w \overset{\Sigma^*}{\sim} u$. Since C is assumed to be self-consistent with respect to $\overset{\Sigma^*}{\sim}$ and $\overset{Q}{\sim}$, it follows that $\delta(q_0, w) \overset{Q}{\sim} \delta(q_0, u)$. This, together with the fact that $\overset{\Sigma^*}{\sim}$ and $\overset{Q}{\sim}$ are property preserving, implies that $P(w, \delta(q_0, w)) = P(u, \delta(q_0, u))$. By assumption 3, we know that $P(u, \delta(q_0, u))$ and thus $P(w, \delta(q_0, w))$. $\qquad\square$

2.4 Self-Consistency Checking

The above definition of self-consistency is very general. It is so general, in fact, that we have not yet discovered a practical method for checking a circuit for self-consistency with respect to any (non-trivial) input and state relations. However, there are more specialized versions for which efficient self-consistency verification can be accomplished using symbolic simulation. We will illustrate in Section 3 that important classes of circuits and properties can be formulated as self-consistency checks.

Our main restriction to the general framework is to limit the input relation to a pointwise extension of a function from Σ to Σ^*. This input function takes an arbitrary input sequence and returns a "simpler" sequence. "Simpler" here refers to the complexity needed to execute such a sequence correctly. Thus, the input function will often return a significantly longer sequence than it is given as input. For example, the "simplified" sequence for DLX, in our initial example, is the given sequence padded with NOPs. More formally, if $s : \Sigma \to \Sigma^*$, then the *input function corresponding to s*, $\mathcal{S} : \Sigma^* \to \Sigma^*$, is defined as:

$$\mathcal{S}(w) = \begin{cases} \epsilon & \text{if } w = \epsilon, \\ \mathcal{S}(\tilde{w})s(\sigma) & \text{if } w = \tilde{w}\sigma. \end{cases}$$

In this context, Definition 1 specializes to:

Definition 4. Assume $\mathcal{C} = (\Sigma, \mathcal{Q}, q_0, \delta)$ is a circuit and $s : \Sigma \to \Sigma^*$ is a function. Let $\mathcal{S}$ be the corresponding input function of s. Given a relation $\overset{\mathcal{Q}}{\sim} \subseteq \mathcal{Q} \times \mathcal{Q}$, we say that $\mathcal{C}$ is *self-consistent with respect to $\mathcal{S}$ and $\overset{\mathcal{Q}}{\sim}$* if and only if

$$\forall w \in \Sigma^*. \ \delta(q_0, w) \overset{\mathcal{Q}}{\sim} \delta(q_0, \mathcal{S}(w)).$$

Of course, the above definition still appears to force us to check the self-consistency condition for infinitely many input sequences. However, in the usual manner we can accomplish the check by induction, as formalized in the following theorem.

Theorem 5. Let $\mathcal{C} = (\Sigma, \mathcal{Q}, q_0, \delta)$ be a circuit and $\mathcal{S} : \Sigma^* \to \Sigma^*$ the input function corresponding to some function $s : \Sigma \to \Sigma^*$. If for every $\sigma \in \Sigma$ and every $q_1, q_2 \in \mathcal{Q}$ we have that $q_1 \overset{\mathcal{Q}}{\sim} q_2$ implies $\delta(q_1, \sigma) \overset{\mathcal{Q}}{\sim} \delta(q_2, s(\sigma))$, then $\mathcal{C}$ is self-consistent with respect to $\mathcal{S}$ and $\overset{\mathcal{Q}}{\sim}$.

Proof. Follows trivially by induction on the length of w. $\qquad\qquad\square$

A more pictorial description of the condition of Theorem 5 is shown in Figure 3. Note that verifying this condition only involves verifying a finite number of cases.

Theorem 5 requires us to check the condition for every pair of states q_1 and q_2 and every input σ. A more accurate (but also notationally more complex) statement would be to require this condition to hold only for all reachable states and all valid inputs given the current state. We will not formalize this notion, but will often make that assumption in actual self-consistency checking.

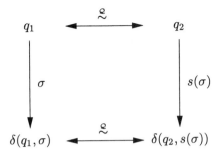

Fig. 3. Illustration of Theorem 5.

3 Examples

In this section, we present two simple examples to illustrate self-consistency checking. We do not purport that these examples represent complex verification results. To the contrary, these examples are simply intended to illustrate the use of self-consistency. In Section 5 we will discuss how we perform self-consistency checking in practice and present verification results from industrial microprocessor designs.

3.1 Pipeline Bypassing

We introduced the DLX pipelining example in Section 1. Knowing how NOPs behave in a simple pipeline, we expect that executing both programs in Figure 2 will result in the same register values. This is because NOPs do not change the architectural state we are verifying. Inserting a NOP after every instruction will essentially remove the functionality of the bypass circuitry from the results [3]. We can then compare the results of executing sequences with and without NOPs to determine the correctness of the bypass network.

This idea is easily scaled to practically sized pipelines. The mapping function S must simply be defined to insert enough NOPs (or some other stalling mechanism for more complex pipelines where NOPs are not sufficient) so that only one instruction is in the pipeline at a time. Additionally, we cannot compare register values until the instructions have exited the pipeline and written back their results. Thus, before comparison we must *flush* the pipeline, which results in all of the outstanding computation in pipe registers being placed in architectural state. For a simple pipeline like our example, flushing can be accomplished by inserting a sufficient number of NOPs. In more complicated pipelines, the hardware itself may be used to force a flush. Burch and Dill [BD94b] introduced an

---

[3] These examples assume a relatively unsophisticated pipeline and properties that are not affected by extra NOPs. More complex implementations require more sophisticated techniques.

automatic technique for flushing, and Burch [Bur96] has subsequently extended it for more complex superscalar processor architectures.

With every proof, we must give appropriate property-preserving definitions for $\mathcal{S}$ and $\overset{\varrho}{\sim}$. For this example, the input function ($\mathcal{S}$) is defined by the pointwise extension of the function $s : \Sigma \to \Sigma^*$:

$$s(op_i) = op_i \text{ NOP NOP NOP NOP}.$$

For the state relation, we only want to compare the final results contained in the register file (after flushing). Consequently, we introduce a projection function Π, that discards all state but the register file. We define $\overset{\varrho}{\sim}$ as:

$$q_1 \overset{\varrho}{\sim} q_2 \iff \Pi(\delta(q_1, \text{NOP NOP NOP NOP})) = \Pi(q_2).$$

Note that a successful self-consistency proof will establish only that the pipeline control logic is correct and not overall correctness. In particular, a certain class of bugs cannot be detected with self-consistency alone. For example, if the ALU is incorrect (but consistently incorrect), self-consistency with respect to the above input function and state relation will not detect the bug. However, this class of bugs can be found by proving properties about the ALU on the simpler set of input sequences with traditional verification techniques. This restricted property verification is typically much easier to perform than for the general case. This technique was formalized in Theorem 3.

3.2 Superscalar Arbitration

Pipeline control logic in superscalar processors is very complex. One of the reasons is that dependencies and other interactions between parallel instructions must be tracked within and between pipelines as they execute. For example, the instructions in Figure 5a cannot be allowed to execute in parallel because of the data dependency. Another example of the arbitration between parallel pipelines is write access to internal memories like register files (Figure 4).

If the instructions in Figure 5a execute in parallel, the register file arbitration logic must only allow the second write to R3. Allowing both writes simultaneously could result in incorrect data. As before, we intuitively expect that executing program 5b will produce the same functional result as executing program 5a. When the NOPs are inserted, the hazard detection logic is not necessary— because there are no hazards to detect. We can therefore use equivalent classes of sequences like these to prove the functional correctness of the arbitration logic.

The input function ($\mathcal{S}$) is defined by the pointwise extension of the function $s : \Sigma \to \Sigma^*$:

$$s((op_i, op_j)) = (op_i, \text{NOP})\,(op_j, \text{NOP})\,(\text{NOP}, \text{NOP}).$$

Again, we only require that the state relation compare the final results contained in the register file. We define $\overset{\varrho}{\sim}$ as:

$$q_1 \overset{\varrho}{\sim} q_2 \text{ if and only if } \Pi(\delta(q_1, (\text{NOP}, \text{NOP}), (\text{NOP}, \text{NOP}))) = \Pi(q_2).$$

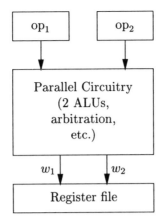

Fig. 4. Example of parallel writes to register file.

```
ADD R3, R1, R2
SUB R3, R4, R5
```

```
ADD R3, R1, R2
NOP
SUB R3, R4, R5
NOP
```

(a)   (b)

Fig. 5. Code sequences with ordering constraints.

We assume for this example that the pipeline is only two stages deep. For a realistic pipeline, more NOPs (or other appropriate stalling mechanisms) would be required.

4   Practical Self-Consistency Checking

As shown in Section 2.4, we can prove self-consistency by establishing the condition in Theorem 5. A naive approach to verifying the condition would be as follows. First, initialize the circuit to an arbitrary state q_1 by using a BDD variable for every state holding element in the circuit. Next, symbolically simulate the circuit for an arbitrary input value to obtain $\delta(q_1, \sigma)$ and $\delta(q_2, s(\sigma))$. Finally, compute $\overset{\circ}{\approx}$ and the implication. Although this approach allows complete self-consistency verification, it is limited to fairly trivial circuits (on the order of 20-100 latches).

The first improvement to the naive approach stems from the following observation. Although the state relation ($\overset{\circ}{\approx}$) is almost always a relation and not a function, it is often the case that if $q_1 \overset{\circ}{\sim} q_2$, large portions of the state q_2 are uniquely determined by q_1. These parts of q_2 can often be computed quite efficiently. Thus, we can often significantly reduce the number of new variables needed to represent q_2, because that representation has been derived from q_1. The second improvement stems from the face that certain input values will not affect any of the state holding elements that are part of the state relation. This observation can be used to reduce the number of variables needed even further. Finally, for some types of circuits, restricting the states and inputs to valid combinations can reduce the number of variables, and at least to a first approximation, the run-time requirements to perform the verification.

With techniques like those described above, we can often significantly increase the size of circuits for which we can perform self-consistency verification. However, the main motivation for self-consistency came from attempts at verifying *extremely* large control oriented circuits with many thousands of latches. For such circuits, complete self-consistency verification is not immediately practical. Consequently, we focus on self-consistency *checking*, as opposed to *verification*.

Although scalar simulation could be used to check self-consistency, we wanted to obtain more confidence in our checking. Thus, we use symbolic simulation for self-consistency checking and the number of variables used in the simulation (i.e., the coverage obtained) is simply determined by the run time requirements and the time (and patience!) available. The complexity of the circuits we are working with prevents us from performing a complete self-consistency *verification*. We can, however, perform a fairly impressive amount of self-consistency *checking*. Using self-consistency this way allows us to cover many simulation cases simultaneously with the symbolic simulator. This is much more effective than traditional simulation but of course less effective than complete verification.

We have implemented this technique using the Voss system [Seg93]. Voss is a general purpose verification platform consisting of a symbolic simulator, built-in BDD support, and a general-purpose functional language (FL). Voss supports mixed scalar and symbolic simulation and is an excellent prototyping environment for these types of applications. In addition, Voss provides interactive circuit debugging support.

5   Results

The examples given in Section 3 are simple, and intended to illustrate the use of self-consistency checking. In reality, the motivation for self-consistency came from the need to verify extremely complex circuits where the specification was absent or at best elusive. We have found this approach to be effective for real-world verification examples.

It was straightforward (but by no means trivial) to derive the property-preserving input transformations S and state-equivalence relations $\overset{\circ}{\sim}$ for our actual circuits. In fact, we often leveraged the circuitry itself to implement the elements of the self-consistency framework. For example, the input transformations S for our case-studies are considerably more complex than just adding

NOPs to the input stream. Even so, the basic approach is the same: modify the input stream in some well-defined and regular manner and in a way that simplifies the operation of the hardware.

The results we report in this section concern the self-consistency check only. Property verification on the simpler equivalent circuits could be approached using existing methods. Neither of our examples had a specification amenable for property checking, and both were of sufficient size that it would have been difficult even if a specification had existed. *The results in this section emphasize the practical usefulness of self-consistency checking even when a specification of a system is incomplete or unavailable.*

The results we report here are circuits from industrial microprocessor designs. Our case-studies involved RTL-level descriptions of large (many thousands of gates) control circuits. Our examples are synthesized from an HDL description into the format used by Voss. Symbolic simulation is used to compute the verification conditions and the results are compared with a library of FL functions. Any differences in the resultant traces are reported as a failure set.

For inconsistencies, another FL function is used to return an element of the failure set, which serves as a specific counterexample. This function can be written so that it returns as many signals de-asserted as possible—a very useful feature for debugging. With unnecessary (to the counter-example) control signals de-asserted, the debugging process is considerably simplified.

Results from our case-studies are shown in Table 1. Although we do not perform complete verification, the number of cases actually covered is still impressive when compared with current simulation standards. Circuit A is a pipelined "stream processor", which annotates a stream of data. The state-equivalence relation was based on the expectation that the behavior of the pipeline be consistent regardless of data alignment. The complexity of the verification was greatly reduced by leveraging surrounding circuitry to compute the state-equivalence relation.

Circuit B is a parallel pipeline arbitrating between several execution threads. We were interested in verifying an extremely complex block of control logic within circuit B. We were unable to characterize the reachable state space of circuit B, so we approximated the reachable state space with the reset state. However, the functional verification was quite thorough, as known bugs were independently discovered by our method.

Circuit	Gates	Latches	Simulation Variables	Execution Time (hr)	Equivalent Simulation Cases
A	8452	2506	49	3	$6 * 10^{14}$
B	72664	11709	144	10	$2 * 10^{43}$

Table 1. Self-consistency checking results.

Because the verification times we report may seem fairly long, it is important to point out that in practice the Voss verification runs are much shorter. Because the symbolic simulation approach is so amenable to partial verification, most of our interactive work is of that nature. Short (under five minutes) runs can reveal significant information. In fact, all of the bugs we have discovered have been revealed through this type of partial verification. The complete (time intensive) verification runs can be performed in batch mode as a safety net.

6   Conclusions

There are obviously many applications of self-consistency checking that we have not yet discovered. We expect that applications which exploit micro-architecture structure will be the most useful.

There is significant work left to be accomplished in constructing reachable state invariants for facilitating the inductive self-consistency proof. We anticipate that it will be fruitful to over-approximate the reachable state in ways that do not result in false negatives.

Ideally, the distance between self-consistency theory (verification) and practice (checking) will be reduced. Obviously, advances in the efficiency of symbolic simulation will aid in the completeness of the self-consistency proofs.

Acknowledgments

We would like to thank our colleagues at Intel Development Labs and Stanford University for providing intellectually stimulating and fruitful research environments.

The first author is supported at Stanford by an NDSEG graduate fellowship. The third author is partially supported by ARPA under contract ARMY DABT63-95-C-0049-P00002.

References

[BD94a]   V. Bhagwati and S. Devadas. Automatic verification of pipelined microprocessors. In *31st ACM/IEEE Design Automation Conference*, 1994.

[BD94b]   J. R. Burch and D. L. Dill. Automatic verification of microprocessor control. In *Computer Aided Verification. 6th International Conference*, 1994.

[Bur96]   J. R. Burch. Techniques for verifying superscalar microprocessors. In *33rd ACM/IEEE Design Automation Conference*, 1996.

[Cyr93]   D. Cyrluk. Microprocessor verification in PVS: A methodology and simple example. Technical Report SRI-CSL-93-12, SRI Computer Science Laboratory, December 1993.

[Gup92]   A. Gupta. Formal hardware verification methods: A survey. *Formal Methods in System Design*, 1:5–92, 1992.

[HP90]   J. L. Hennessy and D. A. Patterson. *Computer Architecture: A Quantitative Approach*. Morgan Kaufmann, 1990.

[Hun85]   W. A. Hunt, Jr. FM8501: A verified microprocessor. Technical Report 47, University of Texas at Austin, Institute for Computing Science, December 1985.

[Seg93]   C. H. Seger. Voss: A formal hardware verification system user's guide. Technical Report 93-45, Department of Computer Science, Univerisity of British Columbia, 1993.

[SGGH91]   J. B. Saxe, S. J. Garland, J. V. Guttag, and J. J. Horning. Using transformations and verification in circuit design. Technical Report 78, DEC Systems Research Center, September 1991.

[Wil95]   R. Wilson. Verification feels strain. *Electronic Engineering Times*, (840):18–22, March 1995.

[Win95]   P. J. Windley. Formal modeling and verification of microprocessors. *IEEE Transactions on Computers*, 44(1):54–72, January 1995.

Inverting the Abstraction Mapping:
A Methodology for Hardware Verification

David Cyrluk*

Dept. of Computer Science, Stanford University, Stanford CA 94305 and
Computer Science Laboratory, SRI International, Menlo Park, CA 94025
cyrluk@cs.stanford.edu

Abstract. Abstraction mappings have become a standard approach to verifying the correctness of processors. When used in a straightforward manner this approach suffers from generating extremely large intermediate terms that have to be simplified.

In an interactive theorem prover the complete expansion of the abstraction mapping is not even possible. Yet, with human guidance it is interactive theorem proving that is applied to examples too large to be handled by automated methods.

We present a methodology for verifying the correctness of processors that aims to limit the size of intermediate terms generated in an interactive proof and to manage the complexity of the proof search.

The main idea of this methodology is that, instead of expanding the abstraction mapping and thereby introduce large implementation terms into the specification, we try to identify sub-terms of implementation terms that can be replaced by specification state variables. This is done by inverting the equations defining the abstraction mapping so to rewrite implementation terms into abstract state variables.

This method has been successfully applied to the verification of the DLX processor. It has also been applied to the verification of the ALU pipeline and Saxe pipeline and has simplified their proofs.

Finally, lessons learned from this methodology can help develop better heuristics employed by automatic methods.

1   Introduction

The use of abstraction mappings [1] for the verification of microprocessors and other sequential hardware circuits is commonplace [3, 7, 11]. Both automated stand-alone tools [3] and automated proof strategies for use in interactive theorem provers [6] have been developed based on the use of abstraction mappings. In [5] we developed a language *GTL2* that is appropriate for specifying the correctness of sequential hardware circuits using abstraction mappings.

---
* This research was partially supported by SRI International, DARPA contract NAG2-703, and NSF grants CCR-8917606, CCR-8915663. Discussions with Deepak Kapur helped immensely in getting this paper written.

A common theme of all of these approaches is to expand the definitions of the specification machine, the implementation machine, and the abstraction mapping to generate large *if-then-else* expressions. These expressions are then compared by performing case analysis on the conditions appearing in them.

In large examples these expressions can get huge. In the PVS verification system [10], the complete expansion of the abstraction mapping for this machine causes PVS to run out of memory! There are over 100 embedded *if-then-else* expressions in the expansion of the abstraction of the register file. It takes over 10 screens to print the entire expansion.

In spite of this size, the automated tool finishes the verification in a matter of seconds. In PVS, larger and more complicated processor verifications have been carried out [11]. How can an automated tool that generates such large expressions verify their equivalence so quickly? Are there systematic methodologies that a user of an interactive proof system can make use of to structure the verification of large circuits?

Since large realistic processors are still beyond the capability of completely automatic tools, a methodology for structuring large verifications is needed.

In this paper we propose a methodology that can give the verifier of large processors some guidance in structuring an interactive proof attempt. We also look at whether this methodology can give insight into what makes the automated tools work so effectively and possibly how they might be improved.

At the highest level, the methodology rests on the observation that the proof of correctness of processors using the abstraction mapping approach comes down to identifying in the implementation machine certain structures that correspond to state elements in the abstract machine. The main example is that the implementation register file along with the pipeline bypass logic behaves like the abstract register file.

The aim of this paper is to translate this very high-level idea into a concrete proof methodology. The main idea is instead of using the abstraction mapping to introduce into the specification machine large implementation terms, we invert the abstraction mapping—hoping to replace implementation terms with simpler specification state variables.

In developing and implementing this methodology several insights became apparent. One is that in choosing conditions to perform case analysis on, we should select conditional formulas in the following order: first formulas that determine when the specification machine stalls, then formulas that appear in the specification machine, and then formulas that appear in the implementation machine. This observation should also be useful to finite state methods that rely on BDDs to encode state; the BDD variables should also be ordered as above. We will discuss additional insights in the conclusion.

In the next section we present the basic framework in which our methodology operates. Section 3 illustrates our methodology on a simple example. Section 4 discusses how our methodology is applied to the verification of the DLX processor. We end with comparisons to other work and conclusions.

2 Preliminaries

The basic framework is illustrated by the conventional commutative diagram shown in Figure 1.

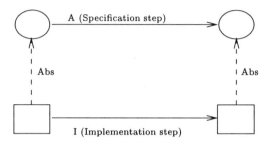

Fig. 1. The commutes diagram

Let V be a state variable of the specification machine and s an arbitrary implementation state. We denote by V_{spec} the term that results from following the left upper path through the commutative diagram. Similarly, V_{impl} denotes the term resulting from following the lower right path:

$$V_{spec} = A(\text{Abs}(s)), \tag{1}$$

$$V_{impl} = \text{Abs}(I(s)). \tag{2}$$

Common to both the stand-alone tools approach and the interactive theorem proving approach is the representation of the specification and implementation machines as transition systems. These transition systems are represented as equations that specify how the state variables are updated by the execution of one machine cycle. These equations are implementation equations in $GTL2$ [5] and can be stated in the following form.

$$\bigwedge_j V_j(\text{next}(s)) = \text{next}_j(s) \tag{3}$$

where V_j are state variables, next is the next state function (either I or A in Figure 1) and next_j is a function that specifies how state variable V_j changes.

Additionally, the abstraction mapping is given as equations that equate state variables in the specification machine with terms involving state variables of the implementation machine. These equations will be of this form:

$$\bigwedge_j V_j(\text{Abs}(s)) = \text{Abs}_j(s), \tag{4}$$

An example abstraction mapping for a three stage pipeline might be:

$$PC_A(\text{Abs}(s)) = PC_I(s),$$

$$REG_A(\text{Abs}(s)) = REG_I(\text{next}(\text{next}(\text{next}(s)))).$$

In practice these equations have the effect of introducing implementation machine terms into V_{spec}. In the above three stage pipeline, where the parts of the abstraction mapping involves executing the implementation machine for three cycles, the terms introduced by the abstraction mapping can get quite large.

Now let us consider inverting the equations describing the abstraction mapping. That is, instead of equations of the form 4 we reverse the equations as follows:

$$\bigwedge_j \text{Abs}_j(s) = V_j(\text{Abs}(s)). \tag{5}$$

The abstraction mapping for our three stage pipeline would then become:

$$PC_I(s) = PC_A(\text{Abs}(s)),$$

$$REG_I(\text{next}(\text{next}(\text{next}(s)))) = REG_A(\text{Abs}(s)).$$

Now the term V_{spec} will only contain specification terms. The inverse abstraction mapping tells us how to transform V_{impl} into V_{spec}, i.e. how to transform terms involving implementation state variables into equivalent specification state variables. The hope is that since the resulting terms, V_{spec} and V_{impl} will be much smaller the verification should be easier and more manageable.

It turns out that this is only partially true. The difficulty is that there is no straightforward way to obtain the inverse abstraction mapping. That is, Equation 5 is not sufficient to invert the abstraction mapping. In particular, in different contexts there may be many terms equivalent to $\text{Abs}_j(s)$. For example, consider the case in our three stage pipeline example when the pipe is filled with three store instructions. In this case, since store instructions do not change the register file, the register file remains unchanged after executing the machine three cycles. Thus, in a context in which the pipe is filled with store instructions we would have to know that we could replace $REG_I(s)$ with $REG_A(\text{Abs}(s))$. Thus, to be complete, new equations that can rewrite all possible implementation terms that are equivalent to $REG_I(\text{next}(\text{next}(\text{next}(s))))$ in different contexts need to be generated.

While in general it is difficult to generate such a complete set of rewrite rules, we have found this method greatly simplifies the verification of the processor verification benchmarks to which we have applied it. When used in the context of an interactive theorem prover, like PVS, the idea of inverting the abstraction mapping provides a powerful methodology to guide the human interaction.

On the other hand, while at this time the difficulty of generating a complete set of equations makes this approach problematic for a completely automatic tool, the ideas presented here can help to provide a better understanding as to how those automatic tools work, why their heuristics are effective, and perhaps how those heuristics can be improved.

In the next section we illustrate our method through the complete verification of the simplest one-stage pipeline.

3 A Simple Example

Our simple one-stage pipeline has *stall*, *dstn*, and *src* inputs. Its internal state consists of a register file, *SREG*. Two functions operate on register files: apply and update. Apply reads from a register file at a source address. Update writes to a register file. The specification of the pipeline is that if the *stall* input is not asserted then the value of *SREG* is updated at the *dstn* input to the current value of *SREG* at the *src* input.

The above specification is given by the following equation:

$$
\begin{aligned}
SREG(\mathrm{step}(as)) = \mathrm{cond}(\mathrm{NOT}\ & stall(as), \\
& \mathrm{update}(SREG(as), \\
& \qquad dstn(as), \\
& \qquad \mathrm{f}(\mathrm{apply}(SREG(as),\ src(as)))), \\
& SREG(as)).
\end{aligned} \tag{S1}
$$

The implementation consists of adding one pipeline stage. This is done by adding the registers *STALLD*, *SRCD*, *DSTND*, and *WBREG*. *STALLD*, *SRCD*, and *DSTND* store the previous cycle's inputs. The write back register *WBREG* stores the contents of the implementation register file, *IREG*, at the previous cycle's *src* input. This value is what will be written to the register file. Note that the equation, I2, that describes how *WBREG* is updated contains the pipeline bypass logic.

The implementation is given by the following equations:

$$
\begin{aligned}
IREG(\mathrm{next}(is)) = \mathrm{cond}(& STALLD(is), \\
& IREG(is), \\
& \mathrm{update}(IREG(is),\ DSTND(is), \\
& \qquad WBREG(is))),
\end{aligned} \tag{I1}
$$

$$
\begin{aligned}
WBREG(\mathrm{next}(is)) = \mathrm{f}(\mathrm{cond}(& STALLD(is), \\
& \mathrm{apply}(IREG(is),\ src(is)), \\
& \mathrm{cond}(src(is) = DSTND(is), \\
& \qquad WBREG(is), \\
& \qquad \mathrm{apply}(IREG(is),\ src(is))))),
\end{aligned} \tag{I2}
$$

$$
SRCD(\mathrm{next}(is)) = src(is), \tag{I3}
$$

$$
DSTND(\mathrm{next}(is)) = dstn(is), \tag{I4}
$$

$$
STALLD(\mathrm{next}(is)) = stall(is). \tag{I5}
$$

Finally, the abstraction mapping is given by these equations:

$$
SREG(\mathrm{Abs}(is)) = IREG(\mathrm{next}(is)), \tag{Abs1}
$$

$$
src(\mathrm{Abs}(is)) = src(is), \tag{Abs2}
$$

$$dstn(\text{Abs}(is)) = dstn(is), \qquad\qquad\qquad (\text{Abs3})$$
$$stall(\text{Abs}(is)) = stall(is). \qquad\qquad\qquad (\text{Abs4})$$

The relationship between apply and update is given by the following equation:

$$
\begin{aligned}
&\text{apply}(\text{update}(reg,\ value,\ dstn\_addr), \\
&\qquad src\_addr) \\
&= \text{cond}(src\_addr = dst\_addr, \qquad\qquad (\text{APP\_UPD}) \\
&\qquad value, \\
&\qquad \text{apply}(reg,\ src\_addr)).
\end{aligned}
$$

3.1 Traditional Method

In this section we illustrate the verification of the above pipeline using the traditional method similar to methods discussed in [3, 6]. This method simply expands the definitions of the specification machine, implementation machine, and abstraction mapping. The resulting terms which contain embedded conditionals are then compared by case splitting on those conditionals and simplified using reasoning about equalities and updates.

In the above example let $SREG_{spec} = \text{SREG}(\text{step}(\text{Abs}(is)))$ and $SREG_{impl} = \text{SREG}(\text{Abs}(\text{next}(is)))$ then the correctness condition corresponding to the commutative diagram is:

$$SREG_{spec} = SREG_{impl}. \qquad\qquad\qquad (6)$$

After rewriting we have:

$SREG_{spec} =$
  cond(NOT $stall(is)$,
        update(cond($STALLD(is)$,
                  $IREG(is)$,
                  update($IREG(is)$, $DSTND(is)$, $WBREG(is)$))),
            $dstn(is)$,
            f(apply(cond($STALLD(is)$,
                    $IREG(is)$,
                    update($IREG(is)$, $DSTND(is)$, $WBREG(is)$))),
                $src(is)$)))),
        cond($STALLD(is)$,
            $IREG(is)$,
            update($IREG(is)$, $DSTND(is)$, $WBREG(is)$)))),

and

$SREG_{impl} =$
  cond(stall(is),
        cond($STALLD(is)$,
            $IREG(is)$,
            update($IREG(is)$, $DSTND(is)$, $WBREG(is)$))),

update(cond($STALLD(is)$,
$\quad\quad\quad$ $IREG(is)$,
$\quad\quad\quad$ update($IREG(is)$, $DSTND(is)$, $WBREG(is)$))),
$\quad\quad$ $dstn(is)$,
$\quad\quad$ cond($STALLD(is)$,
$\quad\quad\quad$ f(apply($IREG(is)$, $src(is)$))),
$\quad\quad\quad$ cond($src(is) = DSTND(is)$,
$\quad\quad\quad\quad$ f($WBREG(is)$),
$\quad\quad\quad\quad$ f(apply($IREG(is)$, $src(is)$))))))).

If we do case analysis on $src(is) = DSTND(is)$ and $stall(is)$ both $SREG_{spec}$ and $SREG_{impl}$ reduce to the same term. Note that we might also have decided to do case analysis on $STALLD(is)$ which would have been unnecessary.

3.2 Inverting the abstraction mapping

We now redo the same proof but use the equations for the abstraction mapping in the reverse direction:

$$IREG(next(is)) = SREG(Abs(is)), \quad\quad\quad \text{(Abs1\_rev)}$$
$$src(is) = src(Abs(is)), \quad\quad\quad \text{(Abs2\_rev)}$$
$$dstn(is) = dstn(Abs(is)), \quad\quad\quad \text{(Abs3\_rev)}$$
$$stall(is) = stall(Abs(is)). \quad\quad\quad \text{(Abs4\_rev)}$$

Applying these rules we get:

$SREG_{spec} =$
$\quad$ cond(NOT $stall(Abs(is))$,
$\quad\quad$ update($SREG(Abs(is))$, $dstn(Abs(is))$,
$\quad\quad\quad$ f(apply($SREG(Abs(is))$, $src(Abs(is))$)))),
$\quad\quad$ $SREG(Abs(is))$),

and (with expanding Abs once before applying the rewrite rules)

$SREG_{impl} = IREG(next(next(is))) =$
$\quad$ cond($stall(Abs(is))$,
$\quad\quad$ $SREG(Abs(is))$,
$\quad\quad$ update($SREG(Abs(is))$, $dstn(Abs(is))$,
$\quad\quad\quad$ f(cond($STALLD(is)$,
$\quad\quad\quad\quad$ apply($IREG(is)$, $src(Abs(is))$),
$\quad\quad\quad\quad$ cond($src(Abs(is)) = DSTND(is)$,
$\quad\quad\quad\quad\quad$ $WBREG(is)$,
$\quad\quad\quad\quad\quad$ apply($IREG(is)$, $src(Abs(is))$)))))))).

There is a slight advantage over the traditional method in that the terms generated are smaller. In a realistic example with a five stage pipeline the terms would be considerably smaller.

Unfortunately we also have a huge disadvantage: the above equation no longer simplifies. No rewrite rules are applicable and no amount of case analysis can reduce the above equation to true.

The reason for this is that when we inverted the abstraction mapping we no longer have a confluent set of rewrite rules. Much like in the Knuth-Bendix completion procedure [8], we need to add equations to make the resulting system canonical.

Along with simple properties of cond, the following three rules can be added to finish the proof:

$$\text{cond}(STALLD(is),$$
$$IREG(is),$$
$$\text{update}(IREG(is), DSTND(is), WBREG(is))) \qquad \text{(ABS1')}$$
$$= SREG(\text{Abs}(is))$$

$$\text{cond}(src\_addr = dst\_addr,$$
$$value,$$
$$\text{apply}(reg, src\_addr)) \qquad \text{(APP\_UPD\_REV)}$$
$$= \text{apply}(\text{update}(reg, value, dstn\_addr),$$
$$src\_addr)$$

$$\text{cond}(p, \text{apply}(then, addr),$$
$$\text{apply}(else, addr)) \qquad \text{(PUSH\_COND)}$$
$$= \text{apply}(\text{cond}(p, then, else), addr)$$

Using APP_UPD_REV

$$\text{cond}(src(\text{Abs}(is)) = DSTND(is),$$
$$WBREG(is),$$
$$\text{apply}(IREG(is), src(\text{Abs}(is))))$$

rewrites to

$$\text{apply}(\text{update}(IREG(is), DSTND(is), WBREG(is)),$$
$$src(\text{Abs}(is)))$$

Thus $SREG_{impl}$ is rewritten to

$$\text{cond}(stall(\text{Abs}(is)),$$
$$SREG(\text{Abs}(is)),$$
$$\text{update}(SREG(\text{Abs}(is)), dstn(\text{Abs}(is)),$$
$$\text{f}(\text{cond}(STALLD(is),$$
$$\text{apply}(IREG(is), src(\text{Abs}(is))),$$
$$\text{apply}(\text{update}(IREG(is), DSTND(is), WBREG(is)),$$
$$src(\text{Abs}(is)))))))).$$

PUSH_COND will lift the embedded applys so that ABS1' rewrites $SREG_{impl}$ to:

cond($stall$(Abs(is)),
   $SREG$(Abs(is)),
    update($SREG$(Abs(is)), $dstn$(Abs(is)),
        f(apply($SREG$(Abs(is)), src(Abs(is)))))).

Now case analysis on $stall$(Abs(is)) finishes the proof. Note also that this is the only choice for case analysis.

The basic lesson from this example that we will apply to more complicated examples is that if we can generate the necessary rules, then by inverting the abstraction mapping we can limit our case analysis to conditions that appear in the specification machine. This should lead to a big savings since there are usually much fewer conditions in the specification than in the implementation.

Put another way, we divide the proof of correctness into two parts. The first part is the inversion of the abstraction mapping and the generation of the new rules. This requires some understanding of the implementation and, as will be seen, may require some user guidance. The second part is more automatic and essentially uses our core hardware strategy from [6] to rewrite using the inverted abstraction mapping, the machine definitions, and the new rules, and then perform case analysis on the resulting terms. If we have generated enough rules the case analysis need only be on specification conditions.

As will be seen in Section 4 we can generate different instances of the inverse abstraction mapping lazily by simplifying the inverse abstraction mapping at the same time that we perform case analysis on the specification machine

3.3 Generating new equations

We now look at how the rules needed above can be generated.

One possible method for generating new rules is to run a conditional Knuth-Bendix completion procedure on the implementation equations, reverse abstraction mapping, and domain equations. We have done this by hand on several small examples, but the cost of blindly running a completion procedure on large examples may quickly exceed the potential benefit.

However, the ideas behind completion can help. Let us look more closely at our simple example. Rule ABS1' is generated by noticing that the addition of the rule Abs1_rev causes $IREG$(next(is)) to rewrite to two terms, the left and right hand sides of equation ABS1'. Thus ABS1' is added to eliminate that source of non-confluence.

We could have continued in this fashion generating many conditional rewrite rules. Instead we note that the goal of our rewriting process is to replace implementation terms with corresponding specification terms. At a high level this corresponds to recognizing in the implementation circuit instances of the specification circuit.

In processor verification examples this corresponds to recognizing instances of the register file distributed (sometimes in a hidden fashion) in the implementation machine. In this example the bypass logic combined with the implementation register file behave as one larger register file. In more complicated examples this observation still holds, although it is somewhat obscured by the presence of additional logic that might stall the pipeline or provide branch prediction, etc.

By using equation APP_UPD_REV we are in essence looking for instances of register files in the implementation machine. In this case the addition of the one rule PUSH_COND allowed us to eventually apply APP_UPD_REV.

In general it is not always so easy to generate a simple rule that will allow us to syntactically recognize applications of APP_UPD_REV. The particular form of the implementation equations I1-I5 were important for the above example to work so simply.

As rule I2 was written it is easy to recognize that it can be rewritten by equation APP_UPD_REV and PUSH_COND to the following rule before the proof is even begun:

$$
\begin{aligned}
\text{WBREG}(\text{next}(is)) = \text{f}(\text{cond}(\text{STALLD}(is), \\
\text{apply}(\text{IREG}(is), \text{src}(is)), \\
\text{apply}(\text{update}(\text{IREG}(is), \\
\text{DSTND}(is), \\
\text{WBREG}(is)), \\
\text{src}(is))))
\end{aligned} \qquad (\text{I2}')
$$

However, I2 could originally have been written in such a way as to make it difficult to transform it into I2'. There are several possible approaches to dealing with this difficulty:

One is to canonize the right hand sides of implementation equations into a form that can be matched against the left hand side of rule APP_UPD_REV. A possible canonical form can be generated by pushing uninterpreted symbols out of cond terms and re-ordering the conditionals that appear so that equality between addresses appear last.

Another possibility is to build into the matching routine enough understanding of cond and updates so that it can match the left hand side of APP_UPD_REV.

In an interactive prover such as PVS the most reasonable method is to have the user help with the identification. In our simple example the user could actually have rewritten equation I2 as the simpler equation I2'.

An intelligent user could have gone even further and defined a new function stall_update with the following definition:

$$
\begin{aligned}
\text{stall\_update}(\textit{stall?, reg, addr, val}) = \\
\text{cond}(\textit{stall?, reg, } \text{update}(\textit{reg, addr, val}))
\end{aligned}
$$

The presence of this function could have aided the user (or automated tool) in recognizing that the right hand sides of equations S1, I1, and I2 were all applications of stall_update.

4   The DLX Processor

We now illustrate how this methodology is applied to the DLX processor. The example that we verified was translated from the description used by Burch and Dill in [3]. Our verification differs from theirs in two respects. Firstly, we deal with stuttering slightly differently. Where they stall the specification machine based on a predicate indicating whether the implementation machine is in a visible state, we use an oracle to specify how many cycles the implementation needs to move from one visible state to the next.

Secondly, their verification is done using their automated tool, whereas our verification is done in PVS, an interactive theorem prover. In some ways this difference makes our task both easier and more difficult. It is easier since the user can guide the theorem prover. It is more difficult because the intermediate goals generated by the prover must be manageable and understandable to the user. Furthermore, the general purpose nature of the theorem prover sometimes results in inefficiencies that a special purpose tool can avoid. In practice this means that the automated tool can generate much larger terms before blowing up.

A separate verification of the DLX was done in [2]. This verification was done by hand and represents an extreme in human guidance. This proof was done by decomposing the pipeline into small stepwise refinements that were separately verified.

Our approach falls in between these two extremes. It provides a guide and methodology for human guidance, but automates most of the proof and as will be seen requires little more insight from the user than was required in our simple toy example.

4.1 Setup

We now give some details of the specification and verification of DLX.

The specification state consists of four state variables: `pc`, `imem`, `reg`, `dmem`. The implementation state consists of seventeen state variables.

The DLX is a five stage pipeline. The default execution of the pipeline assumes that branches are not taken. If it turns out that a branch is taken then the pipeline takes an extra cycle to correct for assuming the wrong next program counter. Additionally the pipeline does not assume a load delay slot. Thus, the pipeline needs to be stalled if the instruction following a load depends on the destination of the load.

The above two stall conditions are defined in the DLX processor with the following equations:

```
branch_taken(s) =
        (NOT bubble_id(rs)) AND
        (   opcode_class(opcode_of(instr_id(rs))) = J
         OR (   opcode_class(opcode_of(instr_id(rs))) = BEQZ
            AND next_a(rs) = 0)),
stall_issue(s) =
        (NOT bubble_ex(s)) AND
        opcode_class(opcode_ex(s)) = LOAD AND
        (NOT bubble_id(s)) AND
        (   rf1_of(instr_id(s)) = dest_ex(rs)
         OR rf2_of(instr_id(s)) = dest_ex(s)).
```

Thus the following predicate defines when an implementation state is visible:

```
visible(s) = NOT branch_taken(s) AND NOT stall_issue(s)
```

This is the negation of the stalling condition used in [3]. In their approach the user also needs to provide this information.

The user also needs to provide an oracle that specifies how many cycles the machine needs to execute to reach the next visible state:

```
num_cycles(vs) =
  IF stall_issue(next(vs))
     THEN IF branch_taken(next(next(vs)))
             THEN 3
             ELSE 2
          ENDIF
  ELSIF branch_taken(next(vs))
     THEN 2
     ELSE 1
  ENDIF.
```

The visible predicate and the num_cycles function are important aides in guiding the proof of correctness. They provide a definition of the context in which specification states and implementation states need to be equal. They also specify when the specification machine stutters. Thus even though they are technically part of the implementation machine they are also implicitly part of the specification machine and provide additional sources of case analysis.

An additional source of up-front user guidance is in transforming the following definitions. a and b are the two inputs to the alu. The bypass logic is embedded in their definition.

```
a(next(rs)) =
  IF rf1_of(instr_id(rs)) = reg0
     THEN 0
  ELSIF rf1_of(instr_id(rs)) = dest_ex(rs)
     THEN result_mem(rs)
  ELSIF rf1_of(instr_id(rs)) = dest_mem(rs)
     THEN result_wb(rs)
     ELSE reg(rs)(rf1_of(instr_id(rs)))
  ENDIF,
b(next(rs)) =
  IF opcode_class(opcode_of(instr_id(rs))) = ALU_IMMED
     THEN short_immed_of(instr_id(rs))
  ELSIF rf2_of(instr_id(rs)) = reg0
     THEN 0
  ELSIF rf2_of(instr_id(rs)) = dest_ex(rs)
     THEN result_mem(rs)
  ELSIF rf2_of(instr_id(rs)) = dest_mem(rs)
     THEN result_wb(rs)
     ELSE reg(rs)(rf2_of(instr_id(rs)))
  ENDIF
```

to

```
a(next(rs)) = select(assign(assign(next_reg(rs), dest_mem(rs),
                                   next_result_wb(rs)),
                            dest_ex(rs), next_result_mem(rs)),
                     rf1_of(instr_id(rs)))),
b(next(rs)) =
   IF opcode_class(opcode_of(instr_id(rs))) = ALU_IMMED
      THEN short_immed_of(instr_id(rs))
      ELSE
         select(assign(assign(next_reg(rs),
                              dest_mem(rs),
                              next_result_wb(rs)),
                       dest_ex(rs),
                       next_result_mem(rs)),
                rf2_of(instr_id(rs)))
   ENDIF
```

where `select` and `assign` are defined as:

```
select(reg, addr) = IF addr = 0
                       THEN 0
                       ELSE apply(reg, addr)
                    ENDIF,
assign(reg, addr, data) = IF addr = 0
                             THEN reg
                             ELSE update(reg, addr, data)
                          ENDIF.
```

This is the same type of transformation required in our simple example where equation I2 was transformed into equation I2'.

The inverse abstraction mapping is given as:

```
pc(vs) = pc(ABS(vs))
reg(next(next(next(next(vs))))) = reg(Abs(vs))
dmem(next(next(next(vs)))) = dmem(Abs(vs))
imem(vs) = imem(Abs(vs))
```

4.2 Proof

Now the proof proceeds essentially in the manner specified by our microprocessor strategy defined in [6]. Case analysis is performed on the `num_cycles` function. This yields different goals for each branch in this function. In each goal the two terms corresponding to the two paths in the commutative diagram are expanded much like in our simple example. Case analysis is then performed on the conditions in the specification side of the commutative diagram.

A method is used lazily to generate additional rules corresponding to specializations of the inverse abstraction mapping in different contexts. The inverse abstraction mapping is kept as a hypothesis in the proof goal. As the proof goal

is split, the left hand side of the inverse abstraction mapping is further simplified
in each context. This simplified inverse mapping is then used to rewrite terms
in the rest of the proof goal that are equivalent to an abstract state variable.

This is enough to finish automatically all but two cases. In the remaining
two cases the conditions that are hidden in the definitions of stall_issue and
branch_taken are brought out and case analysis is performed on them. This still
does not completely finish the proof. Goals of the following form are left:

```
  next_dest_ex(s!1) ≠ rf1_of(apply(imem(Abs(s!1)), pc(Abs(s!1))))
  AND assign(reg_file_term,
              next_dest_ex(s!1),
              next_result_wb(next(next(s!1)))) = reg(Abs(s!1))
IMPLIES
  select(assign(reg_file_term,
                next_dest_ex(s!1),
                next_result_mem(next(s!1))),
          rf1_of(apply(imem(Abs(s!1)), pc(Abs(s!1))))))
  = select(reg(Abs(s!1)), rf1_of(apply(imem(Abs(s!1)),
          pc(Abs(s!1))))).
```

Goals of this form are obviously true, but currently have to be left to the user
to complete. Note that the above goal would have been automatically proven if
we had a rule of the form

```
dstn_addr ≠ src_addr IMPLIES
    select(assign(reg_file_term, dstn_add, value1),
          src_addr)
    = select(assign(reg_file_term, dstn_addr, value2),
            src_addr).
```

This is another rule that is needed to complete the inversion of the abstraction
mapping. It indicates an incompleteness of our lazy method of inverting the
abstraction mapping. Our method only looks at possible boolean contexts, not
at functional contexts. Thus rules that describe how reg(Abs(s!1)) behaves
inside different selects and assigns were never generated.

5   Conclusions, Future Work and Comparison with Other Work

We have presented a methodology for structuring the human guidance of processor proofs in an interactive prover. We have seen how this methodology while
incomplete allows us to automate much of the verification and provide guidance
where human input is needed.

The methodology also gives us insight why the automatic methods work well.
In the method of [3] they generate extremely large *if-then-else* terms represented
in a DAG structure. A large part of this *if-then-else* term is devoted to different
instances of the abstraction mapping being fully expanded. Their method also
employs a heuristic to rewrite conditional terms into register file terms [4]. As
our proof indicates if this is done only a relatively few number of case analysis
(relative to the size of the whole if-then-else) is needed to finish the proof. Our

method also indicates that if the specification is not written right this heuristic may not be useful.

Our method also indicates that a good heuristic is to choose to split on specification conditions before splitting on conditions found in the expanded abstraction mapping. It would be interesting to see how well this heuristic works when in employed in an automatic procedure.

Additionally as the automatic methods tackle larger problems they are likely to hit a point where some form of human guidance is needed. Such guidance is likely to be similar to the type that users of the Boyer-Moore theorem prover use—namely providing additional lemmas. The methodology in this paper can give ideas on what additional lemmas are likely to be helpful.

Finally this is just the first experiments in using this methodology. With more experience we hope to be able to develop additional ideas on how to more fully automate the inversion of the abstraction mapping.

References

1. Martín Abadi and Leslie Lamport. The existence of refinement mappings. *Theoretical Computer Science*, 82(2):253–284, May 1991.

2. E. Boerger and S. Mazzanti. A correctness proof for pipelining in risc architectures. Dimacs technical report, DIMACS, 1996. (To appear).

3. J. R. Burch and D. L. Dill. Automatic verification of pipelined microprocessor control. In David Dill, editor, *Computer-Aided Verification, CAV '94*, volume 818 of *Lecture Notes in Computer Science*, pages 68–80, Stanford, CA, June 1994. Springer-Verlag.

4. Jerry Burch. Personal Communication.

5. D. Cyrluk and P. Narendran. Ground temporal logic—a logic for hardware verification. In David Dill, editor, *Computer-Aided Verification, CAV '94*, volume 818 of *Lecture Notes in Computer Science*, pages 247–259, Stanford, CA, June 1994. Springer-Verlag.

6. D. Cyrluk, S. Rajan, N. Shankar, and M. K. Srivas. Effective theorem proving for hardware verification. In Kumar and Kropf [9], pages 203–222.

7. David Cyrluk. Microprocessor verification in PVS: A methodology and simple example. Technical Report SRI-CSL-93-12, Computer Science Laboratory, SRI International, Menlo Park, CA, December 1993.

8. Nachum Dershowitz and Jean-Pierre Jouannaud. Rewrite systems. In Jan van Leeuwen, editor, *Handbook of Theoretical Computer Science*, volume B: Formal Models and Semantics, chapter 6, pages 243–320. Elsevier and MIT press, Amsterdam, The Netherlands, and Cambridge, MA, 1990.

9. Ramayya Kumar and Thomas Kropf, editors. *Theorem Provers in Circuit Design (TPCD '94)*, volume 910 of *Lecture Notes in Computer Science*, Bad Herrenalb, Germany, September 1994. Springer-Verlag.

10. Sam Owre, John Rushby, Natarajan Shankar, and Friedrich von Henke. Formal verification for fault-tolerant architectures: Prolegomena to the design of PVS. *IEEE Transactions on Software Engineering*, 21(2):107–125, February 1995.

11. Mandayam K. Srivas and Steven P. Miller. Formal verification of the AAMP5 microprocessor. In Michael G. Hinchey and Jonathan P. Bowen, editors, *Applications of Formal Methods*, Prentice Hall International Series in Computer Science, chapter 7, pages 125–180. Prentice Hall, Hemel Hempstead, UK, 1995.

Validity Checking for Combinations of Theories with Equality

Clark Barrett, David Dill, and Jeremy Levitt

Computer Systems Laboratory, Stanford University, Stanford CA 94305, USA

Abstract. An essential component in many verification methods is a fast decision procedure for validating logical expressions. This paper presents the algorithm used in the Stanford Validity Checker (SVC) which has been used to aid several realistic hardware verification efforts. The logic for this decision procedure includes Boolean and uninterpreted functions and linear arithmetic. We have also successfully incorporated other interpreted functions, such as array operations and linear inequalities. The primary techniques which allow a complete and efficient implementation are expression sharing, heuristic rewriting, and congruence closure with interpreted functions. We discuss these techniques and present the results of initial experiments in which SVC is used as a decision procedure in PVS, resulting in dramatic speed-ups.

1   Introduction

Decision procedures are emerging as a central component of formal verification systems. Such a procedure can be included as a component of a general-purpose theorem-prover [8, 11], or as part of a domain-specific verification system [1]. Domain-specific systems for formally verifying the correctness of hardware designs in particular are becoming increasingly important, and they are beginning to be incorporated into industrial electronic design automation tools. As part of a general design tool used by non-experts, it is important that a verification system be automatic. To be useful in practice, it is important that the system be flexible and efficient.

Complex hardware designs are hierarchical; a decision procedure to verify such a design needs to have the flexibility to abstract away lower levels of the hierarchy. While BDDs have been applied to the verification of hardware, their use has been limited to certain application domains. Although automatic and very efficient for certain problems, they do not easily support abstraction. Furthermore, they have difficultly dealing with common operations, such as multiplication. Other procedures use a system of rewrites based on Knuth-Bendix completion to attempt to prove the validity of a formula [7, 3]. However, these systems are limited and often inefficient.

For several years, we have been developing a decision procedure for use in hardware verification. The data structure we use resembles a BDD in that expressions are represented as directed acyclic graphs (DAGs) and reused whenever possible. However, by extending our domain from Boolean formulas to decidable fragments of first-order logic, we obtain a procedure which is well suited for the verification of hardware. We maintain the benefits of automation and efficiency, while overcoming the drawbacks of BDDs. Lower levels of a design hierarchy can be easily abstracted through the use of uninterpreted functions, while linear arithmetic, inequalities and other functions and relations can be naturally represented and reasoned about.

Furthermore, rather than relying on Boolean case-splitting or syntactic rewriting alone, we use a hybrid approach: we begin with a formula α which is an arbitrary Boolean combination of atomic formulas. As in previous procedures [6], we maintain a data structure (called a *context*), which is a database containing a set of logical expressions partitioned into equivalence classes. The initial context contains α and all of its subexpressions, each in its own equivalence class. The procedure begins by extracting an atomic formula β from α; it then asserts β and simplifies α in the new context (i.e. substitutes **true** for β). During this simplification, a set of heuristic rewrites are applied which reduce the size of the DAG and increase the overall performance significantly. The procedure then recursively checks the simplified formula for validity. If the recursive call concludes that the formula is not valid, the procedure halts and produces a counterexample. Otherwise, the procedure restores the pre-assertion context, denies β (i.e. substitutes **false** for β), simplifies α in the new context, and then checks the simplified formula recursively as before.

To guide the simplification process and allow the integration of rewrite rules, we define a total ordering on expressions. Intuitively, the expression ordering determines which of two expressions is the *simplest*. Given any atomic formula, we can quickly find the simplest equivalent expression in the current context. This ordering is *monotonic* with respect to term structure so that simplification of an expression using substitution and rewriting can proceed from the bottom up, substituting the simplest equivalent formula for each sub-formula.

One of the problems that a context must deal with is detecting inconsistent combinations of atomic formulas, such as $x = y$ and $f(x) \neq f(y)$. *Congruence closure* is a well-known method for dealing with such phenomena [4, 12, 9]. Although it is true as was reported in [6] that congruence closure is often unnecessary, we have found practical applications where congruence closure is required. The addition of interpreted functions complicates the congruence closure algorithm. Nelson and Oppen propose a solution in which each theory containing interpreted functions has a separate decision procedure. These procedures communicate by propagating information about equality. Several systems have decision procedures based on Nelson and Oppen's methods for congruence closure [9], including STeP [8], and the Extended Static Checker (ESC) being developed by Nelson and Detlefs et al. at DEC [10]. Our implementation most closely resembles that of Shostak [13, 2]. It provides tightly-coupled decision procedures which interact through a single congruence closure algorithm and is thus more efficient than the scheme of Nelson and Oppen.

The entire system is coded in C++ with due attention to low-level efficiency issues. Both the logic and the procedure are significantly extended from that described by Jones, Dill and Burch in 1995 [6]. Despite the addition of congruence closure, linear arithmetic and other extensions, the new system is as fast or faster than the previous version, depending on the examples used.

The remaining sections are organized as follows. We first give some background and present the basic algorithm in Section 2. Section 3 describes the ordering on expressions. Section 4 discusses our implementation of congruence closure, and Section 5 compares it with Shostak's method. Section 6 discusses extensibility, and Section 7 presents results and conclusions. The Appendix contains some of the more involved proofs.

2 Background

2.1 The Logic

Our logic has the following abstract syntax:

$$
\begin{array}{lcl}
expr & ::= & const \\
& | & function\ symbol\,(expr,\ \ldots,\ expr) \\
& | & \textbf{ite}\,(expr,\ expr,\ expr) \\
& | & (expr\ =\ expr) \\
& | & \textbf{add}\,(rational\ const,(rational\ const,\ expr),\ldots,(rational\ const,\ expr)) \\
const & ::= & \textbf{false} \\
& | & \textbf{true} \\
& | & rational\ constant \\
& | & @symbol
\end{array}
$$

Expressions are classified by *sort*: constants, uninterpreted functions, if-then-else expressions (**ites**), equalities and **add** expressions. The addition of other interpreted functions such as array operations and linear inequalities is discussed in Section 6. Constants include the Boolean constants **true** and **false**, rational constants, and special user-defined constants distinguished by an initial "@". By definition, no two constants can be equal. Variables are uninterpreted functions of arity 0. Boolean expressions are represented in terms of **ites**.

The syntax defines an abstract syntax tree; in the remainder of the paper whenever we refer to an expression, we are referring to its abstract syntax tree. If α and β are two expressions, we say that α is a subexpression of β if α is a subtree of β. We say that α is a child of β if α is a subtree rooted at a child of the root of β. In this case, we also say that β is a parent of α. The depth of an expression is defined to be the height of its tree. Formally,

Definition 1 (Depth) *The depth of an expression α, $D(\alpha)$, is defined recursively as: if α has no children, $D(\alpha) = 0$. Otherwise, $D(\alpha) = max\{D(c) \mid c$ is a child of $\alpha\} + 1$.*

Since **add** expressions are interpreted, they behave a little bit differently. We define the depth of an **add** expression to be the depth of its least simple child. In effect, the least simple child represents the **add** expression when it is compared with other sorts of expressions. More will be said about this in Section 3. It is also necessary to note that the child of an **add** expression in normal form cannot be an **add** expression itself.

The following lemma is easily shown by induction on $D(\beta)$.

Lemma 1. *If α is a subexpression of β and β is not an **add** expression, then $D(\alpha) < D(\beta)$.*

An important concept is that of an *atomic* expression. An atomic expression is defined as an expression containing no **ite** subexpressions.[1]

As was previously stated, expressions are represented by DAGs. Each expression has an entry in a hash table. This ensures that there is one unique copy for every expression and enables sharing of subexpressions in the DAG. Before inserting an expression into the hash table, certain rewrite rules are applied. An example of such a rewrite rule is:

$$\mathbf{ite}(\alpha, \alpha, \beta) \;\rightarrow\; \mathbf{ite}(\alpha, \mathbf{true}, \beta).$$

As mentioned above, a context consists of a set of expressions partitioned into equivalence classes. These classes are maintained using Tarjan's *union* and *find* primitives [2] [14]. Each equivalence class contains a unique representative element called the *equivalence class representative* which is an expression used to represent the class. If expressions α and β are in the same equivalence class in context C, we say that $\alpha \sim_C \beta$.

2.2 Validation Algorithm

A simple validation algorithm (based on the one used in [6]) is shown in Figure 1. Given an expression, we pick an atomic expression (called a *splitter*) that appears as the if-part of an **ite** subexpression and perform a case split. We repeat this recursively, reducing the **ite** to its then-part or else-part on each case

---

[1] Ideally, an atomic expression would be an expression containing no Boolean subexpressions. However, since we do not store explicit type information we can only identify a Boolean expression if it appears in the if-part of an **ite**. Though this theoretically limits the completeness of our algorithm, we have not found it to be a problem for practical applications. The issue can be resolved fairly easily by adding explicit typing or more sophisticated type inference.

[2] In our implementation, (as in [13]) *union*(a, b) is deterministic about which find pointer gets changed; in our case it always sets *find*(b) equal to a. This is different from Tarjan's *union* which sets the *find* of the tree with smaller rank. Although this decreases the theoretical worst-case performance of *find*, the actual impact is negligible since the program spends very little time in the *union/find* code. Also, we do use path-compression (in *find*) which has a much more significant effect than union-by-rank since there are many more calls to *find* than to *union*.

```
Validate(e)                              Assert(e)
  e := Simp(e);                            CASE e OF
  splitter := FindSplitter(e);               a = b : Merge(a, b);
  IF e = true THEN                           ELSE  : Merge(true, e);
    RETURN true;                           ENDCASES
  IF splitter = NULL THEN
    RETURN false;                        Deny(e)
  PushContext;                             CASE e OF
  Assert(splitter);                          a = b : add (a,b) to diseq list
  result := Validate(e);                     ELSE  : Merge(false, e);
  PopContext;                              ENDCASES
  IF result = true THEN BEGIN
    PushContext;                         Merge(e1,e2)
    Deny(splitter);                        union(e1,e2);
    result := Validate(e);
    PopContext;                          NewExpr(e)
  END                                      Perform rewrites on e
  RETURN result;                           IF e not in hash table THEN BEGIN
                                             insert e in hash table
Simp(e)                                      find(e) := e;
  e := Signature(e);                       END
  IF e is of the form a = b AND            RETURN e;
    (a',b') ∈ diseq list
    where find(a')=a and               Signature(e)
          find(b')=b THEN                  RETURN
    RETURN false;                            NewExpr(f(Simp(t_1),...,Simp(t_n)));
  ELSE                                       where e = f(t_1,...,t_n)
    RETURN find(e);
```

Fig. 1. Basic Validation Algorithm. Note that **true** indicates the constant expression, whereas true simply refers to the Boolean value.

split. If the expression reduces to **true** when all the splitters are exhausted, we return true, otherwise we return false.

The algorithm presented in this paper is a refinement of the one shown in Figure 1, so we will start by describing this simpler version first. The purpose of Merge is to join the equivalence classes of the two atomic expressions passed to it. Merge is always called with the first expression simpler than the second; in particular, equality expressions are always rewritten to ensure that the left-hand side is simpler than the right-hand side. We will discuss what it means for one expression to be simpler than the other in the next section. Recall that union makes the first argument the new equivalence class representative for the merged equivalence class. Simp recursively traverses an expression, replacing each subexpression with its equivalence class representative. It also detects equalities which are contradictory. This is done by maintaining a list of disequalities which is updated whenever an equality is denied. The significance of the Signature function will be discussed in Section 4. NewExpr allows for expression sharing by checking if an expression already exists before creating it. One important feature of our implementation is that the function `FindSplitter` is free to choose any atomic Boolean expression, making it easy to add or change splitting heuristics.

The algorithm presented in Figure 1 is sound and complete for Boolean tautology checking but is incomplete for logics which include interpreted and uninterpreted functions. For example, it cannot even validate

$$f(a) \neq f(b) \Rightarrow a \neq b.$$

We will show how to complete the algorithm in Section 4, but first we discuss what it means for one expression to be simpler than another and why this is important.

3   Expression Ordering

We use a set of rules to determine a total ordering $\prec$ on the expressions in our logic. If $\alpha \prec \beta$, we say that α is *simpler* than β. This ordering was designed to have two convenient monotonicity properties. First, it is monotonic with respect to subexpressions. That is, if α is a subexpression of β, then $\alpha \prec \beta$. Second, it is monotonic with respect to substitution, so that if we replace a subexpression with a simpler subexpression, the result is simpler. These properties aid intuition as well as implementation.

As demonstrated by Shostak [13], it is possible to efficiently implement tightly-coupled interacting decision procedures without an ordering on expressions. However, there are several significant benefits to implementing an ordering: without a monotonic ordering, it is possible to have $find(a) = b$ and $find(f(b)) = f(a)$. This kind of behavior is counterintuitive and can increase the difficulty and the complexity of the implementation significantly. On the other hand, monotonicity ensures that whenever we assert that two expressions are equal, the simpler of the two can be substituted for the other throughout the DAG resulting in a more intuitive representation. More importantly, such substitution increases sharing and leads to a more compact representation in the DAG.

Perhaps most importantly, by enforcing that any rewrites which are applied always result in a simpler expression, a monotonic ordering ensures that the heuristic rewrites cannot form an infinite loop. Such rewrites can have a significant impact on performance. On a set of test cases taken from actual verification work, the average speed-up with rewrites is about 2, excluding one large example which has a speed-up of 14.5.

The ordering we use is defined by the following rules (applied in order).

1. Constant expressions are always simpler than non-constant expressions. For arbitrary Boolean, rational and user-defined constants b, r and c, we define $b \prec r \prec c$. For Booleans, we simply have **false** $\prec$ **true**. Rational constants are ordered numerically and user-defined constants are ordered lexicographically.

2. **add** expressions behave like their most complex child when compared with expressions of a different sort (if comparing directly with the most complex child, the child is simpler). When comparing two **add** expressions, their children are compared from most complex to least complex. The first pair of children which are not equivalent determines the ordering.

3. If two expressions have different depths, the one with the smaller depth is simpler.

4. We define uninterpreted functions to be simpler than equalities and equalities to be simpler than **ites**. If two expressions are of the same sort, they are compared as follows: If the two expressions are uninterpreted functions with different function names, the expression with the lexicographically simpler name is the simplest; otherwise, the children of the expressions are compared in order, and the first pair of children which are not equivalent determine the ordering.

Lemma 2. *Rules 1-4 determine a total order on expressions.*

The proof is omitted, but is straightforward and can be accomplished by case splitting. We state the monotonicity properties in two theorems.

Theorem 3. *If α is a subexpression of β then $\alpha \prec \beta$.*

Theorem 4. *If $\alpha' \prec \alpha$, α is a subexpression of β, and β' is the result of replacing α with α' in β, then $\beta' \prec \beta$.*

The proof of the Theorem 3 is in the appendix, and the proof Theorem 4 is omitted.

4 Congruence Closure

A context C is said to be closed with respect to congruences if the following property holds for all expressions α and β in C:

Property 1 *If α and β are expressions of the same sort with the same number of children, and if each pair of corresponding children are in the same equivalence class, then α and β are in the same equivalence class.*

The notion of a *signature* [4] is helpful in ensuring that this property holds. The signature of an expression e denoted $signature(e)$ is defined to be the expression in which each child has been replaced with its equivalence class representative (notice that this is what the function **Signature** in Figure 1 does). Each expression u maintains a pointer to the expression which is its signature. We denote this expression by $sig(u)$. In order to ensure that Property 1 holds, we must enforce that $sig(u) = signature(u)$. Each equivalence class representative maintains a list (called the use-list) of expressions in which the equivalence class representative appears as a child.

```
Simp(e)                                   Deny(e)
  IF e is atomic THEN                       Merge(false, e);
    RETURN find(e);
  ELSE                                    NewExpr(e)
    RETURN find(Signature(e));             Solve, normalize, and rewrite
                                           IF e is in hash table THEN
                                             RETURN e;
Merge(a, b)                              ELSE BEGIN
  IF find(a) ≠ find(b) THEN BEGIN          insert e in hash table
    IF IsConst(a) AND IsConst(b) THEN      sig(e) := e;
      Inconsistent := true;                find(e) := e;
    ELSE BEGIN                             use(e) := {};
      union(a, b);                         FOREACH child c of e DO
      FOREACH u IN use(b)                    use(c) := use(c) ∪ e;
        IF sig(u) = u THEN BEGIN          END
          sig(u) := Signature(u);        RETURN e;
          IF find(u) ≠ find(sig(u)) THEN
            IF u = find(u) THEN
              Merge(sig(u), u);
            ELSE
              Assert(NewExpr(find(u) = find(sig(u))));
        END
    END
  END
END
```

Fig. 2. Modifications to the original algorithm.

Figure 2 shows the modifications needed to implement congruence closure. It also includes modifications necessary for dealing with **add** expressions. NewExpr now puts **add** expressions in a normal form with respect to their children, in addition to performing rewrites. Furthermore, if an equality involves one or two **add** expressions, it is solved so that the most complex child of the **add** expressions appears alone on the right-hand side. This guarantees that when we assert such an equality, the most complex child expression will be replaced with a sum of simpler expressions. Thus, variable elimination occurs automatically. In fact, the interaction between solving and congruence closure is fairly subtle and is one of the reasons that expression ordering was originally introduced; we wanted to guarantee that solving produces a simpler expression.

Whereas the algorithm in Figure 1 is unable to detect inconsistent assertions, detecting inconsistency in the new algorithm is surprisingly simple. A context is inconsistent if and only if there is an equivalence class which contains more than one constant. This is because any other inconsistency will eventually propagate to equating **true** and **false** which are both constants.

The purpose of the additional code in Merge is to maintain the following invariant which we show is equivalent to Property 1. Recall that $u \sim_C v$ means that u and v are in the same equivalence class (i.e. $find(u) = find(v)$).

Theorem 5. *A context C satisfies Property 1 if for each expression u, $u \sim_C$ signature(u).*

Proof: Suppose $u \sim_C$ signature(u) for each expression u in C. Let α and β be any two expressions which are of the same sort and for which corresponding children are in the same equivalence class. By definition, then, signature$(\alpha) =$ signature(β). But we know that find$(\alpha) =$ find$(signature(\alpha))$ and find$(\beta) =$ find$(signature(\beta))$, so find$(\alpha) =$ find(β) and thus $\alpha \sim_C \beta$. $\quad\quad\square$

Since our algorithm creates new signatures as it goes, a concern is whether an infinite number of new signatures could be generated. For example, if we assert $f(x) = x$ and if we make find$(x) = f(x)$, then the signature of $f(x)$ becomes $f(f(x))$. This process could then repeat. Theorem 3 guarantees that this will not happen since it is impossible for the equivalence class representative of an expression to be one of its subexpressions. This is another benefit of monotonicity in our expression ordering.

We claim that with the modifications shown in Figure 2, every pair of expressions in the context will satisfy Property 1 on each call to `Validate`:

Theorem 6. *Each time `Assert` or `Deny` is called from `Validate`, the resulting context is closed with respect to congruences.*

The proof can be found in the appendix.

A fairly significant optimization is to only maintain congruence closure for atomic expressions that are not constants. We can do this because completeness only requires that we know whether an expression is true in a terminating case of `Validate`, and in all such terminating cases the final expression is atomic.

An additional benefit of implementing congruence closure is the ease with which disequalities can be handled. In Figure 1, disequalities between equivalence classes are maintained in a special-purpose disequality table. In the new algorithm, in order to deny $a = b$ (or conceptually, assert $a \neq b$), we simply merge $a = b$ with **false**. Now, if we ever try to equate a and b, the equality will become equal to **true**. As mentioned above, this will result in attempting to merge **true** and **false**, and the inconsistency will be discovered.

5 A Comparison to Shostak's Algorithm

As stated in the introduction, our implementation most closely resembles Shostak's algorithm for congruence closure [13]. Recently, Cyrluk et al. published a complete and rigorous analysis of Shostak's algorithm[2]. For ease of comparison, we have written the code in this paper in a similar fashion. Despite the similarities, there are some significant differences.

First, and perhaps most importantly, Shostak's method requires that expressions be converted to disjunctive normal form. Our algorithm does not require this. Not only does this eliminate the overhead of converting to DNF, but it also gives our technique a tremendous advantage when there is a large amount of sharing in the expression DAG.

Second, we allow the inclusion of heuristic rewrite rules which can significantly improve performance (as discussed earlier).

Third, we implement signatures as actual expressions, rather than as tags associated with expressions. This allows us to compute the congruence closure using only a single loop whereas Shostak's method requires a double loop. The reason for this is as follows. Suppose we merge expressions a and b so that $find(b)$ becomes a. We want to guarantee that every expression whose signature contains b is updated and then merged with any other expressions which have the same signature. A single loop over all parents of b is necessary to update all the signatures. In Shostak's implementation, a second loop is required: if u is a parent of b whose signature has changed, then for each such u, all parents of a are checked to see if any have the same signature as u. Since we represent signatures as expressions, the second loop is unnecessary: we simply need to ensure that each expression is in the same equivalence class as its signature. Since the signatures of the parents of a have not changed, there is no need to revisit them. While compared to Shostak's procedure our algorithm generates many extra expressions when a signature changes multiple times, in addition to making the algorithm simpler these extra expressions are actually required in order to return to previous contexts. To avoid updating these old signatures in the congruence closure computation, we check each expression in the use list to see if it is its own signature. If it is not, we know it is an old signature and skip it.

The cost of checking old signatures is also offset by another advantage which comes from using expressions for signatures. Shostak's code requires that the use list of an equivalence class representative contains all expressions which have a child in that equivalence class. We only require that their signatures be in the use list. Thus if there are multiple expressions with the same signature, we have a single entry in the use list whereas an implementation of Shostak's algorithm will have an entry for each expression.

Finally, another difference is the aggressiveness with which we update and simplify expressions. Shostak's implementation waits until an expression is used in an assertion before adding it to use lists; we add expressions to use lists as soon as they are created and we have found it to be faster for many examples. This may be due to the difference between reducing a DAG and reducing a conjunct of formulas.

Other minor differences include the ability to handle disequalities, and support for user-defined constants.

6 Extensions

An advantage of Shostak's method is that his decision procedure easily accommodates new theories with interpreted functions, as long as they are canonizable and algebraically solvable [13]. While our procedure places slightly more stringent requirement on new theories, new theories are similarly easy to accommodate. We require that each new sort of expression be totally ordered and canonizable and that it be possible to solve an equation over interpreted functions for the most complex variable.[3] These are the requirements we meet to support **add** expressions.

A more complicated extension is adding linear inequalities, since the data structures which exist for congruence closure are not sufficient to store all of the implications of asserting an inequality. We solve this problem by adding additional lists at each unasserted inequality which contain a set of expressions implied by the current context. This approach is slow for expressions which are dominated by inequalities, but quite satisfactory for expressions which contain a mix of inequalities, linear arithmetic, Boolean formulas, and array operations.

At the cost of making the procedure incomplete, it is possible to even add interpreted functions which do not satisfy the requirements for canonizability or solvability. A good example of this is the addition of **read** and **write** as new sorts of expressions which implement basic array operations. **read** takes an array and an address and returns the element at that address. **write** takes an array, an address, and a value, and returns the given array with the new value at the specified address. Instead of providing a complete theory for these operations, we treat them as uninterpreted functions and add an automatic rewrite which reduces **read**(**write**$(s, a1, v), a2)$ to **ite**$(a1 = a2, v, \textbf{read}(s, a2))$.[4] Our algorithm is now incomplete in that it is unable to deal with cases in which **write** expressions are directly equated with other expressions. However, it has been sufficient to deal with most of the verification examples that we have encountered. This demonstrates a further advantage of being able to include rewrites.

7 Results and Conclusions

As a point of comparison, we are experimentally using SVC to assist in proofs done using PVS. Where possible, SVC is used as a decision procedure in place of PVS's internal procedures which are an implementation of Shostak's algorithm. Often, SVC also replaces sequences of Boolean simplifications that are necessary in PVS to put formulas into a disjunctive normal form. Since SVC allows arbitrary Boolean formulas, such simplifications are not required.

So far, our experiments have included proving a fragment of a bounded retransmission protocol verified in PVS by Havelund and Shankar [5] and a simple three stage microprocessor pipeline. The results of running the proofs on a HyperSPARC with 128 MB memory are shown in Table 1 and demonstrate significant speedups even for these small examples.

The speedup in the actual decision procedure is even more significant than shown by the data. In the data for PVS with SVC, the time spent inside of SVC

---

[3] As in [13], equations involving interpreted functions from more than one theory solve the topmost interpreted functions by treating interpreted functions from other theories as variables.

[4] Adding this rewrite requires a slight adjustment to the expression ordering.

Example	PVS	PVS with SVC
protocol	78.77s	16.11s
pipeline	56.65s	19.71s

Table 1. PVS example with and without SVC decision procedure.

is very small (less than one second) compared to the time PVS spends preparing to send the data to SVC. Of course it is unfair to conclude from these data alone that our algorithm is significantly superior to Shostak's since PVS is coded in LISP and SVC is coded in C++. It does, however, provide some insight into how SVC compares with other verification tools.

We have presented an efficient and flexible algorithm for validity checking with equality and uninterpreted functions. The algorithm improves on previous work by combining the efficiency and speed of [6] with the completeness and extensibility of [13]. SVC has been successfully used both as a brute-force hardware verification tool, and as a fast supplemental decision procedure for more general theorem provers.

Some of the future work we envision is improving the implementation of linear inequalities, increasing the number of interpreted functions in the logic, and developing improved heuristics for choosing splitters. It is interesting to note that determining how to choose splitters is very similar to the problem of BDD variable ordering and so we expect there will be a significant advantage in finding good heuristics.

Acknowledgments

We would like to thank Jens Skakkebæk for his invaluable help in putting together and conducting the experiments with PVS, and for uncovering bugs in SVC. We would also like to thank Robert Jones whose initial work on SVC laid the foundation for our continuing effort and who provided much appreciated feedback as we prepared this paper. This research was supported by ARPA contract DABT63-95-C-0049, and by a National Defense Science and Engineering Graduate Fellowship.

Appendix

Proof of Theorem 3: We first show the following lemma.

Lemma 7. *If $D(\alpha) < D(\beta)$ then $\alpha \prec \beta$*

Proof: Let α' be α if α is not an **add** expression and the most complex child of α if α is an **add** expression. Define β' similarly so that the ordering of α and β is determined by the ordering of α' and β' in accordance with Rule 2. Since **add** expressions behave and have the same depth as their most complex child, we know that $D(\alpha') < D(\beta')$ and since **add** expressions cannot contain other **add** expressions as children, we know that neither α' nor β' are **add** expressions. Now β' cannot be a constant since its depth is greater than the depth of α' and constants have 0 depth. Thus if Rule 1 applies, it must be the case that α' is a constant so that $\alpha' \prec \beta'$. If Rule 1 does not apply, then the ordering is determined by Rule 3 which directly implies that $\alpha' \prec \beta'$. Thus $\alpha \prec \beta$. $\square$

We now proceed to prove Theorem 3. If β is not an **add** expression, then by Lemma 1, $D(\alpha) < D(\beta)$ and thus by Lemma 7, $\alpha \prec \beta$. If β is an **add** expression, then there are two cases. If α is a child of β, then by Rule 2, $\alpha \prec \beta$. If α is is a subexpression of a child (recall that the child cannot be an **add** expression), then α has a smaller depth than the child by Lemma 1. Now since the depth of the **add** expression is the depth of its most complex child, it must be the case that $D(\alpha) < D(\beta)$ so that by Lemma 7, $\alpha \prec \beta$. $\square$

Proof of Theorem 6: Initially, every expression is its own signature and its own equivalence class representative, so by Theorem 5, the context satisfies Property 1. Now we must show that this property still holds after a call to `Assert` or `Deny`.

Suppose that we have an arbitrary context C which satisfies Property 1. In the following discussion, we will subscript *find* and signature with the context to which we are referring. Thus we have:

$$\forall\, u \in C.\ u \sim_C signature_C(u).$$

Let $C_0 = C$ and let C_n be the context which results from calling `Assert` or `Deny` from `Validate` so that for $0 \le i \le n$, C_i represents an intermediate context, and all such intermediate contexts are represented by some C_i. We will use three lemmas:

Lemma 8. *If $u \sim_{C_i} v$ then $u \sim_{C_j} v$ where $0 \le i \le j \le n$.*

Proof: Each context is derived from the previous context by either adding an expression or merging two equivalence classes. Thus, once two expressions are in the same equivalence class, they will always be in the same equivalence class.

Lemma 9. *If $i < j$ then $signature_{C_j}(signature_{C_i}(u)) = signature_{C_j}(u)$.*

Proof: Let c be an arbitrary child of u. We know that $c \sim_{C_i} find_{C_i}(c)$. By Lemma 8, we know that $c \sim_{C_j} find_{C_i}(c)$. This means that in context C_j, c and $find_{C_i}(c)$ have the same equivalence class representative. Since c and $find_{C_i}(c)$ are corresponding children in u and $signature_{C_i}(u)$ respectively, we thus conclude that u and $signature_{C_i}(u)$ have the same signature in context C_j. $\square$

Lemma 10. *For arbitrary expressions u and v, as a result of executing* `Assert(NewExpr(find(u) = find(v)))` *in context C_i, it will be the case that $u \sim_{C_j} v$ for some $j > i$.*

Proof: As mentioned above, `NewExpr` rewrites equalities into a normal form in which the left hand side is simpler than the right hand side. Let $\alpha_1 = find_{C_i}(u)$ and $\alpha_2 = find_{C_i}(v)$. These are already normal form expressions (i.e. no rewrites should apply), so as long as neither one is an interpreted function, the result of `NewExpr` will simply be an equality with the simpler of the two on the left and the other on the right. Thus the call to `Assert` leads directly to a call to `Merge` and thus $u \sim_{C_{i+1}} v$. If α_1 or α_2 is an interpreted function such as **add**, `NewExpr` will solve the equality to place the most complicated variable alone on the right hand side of the equation. Thus we will have a new equation of the form $\alpha' = \beta$ where β is a single variable which appears in either α_1 or α_2. Assume without loss of generality that it appears in α_1. When `Assert` is called on $\alpha' = \beta$, we will go through everything on the use list of β and eventually find α_1. We will replace β with α' in α_1 to get α_2 (or a signature which eventually gets put in the same equivalence class as α_2 if other simplifications on α_2 were pending when the assertion took place). Now, since α_1 and α_2 eventually end up in the same equivalence class in some context $j > i$, it must be the case that $u \sim_{C_j} v$. □

We now proceed with the proof of Theorem 6. Suppose that C_n does not satisfy Property 1. Then there exists some expression e such that it is not the case that $e \sim_{C_n} signature_{C_n}(e)$. Let i be the minimum for which $e \in C_i \wedge \neg(e \sim_{C_i} signature_{C_i}(e))$ (obviously $i > 0$). Consider C_{i-1}. It cannot be the case that $e \notin C_{i-1}$, because `NewExpr` ensures that each new expression is both its own signature and its own equivalence class representative. It must be the case, then, that $e \sim_{C_{i-1}} signature_{C_{i-1}}(e)$. Obviously $signature_{C_{i-1}}(e) \neq signature_{C_i}(e)$. This means that for some child e' of e, $find_{C_{i-1}}(e') \neq find_{C_i}(e')$. The only way for this to happen is if C_i is the result of calling $union(a, b)$ where $b = find_{C_{i-1}}(e')$ and $a = find_{C_i}(e')$. In this case, $signature_{C_{i-1}}(e)$ is on the use list of b. Assuming that we maintain sig pointers correctly so that $sig(u) = signature(u)$[5], the body of the loop will be executed with u set to $signature_{C_{i-1}}(e)$. This will occur in some context C_j where $i \leq j$. By Lemma 9, $signature_{C_j}(u) = signature_{C_j}(e)$. The result of the body being executed is either nothing, if $u \sim_{C_j} signature_{C_j}(u)$, or a call to `Merge` which merges u and $signature_{C_j}(u)$, or a call to `Assert` which results in $u \sim C_{j'} signature_{C_j}(u)$ for some $j' > j$ by Lemma 10. In each case, we know that $u \sim_{C_k} signature_{C_k}(u)$ where C_k is the context after executing the body of the loop (clearly $k \leq n$). But by Lemma 9, $signature_{C_k}(u) = signature_{C_k}(e)$. Thus, we have $signature_{C_{i-1}}(e) \sim_{C_k} signature_{C_k}(e)$. And by Lemma 8, we know that $e \sim_{C_k} signature_{C_{i-1}}(e)$. So $e \sim_{C_k} signature_{C_k}(e)$ in contradiction to our assumption that e was not equivalent to its signature in contexts C_i through C_n. □

---

[5] It is easy to see that this is true for atomic expressions, the only relevant case in our optimized algorithm.

References

1. J. R. Burch and D. L. Dill, "Automatic Verification of Microprocessor Control", In Computer Aided Verification, 6th International Conference, 1994.
2. D. Cyrluk, P. Lincoln and N. Shankar, "On Shostak's Decision Procedure for Combinations of Theories", Proceedings of the 13th International Conference on Automated Deduction, New Brunswick, NJ, July 1996, 463-477.
3. A. J. J. Dick, "An Introduction to Knuth-Bendix Completion", The Computer Journal 34(1):2-15, 1991.
4. P. J. Downey, R. Sethi and R. E. Tarjan, "Variations on the Common Subexpression Problem", Journal of the ACM, 27(4):758-771, 1980.
5. K. Havelund and N. Shankar, "Experiments in Theorem Proving and Model Checking for Protocol Verification", In Proceedings of Formal Methods Europe, March 1996, 662-681.
6. R. B. Jones, D. L. Dill and J. R. Burch, "Efficient Validity Checking for Processor Verification", IEEE/ACM International Conference on Computer Aided Design, 1995.
7. D. E. Knuth and P. B. Bendix, "Simple Word Problems in Universal Algebras", In Computational Problems in Abstract Algebra, ed. J. Leech, 263-297, Pergamon Press, 1970.
8. Z. Manna, et al., "STeP: the Stanford Temporal Prover", Technique Report STAN-CS-TR-94, Computer Science Department, Stanford, 1994.
9. G. Nelson and D. C. Oppen, "Simplification by Cooperating Decision Procedures", ACM Transactions on Programming Languages and Systems, 1(2):245-257, 1979.
10. G. Nelson, D. Detlefs, K. R. M. Leino and J. Saxe, "Extended Static Checking Home page", <URL:http://www.research.digital.com/SRC/esc/Esc.html>, 1996.
11. S. Owre, et al., "Formal Verification for Fault-Tolerant Architectures: Prolegomena to the Design of PVS", IEEE Transactions of Software Engineering, 21(2):107-125, 1995.
12. R. E. Shostak, "An Algorithm for Reasoning About Equality", Communications of the ACM, 21(7):583-585, 1978.
13. R. E. Shostak, "Deciding Combinations of Theories", Journal of the ACM, 31(1):1-12, 1984.
14. R. E. Tarjan, "Efficiency of a Good but not Linear Set Union Algorithm", Journal of the ACM, 22(2):215-225, 1975.

A Unified Approach for Combining Different Formalisms for Hardware Verification*

Klaus Schneider and Thomas Kropf

University of Karlsruhe, Department of Computer Science,
Institute for Computer Design and Fault Tolerance (Prof. D. Schmid),
P.O. Box 6980, 76128 Karlsruhe, Germany,
e-mail: {Klaus.Schneider, Thomas.Kropf}@informatik.uni-karlsruhe.de,
http://goethe.ira.uka.de/hvg/

Abstract. *Model Checking as the predominant technique for automatically verifying circuits suffers from the well-known state explosion problem. This hinders the verification of circuits which contain non-trivial data paths. Recently, it has been shown that for those circuits it may be useful to separate the control and data part prior to verification. This paper is also based on this idea and presents an approach for combining various proof approaches like model checking and theorem proving in a unifying framework. In contrast to other approaches, special proof procedures are available to verify circuits with data sensitive controllers, where a bidirectional signal flow between controller and data path can be found. Generic circuits can be verified by induction or by model checking finite instantiations.*

By giving the system 'proof hints', also the verification effort for model checking based proofs can be considerably reduced in many cases. The paper presents an introduction to the different proof strategies as well as an algorithm for their combination. The underlying C@S system also allows the efficiency evaluation of different approaches to verify the same circuits. This is shown in different case studies, demonstrating the trade-off between interaction and verifiable circuit size.

1 Introduction

Formal verification is the task of proving that a given specification holds for a certain design. Although more powerful than traditional simulation, a breakthrough in industrial use has been achieved only after the introduction of binary decision diagrams and symbolic state traversal algorithms [1, 2], leading to powerful techniques like symbolic model checking [3]. Using these approaches, a fully automated verification of significantly large circuits has become possible. However, as only finite state systems can be modeled, the size of the processable circuits is limited, due to the so-called state explosion problem. This restriction does not hold for theorem prover based approaches, where mostly higher-order

---

* This work has been financed by the DFG project Automated System Design, SFB No.358.

logic is used as the underlying formalism [4]. Unfortunately, these approaches require a considerable amount of manual interaction. Thus various approaches have been presented to partially automate the verification by incorporating automated reasoning procedures [5, 6] or by adding abstraction and compositional verification techniques to allow larger systems to be verified than by finite state approaches [7, 8, 9].

Many circuits are composed of a controller and a data path. Recently it has been shown, that it is useful to separate these parts prior to verification [10, 11, 12, 13, 14]. Proceeding this way, finite state approaches may be used for the controller part 'guiding' the verification of the data path, whereas the latter often requires theorem proving techniques. However, these approaches are also not able to provide full automation for circuits with heavy interaction between control and data. This paper presents a new approach which provides a general framework called C@S for exploring the combination of different proving techniques. The system aims at providing many different decision procedures which are invoked as soon as automizable proof goals are detected. Other verification goals are split up interactively into subgoals until they can be finally proved by decision procedures. C@S has been implemented on top of the HOL [15] system and has interfaces to the SMV [16] system and RRL [17]. As a result, C@S currently enriches HOL by the following decision procedures and proof methods:

- linear temporal logic theorem proving and model checking
- CTL model checking as implemented in SMV
- ω-automata
- Pressburger arithmetic
- Büchi's monadic second-order theory of one successor
- inductionless induction as implemented in the RRL system
- invariant rules for eliminating goals with data dependent processing

In order to ease the adaption of proof goals to different proof approaches, a class of higher-order formulas called hardware formulas has been defined as a uniform representation for specifications and implementations [18]. In hardware formulas, time and data is represented separately. Once a proof goal has been transformed in hardware formulas, it may be converted according to the proof method to be used. We have provided an algorithm for automatically choosing the combination of proof procedures for a given circuit.

In case of data dependent loops, invariant based techniques may be used to cut the data dependency. The resulting proof goals without data dependency can then be proved by traditional inductive reasoning or simple rewriting. Thus it is possible to perform inductive proofs for generic circuits with data dependencies. Moreover, those techniques also allow to indicate 'proof hints' to the system that allow to significantly reduce the effort of model checking. Thus it is sometimes possible to verify a generic circuit with a concrete but large desired bit width, not being possible before.

The idea of enriching interactive theorem provers for higher order logic by powerful decision procedures can also be found in the PVS system. However, PVS does not support a proof methodology which is based on a separation of control and data flow and no common formalism like hardware formulas are used. Our specification language is based on linear time temporal logic (LTL), enriched by data equations. This language is similar to the CTL extension used for word level model checking [14]. This approach aims at a full automation, hence only finite bitvectors are possible as abstract data types. The approach of Hojati and Brayton [12] can be seen as complementary to our approach as especially for circuits without full control/data interaction valuable decidability results have been achieved, which can be directly used in C@S. By using invariants we are also able to cope with those circuits, not treated in [12] – the price to be paid for that is interaction. The approach of Hungar, Grumberg and Damm [13] is also based on the idea to separate control and data. They use FO-ACTL (first order version of ACTL), whereas we support full LTL enhanced by data equation expressions. As they use a first-order variant of temporal logics the reasoning process has to be performed using a dedicated calculus involving the construction of first-order semantic tableaux. We try to use a more rigorous separation of control and data such that proving tools for propositional temporal logics may be used in conjunction with e.g. term rewriting systems.

Moreover, to our knowledge this is one of the first approaches, where actually verified case studies are presented, allowing to compare the pros and cons of different approaches to verify the same circuit.

The outline of the paper is as follows: the next two sections describe specification and verification techniques of C@S. Section 4 illustrates the use of some of these strategies by considering different case studies.

2   Specification Language

The key property of a specification is its readability. Hence, abstraction [19] has to be used for the structure of the modules, for the underlying temporal behavior and for the data that is processed in the system in order to lift the descriptions to a level that is appropriate for the designer. Structural abstraction is required for hierarchical verification where at the next level of hierarchy, an implementation description is replaced by its specification. C@S offers all kinds of abstractions, but due to lack of space, only temporal abstraction and data abstraction is considered in this paper.

2.1   Temporal Properties

Temporal properties can be captured by a lot of different formalisms, e.g. linear temporal logic, computation tree logic, μ-calculus, ω-automata or simple arithmetic. Due to its readability and expressiveness, linear temporal logic is in most cases the best choice for specifying the temporal behavior of a system. For this reason, temporal properties are usually specified using this logic [20] in C@S.

Especially for the specification of event-oriented properties, the W operator has proved to be convenient. $[x \ W \ b]$ expresses the fact, that x must hold when the event signal b becomes 1 for the first time. Together with the 'Next-Time' operator X, W builds a 'temporal basis' which allows to express all kinds of temporal relations, as e.g. $Gx = [0 \, W \, (\neg x)]$ (x holds always), $Fx = \neg \, [0 \, W \, x]$ (x holds at least once) or $[x \, U \, b] = [b \, W \, (x \rightarrow b)]$ (x holds until b).

In the following, the W operator is used as a representative to illustrate the reduction to other formalisms. The specification language of C@S has also other built-in temporal operators, which are treated analogously.

In some other cases, other formalisms such as arithmetic or ω-automata are more useful. C@S directly supports temporal logics and ω-automata; a translation procedures for Büchi's monadic second-order theory is currently implemented.

2.2 Abstract Data Types

In order to process higher order data types $\mathbb{C}$ with a digital circuit, a finite number of boolean values is usually collected and interpreted as single unit according to an interpretation function $\Phi_{\mathbb{C}}$. For example, if natural numbers are to be processed, then a tuple $[a_n, \ldots, a_0]$ of boolean values is often interpreted as the natural number $\Phi_{\mathbb{N}}([a_n, \ldots, a_0]) := \sum_{i=0}^{n-1} 2^i \times a_i$. Each specification using the abstract data type $\mathbb{C}$ can therefore be described at three levels of abstractions:

- First, the data operations are formulated on the abstract data type $\mathbb{C}$, as e.g. $\Phi_{\mathbb{N}}([b_2, b_1, b_0]) = \Phi_{\mathbb{N}}([a_1, a_0]) + 1$. This is the most readable version, but the verification of these descriptions requires to implement a special decision procedure for each used data type $\mathbb{C}$.
- The data operations can directly be formulated as propositional formulae on the bits $a_n, \ldots, a_0$, where n has to be a concrete number. Using two bits, the above example is represented as $(b_0 = \neg a_0) \wedge (b_1 = a_1 \oplus a_0) \wedge (b_2 = a_1 \wedge a_0)$. The advantage of this description is that simple tautology checking can be used as decision procedure. However, these descriptions are not readable at all and do not allow n-bit specifications.
- Finally, the data operations may be formulated at the 'bitvector level', where a bitvector is defined to be a boolean list with *an arbitrary length*. Using predefined bitvector operators, the above example looks like $[b_2, b_1, b_0] = $ INC $([a_1, a_0])$, where INC is defined as follows $((B \rhd b)$ appends b at the right hand side of the bitvector B.):

$(\text{INC} \, ([]) = [1]) \wedge$
$(\text{INC} \, ([a_n, \ldots, a_0]) = \text{if } a_0 \text{ then } (\text{INC} \, ([a_n, \ldots, a_1])) \rhd 0) \text{ else } [a_n, \ldots, a_1, 1]$

This level has many advantages: first its readability is comparable to descriptions at the abstract level. The only data type that is really used is 'bitvector', hence it is sufficient to have a decision procedure for bitvectors. This circumvents the problem of writing for each data type a separate decision procedure. For fixed lengths of bitvectors, it is also possible to translate these descriptions to propositional logic such that tautology checking can serve as decision procedure.

3 Verification Strategies

3.1 The Unifying Principle

In general, a specification in C@S may consist of a combination of temporal operators and abstract data type expressions. However, the combination of data abstraction with temporal abstraction leads to highly undecidable theories. Hence, it is useful to separate the temporal and the data part such that they can be tackled with different approaches. For example, the specification $[(C = A + 1) \, \mathsf{W} \, (B = 0)]$ (A, B, C are bitvectors interpreted as natural numbers) is transformed into $[\varphi_1 \, \mathsf{W} \, \varphi_2]$ where φ_1 and φ_2 are new variables with the meaning $\varphi_1 := (C = A + 1)$ and $\varphi_2 := (B = 0)$. These new variables are added by let-expressions similar to those of functional programming languages as follows:

$$\mathsf{let} \; \left[\begin{pmatrix} \varphi_1 \\ \varphi_2 \end{pmatrix} := \begin{pmatrix} C = A + 1 \\ B = 0 \end{pmatrix} \right] \; \mathsf{in} \; [\varphi_1 \, \mathsf{W} \, \varphi_2] \; \mathsf{end}$$

In general, the all specifications in C@S follow the template given below:

$$\mathsf{C\_SPEC}(\mathbf{i}, \mathbf{o}) := \begin{pmatrix} \mathsf{let} \\ \quad \begin{pmatrix} \varphi_1 \\ \vdots \\ \varphi_n \end{pmatrix} := \begin{pmatrix} \Theta_1(\mathbf{i}, \mathbf{o}) \\ \vdots \\ \Theta_n(\mathbf{i}, \mathbf{o}) \end{pmatrix} \\ \mathsf{in} \\ \quad \Phi(\varphi_1, \ldots, \varphi_n) \\ \mathsf{end} \end{pmatrix}$$

The above specification consists of a *temporal abstraction* Φ and the *data equations* $\Theta_1(\mathbf{i}, \mathbf{o}), \ldots, \Theta_n(\mathbf{i}, \mathbf{o})$. Φ is a propositional linear temporal logic formula containing only the propositional variables $\varphi_1, \ldots, \varphi_n$, i.e. no data expressions may occur. The above separation between temporal and data abstraction parts follows the usual distinction of digital circuits into a controller and a data path. However, $\Phi(\varphi_1, \ldots, \varphi_n)$ is not meant to be a specification for the controller.

There are well-known proof procedures for temporal logic (see [20] for an overview) and for proving the consistency of equations with abstract data types [21, 22, 23]. However, it is difficult to combine both classes of proof procedures. In order to combine these approaches, a class of higher-order formulae called hardware formulae have been developed in [18]. Hardware formulae consist of a finite state system (that may however be generic) and a finite set of safety, liveness and fairness properties.

In the non-generic form, hardware formulae correspond directly to ω-automata, which are related to the Staiger/Wagner class [24] of ω-regular languages. Hence, these hardware formulae are decidable and symbolic traversal techniques can be used directly for their proof [18, 25].

On the other hand, hardware formulae can be interpreted as rewrite systems *if only the safety properties are considered*. Hence, a translation of the temporal logic part $\Phi(\varphi_1, \ldots, \varphi_n)$ of a given specification into a hardware formula without fairness constraints allows the use of inductionless induction methods [21, 22, 23] for its verification.

However, the method for translating LTL into hardware formulae [18, 25] generates in some cases additional fairness constraints. Hence, we have to modify the translation procedure to circumvent the generation of these fairness constraints. This is done by providing appropriate invariants for the translation procedure as shown in the following sections.

3.2   Strategies for Verifying Temporal Behavior

In this section, the W operator is used to illustrate different translation procedures of C@S. All presented procedures are able to capture full linear temporal logic. The main strategy is strategy Φ_{HW} for translating verification goals into hardware formulae since it allows the combination with various other strategies[2].

Φ_{IV} can be used to reduce the complexity of the result of Φ_{HW} by providing more information of the system by additional invariants. Moreover, the use of Φ_{IV} eliminates in most cases the occurrence of fairness constraints when LTL specifications are translated into hardware formulae with Φ_{HW} such that the remaining goals can be proved with rewriting techniques. C@S has also other proof procedures for temporal reasoning as $\Phi_{\mathbb{N}}$ and Φ_{μ} which are however only invoked in the interactive mode.

Strategy Φ_{HW} It is straightforward to define various temporal operators by hardware formulae. For example, the W-operator can be defined as follows:

$$\vdash_{def} [x \ W \ b] := \left(\begin{array}{c} \exists q. \ (q = 0) \wedge G \ (Xq = q \vee b) \wedge \\ G \neg b \vee x \vee q \end{array} \right)$$

The strategy Φ_{HW} uses these definitions to translate the temporal abstraction Φ into an equivalent hardware formula $\mathcal{H}_{\Phi}$. In order to capture full LTL, additional universal quantification over input variables and additional fairness constraints are required for hardware formulae [18].

Strategy Φ_{IV} Provided that invariants are given for some occurrences of temporal operators, this strategy eliminates the corresponding temporal operators by invariant rules of C@S. For example, the invariant rule of the W operator is as follows [10]:

$$[x \ W \ b] := \left(\begin{array}{c} \exists J. \\ J \wedge \\ G \ (\neg b \wedge J \rightarrow XJ) \wedge \\ G \ (b \wedge J \rightarrow x) \end{array} \right)$$

The bound variable J is called an invariant and the bound subformulae have the following meaning: The first one states that the invariant J holds at the beginning of the computation, the second one states that if the termination

---

[2] We have set up some WWW pages where some aspects of C@S, as e.g. strategy strategy Φ_{HW} can be tested: http://goethe.ira.uka.de/hvg/cats/

condition b does not hold, the invariant implies the invariant at the next time instance, and the third subformula states that if the termination condition b and the invariant J hold at a certain time $\delta + t_0$, then the condition x holds also at time $\delta + t_0$.

Invariant rules are applied for the following purposes: first, they are used to cut data dependencies between control and data part of the verification goals. Second, they often eliminate the introduction of fairness constraints by Φ_{HW}. Hence, after an application of Φ_{HW} simple rewrite systems are obtained that can be checked by traditional rewrite methods. A third purpose is that even in case that model checking techniques are used afterwards, the complexity of these procedures can be significantly reduced as shown in section 4.3.

Strategy $\Phi_{\mathbb{N}}$ This strategy translates the temporal abstraction Φ into an equivalent arithmetical formula by replacing the temporal operators by equivalent arithmetical expressions. This replacement leads to the translation of LTL into the first order theory of one successor. The W operator can be replaced according to the following theorem:

$$[x \; \mathsf{W} \; b]^{(t_0)} = \exists \delta. \left\{ \left[\forall t.t < \delta \rightarrow \neg b^{(t+t_0)} \right] \wedge b^{(\delta+t_0)} \right\} \rightarrow x^{(\delta+t_0)}$$

This strategy yields in results which are directly related to timing diagrams and allow hence a graphical visualization. The proofs proceed usually manually by using induction and decision procedures such as Preßburger or Skolem arithmetic [27].

3.3 Strategies for Verifying Abstract Data Types

In this section, different strategies for handling the data expressions in verification goals are discussed.

Strategy Θ_B This strategy transforms the equations $\Theta_1, \ldots, \Theta_n$ into propositional formulae $\Theta_1^{prop}, \ldots, \Theta_n^{prop}$ that can be substituted in the temporal abstraction Φ. This strategy does only work with non-generic circuits.

Strategy Θ_{BV} This strategy translates operations on abstract data types such as addition on natural numbers in corresponding operations on bitvectors. If the specification is already given on bitvectors instead of (real) abstract data types, then this strategy is not required.

Strategy Decompose If the circuit is recursively defined, then this strategy decomposes the verification goal into separate subgoals by applying a suitable induction rule which directly reflects the definition of the circuit. According to the induction hypothesis, the circuit structure is replaced by the circuit's specification in the induction step. Note that circuits which are built up non-recursively but use recursively defined modules are not changed by this strategy.

Strategy HOL This strategy allows all rules of the HOL system which is currently used as platform for the implementation of C@S. Hence, this strategy is not specialized for any circuit structure or any verification goal and allows arbitrary manipulations of the goal. It is however the intention to invoke this strategy only if a fully automated proof is not possible. In such cases HOL is used to prove additional lemmas or to change the design hierarchy such that one of the other rules can be applied.

3.4 Combining the Strategies

The above strategies have to be combined specifically for each verification problem. In general, the choice of the suitable strategy can be done according to the algorithm *Verify* below. *imp* is the implementation description, *sp* is the list of specifications for the currently considered module and *inv_list* is a possibly empty list of given invariants.

$$
\begin{aligned}
&FUNCTION\ Verify(imp,sp,inv\_list) \\
&\quad IF\ Recursive\_Definition(imp) \\
&\quad THEN\ MAP\ Verify\ (Decompose(imp,sp,inv\_list)) \\
&\quad ELSE \\
&\qquad let\ sp1 = \Phi_{HW}(\Phi_{IV}(inv\_list,sp)) \\
&\qquad in\ IF\ fixed\_bv\_length(imp) \\
&\qquad\quad THEN\ HW\_Decide(HW\_IMP(imp,\Theta_B(sp1))) \\
&\qquad\quad ELSE \\
&\qquad\qquad IF\ one\_safety(sp1) \\
&\qquad\qquad THEN\ Inductive\_Completion(imp,sp1) \\
&\qquad\qquad ELSE\ HOL(imp,sp,inv\_list)
\end{aligned}
$$

Verify first tries to *decompose* recursively defined circuits by applying induction rules. The induction principle follows directly the structure of the circuit, which is assumed to be well-defined as in the Boyer-Moore theorem prover [28]. According to the induction hypotheses, the implementation descriptions in the induction step(s) are immediately replaced by the specifications and the thereby obtained goals are also fed into *Verify*. If a non-recursive circuit is given to be verified, *Verify* first translates the temporal abstraction of the specification into a hardware formula using strategy Φ_{HW}. If additional invariants are given in *inv_list*, these are used by applying strategy Φ_{IV} right before applying Φ_{HW}.

If only a concrete instantiation of the recursion scheme is to be proved (e.g. a 16-bit adder), then *Verify* translates the data equations into propositional formulae and calls the decision procedure *HW_Decide* based on symbolic model checking for hardware formulae. If this succeeds, the circuit has been verified without any interaction. Otherwise, it is checked, if the acceptance condition of the obtained hardware formula has only a safety property. If this is the case, then the hardware formula can be interpreted as a set of universally quantified equations and hence it can be proved by inductionless induction. The inductionless induction procedure requires some minor interaction for giving appropriate term orderings. In some cases, the inductive completion will run into infinite loops. In these cases, *Verify* will switch to the HOL system and requires further interaction to modify the proof goal such that another call to *Verify* will become more successful.

In general, the more information on the circuit and its specification (i.e. invariants and lemmas on the abstract data types) is given to the system, the more efficient its verification will be. In the next section, some case studies of simple circuits will show in detail how C@S can be used for the verification of digital circuits.

4   Case Studies

In this section, four different circuits are presented to show how different verification strategies are used for their verification. The first case study, a generic arbiter from [16], shows that small and medium-sized circuits can often be automatically verified. The second case study, a switching element for an ATM network [29], shows that the design hierarchy of a circuit can be exploited to reduce the costs of verification. The third case study, an n-bit adder, shows how circuits with abstract data types can be verified automatically. It also shows that the complexity of the verification process can be reduced by giving the system more information by invariants of the algorithm. The invariants express in a formal way the interaction between control and data path of the circuit. Finally, the last case study considers sequential n-bit multipliers in order to show that even the use of invariants does not always allow an easy automation. In these cases, inductive reasoning with invariant rules has to be used for the proof of additional lemmas used in an interactive verification.

4.1   Generic Arbiter

If several components have to share a common resource, there is the problem to avoid conflicts. Usually, arbitration circuits or protocols are used to manage the access of the resource. These arbitration circuits have at least to fulfill the following properties:

- at each point of time, at most one component has access to the resource
- only components have access to the resource, which have requested an access
- after a finite time each request is satisfied

In this section, using C@S the verification of the arbitration circuit given in [16] is presented. This arbiter is generic, i.e. it can be implemented for an arbitrary

Compo-nents	File Size [Bytes]	Time [Sec.]	Storage
10	7249	0.73	10305
20	24859	3.10	10900
30	52869	7.48	11249
40	91279	15.00	14175
50	140089	26.41	18433
60	199299	39.09	23864
70	268909	56.03	30417

Table 1.: Results for the arbiter

number of components. It is assumed that no access of the resource requires more than one point of time and that the components request access to the resource until the request is finally granted. The results for some numbers of components are given in table 1. In practice, it is however unlikely that more than 20 components share the same resource. Hence, the above results are completely sufficient and an interactive inductive proof of the generic n-component circuit is not necessary, although possible. In this example, the translation of the given temporal properties do not differ from the originally given ones of [16]. Hence, we obtain similar runtimes as reported in [16].

4.2 The Fairisle Switching Element

Module	time1	vars	time2
ACK	0.26	0	10.43
DMUX2B4CAll	0.18	8	1.27
ARBITER	0.23	6	1.45
PAUSE16	54.15	16	6.13
PRIORITY	0.56	0	8.38
DECODE_N	1.30	0	25.83
TIMING	0.17	3	0.89
ARBITERS	??	24	15.66
PRIOR_DECODE	156.77	16	27.42

Table 2.: Selected runtimes of the Fairisle switching element

The Fairisle switching element has been developed at the University of Cambridge and is used for the implementation of routing elements in a ATM network. The circuit is implemented in a highly hierarchical manner with 3 main modules and about 9 hierarchy levels. Curzon describes in [29] a verification of the circuit using the HOL system that has been done almost without any decision procedures and that took roughly six weeks. Using the C@S system, a complete automation is not possible since the circuit is too big to be handled by simple decision procedures. However, C@S allows to exploit the hierarchical design since it is possible to replace implementation descriptions at a higher level by previously verified specifications. This hierarchical verifications allows all modules to be verified automatically except for the top level module. This remaining verification step has been done by very simple interactions in the HOL system, which essentially combine the specifications of the three used submodules.

A comparison to Curzon's work is difficult, since Curzon described his specifications in a hardware description language which is not available in the C@S system. Nevertheless, the verification of the switching element took roughly a single man week with the C@S system, where most of the time was spent for describing hardware structure and behavior.

4.3 Von Neumann Adder

The sequential adder of this circuit goes back to John v. Neumann. A simple implementation of the algorithm is shown in figure 1. The behavior of the circuit is as follows: initially, the circuit is ready for new computations. This is indicated by the output rdy. If the circuit is ready at any point of time, and a new computation is requested at req, then the inputs A and B are read into the registers R_A and R_B and the circuit switches into computation mode. In this mode, the circuit performs iterations and ignores further requests. In the computation mode, the output rdy is false. If the computation terminates, then the circuit leaves the computation mode $rdy = 1$ and stores the results until a new computation is requested. Note that, the number of required iteration steps depends on the given numbers and that an upper bound is the length n of the given bitvectors.

The correctness of the above algorithm is based on the following theorem, which guarantees that the sum of R_A and R_B remains unchanged during the iterations:

$$\left(\sum_{i=0}^{n} a_i 2^i\right) + \left(\sum_{i=0}^{n} b_i 2^i\right) = 2 \left(\sum_{i=0}^{n} (a_i \wedge b_i) 2^i\right) + \left(\sum_{i=0}^{n} (a_i \oplus b_i) 2^i\right)$$

A specification of the circuit, whose verification is outlined in detail in the following, is therefore the following: $rdy \wedge req \rightarrow [(C = (A+B))\ \mathsf{W}\ rdy]$. The verification of this specification can be done by various strategies though it is certainly given at bitvector level. Figure 2 shows the runtimes (in seconds) and storage requirements (in BDD nodes) on a Sun Sparc 10 for some strategies.

The simplest strategy that can be used is $\Phi_{HW} \circ \Theta_B$, i.e. translating the data equations into propositional logic and the temporal abstraction into a hardware formula. Finally, symbolic model checking based on binary decision diagrams

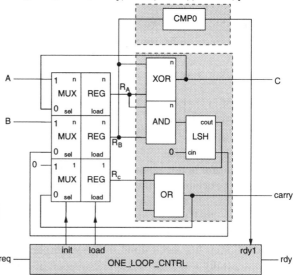

Fig. 1.: Implementation of a sequential adder

(BDDs) [30] is used as a decision procedure. It is well-known that the size of the BDDs is important for the efficiency of the verification and depends crucially on the variable ordering. For the verification of the adder, the following ordering has been used: $req \prec q \prec a_0 \prec b_0 \prec a_1 \prec b_1 \ldots a_{n-1} \prec b_{n-1}$

If the number of iteration steps did not depend on the data, it would be possible to replace the sequential circuit by a sequence of operation units, such that a combinational circuit is obtained which can then be verified by simple tautology checking. Nevertheless, this 'unrolling' can also be used for the verification of the adder, since an upper bound for the number of iterations is known and the values of the registers do not change if further iterations are performed after termination (see figure 2).

In order to enhance the efficiency of the verification, invariants can be used to use the basic design ideas of the circuit. In this case, it is important to observe that the sum of the registers remains unchanged during the iteration, i.e. the formula $J_1 := q \wedge (A{+}B) = (R_A{+}(R_c \triangleright R_B))$ is an invariant of the loop ($(R_c \triangleright R_B)$ means the bitvector which is obtained by adding R_c at the left hand side of R_B). This invariant allows the elimination of all temporal operators. For reasons of readability, define the values R_A°, R_B° and R_c° of R_A, R_{BC} and R_c at the next point of time by $R_A^\circ := (R_A \oplus R_B)$, $R_B^\circ := \mathsf{LSH}\,(R_A \wedge R_B)$ and $R_c^\circ := R_c \vee \mathsf{HD}\,((R_A \wedge R_B))$. Then the following subgoal remains to be proved:

$$(A{+}B) = (R_A{+}(R_c \triangleright R_B)) \rightarrow (A{+}B) = (R_A^\circ{+}(R_c^\circ \triangleright R_B^\circ))$$

Using strategy $\Theta_\mathbb{B}$, the above goal can be translated into pure propositional logic and can be solved afterwards by tautology checking. In order to reduce this complexity, one might think to rewrite the above goal with its assumption, i.e. to replace $(A{+}B)$ by $(R_A{+}(R_c \triangleright R_B))$. Hence, the tautology checker only needs to check that $(R_A{+}(R_c \triangleright R_B))$ is equal to $(R_A^\circ{+}(R_c^\circ \triangleright R_B^\circ))$. However, this formula is not valid, since the assumption also guaranteed that the sum of R_A and $(R_c \triangleright R_B)$ has at least $n + 1$ Bits. This information has been lost in the above term but is necessary for the validity of the specification.

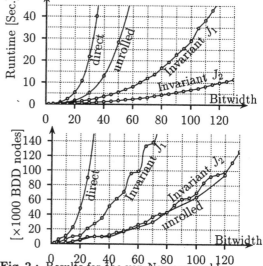

Fig. 2.: Results for the von Neumann adder

Nevertheless, the number of variables can be reduced by the invariant $J_2 := q \wedge ((A{+}B) \sim_\mathbb{N} (R_A{+}(R_c \triangleright R_B))) \wedge (R_c \rightarrow \Phi_\mathbb{N}((R_A{+}R_B)) < 2^n)$. J_2 contains explicitly the knowledge that there is at least one carry, i.e. if $R_c = 1$ holds, then the remaining sum of R_A and R_B has only n bits, otherwise the sum may have $n + 1$ bits. The resulting goal can now be solved by strategy $\Theta_\mathbb{B}$ and additional tautology checking. Figure 2 shows the detailed results of both invariant approaches. It can be seen that the more information is given to the system, the more efficiently the circuit can be verified. However, this observation is not always valid as shown in the next section.

4.4 Sequential Multiplier Circuits

The circuit given in figure 3 implements a russian multiplier with the same controller for the loop as in the previous section. The previously presented strategies can also be used for the verification of this circuit, the results are given in table 3 (runtime again in seconds; storage in number of BDD nodes).

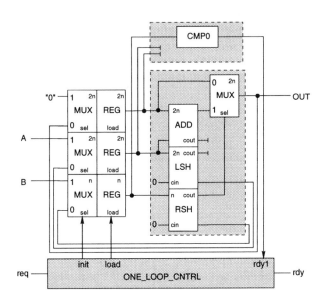

Fig. 3. Implementation of the russian multiplier

The first column shows the results of strategy $\Phi_{HW} \circ \Theta_{\mathbb{B}}$. The corresponding hardware formula of the n-bit circuit has exactly $(n+2)2^{2n}+2^n$ reachable states and can not be verified for more than 8 bits (this required 31 MByte). Using the invariant

$$q \wedge \left((A \times B) \sim_{\mathbb{N}} (R_C + \mathsf{LASTN}_{2n} ((R_A \times R_B)))\right),$$

it also only possible to verify at least 8 bits. Unroling the circuit allows even to verify the 11 bit circuit, but more than 11 bits can not be verified with this approach. The strategies which were quite successful for the adder are not applicable for the verification of this circuit. In contrast to the adder, the complexity of this verification problem stems from the data equations and only minor complexity stems from the control part. It is however well-known, that propositional formulae for multiplication have no variable ordering which gives the corresponding BDDs a non-exponential size. Hence, strategy $\Theta_{\mathbb{B}}$ is not the right choice for this circuit. Instead, the n-bit circuit has to be verified by induction for handling the data equations and invariants for handling the control part. This strategy

Bits	Russian Multiplier						ADD/SHIFT Multiplier			
	direct		Invariant		unrolled		direct		unrolled	
	Time	Storage	Time	Storage	Time	Storage	Time	Storage	Time	Storage
2	0.22	3254	0.11	567	0.17	56	0.24	2805	0.10	41
3	0.38	9725	0.26	3503	0.15	361	0.39	7513	0.13	268
4	0.99	10923	0.74	10303	0.31	1820	1.04	10429	0.27	1553
5	3.32	31024	3.42	47990	0.53	7160	4.02	23286	0.39	6669
6	15.63	116755	20.13	221678	1.20	11049	22.87	87532	1.02	11587
7	107.94	467636	214.38	989823	3.50	39379	138.09	357527	3.13	39768
8	1203.04	1901547	3416.04	4256226	12.78	143568	1599.76	1484104	11.94	153791
9	—	—	—	—	62.86	449337	—	—	64.56	430916
10	—	—	—	—	435.54	1518158	—	—	469.30	1567273
11	—	—	—	—	3468.37	4212190	—	—	4546.16	4812990

Table 3. Results for the multiplier circuits

required a lot of user interaction and cannot be explained here in detail. For example. the following lemmas had to be proved:

$$(y \text{ MOD } 2) = 0 \vdash (2 \times x) \times ((y \text{ DIV } 2)) = x \times y$$
$$(y \text{ MOD } 2) = 1 \vdash x + (2 \times x) \times ((y \text{ DIV } 2)) = x \times y$$

The russian multiplier given in figure 3 is usually not used, since it contains more flipflops than necessary. A more efficient (in terms of chip size) sequential multiplier is given in figure 4. However, this multiplier is slower, since it requires always n iteration steps in contrast to the multiplier of figure 3 that required at most n iteration steps. The savings of chip area lead however not to any savings for the complexity of the verification. The results for the multiplier of figure 4 are given in table 3.

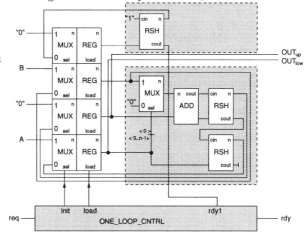

Fig. 4.: Implementation of an ADD/SHIFT multiplier

5 Conclusion

In this paper, we have presented a new tool for the verification of digital circuits. Different decision procedures are coupled by transforming specification and implementation descriptions into a common representation based on higher order logic. Dependent on the class of the verification problem, the most suited solution is selected and the necessary decision procedures are invoked. In contrast to previous approaches, several solution strategies may be combined to verify circuits for which a specification includes complex timing as well as abstract data types. Using different examples, different verification strategies have been demonstrated and compared with regard to the circuit size and necessary amount of interaction.

References

1. R.E. Bryant. Graph-Based Algorithms for Boolean Function Manipulation. *IEEE Transactions on Computers*, C-35(8):677–691, August 1986.
2. O. Coudert, C. Berthet, and J.C. Madre. Verification of sequential machines using boolean functional vectors. In L. Claesen, editor, *IMEC-IFIP Workshop on Applied Formal Methods for Correct VLSI-Design*, 1990.
3. J.R. Burch, E.M. Clarke, K.L. McMillan, D.L. Dill, and L.J. Hwang. Symbolic Model Checking: 10^{20} States and Beyond. *Information and Computing*, 98(2):142–170, June 1992.
4. M.J.C. Gordon. Why higher-order logic is a good formalism for specifying and verifying hardware. In G.J. Milne and P.A. Subrahmanyam, editors, *Formal Aspects of VLSI Design*, pages 153–177. Computer Laboratory, University of Cambridge, 1986.
5. R. Kumar, K. Schneider, and T. Kropf. Structuring and Automating Hardware Proofs in a Higher-Order Theorem-Proving Environment. *International Journal of Formal System Design*, pages 165–230, 1993.
6. S. Owre, J.M. Rushby, N. Shankar, and M.K. Srivas. A tutorial on using PVS for hardware verification. In T. Kropf and R. Kumar, editors, *Proc. 2nd International Conference on Theorem Provers in Circuit Design (TPCD94)*, volume 901 of *Lecture Notes in Computer Science*, pages 258–279, Bad Herrenalb, Germany, September 1994. Springer-Verlag. published 1995.
7. E.M. Clarke, D.E. Long, and K.L. McMillan. Compositional Model Checking. In *Proceedings of Fourth Annual Symposium on Logic in Computer Science*, pages 353–361, Washington D.C., June 1989. IEEE Computer Society Press.
8. E.Clarke, O. Grumberg, and D. Long. Model Checking and Abstraction. In *Proceedings of the Nineteenth Annual ACM Symposium on Principles of Programming Languages*, New York, January 1992. ACM.
9. D.E. Long. *Model Checking, Abstraction, and Compositional Verification*. PhD thesis, Carnegie Mellon University, 1993.
10. K. Schneider, T. Kropf, and R. Kumar. Control-Path Oriented Verification of Sequential Generic Circuits with Control and Data Path. In *Proceeding of the European Design and Test Conference*, pages 648–652, Paris, France, March 1994. IEEE Computer Society Press.
11. M. Langevin, E. Cerny, and R.E. Ladner. An extended OBDD representation for extended FSMs. In *The European Design and Test Conference*, pages 208–303, Paris, February 1994. IEEE Computer Society Press. EDAC94.
12. R. Hojati and R.K. Brayton. Automatic Datapath Abstraction in Hardware Systems. In *Proc. of the International Conference on Computer-Aided Verification*, pages 98–113. Springer Verlag, LNCS, 1995.

13. H. Hungar, O. Grumberg, and W. Damm. What if model checking must be truly symbolic. In P.E. Camurati and H. Eveking, editors, *Correct Hardware Design and Verification Methods*, volume 987 of *Lecture Notes in Computer Science*, pages 1–20, October 1995. IFIP WG10.5 Advanced Research Working Conference CHARME'95, Springer-Verlag.

14. E. Clarke and X. Zhao. Word level symbolic model checking. Technical Report CMU-CS-95-161, Carnegie Mellon University, Pittsburgh, PA 15213, May 1995.

15. M.J.C. Gordon and T.F. Melham. *Introduction to HOL: A Theorem Proving Environment for Higher Order Logic*. Cambridge University Press, 1993.

16. K.L. McMillan. *Symbolic Model Checking*. Kluwer Academic Publishers, Norwell Massachusetts, 1993.

17. D. Kapur and H. Zhang. RRL: a rewrite rule laboratory. In Lusk and Overbeek, editors, *9th International Conference on Automated Deduction*, pages 768–769. Springer-Verlag, 1988.

18. K. Schneider. *Ein einheitlicher Ansatz zur Unterstützung von Abstraktionsmechanismen der Hardwareverifikation*, volume 116 of *DISKI (Dissertationen zur Künstlichen Intelligenz)*. Infix Verlag, Sankt Augustin, 1996. ISBN 3-89601-116-2.

19. T.F. Melham. Abstraction mechanisms for hardware verification. In G. Birtwistle and P.A. Subrahmanyam, editors, *VLSI Specification, Verification, and Synthesis*, pages 129–157, Boston, 1988. Kluwer Academic Publishers.

20. E.A. Emerson. Temporal and Modal Logic. In J. van Leeuwen, editor, *Handbook of Theoretical Computer Science*, volume B, pages 996–1072, Amsterdam, 1990. Elsevier Science Publishers.

21. G. Huet and J.-M. Hullot. Proofs by induction in equational theories with constructors. *Journal of Computer and System Sciences*, 25:239–266, 1982.

22. L. Fribourg. A strong restriction of the inductive completion procedure. In *13th International Colloqium on Automata, Languages and Programming*, pages 105–115, 1986.

23. J.-P. Jouannaud and E. Kounalis. Proofs by induction in equational theories without constructors. *Information and Computation*, 82:1–33, 1989.

24. L. Staiger and K.W. Wagner. Automatentheoretische Charakterisierungen topologischer Klassen regulärer Folgenmengen. *Elektron. Informationsverarb. Kybernet.*, 10:379–392, 1974.

25. K. Schneider. Translating LTL Model Checking to CTL Model Checking. Technical Report SFB358-C2-3/96, Universität Karlsruhe, Institut für Rechnerentwurf und Fehlertoleranz, January 1996.

26. C.A.R. Hoare. An axiomatic basis for computer programming. *Communications of the ACM*, 12:576–580, 1969.

27. M. Machtey and P.Young. *An Introduction to the General Theory of Algorithms*. North-Holland, 1978.

28. R.S. Boyer and J.S. Moore. *A Computational Logic Handbook*. Academic Press, 1979.

29. P. Curzon. Tracking design changes with formal verification. In T.F. Melham and J. Camilleri, editors, *International Workshop on Higher Order Logic Theorem Proving and its Applications*, volume 859 of *Lecture Notes in Computer Science*, pages 177–192, Malta, September 1994. Springer-Verlag.

30. J.R. Burch, E.M. Clarke, K.L. McMillan, and D.L. Dill. Sequential Circuit Verification Using Symbolic Model Checking. In *Proceedings of the 27th ACM/IEEE Design Automation Conference*, pages 46–51, Los Alamitos, CA, June 1990. ACM/IEEE, IEEE Society Press.

Verification Using Uninterpreted Functions and Finite Instantiations

Ramin Hojati[1], Adrian Isles, Desmond Kirkpatrick, Robert K. Brayton

Abstract

One approach to address the state explosion problem in verification of microprocessors with wide datapaths is to model variables as integers and datapath functions as uninterpreted ones. Verification then proceeds by either symbolically simulating this abstract model, or creating a small finite instantiation which contains all possible behaviors. In this paper, we first prove that the reachability problem for models with uninterpreted functions and predicates only of the form $x = y$, where both x and y are integer variables, is undecidable. However, such predicates are generally only needed in the property being checked and not in the model. For properties involving predicates of the forms $x = term$ and $x = y$, we provide complete and partial verification techniques using finite instantiations respectively. Applications of these result to the verification of the control circuitry of superscalar microprocessors are provided, where one can verify various correctness properties using models with one or a few bit integers.

1 Introduction

In designing modern microprocessors, validation engineers begin their work as soon as the design process begins. The validation teams are often as large or larger than the design teams. Currently, validation is done using simulation, where two commonly used techniques are developing ad hoc behavioral simulators for specific parts of the design, and providing test vectors for conventional HDL simulators. This process is costly both in terms of man-power and CPU time. Worse, there is no guarantee that the whole design has been tested, and this problem is getting more severe with the increasing complexity of designs. For example, [Hsu93] reports that in the design of a commercial microprocessor, a more aggressive design had to be abandoned since the verification effort was expected to be too large.

Formal design verification (verification hereafter) is the process of proving properties of designs. Since verification checks all possible behaviors, it provides a guarantee that the verified property holds for the design. Although, verification is still only as good as the properties checked, generally it is much easier (and less time-consuming) to come up with a good set of properties, than to develop a good set of test vectors. The main problem with verification is capacity. Not counting the on-chip cache, current microprocessors contain in the order of tens of thousands of latches. For example, UltraSparc ([Sparc95]) contains 20,000 clocked elements. However, the current BDD-based techniques for verification can at best handle circuits with 100-600 latches[2]. This *verification gap* can be made smaller by using *modular verification*, in which different subsystems such as dispatch units, memory subsystems, bus interfaces, and functional units are verified separately. When verifying a subsystem, other subsystems are replaced (manually) by simple modules which contain the behavior of the detailed modules they replace. Our approach ([HB95]) is to model the circuits using a concurrency model called Integer Combinational Sequential (ICS), and to use abstraction to nullify the effect of large datapaths and memory elements on the state explosion problem. Using ICS models, the datapath variables are modeled as integers, functional units as uninterpreted functions, and memory as infinite memory. ICS models can easily and automatically be extracted from descriptions in hardware description languages such as Verilog, with very little work on the part of the designers.

Given a property, verification of ICS models proceeds using two techniques. For a subset of circuits where certain structural properties hold, verification can be done using *finite*

---

1. This work was supported in part by SRC, under contract DC-324-033.
2. [CZ95] reports a BDD-based verification with 600 latches.

instantiations, where the integer variables are replaced by variables a few bits wide, and the uninterpreted functions are given appropriate interpretations (meanings). By performing this reduction, we generally get a model which can be attacked using current verification techniques (e.g. SMV, HSIS, etc.). If this reduction is not possible, a symbolic simulation algorithm can be used to verify the property directly. Symbolic simulation does not always terminate, but termination is guaranteed in some cases, such as some correctness properties of pipelined circuits.

In modern microprocessors, there are two especially hard-to-design components. One is the memory subsystem and its interfaces; the other is the fetch-issue-execution-writeback unit, which we refer to as the *pipeline control*. Issues such as speculative execution, out-of-order completion, register re-naming, and interrupts make the design of pipeline controls difficult. Since pipeline controls communicate with functional units, these must be modeled also for proving some correctness properties. For example, let $v(r)$ denote the value of register r, and assume we are interested in proving property P: when instruction $r_3 = ADD(r_1, r_2)$, is about to be retired, the value which is written back is $v(r_1) + v(r_2)$. To prove P directly, we need to model the integer functional unit. However, the fact that the functional unit does "addition" is irrelevant for the correctness of pipeline control: it can be modeled as an uninterpreted function[1].

[HB95] introduced a set of abstraction techniques using finite instantiation for circuits without uninterpreted functions. Examples of such circuits are communication protocols, and memory controllers. These results were applied in [HMLB95] to verify the correctness of a model of SUN's SPARC 8 memory subsystem using only one bit for data and a few bits for addresses. In this paper, we extend these results to the case of uninterpreted functions.

We first prove that if ICS models are allowed to contain uninterpreted functions with predicates of the form $x = y$, where x and y are integers, then the *reachability problem* (whether a specific state s is reachable) becomes undecidable. This provides a limit for the applicability of finite instantiation techniques. However, in verifying many properties of pipeline controls, where the functional units are replaced by uninterpreted functions, the predicates operating on data can be abstracted to be unconstrained inputs, taking any value 0 or 1. Therefore, data predicates are not needed in the model, but just in the property.

The syntax and semantics of ICS models is similar to hardware. When a model contains uninterpreted functions, the integer variables take *ICS terms*, which are obtained by applying uninterpreted functions to constants (intuitively unconstrained integer inputs) and smaller terms. We then prove that for ICS models with only uninterpreted functions, an important class of properties of the form "when binary variable b become true, $(x_1 = \alpha_1) \wedge \ldots \wedge (x_m = \alpha_m)$ ", where x_i s are integer variables and α_i 's are ICS terms, can be verified using finite instantiations. This result can be used, for example, to prove the following property Q, using two-bit integers, and appropriate interpretations for the uninterpreted functions. The property Q is: "Assuming r_1 and r_2 contain c_1 and c_2 initially, if the instruction $r_3 = ADD(r_1, r_2)$ (call it I) is issued after any stream of instructions which do not modify r_1 and r_2, the result which is written to r_3 when I is retired is $ADD(c_1, c_2)$ ". Note that c_1 and c_2 can be any arbitrary integers, and all instructions before I cannot modify r_1 and r_2 but can read them. One can also prove that if any instruction after I has as one of its sources r_3 (and no intervening instruction has written to

---

1. Modeling the integer functional units by uninterpreted functions allows for more behavior than the original system. Hence, in theory, there is a possibility of false negatives. In practice, we do not expect this to be a problem.

r_3), then it will receive the value $ADD(c_1, c_2)$, i.e. the bypass circuitry works correctly in this case.

Property Q is only an approximation to property P since no instruction before I is issued is allowed to modify r_1 and r_2. To verify P, our approach is to run the unpipelined machine N in parallel with the implementation M. N halts after it executes I. After M retires I, it is checked that the r_3 registers of M and N are equal. This property has the form "when b becomes true, $x = y$", where x and y are integer variables.

Although we do not have techniques for exact verification using finite instantiations for this property (and there is some evidence that if such techniques exists, they may be expensive), we provide techniques which prove the property for a class of possible bugs. For example, they can catch any bug such that when b becomes true, the ICS terms stored at x and y differ in the number of times some constant or uninterpreted function is used. For example, if x contains $f(a, b)$ and y contains $f(g(a), b)$, then this bug will be uncovered, since the function g does not occur in $f(a, b)$. Verifying this bug requires only one bit for each integer variable. This approach provides a combination of testing and verification (hence false positives are possible), and should be effective in exposing bugs where little data variation is needed. [HYMD95] technique combines verification and simulation by building the graph of the "control part", and then producing a set of test vectors which ensure that all control arcs are traversed. This uncovered many subtle bugs, and confirms that, for exposing bugs in microprocessors, little data variation is needed. Our third technique is related to this in the sense that a very large set of test vectors is covered (specifically all control arcs will be checked), although 100% coverage is not obtained as with formally proving the property.

There is an extensive literature on verifying microprocessors. In most of these approaches ([Cyr93], [BD94], [Sri95]), a pipelined machine is compared to an unpipelined machine by proving that one step of the unpipelined machine is "equivalent" to one step of the pipelined one. These approaches require the user to provide a subset of the possible states of the implementation machine in which the correspondence can be proved. This set has to be large enough to contain all reachable states (otherwise verification is not complete), but it should not be too large to contain unreachable states in which the machine may behave unpredictably. Providing such an invariance, in theory, is equivalent to providing the set of reachable states, and therefore can be very challenging for complicated machines. These techniques have not so far been applied to the superscalar processors. [BaD94] also requires substantial user intervention, by asking the users to provide β-relation, and restricting the circuits to be k-definite. On the other hand, symbolic simulation algorithm of [Cor93], [SS95], and [HB95] are nearly automatic. However, [Cor93] does not give an algorithm for pipelined machines; [SS95] requires some user intervention, and their algorithm may not terminate and can give false negative results; and [HB95]'s BDD-based algorithm does not guarantee termination. To our knowledge, there is no previous literature on verification using uninterpreted functions and finite instantiations.

The paper is organized as follows. In section 2, the ICS models, and their operational semantics are reviewed. Section 3 contains the technical part, proving the three main results: undecidability of ICS models with the predicate $x = y$, verification of properties involving $x = term$, and partial verification of properties involving the predicate $x = y$. Section 4 gives an application to the correctness of superscalar microprocessors. Section 5 concludes the paper and discusses future research directions.

2 Integer Combinatorial/Sequential (ICS) Concurrency Model

In this section, the integer combinational/sequential concurrency model is reviewed; more detail can be found in [HB95]. ICS is designed to represent systems composed of control, datapath, and memory. Its syntax and semantics is similar to hardware systems, with the addition of some machinery to reason about non-deterministic gates, integers functions and predicates, and how they affect the state spaces.

2.1 Syntax

The primitives are: variables, tables, interpreted functions and predicates, uninterpreted functions and predicates, constant creators, latches, and memory functions.

Variables. Variables are of two types: *finite* and *integer*. Finite variables take values from some finite domain; integer variables take integer values $(0, 1, 2, \ldots)$.

Tables. A table is a relation defined over a set of finite variables, divided into inputs and outputs. A table is a *function* if for every possible input tuple there is at most one output tuple (incompletely specified functions are allowed). Otherwise it is a *relation*. If a table has only one binary output, and is a function, then it is a *predicate*.

Interpreted Functions and Predicates. A predefined set of functions and relations over integers is built in. The interpreted functions are: $y := x$, $y := if(b, x)$, $z := mux(b, x, y)$, $z := x + y$, $y := x + c$, where x, y, and z are integer variables, b a binary variable, and c a numeral $(0, 1, 2, \ldots)$. The interpreted predicates are $y = x$ (equality), $y < x$, $x = c$, $(x \bmod m) = r$, and $(x \bmod m) < r$, where r and m are given integers.

Uninterpreted Functions and Predicates. These are a set of function and predicate symbols with their arities and domain variables given. For example, $f(x_1, x_2)$ may be the specification of an uninterpreted integer function defined over binary variable x_1 and integer variable x_2. Predicates of the form $x = term$, where x is an integer variable, and $term$ is an ICS term are also allowed. An *ICS term* is built recursively from numerals, constants, interpreted and uninterpreted functions. Therefore, numerals and constants are ICS terms, and if f is an n-ary function and $t_1, \ldots, t_n$ are ICS terms, then $f(t_1, \ldots, t_n)$ is also an ICS term. Note that ICS terms do not involve any variables or predicates. Examples of ICS terms are c_0, $f(c_0, c_1)$, $g(15, f(c_2, c_3))$, and $c_1 + c_2$.

Constant Creators. A constant creator is a special element with no input, and an integer output. Intuitively, it is a higher-order function, which creates a new constant (i.e. a function with no argument) each time called. Intuitively, a constant creator models an unconstrained integer input.

Latches. A latch is defined on two variables over the same domain: input (or *next state*) and output (or *present state*). Present and next state variables may overlap, e.g. when an output of a latch is an input to another. Every latch has a set of initial values, which are a subset of the domain of its variables. If the latch is integer-valued, then the initial value set can either be a finite set of numerals, or a given constant. Predicates can be used in combination with constants to create an infinite set of initial values. For example, to declare that the initial set of a latch is all integers greater than 5, we let the initial value be some constant c_0. In the first state, the output of the latch is input to a predicate $x > 5$, and the machine continues only if the predicate holds. Hence, only those behaviors are allowed where the initial value of the latch is an integer greater than 5.

Memory Functions. Two functions *read* and *write* are provided with their usual interpretation; *read* is a binary function of a memory and a location; *write* is a ternary function, whose arguments are a memory, a location, and a value. Location and value are variables in the model. Reading a location which has not been written, returns a new constant (like a constant creator).

Definition A *generalized gate* is a table, an interpreted or uninterpreted function or predicate, or a constant creator.

Definition **Data movement operations** are $x := y$, $z := mux(b, x, y)$, and $y := if(b, x)$, where x, y, z are integer variables, and b is binary.

Every model has only a finite number of variables, latches, and generalized gates. Every variable is the output of exactly one generalized gate or latch. Hence, every input to a generalized gate or latch is the output of some other generalized gate or latch; ICS models are closed. A variable can be input to many generalized gates or latches.

Definition A **state** is a triple $(latch, memories, predicates)$, where,

a. *latch* is an assignment of values to the latches. For finite valued latches, the value comes from the domain. For integer valued latches, the value is an ICS term.

b. *memories* is a set of memory elements, where a **memory** is a set of pairs of ICS terms, where the first denotes a location and the second a value.

c. *predicates* is a set of **atomic formulas**, where an atomic formula is any interpreted or uninterpreted predicate applied to ICS terms. Examples are $c_0 < f(c_1)$ and $P(f(c_0, c_1), g(c_0))$. Note that $(c_0 < f(c_1)) \wedge (P(f(c_0, c_1), g(c_0)))$ is not an atomic formula. Intuitively, *predicates* at a state s, is the set of assumptions made to reach s.

Definition Given two ICS terms t_1 and t_2, and two sets of atomic formulas $P = \{P_1, ..., P_n\}$ and $Q = \{Q_1, ..., Q_m\}$, t_1 *is equal to* t_2 *subject to P and Q,* denoted as $t_1|_P = t_2|_Q$ iff the formula $(P_1 \wedge ... \wedge P_n \wedge Q_1 \wedge ... \wedge Q_m) \Rightarrow (t_1 = t_2)$ is valid. For example, if $t_1 = x$, $P = \{x > 7, x \le 8\}$, $t_2 = 8$, and $Q = \varnothing$, then $t_1 = t_2$ subject to P and Q since $(x > 7 \wedge x \le 8) \Rightarrow (x = 8)$ is valid. The equality of two ICS terms can be decided using the algorithms of [Sho79] and [Sho82].

Notation If $P = \{P_1, ..., P_n\}$ is a set of predicates, we will use P to denote $P_1 \wedge ... \wedge P_n$. For example, $P \to (b = 1)$ is the formula $(P_1 \wedge ... \wedge P_n) \to (b = 1)$.

Definition Let states $s_1 = \left(L_1, \left(M_1^1, ..., M_n^1\right), P_1\right)$ and $s_2 = \left(L_2, \left(M_1^2, ..., M_n^2\right), P_2\right)$ be given. $s_1 = s_2$ if the following hold.

a. Let l_1^i and l_2^i denote the values of the i-th latch in L_1 and L_2, respectively. If the i-th latch is finite, then $l_1^i = l_2^i$; otherwise, $l_1^i|_{P_1} = l_2^i|_{P_2}$ must hold.

b. Let $M_k^1[i] = \left(a_k^1[i], v_k^1[i]\right)$ denote the i-th address/value pair in M_k^1, the k-th memory element of s_1. Then, for each $M_k^1[i]$ there exists $M_k^2[j]$ such that $a_k^1[i]|_{P_1} = a_k^2[j]|_{P_2}$ and $v_k^1[i]|_{P_1} = v_k^2[j]|_{P_2}$. Similarly, for each $M_k^2[j]$ there must exist $M_k^1[i]$ such that $a_k^1[i]|_{P_1} = a_k^2[j]|_{P_2}$ and $v_k^1[i]|_{P_1} = v_k^2[j]|_{P_2}$.

c. $P_1 \equiv P_2$, i.e. $P_1 \Rightarrow P_2$ and $P_2 \Rightarrow P_1$.

Definition An **initial state** is a state $(latch_{init}, \varnothing, \varnothing)$, where $latch_{init}$ is an assignment of an initial value to each latch.

2.2 Operational Semantics of ICS

The operational semantics of ICS describes how a transition between two states of an ICS model occurs.

Definition A **gate graph** G is a directed graph where each node is a generalized gate. $(u, v) \in G$ if some output variable of the generalized gate u is an input to the generalized gate v. A cyclic gate graph is said to contain a **combinational loop (or cycle)**.

Remark For an acyclic gate graph G, a root node either has no inputs or is a latch.

Our operational semantics is restricted to acyclic graphs and is defined in terms of a **configuration** (or **state) graph** and its transition relation. Every node in the configuration graph is a pair (s, n), where s is a state of the model, and n represents the number of constants created so far. An initial state of the configuration graph is of the form (s_{init}, k), where s_{init} is an initial state of the model, and k the number of constants in s_{init}. An edge (transition) is defined by the following algorithm which, given a state $u = ((L, M, P), n)$ in the configuration graph, assigns a new value to all next state variables (creating L'), and creates a new memory M', a new set of predicates P', and a new counter n'. In this case, we say there is an edge (u, v) in the configuration graph, where $v = ((L', M', P'), n')$. State transitions are computed as follows.

0. Let $P' = P$, $n' = n$.

1. Let a user-given total order on all memory operations be given. Choose a topological sort O of the gate graph consistent with this memory order.

2. Assign the values given by L to the outputs of the latches.

3. Assign values to the outputs of each generalized gate consistent with its inputs, processing the generalized gates in the topological order O. More precisely, let a generalized gate $g(i, o)$ be given, where i represents the inputs to the gate and o its output.

a. If g is a table representing the relation $R(i, o)$, then $(i, o) \in R$.

b. If g is an integer function, then $o = g(i)$, where o and i are ICS terms.

c. If g is an integer predicate, if $P' \to (g = 0)$ is valid, let $g = 0$; if $P' \to (g = 1)$ is valid, let $g = 1$; otherwise let $o = 0$ or $o = 1$, and $P' = P' \cup \{g = o\}$.

d. If g is a constant creator, then $o = c_{n'}$, where $c_{n'}$ is a fresh constant. Let $n' = n' + 1$.

Step 3 is referred to as **value propagation**. Note that the configuration graph is finite-branching, i.e. for every state, there are a finite number of next states. It is possible that a table is not complete, i.e. there are inputs for which there are no outputs. Then, the set of values assigned to an output of a table may be empty. The empty values propagate, i.e. if one of the inputs to a table is empty, then the output is empty as well.

2.3 ICS Models and Verification

ICS models can be used to model hardware systems. Fairness constraints, which rule out some unwanted behavior (introduced due to abstraction), can be placed on finite latches. In order to verify a property, the property is written as an ICS model, and it is checked that the language of the system is contained in the language of the property. A string in the language of an ICS model M is obtained by traversing a path in the graph of the operational model of M, such that the set of infinitely occurring states of the finite latches satisfy the fairness constraints. The language containment must hold for any interpretation given to uninterpreted functions and constants, i.e. for all possible definitions for the functions and valuations for the constants. In some cases, the property can be complemented automatically. In other cases, the user must do this by hand. The complement of the property is then composed with the system, and it is checked that the language of the composed system is empty. Hence, the verification problem reduces to checking whether the language of an ICS model is empty. Again, emptiness must hold for all interpretations given to uninterpreted functions and constants.

3 Main Results

[HB95] showed that language emptiness of ICS models with only constant creators, data movement operations, and the predicate $x = y$ can be decided using finite instantiations. In this section, we prove the problem is undecidable if we allow, in addition, uninterpreted functions (and even disallow constant creators). However, in proving correctness of microprocessors, usually the integer predicates $x = y$ and $x = term$, operating on data values, are not used in the system; rather, they are used in the property. It is then proved that an important class of properties of the form "when binary variable b becomes true, $(x_1 = \alpha_1) \wedge ... \wedge (x_m = \alpha_m)$ ", can be verified using finite instantiations. In section 4, we give an application of these results to the correctness problem of microprocessors.

3.1 Undecidability of ICS Models with Predicate $x = y$

Theorem 1 The reachability problem for ICS models involving data movement operations, uninterpreted functions and the predicate $x = y$ is undecidable (note no constant creators are allowed).

Proof We reduce the halting problem for two-counter machines to our problem. A two-counter machine is a FSM with two integer variables called counters [HU79]. The counters can be incremented, decremented, and tested against zero. A transition can be represented as $(s, cond_1, cond_2, op_1, op_2, s')$, where,

a. s and s' are the current and next states respectively.

b. $cond_1$ and $cond_2$ are conditions on whether counters 1 and 2 are zero, not zero, or don't care. The transition is taken only if the conditions are satisfied in state s .

c. op_1 and op_2 are the operations performed on counters 1 and 2. Increment, decrement, and no-op are the possible choices.

For a two-counter machine M , we create an ICS model N with one uninterpreted function f representing increment, and a latch initialized to a constant c_0 and always holding that value.

1. For each counter, we define an increment circuit which operates as follows. Let I be the input to the increment circuit. When the circuit is enabled, it sets temporary variables y and z to $f(I)$ and c_0 respectively. The circuit continues only if $y \neq z$. This is accomplished by using an incompletely specified function which has an output only if $y \neq z$. If $z \neq I$, the circuit then lets $z = f(z)$, and repeats the process. If $z = I$, the circuit outputs $f(I)$. Note that we have ensured that the only possible executions for N are those where no value between c_0 and I is equal to $f(I)$.

2. To decrement a counter, we define a circuit which operates as follows. When enabled, it first checks that its input I is not c_0 . If so, it returns c_0 , otherwise it sets a temporary variable y to c_0 , and compares $f(y)$ to I . If the two are equal, it outputs y . If not, it lets $y = f(y)$, and repeats the process. It finally outputs a y for which $I = f(y)$.

3. To test a variable I against 0, N just checks whether $I = c_0$.

N simulates M by checking for a transition $(s, cond_1, cond_2, op_1, op_2, s')$, that $cond_1$ and $cond_2$ hold, and then performs op_1 and op_2 (using the subcircuits defined above), and moves to s' . Let $f^j(I) = f(f^{j-1}(I))$. We note the following during the operation of N .

1. The increment subcircuit does not create an inconsistency, i.e. there are some interpretations for which all the predicates of the form $f^j(c_0) \neq f^k(c_0)$ (which this subcircuit creates) can be satisfied. One such interpretation is having integers as the domain, with f replaced by the increment operation on integers.

2. The decrement subcircuit always terminates. The reason is that the value at the input of a decrement subcircuit is always some formula $f^n(c_0)$, for some n, since these are the only values created by N. Also, note that if the input to a decrement subcircuit is $f^n(c_0)$, then the output of the circuit is $f^{n-1}(c_0)$, since when N created $f^n(c_0)$ it ensured that for all $0 \leq i \leq n-1, f^i(c_0) \neq f^n(c_0)$.

3. During the execution of N, the test $I = c_0$ returns true iff I is actually the formula c_0. To prove this, assume for some value i, $f^i(c_0) = c_0$. This is not possible since when $f^i(c_0)$ was created, the increment circuitry ensured that $f^i(c_0) \neq c_0$.

It follows that if 1) M is at some state s, and N is at its corresponding state s, and 2) if the two counters of M take i and k respectively, and the two counters of N take $f^i(c_0)$ and $f^k(c_0)$ respectively, and 3) if M makes a transition $(s, cond_1, cond_2, op_1, op_2, s')$, with counters taking on i' and k' respectively, then N will eventually end up in its state s' with counters taking $f^{i'}(c_0)$ and $f^{k'}(c_0)$ respectively. Since N merely simulates M, the reverse is also true. Hence, M falls into its halt state iff N reaches a corresponding halt state (QED).

Note that the proof of 1 shows that reachability problem for ICS models becomes undecidable when in addition to data movement operations and the predicate $x = y$, only one uninterpreted function (f), and one constant (c_0) are allowed.

3.2 Verification of Properties Involving Predicate $x = term$

Theorem 2 Let M be an ICS model with constant creators, data movement operations, and uninterpreted functions. Assume 1) the integer latches in M take only constants as their initial values[1]; 2) property P is "when b becomes true, $(x_0 = \alpha_0) \wedge \dots \wedge (x_{m-1} = \alpha_{m-1})$ ", where b and x_i 's are binary and integer variables of M respectively, α_i 's are ICS terms, and the constants in the α_i 's refer to the initial values of integer latches; 3) n is the number of all subformulas in the α_i 's; 4) M_{n+1} is a finite instantiation of M where the uninterpreted functions and the constants in the α_i 's are replaced by appropriate functions (constructed below) and constants on $n+1$ values (i.e. $\lceil \log(n+1) \rceil$ bits), and the constant creators are allowed to take values only from the set $0, \dots, n$. Let P_{n+1} be the property P which results from this interpretation. Then, P holds for M iff P_{n+1} holds for M_{n+1}.

Proof Recall that a state s in an ICS model is a tuple (L, Mem, P), where L is an

---

1. Not allowing numerals as initial values of integer latches does not seem to be a severe restriction, since numerals are useful when interpreted integer functions are allowed, which is not the case here.

assignment of values of all latches, *Mem* is a set of memory elements, and P is a set of predicates. Hence, $f(a)$ may be equal to $f(b)$ in s if $a = b$ is a true predicate in s. Since there are no other integer predicates in M, two ICS terms are equal in any state of M iff they are exactly the same formula. The idea behind the proof is that if a formula β contains a subformula which does not occur in some other formula γ, then β and γ are not equal in any state of M. Hence, the set of all formulas can be divided into a finite set of equivalence classes: those which occur as a subformula in some α, and all other formulas.

Let $\{\beta_0, ..., \beta_{n-1}\}$ denote the set of all subformulas of α_i's, where $\beta_i = \alpha_i$ for $0 \leq i \leq m - 1$. To define M_{n+1}, we give an interpretation $\mathcal{f}$ to each uninterpreted function f of M, and the constants which occur in the α_i's. If a constant c_l occurs as subformula β_k, then let c_l take the value k in M_{n+1}. For each uninterpreted function f, let $\mathcal{f}(k) = l$ if $f(\beta_k) = \beta_l$. Otherwise let $\mathcal{f}(k) = n$. Similarly define other uninterpreted functions of different arities, where if one of the argument is n, then the output is also n. Note that the value n propagates. Note that each predicate $x_i = \alpha_i$ in M becomes $x_i = i$ in M_{n+1}.

Assume P holds for M. Hence, by definition, P holds for all interpretations given to constants and uninterpreted functions. A set of such interpretations is represented by M_{n+1} and P_{n+1}. These interpretations are produced by defining the uninterpreted functions as we did above, and allowing the constants to take values only from the domain $0, ..., n$. Hence, it follows that P_{n+1} holds for M_{n+1}.

Conversely, assume P does not hold for M. Then, there exists a run $r = s_0, a_0, s_1, a_1, ..., a_{k-1}, s_k$ in M such that b holds in s_k but for some i, $x_i \neq \alpha_i$. We will build a run $\hat{r} = \hat{s}_0, \hat{a}_0, \hat{s}_1, \hat{a}_1, ..., \hat{a}_{k-1}, \hat{s}_k$ in M_{n+1} such that b holds in $\hat{s}_k$ and $x_i \neq i$, i.e. P_{n+1} does not hold for M_{n+1}. We do so by ensuring the following two invariants hold of all variables in any transitions (s_i, a_i, s_{i+1}) and $(\hat{s}_i, \hat{a}_i, \hat{s}_{i+1})$.

a. All finite (non-integer) latches take the same values in (s_i, a_i, s_{i+1}) and $(\hat{s}_i, \hat{a}_i, \hat{s}_{i+1})$.

b. If some integer variable z takes an ICS term β_k in (s_i, a_i, s_{i+1}), then z takes the value k in $(\hat{s}_i, \hat{a}_i, \hat{s}_{i+1})$. Otherwise, z takes the value n. We will show how s_0, a_0, s_1 can be constructed satisfying the above invariances; the other transitions are similar. Let O_0 be the topological ordering of the gates used in the transition (s_i, a_i, s_{i+1}) in M. We will use the same ordering to generate the transition $(\hat{s}_i, \hat{a}_i, \hat{s}_{i+1})$ in M_{n+1}. The roots of O_0 are either latches (taking one of their initial values) or constant creators. For finite latches, let them take the same initial value in s_0 as in $\hat{s}_0$. Since the constants in the α_i's can only refer to the initial values of integer latches, if the initial value of a latch l is a constant and it occurs as a subformula β_k in one of the α_i's, assign it k. Otherwise, assign it n. Assign n to the constant creators. Now, by induction, assume the above two invariants hold, before some generalized gate g in M_{n+1} is processed. We show they hold for the output of g.

Case 1. If g is a finite table, then its inputs are the same as those in s_0, a_0, s_1. Hence, its output can be assigned the same value as in $\hat{s}_0$, $\hat{a}_0$, $\hat{s}_1$.

Case 2. If g is a data movement element, since the invariants hold for its input, they will hold for its output.

Case 3. If g is an uninterpreted function f, by the second invariant, and the definition of f, the second invariance will hold for the output of g.

It could be the case that processing gate g will affect whether the invariants hold for other variables besides the output of g. For example, if integer predicates were allowed, two formulas which were not equal before the predicate was processed, could become equal afterwards. This is not a problem in this case, since M does not have any integer predicates. Continuing as above, the run $\hat{r} = \hat{s}_0, \hat{a}_0, \hat{s}_1, \hat{a}_1, ..., \hat{a}_{k-1}, \hat{s}_k$ can be constructed (QED).

3.3 Partial Verification of Properties Involving Predicate $x = y$

In the following, let M be an ICS model with constant creators, data movement elements, uninterpreted functions, and restrict the initial values of the integer latches to be constants. Let property P be "when b becomes true, $x = y$ for integer variables x and y".

Definition Let α be an ICS term. Let $f_0, ..., f_m$ be the uninterpreted functions occurring in α. Given numerals $a_0, ..., a_m$, the **linear expansion** of α with respect to $a_0, ..., a_m$, $lin(\alpha)$, is obtained by adding all the symbols occurring in α, with each f_i having value a_i.

For example, given uninterpreted function f and g with corresponding values $(1, 0)$, the linear expansion of ICS term $f((g(a, b)), f(a))$ is $1 + 0 + a + b + 1 + a$ which is $2 + 2a + b$.

Lemma 3.1 Let α be an ICS term, involving uninterpreted functions $f_0, ..., f_m$ and constants $c_0, ..., c_n$. Let $a_0, ..., a_m$ be a set of numerals. If each f_i is given the interpretation, which returns the sum of its arguments plus a_i, then the value of α under this interpretation is $lin(\alpha)$.

Proof Clear from the definitions (QED).

Lemma 3.2 Assume P does not hold for M, and let k be a given integer. Assume for some run $r = s_0, a_0, ..., a_{n-1}, s_n$,

a. in s_n, b becomes true, $x = \alpha$, $y = \beta$, $\alpha \neq \beta$,

b. for some constant c_i occurring in α or β, the coefficients of c_i in $lin(\alpha)$ and $lin(\beta)$ with respect to $0, 0, ..., 0$ (denoted by n_{i_α} and n_{i_β}) have different residues modulo k.

Let M_k be a finite instantiation of M, obtained by letting the constant creators take values 0 and 1, and each uninterpreted function f_i returning the sum of its arguments modulo k. Then, P does not hold for M_k either.

Proof Since there are no integer predicates, the values which the integer variables take have no impact on the finite variables. Hence, there exists a run $\hat{r} = \hat{s}_0, \hat{a}_0, ..., \hat{a}_{n-1}, \hat{s}_n$ in M_k which agrees with r on the values of finite variables. By the definition of f_i's and lemma 3.1, each integer variable z in $\hat{s}_j, \hat{a}_j$ can take the value $lin(z)$, where z denotes the value of z in s_j, a_j. Let all constants be 0, except for c_i, which is 1. It follows that x and y will take n_{i_α} and n_{i_β} respectively in $\hat{s}_n$. These values are not equal by assumption. Hence, P does not hold for M_k (QED).

An example of the application of lemma 3.2 is the following. Assume, for some run $x = \alpha$, $y = \beta$, $\alpha = f(a, b)$ and $\beta = f(a, c)$. Then $lin(\alpha) = a + b$, and $lin(\beta) = a + c$. Since the coefficients of b (and c) are not equal modulo 2, P will not hold of M_2 (note that the coefficient of b in $a + b$ is 1, but 0 in $a + c$. To answer the question of how many k's should be tried, note that if two integers a and b have the same residues modulo three primes P_1, P_2, P_3, $a = b$ and $b < a$, then $a = b + n(P_1 \times P_2 \times P_3)$ for some n. It is expected that even one bit integers, for which $k = 2$, will expose many of the bugs, since most bugs are control bugs, and should occur under many different data combinations.

Notation For an ICS term α, let α_f denote the number of times f occurs in α. For an uninterpreted function f, let $S_f(a)$ denote a function which adds a to the sum of the arguments of f. Hence, $S_f(1)$ denotes a function which increments the sum of f's arguments.

Lemma 3.3 Let P not hold for M. Let k be an integer. Let there be a run $r = s_0, a_0, ..., a_{n-1}, s_n$, such that in s_n, $b = 1$, $x = \alpha$, $y = \beta$, $\alpha \neq \beta$, and for some uninterpreted function f of M, $\alpha_f \neq \beta_f$ modulo k. Let M_k be a finite instantiation of M, obtained by letting the constant creators take value 0, f returning $S_f(1)$ modulo k, and each other uninterpreted function $g \neq f$ returning $S_g(0)$ modulo k. Then, P does not hold for M_k.

Proof The proof follows by noticing that in M_k, one can create a run in which when b becomes true x takes α_f modulo k and y takes β_f modulo k, which are not the same by assumption (QED).

Lemma 3.4 If techniques of lemmas 3.2 and 3.3 do not uncover any bugs in the verification of P, assuming that sufficiently many k's have been tried, the only possible bugs are where when $b = 1$, x and y take unequal ICS terms, but the number of occurrences of every symbol is exactly the same.

Proof Follows directly from lemmas 3.2 and 3.3 (QED).

An example of the situation in lemma 3.4 is $x = f(g(a, b), a)$ and $y = f(a, g(a, b))$. A simple technique which seems to produce good results is to have each uninterpreted functions return a weighted sum modulo k of its arguments, where the i-th argument is multiplied by i. For example, in this case, $x = (a + 2b) + 2a = 3a + 2b$ and $y = a + 2(a + 2b) = 3a + 4b$, where the coefficients of b are not the same. By creating finite instantiations where the constant creators are allowed to take values $(0, 1)$, and the uninterpreted functions are defined as above, such bugs can be uncovered.

Remark We have not shown in this section that our choice of uninterpreted functions for the finite instantiations is better than merely down-scaling a given concrete design (which already has meanings for the uninterpreted functions). However, by choosing the finite instantiations carefully we are able to *prove* that various classes of bugs are caught, and hence improve the confidence of the designers that good coverage is achieved.

4 Applications to Superscalar Microprocessors

We present an application of the results of the previous section to some correctness problems of superscalar microprocessors (see [Joh91] and [HP90] for background information). Superscalar microprocessors take advantage of instruction parallelism by allowing multiple instructions to be issued and completed in any cycle. Many commercial microprocessors designed in the past few years are superscalar. Throughout this section, we consider an imaginary processor M, which issues instructions out-of-order, allows out-of-order completion of instructions, does register renaming, allows speculative execution, and retires (writes back) the instructions in order. In all superscalar microprocessors we know of, results are retired in-order since precise interrupts and speculative execution (where a branch's outcome in predicted in advance) need to be supported.

When an instruction is issued, it enters a reorder buffer ([SP86]). The *reorder buffer* is a table consisting of a set of rows corresponding to different instructions and a set of columns corresponding to information about the instructions. The rows of the table are implemented as a circular FIFO, with two pointers: head and tail, pointing to the first and last entries. The columns consist of an entry number (or tag), destination register, result, exception information, valid bit saying whether the result is valid, and the program counter of the instruction. When an instruction is issued, it is given the entry pointed to by the tail of the reorder buffer, and the instruction is sent to the reservation stations of the functional units for execution.

A *reservation station* is a buffer which contains all instructions waiting to be executed in a functional unit. Each entry in a reservation station contains the following fields: the instruction's tag, operation to be performed on the two sources, values of sources 1 and 2 or tags for the instructions which will produce them, and 2 bits stating whether sources 1 and 2 are valid. When an instruction is issued, if one of its arguments, say register r, has not been computed, then the tag for the last instruction in the reorder buffer which will write r is saved in the instruction's entry at its reservation station. When an instruction finishes, the result is written to any reservation station which is waiting for this result by comparing the tag of the instruction to the tags at the reservation stations.

When the instruction at the head of the reorder buffer contains valid information, and there are no exceptions, then the result is written to the destination register. If there is an exception associated with the instruction, then the writing of results is stopped, and the corresponding exception handler is called. The program counter of the instruction is used to resume execution after the exception handler returns. To implement speculative execution, if the wrong branch was taken after a jump instruction, all entries after the jump instruction are deleted from the reorder buffer.

Consider an instruction I, $r_3 = ADD(r_1, r_2)$, where r_1 and r_2 are source registers, r_3 the destination, and the values stored at r_1 and r_2 are constants a and b. A property P one may want to verify is that if I is issued with tag number t, when t is at the top of the reorder buffer, then the value stored under the result entry in the reorder buffer is $ADD(a, b)$, provided that there are no exceptions associated with I. To use the results of the previous section, several assumptions are made.

1. The number of registers and the size of the memory is finite.

2. The predicates operating on data can take any value $(0,1)$ at any point in time. This abstraction might prevent us from proving P by allowing too much behavior only if the correctness proof somehow depends on always making the same choice for a predicate Q, i.e. if at some point $Q(x) = 1$, then at a later point, $Q(x)$ should be 1 as well. This would imply that the pipeline control remembers the value of $Q(x)$, which does not appear to be the case in general.

3. We assume that the instructions are generated non-deterministically, but no instruction before I is issued is allowed to modify r_1 or r_2. The initial values for r_1 and r_2 are set to be the constants a and b respectively. To catch errors associated with speculative execution, a program buffer is provided. The *program buffer* is implemented as a circular FIFO, and has a tail and a head pointer. The tail pointer marks the location where the next instruction is retired. The head pointer points to where the current instruction is being read. As soon as an instruction is retired, a new instruction is produced (non-deterministically) in its place. An

instruction has four fields: operation, sources 1 and 2, and destination.

4. Each functional unit is replaced by an uninterpreted function.

The verification of this property can then be done by replacing each integer variable by two bit variables, since $ADD(a, b)$ has three subformulas: a, b and $ADD(a, b)$. Each functional unit should also be replaced by the function described in the proof of theorem 2.

The process of verification using a tool such as [HSIS94] would be as follows (we have not yet implemented the techniques presented in this paper, therefore the analysis is a 'paper and pencil' analysis). The Verilog description of the reorder buffer and its interface to the functional units are chosen for verification by the user. The description is compiled into the intermediate format BLIF-MV ([HSIS94]). The compiler is instructed to use library functions for integer operations such as addition (as opposed to creating a circuit implementing addition), and to leave the functional units undefined. For example, $z := x + y$ for 32 bit integers is compiled into $HSIS - add - 32(x, y, z)$. For our Verilog compiler ([Che94]), the user needs to identify the uninterpreted functions. The user then inputs the property as an automaton or a CTL formula. The BLIF-MV file and the property are read, and are used to automatically extract an ICS model. It is then checked whether the property can be verified using finite instantiations. If so, the finite instantiation is created automatically and the standard verification routines are called.

For example, if the property is "when the result of instruction $ADD(r_1, r_2)$ is written back, $ADD(a, b)$ is saved", then a finite instantiation with two bits is created. Formulas a, b and $ADD(a, b)$ are represented by 0,1, and 2 respectively. Value 3 is used to denote all other formulas. If there is an error, the result has to be changed to be readable by the user. Specifically, the integer values need to be changed to formulas. One solution is to use formulas a, b and $ADD(a, b)$ for 0, 1, and 2 respectively, and use some special symbol such as *other* to denote other formulas. The user can then use this error trace to correct the design.

To get an idea of how large (in terms of the number of latches) the reduced systems can get, assume there are 4 registers, 4 memory locations, 4 instructions in the program buffer, 1 bit for exception information, 2 bits for data, and the tags of the reorder buffer are the locations of each entry in the buffer and therefore are not stored in the buffer, then the reorder buffer requires 8 bits per entry. If we further assume there is 1 bit for the operation in each functional unit and 2 bit tags, then each entry in the reservation stations requires 9 latches. Assuming that 8 different instructions are modeled, the program buffer requires 9 latches per entry. If there are 4 entries in the program and reorder buffers, and 2 reservation stations each holding 2 entries, we get a total of 128 latches: 8 for registers, 8 for memory, 40 for the program buffer (36 for entries and 4 for pointers), 36 for the reorder buffer (32 for entries and 4 for pointers), and 36 for reservation stations. In general, let m denote the number of bits for memory locations, r the number of bits for registers, d the number of bits for data, x the number of bits for exception information, f the number of functional units, e the number of nurtures in each reservation station, o the number of bits in the operations at each functional unit, i the number of bits required to represent an instruction type, p and b are the number of bits for the sizes of the program and reorder buffers. Then, the total number of latches is given by

$$d(2^r + 2^m) + [2^p(i + 3max(r, m)) + 2p] + [2^b(r + d + x + 1 + p) + 2b] + fe(\log b + 2d + 2 + o).$$

Another interesting property is that when the result of I is computed, any instruction in the reservation stations waiting for the result of I, will receive $ADD(a, b)$. This property tests the by-pass circuitry. Yet, a stronger property to verify is that for any stream of instructions, even those which modify r_1 and r_2 before I is issued, when I is retired, the value written back is $ADD(v(r_1), v(r_2))$, where $v(r_1)$ and $v(r_2)$ denote the values of registers r_1 and

r_2 when I was issued. To verify this property, one solution is to run an instruction level interpreter N in parallel with M. N merely executes the functional specification of each instruction. It duplicates the memory and registers of M, and stops right after it executes I. After M writes back I, it is checked that the r_3 registers of M and N are equal. Since the instruction stream can write over r_1 and r_2, when I is issued, the values of registers r_1 and r_2 are not known. Therefore, to verify this property, techniques of section 3.3 are needed.

5 Comparison to Selected Previous Work

5.1 Comparison to [LPP70]

[LPP70] presents a model called *Ianov's program schemas* which is close in flavor to our own ICS models. Ianov's schemas are sequential deterministic programs with simple computation and control. For computation, only uninterpreted functions are allowed, and for control only jumps based on uninterpreted predicates can be used. The domain of the variables is uninterpreted (for ICS models it is the set of integers).

As it turns out most of their work is orthogonal to ours. They did not define an operational model for Ianov's schemas, neither prove any finite instantiations results for them. There is one undecidability result of theirs which is similar to our own undecidability result (theorem 1). In theorem 1, we allow one uninterpreted unary function, one uninterpreted constant, and the interpreted equality predicate. In theorem 4.1 of [LPP70], they allow one uninterpreted unary function, one uninterpreted constant, and one uninterpreted unary predicate, and prove that the divergence (and equivalently halting) problems for such program schemas is undecidable. These are subtle differences and due to that the proofs are completely different: their reduction is from the emptiness check for two-headed ω-automata, whereas ours is from halting of two-counter machines.

On the other hand, some of their results are very interesting from our perspective. For example, they have some results regarding finite interpretations which apply to ICS models as well when integers are finite. They also consider some special cases which may be helpful in understanding how our ICS reachability algorithm can be optimized.

5.2 Comparison to [CN94]

[CN94] introduces a temporal logic with uninterpreted functions and predicates, the *ground logic*. This logic can be used to represent the transition relation of ICS models. [Cyr95] has shown that the validity problem for this logic with two uninterpreted functions, equality symbol, and a constant symbol is undecidable. Note that theorem 1 assumes only one uninterpreted function symbol.

5.3 Comparison to [BD94]

[BD94] uses an efficient temporal logic similar to a subset of ground temporal logic of [CN94] to show that a pipelined circuit implements its architectural specification. The check is set up so that the reasoning can be done only on the transition relation. Therefore, the user needs to specify a superset of the reachable states on which the two descriptions behave the same. In effect, they trade costly reachability computation with user intervention and expressiveness (they are not able to prove general properties). The relationship between [BD94]'s techniques and ICS reachability is similar to the relationship between symbolic simulation ([SB95]) (which allows restricted properties provable by arguing only about the transition relation), and model checking and language containment, which allow more expressive properties, but as a result need to build the set of reachable states. [BD94] extend their techniques in [Bur96] to a very simple superscalar processor, which does not have a reorder buffer and does not perform speculative execution.

6 Conclusions and Future Directions

The finite instantiation techniques are a way of addressing state space explosion due to wide datapaths, by reducing the size of the datapaths to a few bits, many times with no or little sacrifice in verification's accuracy. The results of [HB95] did not deal with uninterpreted functions. Although, these results were enough to verify memory systems and other communication protocols using finite instantiations, they are not powerful enough to handle some correctness properties of pipeline controls in microprocessors. In this paper, we extended these results to uninterpreted functions, and applied the techniques to the correctness of pipeline controls. We also showed the limits of the finite instantiation methods by proving that the problem becomes undecidable when uninterpreted functions and the predicate $x = y$ are used simultaneously.

An important research area is studying the ramification of considering infinite memory. A simple case where the memory is never read by the machine but only read once by the property was dealt with in [HMLB95]. Also, in cases where finite instantiation cannot be applied, efficient symbolic execution algorithms are needed. The works of [Cor93], [BD94], and [HB95] are relevant references for this case.

References

[BaD94] V. Bhagwati, S. Devadas, "*Automatic Verification of Pipelined Microprocessors*", Proceedings of 31st Design Automation Conference, 1994.

[But96] J. Burch, "*Techniques for Verifying Superscalar Microprocessors*", Design Automation Conference, 1996.

[BD94] J. Burch, D. Dill, "*Automated Verification of Pipelined Micro-processors*", Computer-Aided Verification, 1994.

[CZ95] Ed C. Clarke, X. Zhao, "*Word Level Model Checking, A New Approach for Verifying Arithmetic Circuits*", Technical Report, Carnegie Melon University, May 1995.

[CB94] Szu-Tsung Cheng and Robert K. Brayton, "*Compiling Verilog into Automata*", University of California at Berkeley", Memorandum UCB/ERL M94/37, 1994.

[Cor93] F. Corella, "*Automatic High-Level Verification Against Clocked Algorithmic Specifications*", Proceedings of the IFIP WG10.2 Conference on Computer Hardware Description Languages and their Applications, Ottawa, Canada, Apr. 1993. Elsevier Science Publishers B.V.

[Cyr93] David Cyrluk, "*Microprocessor Verification in PVS: A Methodology and Simple Example*", Technical Report SRI-CSL-93-12, Computer Science Laboratory, SRI International, December 1993.

[CN94] D. Cyrluk, P. Narendran, "*Ground Temporal Logic: A Logic for Hardware Verification*", Computer-Aided Verification, 1994.

[Cyr95] D. Cyrluk, private communication, 1995.

[HP90] John L. Hennessy, David A. Patterson, "*Computer Architecture A Quantitative Approach*", Morgan Kaufmann Publishers, 1990.

[HYMD95] Richard C. Ho, Han Yang, Mark A. Horowitz, David L. Dill, "*Architecture Validation for Processors*", Proceedings of the 22nd Annual Intl. Symposium on Computer Architecture, June 1995.

[HB95] R. Hojati, R. K. Brayton, "*Automatic Datapath Abstraction of Hardware Systems*", Conference on Computer-Aided Verification, 1995.

[HMLB95] R. Hojati, R. Mueller-Thuns, P. Loewenstein R. K. Brayton, "*Automatic Verification of Memory Systems which Execute Their Instructions Out of Order*", Conference on Hardware Description Languages and Their Applications, 1995.

[HSIS94] A. Aziz, F. Balarin, S. T. Cheng, R. Hojati, T. Kam, S. C. Krishnan, R. K. Ranjan, T. R. Shiple, V. Singhal, S. Tasiran, H.-Y. Wang, R. K. Brayton and A. L. Sangiovanni-Vincentelli, "*HSIS: A BDD-Based Environment for Formal Verification*", Design Automation Conference, 1994.

[Hsu93] Peter Yan-Tek Hsu, "*Design of the R8000 Microprocessor*", IEEE Micro 1993. Also available at http://www.mips.com under R8000 microprocessor.

[HU79] John E. Hopcroft, Jeffery D. Ullman, "*Introduction to Automata Theory, Languages, and Computation*", Addison-Wesley, 1979.

[Joh91] Mike Johnson, "*Superscalar Microprocessor Design*", Prentice Hall, 1991.

[LPP70] D.C. Luckham, D.M.R. Park, and M.S. Patterson, "*On Formalized Computer Programs*," Journal of Computer and System Sciences, 4, 3, pp. 220-249, June 1970.

[Sparc95] A. Charnas, et al. "*A 64b Microprocessor with Multimedia Support*", International Solid-State Circuits Conference, pp178-179, Feb, 1995.

[SB95] C. H. Seger, R. E. Bryant, "*Formal Verification by Symbolic Evaluation of Partially-Ordered Trajectories*", Formal Methods in System Design, 6:147-189, 1995.

[SS95] Toru Shonai, Tsuguo Shimizu, "*Formal Verification of Pipelined and Superscalar Processors*", Conference on Hardware Description Languages, Tokyo, Japan, August 1995.

[SP86] James E. Smith and Andrew R. Pleszkun, "*Implementing Precise Interrupts in Pipelined Processors*", IEEE Transactions on Computers, Vol. 37, No. 5, May 1986.

[Sri95] Mandayam K. Srivas, Steven P. Miller, "*Applying Formal Verification to a Commercial Microprocessor*", Conference on Hardware Description Languages, Tokyo, Japan, August 1995.

Formal Verification of the Island Tunnel Controller Using Multiway Decision Graphs

Z. Zhou[4], X. Song[1], S. Tahar[2], E. Cerny[1], F. Corella[3] and M. Langevin[4]

[1] D'IRO, Université de Montréal, Canada, {song,cerny}@iro.umontreal.ca
[2] Dept. of ECE, Concordia University, Canada, tahar@ece.concordia.ca
[3] Hewlett-Packard Company, USA, fcorella@hprpcd.rose.hp.com
[4] Nortel Technology, Canada, {zzhou,mlangevin}@nortel.ca

Abstract. *Multiway Decision Graphs* (MDGs) have recently been proposed as an efficient representation tool for RTL designs. In this paper we demonstrate the MDG-based formal verification technique on the example of the Island Tunnel Controller. We also provide comparative experimental results for the verification of a number of properties using two well-known ROBDD-based verification tools SMV (Symbolic Model verifier) and VIS (Verification Interacting with Synthesis). Finally, we study in detail the non-termination problem of the abstract state enumeration and present an solution.

1 Introduction

For automated hardware verification, ROBDDs [1] have proved to be a powerful tool [4, 10]. However, since they require a Boolean representation of the circuit, the size of an ROBDD grows, sometimes exponentially, with the number of Boolean variables. Therefore, ROBDD-based verification cannot be directly applied to circuits with complex datapaths.

There are two different ways to deal with this problem. One is to reduce the complexity of hardware designs by means of data abstraction [6, 14]. Another way is to represent hardware designs at different levels of abstraction, and verify the designs hierarchically. Recently, a number of ROBDD extensions such as BMDs [2] and HDDs [7] have been developed to represent arithmetic functions more compactly than ROBDDs. There also emerged a number of methods [8, 5, 11] which verify the overall functionality of Register-Transfer Level designs at an abstract level, using *abstract* variables to denote data signals and *uninterpreted* function symbols to denote data operations.

We have proposed a new class of decision graphs, called *Multiway Decision Graphs* (MDGs) [8], that comprises, but is much broader than, the class of ROBDDs. The underlying logic of MDGs is a subset of many-sorted first-order logic with a distinction between *concrete* and *abstract* sorts. A concrete sort has an enumeration while an abstract sort does not. Hence, a data signal can be represented by a single variable of abstract sort, rather than by a vector of Boolean variables, and a data operation can be viewed as a black box and represented by an uninterpreted function symbol. MDGs are thus much more compact than ROBDDs for designs containing a datapath.

In [8] we described a basic set of MDG operators and a reachability analysis algorithm based on *abstract implicit enumeration*. The reachability algorithm verifies whether an invariant holds in all reachable states of an *abstract state machine* [9]. One application of the algorithm is the verification of observational equivalence of synchronous circuits.

Burch and Dill [5, 13] proposed a validity checking algorithm for processor verification which is also based on the use of abstract sorts and uninterpreted function symbols. A logic expression representing the correctness statement is generated using symbolic simulation. The algorithm is then used to check its validity. With carefully chosen heuristics for avoiding exponential case splitting, the authors verified a subset of a RISC pipeline processor DLX [5] and a protocol processor (PP) [13]. Cyrluk and Narendran [11] defined a first-order temporal logic – Ground Temporal Logic (GTL) which also uses uninterpreted function symbols. Using a decidable fragment of GTL, they can automate the verification in the PVS theorem prover. These methods, however, are not applicable to verification problems that require state-space exploration, because they can only deal with paths of fixed length. They cannot represent sets of states and compute fixpoints on sets of states. MDGs, on the other hand, provide a tool for both validity checking and verification based on state-space exploration.

In this paper we present a case study of MDG-based hardware verification using the Island Tunnel Controller (ITC) [12] as an example. This example was originally used to illustrate the notation of a heterogeneous logic system supporting diagrams as logic entities [12]. However, no verification experiments were performed. Although the example is small and does not represent the scale of designs we can verify, it is ideal for illustration purposes since it covers most current MDG applications. Moreover, the example is appropriate in size to be used to illustrate in detail the state generalization techniques for dealing with the non-termination problem of our reachability analysis procedure.

The rest of the paper is organized as follows: We briefly review the Multiway Decision Graphs in Section 2 and introduce the Island Tunnel Controller example in Section 3. In Section 4, we describe the hardware modeling of the ITC at an abstract level. In Section 5, we explain the verification of the ITC example including combinational verification, invariant checking and behavioral equivalence checking. We also present experimental results, including a comparison with results obtained using SMV and VIS. Section 6 studies in detail the non-termination problem of the abstract state enumeration and presents several solutions. Finally, Section 7 concludes the paper.

2 MDGs and MDG-based Verification Approaches

The formal logic underlying MDGs is a many-sorted first-order logic, augmented with the distinction between *abstract sorts* and *concrete sorts*. This is motivated by the natural division of datapath and control circuitry in RTL designs. Concrete sorts have *enumerations* which are sets of *individual constants*, while abstract sorts do not. Variables of concrete sorts are used for representing control signals, and variables of abstract sorts are used for representing datapath signals. Data operations are represented by uninterpreted function symbols. An n-ary function symbol has a type $\alpha_1 \times \ldots \times \alpha_n \to \alpha_{n+1}$, where $\alpha_1 \ldots \alpha_{n+1}$ are sorts. The distinction between abstract and concrete sorts leads to a distinction between three kinds of function symbols. Let f be a function symbol of type $\alpha_1 \times \ldots \times \alpha_n \to \alpha_{n+1}$. If α_{n+1} is an abstract sort then f is an *abstract function symbol*. If all the $\alpha_1 \ldots \alpha_{n+1}$ are concrete, f is a *concrete* function symbol. If α_{n+1} is concrete while at least one of $\alpha_1 \ldots \alpha_n$ is abstract, then we refer to f as a *cross-operator*; cross-operators are useful for modeling feedback signals from the datapath to the control circuitry.

A *multiway decision graph* (MDG) is a finite, directed acyclic graph (DAG). An internal node of an MDG can be a variable of a concrete sort with its edge labels be the *individual constants* in the enumeration of the sort; Or it can be a variable of abstract sort and its edges are labeled by abstract terms of the same sort; Or it can be a *cross-term* (whose function symbol is a cross-operator). An MDG may only have one leaf node denoted as **T**, which means all paths in the MDG are true formulas. Thus, MDGs essentially represent relations rather than functions. Just as Bryant's ROBDDs [1] must be *reduced* and *ordered*, MDGs must also be reduced and ordered, and obey a set of other well-formedness conditions given in [8]. We developed a set of MDG algorithms for computing *disjunction, relational product* (*Conjunction* followed by *existential quantification* [4]) and *pruning-by-subsumption*. A detailed description of the algorithms can be found in [8].

A state machine is described using finite sets of input, state and output variables, which are pairwise disjoint. The behavior of a state machine is defined by its transition/output relations, together with a set of initial states. An *abstract description* of the state machine, called *abstract state machine* (ASM) [9], is obtained by letting some data input, state or output variables be of an abstract sort, and the datapath operations be uninterpreted function symbols. Just as ROBDDs are used to represent sets of states, and transition/output relations for finite state machines, MDGs are used to compactly encode sets of (abstract) states and transition/output relations for abstract state machines. We thus lift the implicit enumeration [4, 10] technique from the Boolean level to the abstract level, and refer to it as *implicit abstract enumeration* [8]. Starting from the initial set of states, the set of states reached in one transition is computed by the relational product operation. The frontier-set of states is obtained by *pruning* (removing) the already visited states from the set of newly reached states using pruning-by-subsumption. If the frontier-set of states is empty, then a least fixed point is reached and the reachability analysis procedure terminates. Otherwise, the newly reached states are merged (using disjunction) with the already visited states and the procedure continues the next iteration with the states in the frontier-set as the initial set of states.

One variation of the reachability analysis is invariant checking and the verification of observational equivalence of two ASMs verified as an invariant of the product ASM. When an invariant is violated at some stage of the reachability analysis, a counterexample is given as a sequence of input-state pairs leading from the initial state to the faulty behavior. We also have an equivalence checking procedure for combinational circuits that takes advantage of the canonicity of MDGs [8]. The MDG operators and verification procedures are packaged as MDG tools [18].

3 The Island Tunnel Controller (ITC)

The Island Tunnel Controller (ITC) example was originally introduced by Fisler and Johnson [12]. There is a one lane tunnel connecting the mainland to a small island, as shown in Figure 1. At each end of the tunnel, there is a traffic light. There are four sensors for detecting the presence of vehicles: one at tunnel entrance (ie) and one at tunnel exit on the island side (ix), and one at tunnel entrance (me) and one at tunnel exit on the mainland side (mx). It is assumed that all cars are finite in length, that no car gets stuck in the tunnel, that cars do not exit the tunnel before entering the tunnel, that cars do not leave the tunnel entrance without traveling through the tunnel, and that there is sufficient distance between two cars such that the sensors can distinguish the cars.

In [12], an additional constraint is imposed: "at most sixteen cars may be on the island at any time". The number "sixteen" can be taken as a parameter and it can be any natural number. Thus the constraint may be read: "at most n $(n \geq 0)$ cars may be on the island at any time". With ROBDD-based verification, the performance depends on the particular instance of n, as we shall see in Section 5.3

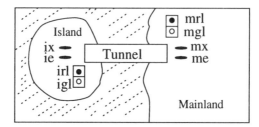

Fig. 1. The Island Tunnel Controller

Fisler and Johnson [12] proposed a specification of ITC using three communicating controllers. Their state transition diagrams are shown in Figure 2 (Figures 2(a), (b) and (c) are taken from [12] except that 16 is replaced by n). The island light controller (ILC) (Figure 2(a)) has four states: *green, entering, red* and *exiting*. The outputs *igl* and *irl* control the green and red lights on the island side, respectively; *iu* indicates that the cars from the island side are currently occupying the tunnel, and *ir* indicates that ILC is requesting the tunnel. The input *iy* requests ILC to release control of the tunnel, and *ig* grants control of the tunnel. A similar set of signals is defined for the mainland light controller (MLC). The tunnel controller (TC) processes the requests for access issued by the ILC and MLC. The island and the tunnel counters keep track of the numbers of cars currently on the island and in the tunnel, respectively. For the tunnel counter, at each clock cycle, the count tc is increased by 1 depending on signals $itc+$ and $mtc+$ or decremented by 1 depending on $itc-$ and $mtc-$ unless it is already 0. The island counter operates in a similar way, except that the increment and decrement signals are $ic+$ and $ic-$, respectively.

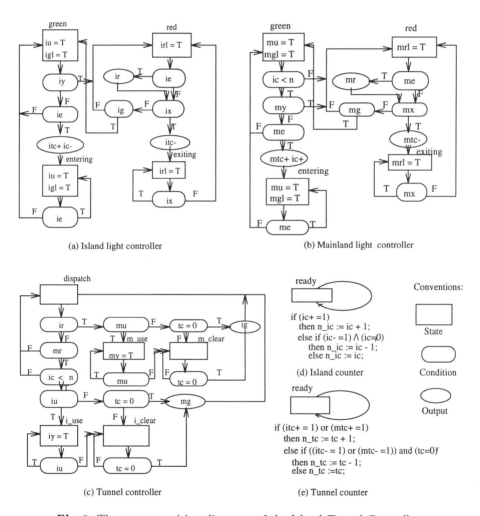

(a) Island light controller

(b) Mainland light controller

(c) Tunnel controller

(d) Island counter

if (ic+ =1)
  then n_ic := ic + 1;
  else if (ic- =1) ∧ (ic≠0)
    then n_ic := ic - 1;
    else n_ic := ic;

(e) Tunnel counter

if (itc+ = 1) or (mtc+ =1)
  then n_tc := tc + 1;
  else if ((itc- = 1) or (mtc- =1)) and (tc=0)
    then n_tc := tc - 1;
    else n_tc :=tc;

Conventions:

State

Condition

Output

Fig. 2. The state transition diagrams of the Island Tunnel Controller

4 Hardware Description of the Island Tunnel Controller

We take as *specification* the ITC state transition diagrams in Figure 2. In this section, we show how they are modeled as abstract state machines.

Both the island and the tunnel counters have each only one control state, *ready*, hence no control state variable is needed. An abstract state variable ic (tc) represents the current count number. At each clock cycle, the count is updated according to the control signals. In this abstract description of a counter, the count ic (tc) is of abstract sort, say $wordn$ for n-bit words. The control signals $(ic+, ic-$, etc.) are of *bool* sort with the enumeration $\{0, 1\}$. The uninterpreted function inc of type $[wordn \to wordn]$ denotes the operation of increment by 1, and dec of the same type denotes decrement by 1. The cross-function $equz(tc)$ of type $[wordn \to bool]$ represents the condition "$tc = 0$" and models the feedback from counter to the control circuitry. Figure 3 shows the MDG of the transition relation of the tunnel counter for a specific variable order.

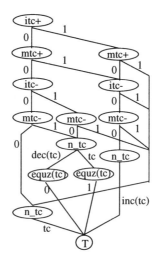

Fig. 3. Transition relation MDG of the tunnel counter

Each of the controllers can have a single control state variable which takes all the possible states as its values. Thus the enumeration of those states constitutes the (concrete) sort of the variable. Let is, ms and ts be the control state variables of the three controllers ILC, MLC and TC. We assign the variables is and ms (and also their next state variables n_is and n_ms) the sort mi_sort having the enumeration $\{green, red, entering, exiting\}$. Similarly, we let variables ts and n_ts be of sort ts_sort which has the enumeration $\{dispatch, i\_use, m\_use, i\_clear, m\_clear\}$. All other control signals (ie, ix, me, mx, etc) are of sort *bool*. The condition "$ic < n$" is represented by the cross-function $lessN(ic)$ of type $[wordn \to bool]$.

We consider an *implementation* as a netlist of components connected by signals. It is described using predefined component definitions in MDG–HDL which is the input language for our tools. The specification is also modeled as an ASM and represented using MDGs, which is compiled from the MDG-HDL descriptions.

5 Verification of the Island Tunnel Controller

In this section, we explain the various verification experiments performed on the ITC example using our MDG tools. All experiments (including those using SMV and VIS) were done on a Sun SPARCstation 10. In the result tables, column *Time* is the CPU time in seconds used for compiling the circuit descriptions and for the reachability analysis, including counterexample generation if necessary. Column *Mem* is the memory allocated in megabytes. Column *#Nodes* is the number of total MDG (or ROBDD) nodes generated.

5.1 Equivalence Checking

In [12], two ILC implementations are given, denoted as I_1 and I_2 where the later is a simplified version of I_1. We first compared I_1 and I_2 of the ILC against the behavioral specification (Figure 2(a)). We found one design error in I_1 and two errors in I_2. For more details, refer to [19]. All the errors were corrected in half an hour with the help of the counterexample facility. These errors were confirmed by the designer. Since MLC is similar to ILC, we also derived the implementations for MLC from the ILC implementations, and compared them with the MLC behavioral specification (Figure 2(b)).

The execution time and resource requirements of the experiments are given in Table 1. Rows *Faulty I_1 vs. Spec.* and *Corrected I_1 vs. Spec.* are the verifications of the implementation I_1 with and without the original errors against the behavioral specification. Similarly for rows *Faulty I_2 vs. Spec.* and *Corrected I_2 vs. Spec..*.

5.2 Combinational Verification

For both MLC and ILC, it is possible to check the equivalence of implementation I_1 with the simplified version I_2 using combinational verification, since the registers in the two circuits have one-to-one correspondence. For combinational verification, we compare two circuits having the same input, output and state variables (if any). For each circuit, we derive an MDG relating each output and state variable to the inputs and state variables (These MDGs collectively represent the output and transition relations). Then, using the canonicity property of MDGs, we simply check that corresponding MDGs for the two circuits have the same MDG identifier. The experimental results for comparing the corrected I_1 and I_2 are given in the last row in Table 1. The execution time is essentially the compilation time for constructing the transition and the output relation MDGs.

Table 1. Equivalence checking and combinational verification

	Island Light Controller			Mainland Light Controller		
	Time(sec)	Mem(MB)	#Nodes	Time(sec)	Mem(MB)	#Nodes
Faulty I_1 vs. Spec.	3.4	1.5	1015	5.0	1.7	1361
Corrected I_1 vs. Spec.	3.4	1.5	988	5.0	1.6	1292
Faulty I_2 vs. Spec.	2.8	1.2	833	3.6	1.5	992
Corrected I_2 vs. Spec.	2.7	1.2	806	3.6	1.6	988
Corrected I_1 vs. I_2	2.7	1.4	783	3.7	1.5	1108

5.3 Checking Invariants on the ITC Specification

Property checking is useful for verifying that a specification satisfies certain requirements. We list below three example properties (invariants) that we verified[5]. We also provide their CTL (Computational Tree Logic) formulas which are used for invariant checking by the ROBDD-based verification tools SMV and VIS.

P1: Cars never travel both in directions in the tunnel at the same time.
   AG (! ((igl=1) & (mgl=1))).
P2: The tunnel counter is never signaled to increment simultaneously by ILC and MLC. AG (!(($itc+$ = 1) & ($mtc+$ = 1))).
P3: The island counter is never signaled to increment and decrement simultaneously. AG (! (($ic-$ = 1) & ($ic+$ = 1))).

For the purpose of comparison, we first show the experimental results for the verification of the above example invariants (P1, P2 and P3) using FSM-based methods. For MDG tools, the counts tc and ic are now assigned a concrete sort according to the counter width which is determined by the instantiation of the constraint, i.e., the maximum number of cars that are allowed on the island. In addition, we also used two well-known ROBDD-based tools: SMV (V2.4.4) [16] and VIS [3].

Table 2 shows the results for checking the conjunction of P1, P2 and P3 on various counter widths. For the SMV columns, the *Time* is the *user* time, while for VIS and MDG columns, it is the *elapsed* time including loading the Verilog or MDG-HDL description file, compilation and invariant checking. For SMV and VIS, we used the node ordering generated by the systems, while for MDG tools, we used manual ordering since no heuristic ordering algorithm is available for the moment. Many different factors influence the experimental results shown in the table. The three tools use different integer encodings, different variable ordering, and different partitioning of the transition relation. But notwithstanding these differences, the table clearly shows the following: (i) Time increases exponentially with the counter width for concrete representations of the problem, and

---

[5] Fisler and Johnson [12] proposed a set of properties that the ITC design should satisfy. Currently, we only consider the verification of invariants.

(ii) the MDG figures are substantially greater than the others for concrete representations. This is because the MDG data structure and the MDG algorithms are far more complicated than those of ROBDDs.

However, MDG tools are able to verify a parameterized implementation having n bits for the counter width, which is not the case for SMV and VIS. The last row in Table 2 gives the results for verifying the properties when we model the design as an ASM rather than an FSM using MDGs. The verification is performed quite efficiently in time independent of the datapath width.

Table 2. ITC Invariant checking. ('–' means that the verification did not terminate in certain amount of time, and '*' means that the verification is not possible.)

Counter Width	SMV			VIS			MDG		
	Time (sec)	Mem (MB)	#Nodes	Time (sec)	Mem (MB)	#Nodes	Time (sec)	Mem (MB)	#Nodes
4 bits	1.2	1.2	10043	15.4	0.5	6492	430	8	19670
5 bits	4.1	1.2	10463	18.9	0.5	3887	810	10	27668
6 bits	16.7	1.2	11240	44.5	0.6	8902	1719	15	41751
7 bits	79.7	1.2	15047	429.9	1.2	33447	5486	26	69911
8 bits	360	1.6	29474	1686	2.4	43428	–	–	–
9 bits	1564	2.1	59292	7584	5.1	128426	–	–	–
10 bits	6263	3.2	117890	31255	9.9	327090	–	–	–
11 bits	–	–	–	–	–	–	–	–	–
n bits	*	*	*	*	*	*	55	2.7	4329

It may be argued that the data abstraction method [6, 14] is sufficient to imply the correctness of this ITC example, i.e., we reduce n to a small number encoded by a few bits, e.g., 2 bits (4), 4 bits (16), etc. Yet in general, the equivalence of the reduced circuit against the original one is not verified mechanically. Also, it is not always obvious how to construct an appropriate data abstract function, or such data abstraction may not even be possible. One such example is the 4×4 Fairisle ATM switch fabric recently verified using MDGs [15, 17], where the datapath contains mixed data/control information. In general, if the control information needs n bits, then it is impossible to reduce the word width to less than n. Hence, in this case ROBDD-based datapath reduction technique is no more practical. On the other hand, using the MDG-based approach, we naturally allow the abstract representation of data while the control information is extracted from the datapath using cross-functions.

Invariant checking is based on reachability analysis of state machines. However, the reachability analysis of ASMs may not terminate in general [8] which is the case for the ITC example. To deal with this problem, we developed a heuristic state generalization method to be explained next.

6 Termination of Abstract State Enumeration

6.1 Review

An abstract state machine may have an infinite number of states due to the abstract variables and the uninterpreted nature of function symbols. Thus a least fixed point may not be reached in state enumeration. In [8], the non-termination problem is explained and a solution is given to a class of interesting problems known as *processor-like* circuits by generalizing the initial state (i.e., by replacing abstract constants with abstract variables as initial values of some of the registers). Informally, a processor-like circuit usually starts from a *ready* state, performs data operations in one or more cycles and then returns to the *ready* state. We first briefly review the initial state generalization method using the tunnel counter (Figure 2(e)), which is a processor-like circuit, as an example.

Let the initial state of the tunnel counter be a generic constant zero of the abstract sort *wordn*. Figure 4(a) shows the MDGs N_0, S_1 and N_1 representing the set of initial states, the set of states reached in one transition, and the frontier-set of states, respectively. There are three paths in S_1, i.e., there are three abstract states. The state $tc = $ zero is subsumed by N_0 and is thus pruned. The state $(tc = dec(\text{zero})) \wedge (equz(\text{zero}) = 0)$ is, in fact, an unreachable state. If we rewrite $equz(\text{zero})$ to an individual constant 1 using the rewrite rule $equz(\text{zero}) \rightarrow 1$, it yields a contradiction (1=0) and the path can be eliminated. The frontier-set of states N_1 thus contains the only state $tc = inc(\text{zero})$. If we continue the reachability analysis with N_1, at the k-th iteration we will have the state $tc = inc^k(\text{zero})$ (where $inc^k(\text{zero})$ is a short hand for $inc(....(inc(\text{zero}))...))$. This illustrates the non-termination problem due to the fact that the terms that label the edges can be arbitrarily large and hence arbitrarily many.

The non-termination problem can be avoided by *generalizing* the initial value zero of the state variable tc to a *fresh* [6] *abstract* variable a. Then, as shown in Figure 4(b), the states $tc = inc(a)$ and $(tc = dec(a)) \wedge (equz(a) = 0)$ in S_1 become instances of $tc = a$ under substitutions $inc(a)/a$ or $dec(a)/a$, respectively. They thus can be pruned and the state exploration procedure terminates right away (the MDG $N_1 = \mathbf{F}$ represents an empty set). Note that the method may have false negatives since the reachable state space is enlarged by the state generalization. For instance, the state $(tc = dec(a)) \wedge (equz(a) = 0)$ becomes a reachable state which covers the previously unreachable state $(tc = dec(\text{zero})) \wedge (equz(\text{zero}) = 0)$.

6.2 State Generalization for General Cases

The complete ITC specification is composed of 5 communicating state machines (Figure 2). Among them, the tunnel and the island counters are typical processor-like circuits. However, the composed ASM is no more a processor-like circuit, and the initial state generalization technique is not applicable directly.

---

[6] A *fresh* variable is disjoint with all other variables.

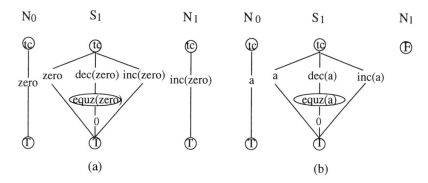

Fig. 4. Non-termination problem and initial state generalization for the tunnel counter.

In the following, we present a new state generalization technique which solves the non-termination problem for a larger class of circuits. To simplify the presentation, we only consider a specification consisting of the ASMs in Figures 2(a), (b), (c) and (e).

Initially, we assume there are no cars in the tunnel. Therefore, the tunnel counter is reset to **zero**. To be safe, the lights on both the island and the mainland sides are reset to *red*, and the tunnel controller is in the state *dispatch*, ready to take requests. The set of initial states thus includes only one state ($is = red$) $\land$ ($ms = red$) $\land$ ($ts = dispatch$) $\land$ ($tc = $ **zero**). In Figure 5, we show a small fraction of the state transition diagram of the composed machine related to the transitions of the ILC between the states *red* and *green*, and the cycles between *green* and *entering*. The Mainland Light Controller and the Tunnel Controller keep their initial state values.

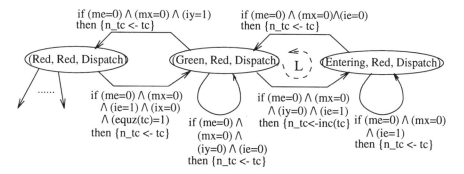

Fig. 5. A fraction of the state transition diagram for the composed machine.

To illustrate the failure of the initial state generalization method on this example, we again try the above method by setting the initial state to ($is = $

$red) \wedge (ms = red) \wedge (ts = dispatch) \wedge (tc = a)$ where a is an abstract variable of sort $wordn$. Figure 6 shows the MDGs representing the set of newly reached states $(S_i, i = 1, 2, ...)$ and the frontier-set of states $(N_i, i = 0, 1, 2, ...)$ in three iterations following the transitions depicted in Figure 5.

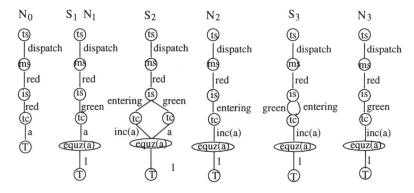

Fig. 6. State enumeration using initial state generalization technique.

Starting from N_0, after three transitions, the set S_3 of newly reached states contains two states: p : $(is = green) \wedge (ms = red) \wedge (ts = dispatch) \wedge (tc = inc(a)) \wedge (equz(a) = 1)$ and q : $(is = entering) \wedge (ms = red) \wedge (ts = dispatch) \wedge (tc = inc(a)) \wedge (equz(a) = 1)$. It can be easily checked that q is an instance of the only state in N_2. Thus, it can be pruned. On the other hand, p is not an instance of any visited states. In the visited states, only the state in N_1 matches p with the same values for control state variables. However, for the abstract state variable and the cross-term, we have a pair of contradictry substitutions $inc(a)/a$ and a/a, which means no valid substitution exists. Therefore, p is considered as a new state in N_3. If we continue the reachability analysis with N_3, the value of tc will become $inc(...inc(a)...)$ with an unbounded number of inc.

A closer examination of this non-termination problem shows that the iterations from N_1 to N_3 represents the loop L (Figure 5) with a data operation inc on the abstract state variable tc. This loop, in fact, resembles the ASM of a processor-like circuit. The state $(is = green) \wedge (ms = red) \wedge (ts = dispatch) \wedge (tc = a) \wedge (equz(a) = 1)$ in N_1 can thus be considered as the initial (entry) state of loop L. However, N_1 is not a generalized state because of the assumption $equz(a) = 1$, which is the reason for non-termination.

The above analysis leads to the observation of a *processor-like loop*. A processor-like loop starts from a control state and returns to this control state after one or more transitions, with data registers updated according to the data operations. Note that a processor-like circuit is represented by an ASM having one and only one processor-like loop. This suggests that it is the entry state of a processor-like loop that should be generalized rather than the initial state of the state machine. In the above example, we have to generalize the state

$(is = green) \wedge (ms = red) \wedge (ts = dispatch) \wedge (tc = a) \wedge (equz(a) = 1)$ in N_1 instead of the states in N_0. Figure 7 shows the state enumeration procedure using the extended method. Starting from N_0, S_1 is reached in one transition: $\{(is = green) \wedge (ms = red) \wedge (ts = dispatch) \wedge (tc = \textsf{zero}) \wedge (equz(\textsf{zero}) = 1)\}$, which can be simplified to $\{(is = green) \wedge (ms = red) \wedge (ts = dispatch) \wedge (tc = \textsf{zero})\}$ using the rewrite rule $equz(zero) \rightarrow 1$. As this state is the entry state of the processor-like loop L, we generalize the constant value of tc to a *fresh abstract* variable a and remove the constraint $equz(a) = 1$. Thus the frontier set of states becomes $N_1' = \{(is = green) \wedge (ms = red) \wedge (ts = dispatch) \wedge (tc = a)\}$. After two transitions, the frontier-set N_3 becomes $\{(is = green) \wedge (ms = red) \wedge (ts = dispatch) \wedge (tc = inc(a))\}$, where the only state in this set is subsumed by N_1' under substitution $inc(a)/a$. The procedure thus terminates.

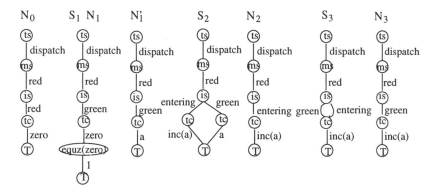

Fig. 7. State enumeration using the extended state generalization technique.

It remains to determine now when and on which state variables the generalization is to be performed, i.e., how to identify processor-like loops and perform the generalization operation. For some circuits, e.g., simple microprocessors, it is possible to identify all processor-like loops by inspection and to manually perform state generalization on the entry states of the loops. However, in general, to find the entry states of all processor-like loops could be very difficult. In the next subsection, we propose a heuristic method.

6.3 Heuristic State Generalization

We developed a heuristic method for state generalization based on the following observation: reachability analysis terminates if we generalize any state within a processor-like loop. Once we generalize a state in a loop, it covers all the abstract states having the same control state values, thus guaranteeing termination.

One method is to perform generalization on every abstract state variable at each clock cycle, i.e., to replace every term that labels an edge issuing from an abstract node with a fresh abstract variable. However, the reachable state space is unnecessarily enlarged since states that are not within processor-like loops are also generalized. As a trade-off, we propose a heuristic solution to this problem: After a certain number of state transitions (specified by the user), if the MDG size of the frontier-set keeps increasing, the value of each state variable in the MDG is generalized. With this heuristic, the state to be generalized is more likely to be within a processor-like loop.

Termination of the abstract state enumeration can be obtained at the cost of false negatives introduced by the state generalization. If the reachability analysis succeeds, we know that the invariant holds even for the enlarged set of reachable states, but if it does not, then we have to examine, e.g., by simulation, whether the counterexample thus produced corresponds to a *real* design error. The heuristic method performs state generalization quite blindly at first, and postpones any manual analysis to the examination of counterexamples, if any.

The results in Table 2 (the last row) are obtained using the heuristic state generalization technique on the complete ITC specification composed of the five ASMs of Figure 2.

7 Conclusions

In this paper, we demonstrated the feasibility of the MDG-based hardware verification at the RT level on a non trivial example–the Island Tunnel Controller. We performed various verification experiments on the example including combinational verification, behavioral equivalence checking and invariant property checking. Using the counterexample facility of the MDG tools, we also showed our ability to identify design errors that were present in the original implementations. Furthermore, we gave a comparative evaluation of the results from invariant checking with the ROBDD-based tools SMV and VIS, and we showed the strength of MDG approach by handling arbitrary data widths. Finally, we studied in detail the non-termination problem of abstract state enumeration and presented a heuristic solution.

The MDG tools are capable of dealing with complex designs. The interested readers may wish to refer to [15, 17] where a case study is presented about the verification of an ATM switch fabric. We are currently developing a model checking algorithm for a restricted first-order temporal logic.

References

1. R. E. Bryant. Graph-based algorithms for boolean function manipulation. *IEEE Transactions on Computers*, 35(8):677–691, August 1986.
2. R. E. Bryant and Y. Chen. Verification of arithmetic circuits with binary moment diagrams. In *32nd ACM/IEEE Design Automation Conference (DAC'95)*. San Francisco, California, June 1995.
3. R. K. Brayton et. al. VIS: A system for verification and synthesis. In *Proc. 8th International Conference on Computer-Aided Verification (CAV'96)*. New Brunswick, New Jersey, USA, July 1996.

4. J. R. Burch, E. M. Clarke, D. E. Long, K. L. McMillan and D. L. Dill. Symbolic model checking for sequential circuit verification. *IEEE Transactions on Computer-Aided Design*, 13(4):401–424, April 1994.

5. J. R. Burch and D. L. Dill. Automatic verification of pipelined microprocessor control. In: D. L. Dill, editor, *Computer Aided Verification*. Lecture Notes in Computer Science 818, Springer Verlag, 1994.

6. E. M. Clarke, O. Grumberg, and D. E. Long. Model checking and abstraction. In *Proc. 19th ACM Symp. on Principles of Programming Languages*. January 1992.

7. E. Clarke, M. Fujita and X. Zhao. Hybrid decision diagrams. In *Proc. IEEE Inter. Conf. on Computer-Aided Design (ICCAD'95)*. San Jose, California, USA, Nov. 1995.

8. F. Corella, Z. Zhou, X. Song, M. Langevin and E. Cerny. Multiway decision graphs for automated hardware verification. IBM technical report RC19676, July 1994. To appear in the journal *Formal Methods in System Design*.

9. F. Corella, M. Langevin, E. Cerny, Z. Zhou and X. Song. State enumeration with abstract descriptions of state machines. In *Proc. IFIP WG 10.5 Advanced Research Working Conference on Correct Hardware Design and Verification Methods (Charme'95)*. Frankfurt, Germany, October 1995.

10. O. Coudert, C. Berthet and J. C. Madre. Verification of synchronous sequential machines based on symbolic execution. In J. Sifakis, editor, *Automatic Verification Methods for Finite State Systems*. Lecture Notes in Computer Science 407, Springer Verlag, 1989.

11. D. Cyrluk and P. Narendran. Ground Temporal Logic: A logic for hardware verification. In: D. L. Dill, editor, *Computer Aided Verification*. Lecture Notes in Computer Science 818, Springer Verlag, 1994.

12. K. Fisler and S. Johnson. Integrating design and verification environments through a logic supporting hardware diagrams. In *Proc. IFIP Conference on Hardware Description Languages and their Applications (CHDL'95)*. Chiba, Japan, Aug. 1995.

13. R. B. Jones and D. L. Dill. Efficient validity checking for processor verification. In *Proc. IEEE International Conference on Computer-Aided Design (ICCAD'95)*. San Jose, California, USA, November 1995.

14. D. E. Long. *Model Checking, Abstraction, and Compositional Verification*. PhD thesis, Carnegie Mellon University, 1993.

15. M. Langevin, S. Tahar, Z. Zhou, X. Song and E. Cerny. Behavioral Verification of an ATM switch fabric using implicit abstract state enumeration. In *Proc. IEEE Inter. Conf. on Computer Design (ICCD'96)*. Austin, Texas, USA, Oct. 1996.

16. K. L. McMillan. *Symbolic model checking*. Kluwer Academic Publishers, Boston, Massachusetts, 1993.

17. S. Tahar, Z. Zhou, X. Song, E. Cerny and M. Langevin. Formal verification of an ATM switch fabric using multiway decision graphs. In *Proc. IEEE Sixth Great Lakes Symposium on VLSI*. Ames, Iowa, USA, March 1996.

18. K.D. Anon, N. Boulerice, E. Cerny, F. Corella, M. Langevin, X. Song, S. Tahar, Y. Xu, Z. Zhou. MDG tools for the verification of RTL designs. In *Proc. 8th International Conference on Computer-Aided Verification (CAV'96)*. New Brunswick, New Jersey, USA, July 1996.

19. Z. Zhou, X. Song, S. Tahar, E. Cerny, F. Corella and M. Langevin. Formal verification of the Island Tunnel Controller using Multiway Decision Graphs. Technical Report 1042, D'IRO, Université de Montréal, Montréal, Canada, July 1996.

VIS

Robert K. Brayton,[1] Gary D. Hachtel,[2] Alberto Sangiovanni-Vincentelli,[1]
Fabio Somenzi,[2] Adnan Aziz,[3] Szu-Tsung Cheng,[1] Stephen A. Edwards,[1]
Sunil P. Khatri,[1] Yuji Kukimoto,[1] Abelardo Pardo,[2] Shaz Qadeer,[1]
Rajeev K. Ranjan,[1] Shaker Sarwary,[4] Thomas R. Shiple,[1] Gitanjali Swamy,[1]
Tiziano Villa[1]

[1] Department of EECS, University of California, Berkeley, CA 94720
[2] Department of ECE, University of Colorado, Boulder, CO 80309
[3] Department of ECE, University of Texas at Austin, TX 78712
[4] Lattice Semiconductor, Milpitas, CA 95035

1 Introduction to VIS

VIS (Verification Interacting with Synthesis) is a tool that integrates the verification, simulation, and synthesis of finite-state hardware systems. It uses a Verilog front end and supports fair CTL model checking, language emptiness checking, combinational and sequential equivalence checking, cycle-based simulation, and hierarchical synthesis.

VIS was designed to maximize performance by using state-of-the-art algorithms, and to provide a solid platform for future research in formal verification. VIS improves upon existing verification tools by:

1. providing a better programming environment,
2. providing new capabilities, and
3. improving performance in some cases.

In addition, software engineering methods were used in the design of VIS in order to encourage further development by a diverse set of researchers. In particular, we provide extensive documentation that is automatically extracted from the source files for browsing on the World Wide Web.

In this tutorial, we will briefly describe each of the capabilities built into VIS and then concentrate on how to use and control VIS to make maximum use of its potential.

1.1 History

Many first generation tools for automatic formal verification were based on two theoretical approaches. The first is temporal logic model checking, where the properties to be checked are expressed as formulas in a temporal logic, and the system is expressed as a finite state system. In particular, Computational Tree Logic (CTL) model checking is a technique pioneered by Clarke and Emerson to verify whether a finite state system satisfies properties expressed as formulas in a branching-time temporal logic called CTL. SMV, a system developed at Carnegie Mellon University, belongs to this class of tools.

Certain properties are not expressible in CTL, but they can be expressed as ω-automata. The second approach, language containment, requires the description of the system and properties as ω-automata, and verifies correctness by checking that the language of the system is contained in the language of the property. Note that certain types of CTL properties involving existential quantification are not expressible by ω-automata. COSPAN, a system developed at Bell Labs, offers language containment.

A combination of both approaches is offered by the HSIS [2] system, which was developed at the University of California, Berkeley. Our experience with verification tools (in particular HSIS) led to the conclusion that sometimes, the simpler and more limited the approach, the more efficient it can be. A number of design decisions that we made for HSIS made it unacceptably slow for some large examples. With these problems in mind, we set about writing a tool that was more efficient, easily extendible, and offered a good programming environment, in order that it can be more easily upgraded in the future as more efficient algorithms are developed.

VIS also has the capability to interface with SIS to optimize logic modules; hence, VIS is an integrated system for hierarchical synthesis, as well as verification. We plan to pursue research on the interaction between verification and synthesis in the future; hence the name VIS, verification interacting with synthesis.

1.2 Overview of VIS

Fig. 1 presents of an overview of VIS. It has three main parts: a front-end to read

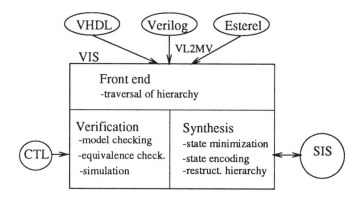

Fig. 1. Block diagram of VIS.

and traverse a hierarchical system described in BLIF-MV, which may have been compiled from a high-level language like Verilog; a verification core, to perform model checking of Fair CTL and test language emptiness; and a path to SIS, to optimize parts of the logic.

1.3 VIS Philosophy

We decided to offer limited but efficient capabilities. We felt that in the future, it would be easy to add more features, as they are required, using a well defined programming interface. In line with this **keep it simple** philosophy, VIS provides the following verification capabilities.

- Only CTL formulas can be checked. Language containment may be handled in a later release. However, we do handle language emptiness checks.
- Fairness constraints must be of Büchi type, i.e., sets of states that must be visited infinitely often. However, the internal VIS data structures do have the capability to support more complicated fairness constraints.

VIS can interact with SIS to assist the task of verification by simplifying parts of the system. Another objective is to support a full-fledged hierarchical synthesis flow, that translates a Verilog description into an optimized multi-level circuit at the gate level. Unlike existing logic optimization systems like SIS, VIS can support hierarchical synthesis.

2 Describing Designs for VIS

VIS operates on an intermediate format called BLIF-MV, which is an extension of BLIF, the intermediate format for logic synthesis accepted by SIS [4]. VIS includes a stand-alone compiler from Verilog to BLIF-MV, called VL2MV [3], which supports a synthesizable subset of Verilog. VL2MV extracts a set of interacting finite state machines that preserves the behavior of the source Verilog program defined in terms of simulated results. Two new features have been added to Verilog:

1. Nondeterminism. A nondeterministic construct, $ND, has been added to specify nondeterminism on wire variables; this is the only legal way to introduce nondeterminism in VIS. Internally, $ND introduces pseudo-inputs which can non-deterministically take any allowed value.
2. Symbolic variables. Sometimes it is desirable to specify and examine the value of variables symbolically, rather than having to explicitly encode them. VL2MV extends Verilog to allow symbolic variables using an enumerated type mechanism similar to the one available in the C programming language.

Currently, a translator from a subset of VHDL to BLIF-MV is being developed and a path from Esterel to BLIF-MV is available through the POLIS system at Berkeley.

3 Verification In VIS

When a BLIF-MV description is read into VIS, it is stored as a "hierarchy" tree, which is a hierarchical description of the design ; it consists of modules that in turn consist of sub-modules. The functionality of a module is represented by a network of gates with arbitrary functionality and latches. The *print_hierarchy_stats* command in VIS prints hierarchy information, and the *print_models* command lists statistics on all the models in the hierarchy. Other useful print commands are *print_io* and *print_latches*.

The hierarchy in VIS can be traversed in a manner similar to traversing directories in UNIX, i.e., a desired node in the tree can be reached by walking up and down with the *cd* command. At any node, simulation, verification and synthesis operations can be performed. The command *pwd* prints the name of the current node. The command *ls* lists all the nodes (submodules) in the current node; *ls -R* lists all the nodes in the current subtree.

We begin by describing the steps involved in converting this hierarchy description into an internal FSM representation. The compound *init_verify* command executes the entire sequence of required initialization commands.

3.1 Flatten

The first step towards verification consists of "flattening" this hierarchical description into a single network (netlist of multi-valued logic gates and latches). The output of the design is a function of its inputs and the latches; this functionality is defined by the functions at the gates, and their interconnections. The *flatten_hierarchy* command creates this network, and the *print_network* command can be used to print it. Other related commands are *print_network_stats* command that prints statistics about the network, and *test_network_acyclic* command that checks the network for combinational cycles.

Note that when a node in the hierarchy is arrived at for the first time during the traversal of the hierarchy using *cd*, there is no network for that node until *flatten_hierarchy* is called for that node.

Also *flatten_hierarchy* automatically checks each table in the network for being deterministic (any non-determinism in the description is taken care of by pseudo-inputs) and completely specified. Since this checking takes some time, it can be turned off safely using the option *flatten_hierarchy -b*, after a BLIF-MV file has been checked once.

3.2 Ordering and Partitioning

The next step towards verification consists of converting this network representation into a functional description that represents the output and next state variables as a function of the inputs and current state variables. We use the BDD (binary decision diagram) and its extension the MDD (multivalued decision diagram) to represent boolean and discrete functions. Before creating the MDDs, it is necessary to order the variables in the support of the MDD. This is accomplished by the *static_order* command, which gives an initial ordering. To undo the current ordering, reinvoke the command *flatten_hierarchy*. At any stage the current variable ordering can be written out to a file using the *write_order* command.

The *build_partition_mdds* command computes the transition function MDDs. Depending on the partitioning method selected, the MDDs for the combinational outputs (COs) are built in terms of either the combinational inputs (CIs) or some subset of intermediate nodes of the network. The MDDs built are stored in a DAG called a "partition". The vertices of a partition correspond to the CIs, COs, and any intermediate nodes used. Each vertex has a multi-valued function (represented by an MDD) expressing the function of the corresponding network node in terms of the partition vertices in its transitive fanin. Hence, the MDDs of the partition represent a partial collapsing of the network. The *inout* method represents one extreme where no intermediate nodes are used, and *total* represents the other extreme where every node in the network has a corresponding vertex in the partition. The *frontier* method creates vertices for intermediate nodes as necessary in order to control the BDD size. The BDD size at which a new vertex is created is controlled by the parameter *partition_threshold*. This partition method encompasses the other two (a large value of partition_threshold leads to "inout" method and a zero value for partition_threshold leads to "total"). The partition graph can be printed to a file with the *print_partition* command. Another related command is the *print_partition_stats* command that prints statistics on the partition graph.

3.3 Dynamic Re-ordering

Dynamic ordering of variables may be enabled and disabled using the *dynamic_var_ordering* command. Dynamic ordering is a technique to reorder the MDD variables to reduce the size of the existing MDDs. Available methods for dynamic reordering are *window* and *sift*. Dynamic ordering (especially using *sift*) may be time consuming, but can often reduce the size of the MDDs dramatically.

Dynamic ordering is best invoked explicitly (using the *dynamic_var_ordering -f <method>* option) after the *build_partition_mdds* and *print_img_info* commands. If dynamic ordering finds a good ordering, then you may wish to save this ordering (using *write_order <file>*) and reuse it (using *static_order -s <method> <file>*). With option *dynamic_var_ordering -e <method>* dynamic ordering is automatically enabled once a certain threshold on the overall MDD size is reached. Enabling dynamic ordering may slow down the verification, but it can make the difference between completing and not completing a verification task.

3.4 FSM Traversal and Image Computation

FSM traversal is the core computation in design verification. Efficient traversal requires grouping the MDDs, in a manner optimal for traversal. To traverse the FSM, the present state, input, and next state variables are organized for easy manipulation. All this information is included in an FSM data structure created in the *compute_reach* command. This also invokes traversal of the entire reachable state set of the FSM representing the design, and may be invoked with different verbosity options to get varying amounts of traversal information. On subsequent calls to *compute_reach*, the reachability computation is not reperformed, but statistics can be printed using *-v*.

The reachability computation makes extensive use of image computation. There are several user-settable options that affect the performance of image computation. The documentation for the *set* command lists these options. Use the command *set image_method* to change the image computation method, and then re-initialize verification (starting at the *flatten_hierarchy* command [5]). The *print_img_info* prints current image information. Notice that while *print_partition _stats* prints information on the next state *functions*, *print_img_info* prints information on the next state transition *relations*. The command *print_img_info* creates transition relations from transition functions by clustering several functions together. The result is a partitioned transition relation. It is often a good idea to force dynamic variable reordering (for instance, *dynamic_var_ordering -f sift*) at this point to reorder these relation MDDs. The reachability computation is an optional step of the model checking algorithm; unreachable states may be used as don't cares to minimize the BDD representation.

3.5 Fairness Constraints

Fairness constraints are used to restrict the behavior of the design. Each fairness condition specifies a set of states in the machine, and requires that in any acceptable behavior these states must be traversed infinitely often (i.e., these states must be on a cycle). Such constraints are called "Büchi fairness" constraints. Fairness constraints are stored in fairness files (with extension .fair by convention). A fairness file is read in by the *read_fairness* command. Active fairness conditions can be displayed by means of *print_fairness*. The *reset_fairness* command is used to reset the fairness constraint to "true"; by default, there is one fairness condition that contains all states.

Fairness constraints remove unwanted behavior from a system. They are a powerful tool, but should be used with care, because it is easy to make a faulty system pass wanted properties by a careless use of fairness constraints. Also, it has been observed that verification is slowed down considerably when there are many fairness constraints.

3.6 Language Emptiness

The language of a design is given by sequences over the set of reachable states that do not violate the fairness constraint. If the language is empty, we know that the system does not exhibit any legal behavior. VIS supports the command *lang_empty* as an alias for model checking the formula *EG true*. This is relevant when verification is done by adding a component which monitors the design and has fair runs precisely when the observed behavior does not satisfy the specification. Before invoking model checking, *lang_empty* can also be used to ensure

---

[5] Whenever a hierarchy is reinitialized, the option *flatten_hierarchy -b* can be used safely for efficiency.

that the system is non-trivial. This is pertinent because the fairness constraints specified may make the entire system "unfair", and an empty system passes all universal properties.

3.7 Model Checking, Invariants and Abstraction

The *model_check* command calls CTL model checking in VIS. Formulas to be checked are specified in a file which is an argument to the command.

An important class of CTL formulas is *invariants*. These are formulas of the form $AG\ f$, where f is a quantifier-free formula. The semantics of $AG\ f$ is that f is true in all reachable states. The command *check_invariant* implements an algorithm that is specialized for these formulas.

When performing model checking and checking invariant properties, one can use the reduce option -*r*, to perform model checking on a "pruned" FSM, i.e., one where parts of the network that do not affect the formula (directly or indirectly) have been removed.

This mechanism can be combined with the abstraction mechanism available through the command *flatten_hierarchy <file>*. *<file>* contains the names of variables to "abstract". For each variable **x** appearing in *<file>*, a new primary input node named **x$ABS** is created to drive all the nodes that were previously driven by **x**. Hence, the node **x** will not have any fanouts; however, **x** and its transitive fanins will remain in the network. Abstracting a net effectively allows it to take any value in its range, at every clock cycle. This mechanism can be used to perform manual abstractions.

3.8 Error Tracing

VIS produces a debug trace to help the designer understand the cause of the failure. Common corrective actions are the correction of an error in the original system description or addition of fairness constraints.

If model checking or language emptiness checks fail, VIS reports the failure with a counterexample, i.e., an error trace of sample "bad" behavior (behavior seen in the system that does not satisfy the property - for model checking, or invalid behavior seen in the system - for language emptiness). This is called the "debug" trace. Debug traces list a set of states that are on a path to a fair cycle and fail the CTL formula.

3.9 Combinational and Sequential Equivalence

In VIS it is also possible to check the equivalence of two networks. The command *comb_verify* verifies the combinational equivalence of two flattened networks. In particular, any set of functions (the roots), defined over any set of intermediate variables (the leaves), can be checked for equivalence between two networks. Roots and leaves are subsets of the nodes of a network, with the restriction that the leaves should form a complete support for the roots. The correspondence between the roots and the leaves in the two networks is specified in a file.

The default option assumes that the roots are the combinational outputs and the leaves are the combinational inputs. Two networks are declared combinationally equivalent if they have the same outputs for all combinations of inputs and pseudo-inputs. An important usage of *comb_verify* is to provide a sanity check when using SIS to re-synthesize portions of a network, as explained in Section 3.11.

The command *seq_verify* tests the sequential equivalence of two networks. In this case the set of leaves has to be the set of all primary inputs. This produces the constraint that both networks should have the same number of primary inputs. The set of roots can be an arbitrary subset of nodes. Moreover, no pseudo-inputs should be present in the two networks being compared. Sequential verification is done by building the product finite state machine. The command *seq_verify* verifies whether any state, where the values of two corresponding roots differ, can be reached from the set of initial states of the product machine. If this happens, a debug trace is provided.

3.10 Simulation

Simulation, although not "formal verification", is an alternate method for design verification. After the command *build_partition_mdds* is invoked, the network can also be simulated. In VIS we provide internal simulation of the BLIF-MV description generated by VL2MV, via the *simulate* command. Thus, VIS encompasses both formal verification and simulation capabilities. *simulate* can generate random input patterns or accept user-specified input patterns.

Any level of the specified hierarchy may be simulated.; the user may traverse the hierarchy to reach the relevant level via the *cd* command and simulate only that part. The *init_verify* command must be called to set up the appropriate internal data structures before simulation.

3.11 Synthesis in VIS

VIS can interact with SIS in order to optimize the existing logic. There are two possible goals/scenarios:

1. Synthesis for verification : Synthesis can be used to optimize the logic that represents the system, for simpler verification.
2. Front-end to synthesis : Files described in Verilog and compiled into *blif_mv* (using VL2MV or another tool) can be synthesized by using VIS and SIS together.

A key fact is that only the current level of the hierarchy is sent to SIS, and not the subtree rooted at the current node. [6] Modules at a lower level are treated as external and the boundary variables are carefully preserved, by reintegrating

---

[6] One would need a flattening routine different from the one which starts the verification flow already in VIS, and such a routine to flatten for synthesis is not yet available.

their multi-valued status after the optimization step in SIS (SIS requires that boundary variables are completely encoded, i.e., are binary variables).

4 Concluding Remarks

We have described VIS a verification and synthesis tool, which offers a better programming environment, new capabilities, and improved performance over existing verification tools. VIS has been implemented in the C programming language, and it has been ported to many different operating systems and architectures. We have tested on VIS on the sequential circuits from the ISCAS benchmark set and some industrial designs.

One of the key goals of VIS is to serve as a platform for developing new verification algorithms. We have used object-oriented programming style of SIS as our paradigm. VIS is composed of 18 packages; each exports a set of routines for manipulating a particular data structure, or for performing a set of related functions (e.g., there are packages for model checking, variable ordering, and manipulating the network data structure). New packages can be added easily. This wealth of exported functions can be used by future programmers to quickly assemble new algorithms. All functions adhere to a common naming convention so that it is easy to find functions in the documentation.

For more information about VIS or to get a copy, visit the VIS home page [5].

Acknowledgments

We would like to thank Adrian Isles, Sriram Rajamani, and Serdar Tasiran for their assistance in developing VIS.

The development of VIS was supported in part by the SRC under contract 96-DC-324 and by generous financial contributions from BNR, Cadence, California Micro, Fujitsu, and Motorola.

References

1. Kenneth L. McMillan. Symbolic Model Checking. Kluwer Academic Publishers, 1993.
2. R. K. Brayton et al. HSIS: A BDD based system for formal verification. Proc. of Design Automation Conference, 1994.
3. S.-T. Cheng. Compiling Verilog into automata. Tech. Rep. UCB/ERL M94/37, May 1994.
4. E.M. Sentovich et al. SIS: a system for sequential circuit synthesis. Tech. Rep. M92/41, May 1992.
5. VIS Home Page : http://www-cad.eecs.berkeley.edu/~vis

PVS: Combining Specification, Proof Checking, and Model Checking*

N. Shankar

Computer Science Laboratory, SRI International, Menlo Park CA 94025 USA
shankar@csl.sri.com
URL: http://www.csl.sri.com/pvs.html
Phone: +1 (415) 859-5272   Fax: +1 (415) 859-2844

PVS is an automated tool for specification and verification that is designed to exploit the synergies between language and deduction, automation and interaction, and theorem proving and model checking. PVS strives for a balance between automation and interaction by using decision procedures to simplify the tedious and obvious steps in a proof leaving the user to interactively supply the high-level steps in a verification. Decision procedures for BDD-based propositional simplification and model checking, equality, and linear arithmetic are integrated into PVS. The utility of the language and deductive features of PVS have been demonstrated in a number of examples, including the specification and partial verification of the Rockwell-Collins AAMP5 processor design [9], and the verification of an SRT divider [13]. This tutorial concentrates on the integration of individual verification techniques available in PVS.

1   The PVS Language

The PVS specification language is based on classical, simply typed higher-order logic augmented with subtypes and dependent types. The use of higher-order logic in PVS allows variables to not only range over datatypes such as numbers, lists, and trees, but also over predicates, functions, functions of functions, etc. The strong type system is needed to disallow self-application of functions and predicates which can lead to inconsistencies such as those related to Russell's paradox. Strong typing is also useful as a way of quickly detecting common specification errors by means of a typechecker. Subtyping in PVS can be used to define refined forms of types such as those for even numbers, prime numbers, injective functions, and order-preserving maps. Typechecking is undecidable for the PVS type system. The PVS typechecker automatically checks for simple type correctness and generates proof obligations corresponding to predicate subtypes. These proof obligations can be discharged through the use of the PVS proof

---

* The development of PVS was funded by SRI International through IR&D funds. Various applications and customizations have been funded by NSF Grant CCR-930044, NASA, ARPA contract A721, and NRL contract N00015-92-C-2177. The PVS system is the result of developed as a collaborative effort involving Sam Owre, John Rushby, Mandayam Srivas, David Cyrluk, Pat Lincoln, Sree Rajan, Judy Crow, among many others.

checker. PVS also has parametric theories so that it is possible to capture, say, the notion of sorting with respect to arbitrary sizes, types, and ordering relations. By exploiting subtyping, dependent typing, and parametric theories, researchers at NASA Langley Research Center and SRI have developed background libraries for basic concepts such as bit vectors, finite sets, and cardinalities. Paul Miner at NASA has developed a specification of portions of the IEEE 854 floating-point standard in PVS [10].

The PVS specification language has a number of features that exploit the interaction between theorem proving and typechecking. Conversely, type information is used heavily within a PVS proof so that predicate subtype constraints are automatically asserted to the decision procedures, and quantifier instantiations are typechecked and can generate TCC subgoals during a proof attempt. The practical experience with PVS has been that the type system does rapidly detect a lot of common specification errors. These language features have been used to obtain readable hardware descriptions in which certain simple semantic errors (e.g., out-of-bounds array references) are guaranteed to be absent.

2   The PVS Proof Checker

PVS has an interactive proof checker where the user constructs a proof by directing the system to apply specific inference steps. PVS has a powerful base of primitive inference rules so that the use of decision procedures and rewriting are built into the core of the system. PVS has a strategy language for defining strategies in a manner similar to the metalanguage tactics and tacticals of LCF [5] and related systems such as HOL [6] and Nuprl [2].

Decision Procedures. PVS employs decision procedures including the congruence closure algorithm for equality reasoning along with various decision procedures for various theories such as linear arithmetic, arrays, and tuples, in the presence of uninterpreted function symbols [14]. PVS does not merely make use of decision procedures to prove theorems but also to record type constraints and to simplify subterms in a formula using any assumptions that *govern* the occurrence of the subterm. These governing assumptions can either be the test parts of surrounding conditional (IF-THEN-ELSE) expressions or type constraints on governing bound variables. Such simplifications typically ensure that formulas do not become too large in the course of a proof. Also important, is the fact that automatic rewriting is tightly coupled with the use of decision procedures, since many of the conditions and type correctness conditions that must be discharged in applying a rewrite rule succumb rather easily to the decision procedures.

Symbolic Model Checking. This approach to verification has been described in McMillan [8] and Burch, Clarke, McMillan, Dill, and Hwang [1]. The integration of symbolic model checking as a decision procedure in PVS is described below in Section 4.

Strategies. The PVS proof checker provides powerful primitive inference steps that make heavy use of decision procedures, but proof construction solely in terms of even these inference steps can be quite tedious. PVS therefore provides a language for defining high-level inference strategies. This language includes recursion, a `let` binding construct, a backtracking `try` strategy construction, and a conditional `if` strategy construction.

Typical strategies include those for heuristic instantiation of quantifiers, repeated skolemization, simplification, rewriting, and quantifer instantiation, and induction followed by simplification and rewriting. There are about a hundred strategies currently in PVS but only about thirty of these are commonly used. The others are used as intermediate steps in defining more powerful strategies. The use of powerful primitive inference steps makes it possible to define a small number of robust and flexible strategies that usually suffice for productive proof construction. Such strategies help in the automated or nearly automated verification of a variety of hardware examples [3, 4, 9, 13].

3 Example Illustrating Rewriting Combined with Decision Procedures

Figure 1 shows a simple 3-stage pipelined processor [1]. Each instruction consists of an opcode `opcode`, two source register addresses `src1` and `src2`, and a destination register address `dstn`. The first stage loads the source register contents into the operand registers `opreg1` and `opreg2`. The second stage applies the ALU operation given by the `opcode` to the contents of the operand registers and stores the result in the write-back register `wbreg`. The final stage writes the contents of `wbreg` into the `dstn` address in the register file. The machine can nondeterministically introduce stall signals in order to implement delayed branching. The hardware must ensure that stalled instructions are not executed and that the outputs of ALU and `wbreg` are properly loaded into the operand registers in case an operand of a new instruction is the result of one of the instructions already in the pipeline.

The state of the machine is modelled as a collection of signals, one per wire, so that at time tick t for $t \geq 0$, the value of signal `wbreg`, for instance, is `wbreg(t)`. The state transitions of the system can be modelled as equations that describe the value of a signal at time t + 1 in terms of the values of various signals at time t. The main correctness property is stated as asserting that for an unstalled instruction at time t, the destination register of the register file at time t+3 registers the result of applying the ALU operation given by the opcode to the contents of the operand registers.

```
correctness: THEOREM (FORALL t:
  NOT(stall(t)) IMPLIES
      regfile(t+3)(dstn(t)) =
        aluop(opcode(t), regfile(t+2)(src1(t)),
              regfile(t+2)(src2(t))) )
```

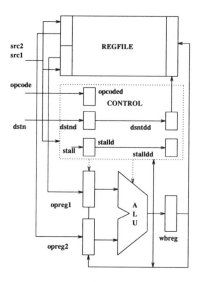

Fig. 1. A 3-Stage Pipelined Processor

This correctness property is proved in about five steps by first symbolically executing the machine to obtain an equality between two large conditional expressions representing the possible values of the left-hand side and the right-hand side of the correctness statement. By case-splitting on the conditions in the conditional expression, and applying BDD-based boolean simplification, we get seventeen subgoals which are easily discharged by the PVS decision procedures. This example can be verified for arbitrary datapath widths in PVS in under a minute of CPU time whereas a verification of this example for a fixed datapath width using model checking takes several hours [1]. The pipeline example described above is unrealistically small but it illustrates the need for integrating rewriting, arithmetic and equality simplification, and BDD-based propositional simplification in order to provide effective deductive support for hardware verification.

4 Integrating Model Checking and Theorem Proving

In the theorem proving approach to program verification, one verifies a property P of a program M by proving $M \supset P$. The model checking approach verifies the same program by showing that the state machine for M is a satisfying model of P, namely $M \models P$. For control-intensive approaches over small finite states, model checking is very effective since a more traditional Hoare logic style proof involves discovering a sufficiently strong invariant. These two approaches have traditionally been seen as incompatible ways of viewing the verification problem.

In recent work [12], we were able to unify the two views and incorporate a model checker as decision procedure for a well-defined fragment of PVS.

Our integration uses Park's relational mu-calculus [1, 11] as a medium for communicating between PVS and a model checker for the propositional mu-calculus. The idea here is that a Kripke model is captured by defining a state type, typically given by a record type construction, and a next-state relation on this state type which defines the state transitions or the edges in the Kripke model. Temporal properties such as those expressible in the branching-time temporal logic CTL can be characterized by fixpoints computed over such a Kripke model using the least and greatest fixpoint operators mu and nu over monotone predicate transformers. Such fixpoint operators can easily be defined using the higher-order logic of PVS. A CTL formula characterizes a predicate over states in this Kripke model. CTL formula are built from the propositional atoms using the propositional connectives and the modalities EX, EG, and EU. With respect to the Kripke model, the CTL formula EX(N, f) holds on those states that have a single edge leading to a state where the formula f holds. The CTL formula EG(N, f) holds on those states that have an infinite outgoing path of edges along which the formula f always holds. The CTL formula EU(N, f, g) holds on those states that have a path of zero or more edges leading to a state where g holds so that f holds along each intermediate state on the path. The other CTL modalities can be defined in terms of the above modalities. In particular, AG(N, f) holds of those states that do not have a path to any state where the formula f is falsified.

```
N: VAR [state, state->bool]

EX(N,f)(u):bool =  (EXISTS v: (f(v) AND N(u, v)))

EG(N,f):pred[state] = nu (LAMBDA Q: (f AND EX(N,Q)))

EU(N,f,g):pred[state] = mu (LAMBDA Q: (g OR (f AND EX(N,Q))))
```

Model checking CTL properties of a given transition system is then achieved by translating the CTL operators into the mu-calculus definitions, then binary coding the finite state type so that the result can be processed by a BDD-based boolean mu-calculus checker.

We can also prove theorems about the CTL operators using their fixpoint characterizations. In particular, we can prove theorems concerning the use of property-preserving abstractions to reduce the state space of a system to manageable levels. Such an abstraction has been used to verify a complicated communication protocol by means of a combination of theorem proving and model checking [7].

CTL cannot express fairness but we can once again define fair versions of the CTL operations in the mu-calculus. For example, the operator fairEG(N, f)(Ff) is the predicate that holds of a state u such that there is a *fair path*, i.e., one where the predicate Ff holds infinitely often, such that the predicate f holds on every state on the path. This is expressed in the mu-calculus by the definition shown below.

```
fairEG(N, f)(Ff): pred[state] =
  nu(LAMBDA P: EU(N, f, f AND Ff AND EX(N, P)))
```

Fair variants of the other connectives can be defined using `fairEG`. Multiple fairness conditions, i.e., using a list of predicates in place of the single fairness predicate `Ff`.

One direct use of fairCTL model checking is that it can be employed for model checking linear-time temporal logic LTL. In joint work with Schuermann and Rajan, we have incorporated LTL model checking into PVS. LTL formulas characterize sets of computation traces, unlike CTL formulas which characterize sets of states in the Kripke model. Various notions of fairness are easily captured in LTL. Many popular verification formalisms are based on LTL. Other forms of finite-state verification such as bisimulation and language inclusion can also be verified in terms of the mu-calculus. Transition systems can be described both synchronously in the style of SMV [8] or asynchronously as in the mutual exclusion example described below.

5 Example: Peterson's Mutual Exclusion Algorithm

Peterson's 2-process mutual exclusion algorithm ensures that two processes P and Q are never simultaneously in their critical section. In order to model this algorithm in PVS, we first identify the state space as a record datatype consisting of P's program counter pcp, a boolean `turn` variable, and Q's program counter pcq. Each process can either be in a `sleeping`, `trying` or `critical` state according to the value of the program counter field of the state. This is formalized in PVS in the theory `mutex` as shown.

```
PC : TYPE = {sleeping, trying, critical}

state : TYPE =
          [# pcp : PC,
             turn: bool,
             pcq : PC #]
```

The behavior of each process is modelled as a pair $\langle I, N \rangle$ where I is an initialization predicate on `state` characterizing the valid initial states of the process, and N is a next-state relation between states. For process P, the initialization predicate is given by I_P and its next-state relation is given by G_P. Initially, process P is in its `sleeping` state. In each transition, P either goes from `sleeping` to `trying` by setting `turn` to FALSE, or from `trying` to `critical` by checking whether `turn` is TRUE or Q is `sleeping`, or exits the `critical` state by setting `turn` to FALSE. Process Q has a symmetrical set of transitions. The composed system is described by conjoining the initialization predicates of P and Q, and by taking the disjunction of the two transition relations in order to obtain the combined transition relation representing the interleaving of P and Q transitions.

```
I_P(s) : bool = (sleeping?(pcp(s)))

G_P(s0, s1): bool =
  (  (s1 = s0)                              %stutter
    OR (sleeping?(pcp(s0)) AND             %try
          s1 = s0 WITH [pcp := trying,
                         turn := FALSE])
    OR (trying?(pcp(s0)) AND               %enter critical
        (turn(s0) OR sleeping?(pcq(s0))) AND
        s1 = s0 WITH [pcp := critical])
    OR (critical?(pcp(s0)) AND             %exit critical
        s1 = s0 WITH [pcp := sleeping,
                       turn := FALSE ]))

I_Q(s) : bool = (sleeping?(pcq(s)))

G_Q(s0, s1): bool = ...

I(s) : bool = (I_P(s) AND I_Q(s))

G(s0, s1) : bool = (G_P(s0, s1)
                    OR G_Q(s0, s1))
```

The main safety property of a two-process Peterson algorithm is very easily verified by means of model checking. The correctness of an N-process version of this algorithm can be reduced to that of the two-process case by means of induction and abstraction. This combination of induction, abstraction, and model checking can be applied to a number of other parametric N-process algorithms.

```
safe(s) : bool = NOT (critical?(pcp(s))
                      AND critical?(pcq(s)))

safety: LEMMA
  I(s) IMPLIES
      AG(G, safe)(s)
```

6   Conclusions

We claim that no single technique such as rewriting, BDDs, or model checking is effective for all aspects of hardware verification. Many examples need the careful integration of these techniques. We have shown some simple examples to illustrate the integration available in PVS. This combination of techniques has been applied to some larger examples such as an SRT divider and Rockwell-Collins AAMP series of processors. The automation available in PVS on these examples can be further improved through the use of more decision procedures (e.g., bit vectors) and better verification methodologies (e.g., abstraction, induction).

References

1. J. R. Burch, E. M. Clarke, K. L. McMillan, D. L. Dill, and L. J. Hwang. Symbolic model checking: 10^{20} states and beyond. *Information and Computation*, 98(2):142–170, June 1992.
2. R. L. Constable, S. F. Allen, H. M. Bromley, W. R. Cleaveland, J. F. Cremer, R. W. Harper, D. J. Howe, T. B. Knoblock, N. P. Mendler, P. Panangaden, J. T. Sasaki, and S. F. Smith. *Implementing Mathematics with the Nuprl Proof Development System*. Prentice-Hall, Englewood Cliffs, NJ, 1986.
3. D. Cyrluk. Inverting the abstraction mapping: A methodology for hardware verification. In *Proceedings of Formal Methods in Computer Aided Design (FMCAD '96)*, 1996. This volume.
4. D. Cyrluk, S. Rajan, N. Shankar, and M. K. Srivas. Effective theorem proving for hardware verification. In Ramayya Kumar and Thomas Kropf, editors, *Theorem Provers in Circuit Design (TPCD '94)*, volume 910 of *Lecture Notes in Computer Science*, pages 203–222, Bad Herrenalb, Germany, September 1994. Springer-Verlag.
5. M. Gordon, R. Milner, and C. Wadsworth. *Edinburgh LCF: A Mechanized Logic of Computation*, volume 78 of *Lecture Notes in Computer Science*. Springer-Verlag, 1979.
6. M. J. C. Gordon and T. F. Melham, editors. *Introduction to HOL: A Theorem Proving Environment for Higher-Order Logic*. Cambridge University Press, Cambridge, UK, 1993.
7. Klaus Havelund and N. Shankar. Experiments in theorem proving and model checking for protocol verification. In *Formal Methods Europe FME '96*, number 1051 in Lecture Notes in Computer Science, pages 662–681, Oxford, UK, March 1996. Springer-Verlag.
8. Kenneth L. McMillan. *Symbolic Model Checking*. Kluwer Academic Publishers, Boston, MA, 1993.
9. Steven P. Miller and Mandayam Srivas. Formal verification of the AAMP5 microprocessor: A case study in the industrial use of formal methods. In *WIFT '95: Workshop on Industrial-Strength Formal Specification Techniques*, pages 2–16, Boca Raton, FL, 1995. IEEE Computer Society.
10. Paul S. Miner. Defining the IEEE-854 floating-point standard in PVS. Technical Memorandum 110167, NASA Langley Research Center, 1995.
11. David Park. Finiteness is mu-ineffable. *Theoretical Computer Science*, 3:173–181, 19.
12. S. Rajan, N. Shankar, and M.K. Srivas. An integration of model-checking with automated proof checking. In Pierre Wolper, editor, *Computer-Aided Verification, CAV '95*, volume 939 of *Lecture Notes in Computer Science*, pages 84–97, Liege, Belgium, June 1995. Springer-Verlag.
13. H. Rueß, N. Shankar, and M. K. Srivas. Modular verification of SRT division. In Rajeev Alur and Thomas A. Henzinger, editors, *Computer-Aided Verification, CAV '96*, number 1102 in Lecture Notes in Computer Science, pages 123–134, New Brunswick, NJ, July/August 1996. Springer-Verlag.
14. Robert E. Shostak. Deciding combinations of theories. *Journal of the ACM*, 31(1):1–12, January 1984.

HOL Light: A Tutorial Introduction

John Harrison

Åbo Akademi University, Department of Computer Science
Lemminkäisenkatu 14a, 20520 Turku, Finland

Abstract. HOL Light is a new version of the HOL theorem prover. While retaining the reliability and programmability of earlier versions, it is more elegant, lightweight, powerful and automatic; it will be the basis for the Cambridge component of the HOL-2000 initiative to develop the next generation of HOL theorem provers. HOL Light is written in CAML Light, and so will run well even on small machines, e.g. PCs and Macintoshes with a few megabytes of RAM. This is in stark contrast to the resource-hungry systems which are the norm in this field, other versions of HOL included. Among the new features of this version are a powerful simplifier, effective first order automation, simple higher-order matching and very general support for inductive and recursive definitions.

Many theorem provers, model checkers and other hardware verification tools are tied to a particular set of facilities and a particular style of interaction. However HOL Light offers a wide range of proof tools and proof styles to suit the needs of various applications. For work in high-level mathematical theories, one can use a more 'declarative' textbook-like style of proof (inspired by Trybulec's Mizar system). For larger but more routine and mechanical applications, HOL Light includes more 'procedural' automated tools for simplifying complicated expressions, proving large tautologies etc. We believe this unusual range makes HOL Light a particularly promising vehicle for floating point verification, which involves a mix of abstract mathematics and concrete low-level reasoning. We will aim to give a brief tutorial introduction to the system, illustrating some of its features by way of such floating point verification examples. In this paper we explain the system and its possible applications in a little more detail.

1 The evolution of HOL Light

In verification tasks, low-level proof checkers are often too tedious to use, while fully automatic provers are seldom effective. The ideal seems to be a judicious combination of interaction and automation. Edinburgh LCF [4] was an influential methodology for achieving this sort of balance. Low-level primitive inference rules are provided which produce theorems, and by the use of an abstract type, it is ensured that theorems can *only* be created by these rules. However the user is free to write arbitrary proof procedures in the metalanguage ML which decompose to these inferences; such procedures can in principle implement practically any automated proof method (e.g. resolution, Presburger arithmetic), and they will be correct per construction, in the sense that they cannot produce a false 'theorem'. Experience suggests that this can usually be done reasonably efficiently. Hence LCF systems offer a unique combination of reliability and programmability.

Paulson [8] and Huet improved Edinburgh LCF, and based on their implementation, Gordon produced a new system based on classical higher order logic rather than Scott's Logic of Computable Functions. This system, HOL, was specifically designed for hardware verification, the use of higher order logic for this purpose having been advocated by Hanna [5] and by Gordon himself [2]. HOL became increasingly popular in the late 80s and early 90s, attracting many users worldwide in academia, industry and military and government agencies. However the HOL88 version was a conceptually complicated and inefficient system, with many of the primitive term operations coded in LISP rather than ML. A new version, hol90, corrected these deficiencies, but did not greatly improve the higher level organization or the theorem-proving technology; HOL proofs are often long and difficult compared with those in other contemporary systems like Isabelle [9] and PVS [7]. Moreover hol90 is at present tied to the New Jersey SML compiler, which is notoriously memory-hungry; this means that it is still slow or even unusable on typical machines.

HOL Light represents an attempt to improve the HOL system more radically, while retaining its traditional strengths. The system is small and light, using the excellent CAML Light interpreter [12] developed at INRIA Roquencourt. The logical development has been cleaned up, and new and powerful proof techniques added. It even supports a new proof *style*, based on Mizar, which allows many proofs to be written more elegantly. Nevertheless, we want to stress that although HOL Light is a recent development, it can be seen as the culmination of a successful line of research spanning about 20 years.

2 Highlights of HOL Light

HOL Light has the following features:

1. It is open. The entire system is coded in ML and all the source code is freely available. Because ML is a fairly readable and high-level language, the implementation often looks quite close to an abstract algorithm description. Hence users can see what is happening inside and gain real confidence and understanding; the system is not just a mysterious black box.

2. It is sound and coherent. The LCF methodology ensures that all extensions with custom proof procedures are correct by construction. Apart from the benefits of soundness, this also helps to make the structure of the system logically clean and comprehensible, rather than its being a complicated entwining of different proof procedures.

3. It is extensible. If users want to implement special purpose tools and decision procedures, they can implement them, and the LCF methodology ensures that they cannot produce false results. The system source code contains numerous examples of special proof procedures, ranging from the simple to the complex.

4. It is an easy matter to write special interface code and connect HOL Light to other systems. For example, we have used the Maple computer algebra system to perform factorization and integration steps, and an implementation of Stålmarck's algorithm [10] to prove tautologies; HOL Light can check the validity of these results internally, maintaining the usual guarantee of correctness even in a mixed system.

5. It is small and lightweight. The system does not require a state-of-the-art workstation to run. CAML Light is very economical and can work quite well on small machines such as PCs and Macintoshes with just a few megabytes of free memory. Nevertheless, we aim to show that the system is not merely a toy, but is capable of being used for real verification tasks.

6. Different proof styles are supported. Though the 'machine code' of the system is a simple set of forward inferences, one can interact with the system in various different high-level ways to prove theorems. For example, one can prove theorems in a backward fashion using tactics, or write a more orthodox mathematical proof in the style of Mizar [11]. A facility for window inference, useful for program refinement and other transformational design methodologies, is under development. All these styles can be intermixed freely in the same script, and even in the same proof.

7. Special-purpose proof procedures are available. In many situations, the user will find that the system already contains tools to handle tricky steps. For example, there is code to automate the definition of inductive relations and types as well as recursive functions with arbitrary well-founded measures. Support for pure logic includes several complementary tautology-proving modules and proof search procedures for first order logic, while there are several domain-specific tools such as decision procedures for linear arithmetic (over naturals, integers and reals).

3 HOL Light in hardware verification

The advantages of theorem-proving techniques over model-checking for hardware (and software) verification are:

1. Specifications can be written in an elegant way using the normal resources of mathematics.

2. Any supporting mathematical theories required can be called upon, after themselves being formally proved, in the verification effort.

We will demonstrate how HOL Light can be used for a particular example: we plan to show the proof of correctness of a CORDIC algorithm for evaluating floating point natural logarithms. Though only a single example, we hope it illustrates many interesting features of the HOL Light system, and the use of theorem provers for verification in general. Floating point correctness is a topical issue, with the Pentium fiasco still a recent memory; moreover, it seems that an incorrect treatment of overflow in a floating-point to integer conversion routine was responsible for the destruction of the European Space Agency's Ariane 5 rocket on its maiden flight.

The first step is to establish what it means for the floating point algorithm to be 'correct'. The natural idea is to specify that the result, when interpreted as a real number, corresponds closely to the true mathematical value. We do indeed choose such a specification, but point out how there are several subtly different ways of making 'corresponds closely' more precise. For the elementary algebraic operations, including square roots, the IEEE Standard [6] mandates that the result should be *the closest representable* value to the true mathematical value (with choices between equally close values resolved using a convention of 'round to even'). For transcendental functions, this is very difficult to achieve in practice, because of a phenomenon known as the 'table maker's dilemma'. We explain two alternative forms of correctness specification that we consider more suitable for the transcendental functions. It is against the second of these that our verification is performed.

The next task is to model the floating point algorithm itself inside HOL Light's logic. It is possible to use a gate-level circuit description, but a more transparent description written in a formalized 'pseudocode' is used here. This gives a good opportunity to discuss techniques which have been used for semantically embedding idealized programming languages [3] and real hardware description languages [1] in HOL. These techniques use a special kind of denotational interpretation of the language as ordinary logical and mathematical constructs, and allow a rather direct means of formal reasoning about programs. Our system description is, therefore, both readable and rigorous.

The verification rests on a number of properties of logarithms. To give a simple example, we use the fact that $ln(1 + x) \leq x$ whenever $x \geq 0$. The ultimate foundation for these results is a rigorously developed theory of real analysis, based on a definitional construction of the real numbers from the natural numbers. We cannot give many details here, but we do hope to give some illustrations of how HOL Light can be used in this more abstract field. In particular, its Mizar-style proofs allow one to write proofs in a rather elegant style (compared with the arcane tactic scripts of most systems, other HOL proof styles included). At the same time, the automated facilities of HOL Light can be called on when required, e.g. to avoid directing routine algebraic steps in detail.

The overall proof is quite intricate, and though it can be rerun in a few minutes, we will concentrate on the general feel of the theorem proving facilities provided by HOL Light. We show how the simplifier can be used extensively to avoid complex and tedious chains of reasoning. Indeed, in a related verification effort of a square root algorithm, the central mathematical invariant property could be proved completely automatically. However we stress that low-level facilities are always there to provide a finer degree of control.

Finally, we show that certain precomputed constants, required for a real implementation, can be calculated in HOL Light using the normal inference rules of the logic. Though slow, this is acceptable in such special circumstances, and gives a cast-iron guarantee on the error bounds.

4 Conclusions

We hope to illustrate that HOL Light can be applied to realistic verification tasks. Moreover, the method of embedding other formalisms, together with the inherent programmability, make HOL Light an excellent core for application-specific reasoning systems.

References

1. R. Boulton et al. Experience with embedding hardware description languages in HOL. In V. Stavridou, T. F. Melham, and R. T. Boute, editors, *Proceedings of the IFIP TC10/WG 10.2 International Conference on Theorem Provers in Circuit Design: Theory, Practice and Experience*, volume A-10 of *IFIP Transactions A: Computer Science and Technology*, pages 129–156, Nijmegen, The Netherlands, 1993. North-Holland.
2. M. J. C. Gordon. Why higher-order logic is a good formalism for specifying and verifying hardware. Technical Report 77, University of Cambridge Computer Laboratory, New Museums Site, Pembroke Street, Cambridge, CB2 3QG, UK, 1985.
3. M. J. C. Gordon. Mechanizing programming logics in higher order logic. In G. Birtwistle and P. A. Subrahmanyam, editors, *Current Trends in Hardware Verification and Automated Theorem Proving*, pages 387–439. Springer-Verlag, 1989.
4. M. J. C. Gordon, R. Milner, and C. P. Wadsworth. *Edinburgh LCF: A Mechanised Logic of Computation*, volume 78 of *Lecture Notes in Computer Science*. Springer-Verlag, 1979.
5. F. K. Hanna and N. Daeche. Specification and verification using higher-order logic: A case study. In G. Milne and P. A. Subrahmanyam, editors, *Formal Aspects of VLSI Design: Proceedings of the 1985 Edinburgh Workshop on VLSI*, pages 179–213, 1986.
6. IEEE. Standard for binary floating point arithmetic. ANSI/IEEE Standard 754-1985, The Institute of Electrical and Electronic Engineers, Inc., 345 East 47th Street, New York, NY 10017, USA, 1985.
7. S. Owre, J. M. Rushby, and N. Shankar. PVS: A prototype verification system. In D. Kapur, editor, *11th International Conference on Automated Deduction*, volume 607 of *Lecture Notes in Computer Science*, pages 748–752, Saratoga, NY, 1992. Springer-Verlag.
8. L. C. Paulson. *Logic and computation: interactive proof with Cambridge LCF*. Number 2 in Cambridge Tracts in Theoretical Computer Science. Cambridge University Press, 1987.
9. L. C. Paulson. *Isabelle: a generic theorem prover*, volume 828 of *Lecture Notes in Computer Science*. Springer-Verlag, 1994. With contributions by Tobias Nipkow.
10. G. Stålmarck. System for determining propositional logic theorems by applying values and rules to triplets that are generated from Boolean formula. United States Patent number 5,276,897; see also Swedish Patent 467 076, 1994.
11. A. Trybulec. The Mizar-QC/6000 logic information language. *ALLC Bulletin (Association for Literary and Linguistic Computing)*, 6:136–140, 1978.
12. P. Weis and X. Leroy. *Le langage Caml*. InterEditions, 1993. See also the CAML Web page: http://pauillac.inria.fr/caml/.

A Tutorial on Digital Design Derivation Using DRS [*]

Bhaskar Bose[†], M. Esen Tuna, and Venkatesh Choppella

Derivation Systems, Inc.
5963 La Place Court, Suite 208
Carlsbad, CA 92008, USA
[†]bose@derivation.com

Abstract. This paper presents a tutorial on digital design derivation using DRS. The DRS system is an integrated formal system for the design of verified hardware. The underlying approach employs a derivation methodology in which a series of correctness preserving transformations are applied to high-level specifications in order to synthesize hardware descriptions. In this paper, we sketch the key steps in the derivation of an example circuit. The example illustrates several aspects of DRS and serves as an introduction to the derivational paradigm of synthesis.

1 Introduction

DRS (Derivational Reasoning System) is an integrated formal system for the development and analysis of verified digital hardware systems. The system provides design engineers with the ability to synthesize verified hardware from high-level specifications. DRS integrates an executable specification language, a polymorphic type inference system and a powerful derivation engine with existing theorem provers and logic synthesis tools to provide a formal framework for design. The derivation methodology has been applied to several non-trivial designs including a 32-bit general purpose microprocessor [1], and a fault-tolerant clock synchronization circuit [6].

2 Design Derivation

The primary mode of reasoning in DRS is *derivation*. Derivation is a form of mathematical proof that deals with correct-by-construction reasoning [3, 4, 5]. A series of correctness preserving transformations are used to derive an implementation from a specification. A key advantage of this approach is that it obviates the need for post-factum verification. This reduces the goal of design verification to proving the correctness of the transformations. The correctness of the transformations is established independently when the transformation is

---

[*] Research reported herein was supported, in part, by The National Aeronautics and Space Administration.

developed, rather than by the designer. The designer simply applies the transformation to a design to achieve some desired goal. In addition, derivation provides the designer the ability to explore the design space while maintaining the rigorous integrity of the specification. In this respect, while deductive (conventional theorem prover based) verification attempts to formalize a particular design, derivation attempts to formalize the design process.

3 Derivation of a Fibonacci Sequence Generator

DRS is an interactive formal system. This interaction provides the designer with direct control of the design process. In this respect, the system resembles a proof-checker in the sense that it automates the algebra needed for circuit synthesis, but requires interactive guidance to perform a derivation. The user interacts with the system through a series of transformation commands that is used to refine the specification from an abstract description to a concrete implementation. In DRS, a sequence of transformations is applied to an initial specification defining a *derivation path* towards an implementation satisfying an intended set of design constraints. Design tactics and constraints imposed by the designer sketch a complex design space with many possible paths between specification and implementation. In practice, however, the derivation path has distinct phases.

Consider the example of a Fibonacci sequence generator. The behavioral specification is obtained from the iterative definition of the Fibonacci function

$$fib(n) = g(n, 1, 1) \text{ where}$$
$$g(x, y, z) = \text{if } lt?(x, 2) \text{ then } y \text{ else } g(dec(x), z, add(y, z)).$$

The specification is an abstract algorithmic description that defines the functionality of the circuit by specifying a sequence of operations and control decisions. The specification describes what operations must occur, but not how they are implemented in hardware.

The first phase in the derivation is to apply a series of transformations at the behavioral level. A class of transformations, called *behavioral transformations*, manipulate the behavioral specification. These transformations usually involve manipulating control and architecture in a tightly integrated relation. Some examples include transformations to achieve a desired scheduling of operations and transformations to move operations between control and architecture. The above specification indicates that the operations *add* and *dec* are executing in parallel. For the sake of the example, let us assume that our final hardware implementation is constrained to use a single arithmetic logic unit. In this phase the *dec* and *add* operations are serialized so that they may be combined into a single logic unit during structural refinement later in the derivation. This serialization (over time) is achieved by introducing an intermediate function h, and splitting the calls to *dec* and *add* across the calls to g and h resulting in

$$g(x, y, z) = \text{if } lt?(x, 2) \text{ then } y \text{ else } h(dec(x), y, z)$$
$$h(x, y, z) = g(x, z, add(y, z)).$$

From a suitable behavior description, DRS automatically builds an abstract *structural* description, denoted by a set of recurrence equations, representing an initial estimation of architecture:

$$w = \mathbf{reg}(g, select(status, g, h, g)) \qquad status = [w, lt?(x, 2)]$$
$$x = \mathbf{reg}(n, select(status, x, dec(x), x)) \qquad rdy = and(equal?(w, g), lt?(x, 2))$$
$$y = \mathbf{reg}(1, select(status, y, y, z)) \qquad ans = y$$
$$z = \mathbf{reg}(1, select(status, z, z, add(y, z)))$$

where

$$select([s, p0], v0, v1, v2) = \mathbf{case}\ s$$
$$g : \mathbf{if}\ p0\ \mathbf{then}\ v0\ \mathbf{else}\ v1$$
$$h : v2$$

The block diagram denoting the initial structural description is shown in Figure 1. (⇥ denotes the select function. For clarity only the inputs to the selector are shown, and the status signal is omitted. ⊔ represents a register unit.)

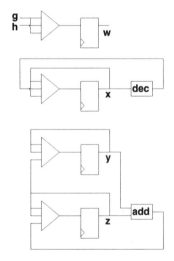

Fig. 1. Initial Structural Description

The structural description defines the components in a circuit and their connectivity. The specification expresses logical behavior and physical organization, but does not address electrical characteristics of the circuit. Timing is coordinated by storage elements whose behavior in turn is governed by an external synchronizing clock. Each variable now denotes an infinite sequence of values over time. $\mathbf{reg}(v, S)$ denotes the sequence of values $< v, S^0, S^1, \ldots >$ where the "reg" function introduces a delay and is interpreted as a register. The construction guarantees that given a particular sequence of input events, the structural description produces the same output-event sequence as does the original behavior specification.

The second phase in the derivation is to refine the structural description to an architecture. A class of transformations, called *structural transformations*, decomposes the description into a system of modules encapsulating signals as co-processes, isolating components of the specification for verification, mapping to existing hardware components, or further algebraic refinement. In the example, the architecture is refined by subsuming the *dec* and *add* operations by a single component. The transformation is valid since these two operations do not occur simultaneously, as imposed by the earlier serialization of these two operations. The results of factoring *dec* and *add* are the synthesis of an abstract component,

$$alu(inst, op\_a, op\_b) = \textbf{case } inst$$
$$\text{nop} : ?$$
$$\text{dec} : dec(op\_a)$$
$$\text{add} : add(op\_a, op\_b)$$

and the derivation of four equations for the signals *alu_out*, *inst*, *op_a*, and *op_b* to communicate with the factored component. The ? symbol denotes a *don't care* value. The original occurrences of *add* and *dec* are replaced with the output of the factored component. The resulting system of equations is

$$w = \textbf{reg}(g, select(status, g, h, g)) \qquad status = tuple(w, lt?(x, 2))$$
$$x = \textbf{reg}(n, select(status, x, alu\_out, x)) \qquad op\_a = select(status, ?, x, y)$$
$$y = \textbf{reg}(1, select(status, y, y, z)) \qquad op\_b = select(status, ?, ?, z)$$
$$z = \textbf{reg}(1, select(status, z, z, alu\_out)) \qquad inst = select(status, \text{nop}, \text{dec}, \text{add})$$
$$rdy = and(equal?(w, g), lt?(x, 2)) \qquad alu\_out = alu(inst, op\_a, op\_b)$$
$$ans = y.$$

The third phase of the derivation is to introduce a lower-level representation. A class of transformations, called *projection transformations*, introduces a lower-level representation. In the example, the architecture is still abstract in the sense that signals represent integer values. To obtain a concrete binary description, these signals are instantiated with bit-vectors of appropriate width. Type declarations are used to project each variable, constant, and operator to a binary representation. For instance, the projection of

$$y = \textbf{reg}(1, select(status, y, y, z))$$

to a binary representation of three bits is rewritten as

$$y_0 = \textbf{reg}(\text{T}, select(status, y_0, y_0, z_0))$$
$$y_1 = \textbf{reg}(\text{F}, select(status, y_1, y_1, z_1))$$
$$y_2 = \textbf{reg}(\text{F}, select(status, y_2, y_2, z_2)).$$

The constant 1 and signal z are also projected to their respective equivalent bit representations. At this stage the entire design is at the binary level. Subsequent transformations impose a logical and physical ordering on the design to map to a particular target technology. Algebraic transformations provide a powerful approach since the massive restructuring and decomposition necessary in reorganizing the design represent purely syntactical manipulations. Ultimately, this formal development produces a hierarchy of boolean subsystems, which are then partitioned into synthesizable subsystems. These boolean subsystems are then passed to logic synthesis tools to generate hardware realizations.

4   Conclusion

The DRS system is based on the philosophy that design is a reasoning process that involves analysis, deduction and generation. Although, derivation is the primary mode of proof in DRS, the system integrates with existing verification systems at several levels. In DRS, verification is necessary to establish the correctness of the specification and representations. In addition, design optimizations that are more easily handled using either mechanical theorem proving techniques or model checking are employed throughout the derivation. For example, suppose we wanted to substitute the derived *alu* specification with an efficient technology dependent implementation. Fully automatic OBDD [2] verification techniques would be sufficient to establish the equivalence between the derived *alu* and its optimized implementation.

The idealized design environment consists of multiple formal systems with secure interaction between them. Proofs in one system are interpreted as valid in another, eliminating the need to re-validate proofs across system boundaries. With this approach the designer is able to employ the most effective tool for a particular design context without sacrificing confidence in the correctness of the design.

References

1. Bhaskar Bose. *DDD-FM9001: Derivation of a Verified Microprocessor*. PhD thesis, Indiana University, December 1994.
2. Randal E. Bryant. Graph-Based Algorithms for Boolean Function Manipulation. In *IEEE Transactions on Computers*, volume C-35, pages 677–691, August 1986.
3. R. Burstall and J. Darlington. A transformation system for developing recursive programs. *Journal of the ACM*, 24:44–67, 1977.
4. Steven D. Johnson. *Synthesis of Digital Designs from Recursion Equations*. The MIT Press, Cambridge, 1984.
5. Steven D. Johnson. Manipulating logical organization with system factorizations. In M. Leeser and G. Brown, editors, *Hardware Specification, Verification and Sythesis: Mathematical Aspects, Lecture Notes in Computer Science*, volume 408, pages 260–281. Springer, Berlin, 1989.
6. Paul S. Miner, Shyamsundar Pullela, and Steven D. Johnson. Interaction of formal design systems in the development of a fault-tolerant clock synchronization circuit. In *13th Symp. on Reliable Distributed Systems*, October 1994.

ACL2 Theorems About Commercial Microprocessors

Bishop Brock, Matt Kaufmann* and J Strother Moore

Computational Logic, Inc., 1717 West Sixth Street, Austin, TX 78703-4776, USA**

Abstract. ACL2 is a mechanized mathematical logic intended for use in specifying and proving properties of computing machines. In two independent projects, industrial engineers have collaborated with researchers at Computational Logic, Inc. (CLI), to use ACL2 to model and prove properties of state-of-the-art commercial microprocessors prior to fabrication. In the first project, Motorola, Inc., and CLI collaborated to specify Motorola's complex arithmetic processor (CAP), a single-chip, digital signal processor (DSP) optimized for communications signal processing. Using the specification, we proved the correctness of several CAP microcode programs. The second industrial collaboration involving ACL2 was between Advanced Micro Devices, Inc. (AMD) and CLI. In this work we proved the correctness of the kernel of the floating-point division operation on AMD's first Pentium-class microprocessor, the AMD5$_K$86. In this paper, we discuss ACL2 and these industrial applications, with particular attention to the microcode verification work.

1 ACL2

ACL2 stands for "A Computational Logic for Applicative Common Lisp." ACL2 is both a mathematical logic and system of mechanical tools which can be used to construct proofs in the logic. The logic, which formalizes a subset of Common Lisp, is a high level programming language which can be executed efficiently on many host platforms. Thus, programmers can define models of computational systems and these models can be executed ("simulated") to test them on concrete data. But because the language is also a formal mathematical logic it is possible to reason about the models symbolically. Indeed, it is possible to prove theorems establishing properties of the models and to check these proofs with mechanical

---

* Matt Kaufmann's address is now Motorola @ Lakewood, P.O. Box 6000, MD F52, Austin, TX 78762

** The theorem prover used in this work was supported in part at Computational Logic, Inc., by the Defense Advanced Research Projects Agency, ARPA Order 7406, and the Office of Naval Research, Contract N00014-94-C-0193. The views and conclusions contained in this document are those of the author(s) and should not be interpreted as representing the official policies, either expressed or implied, of Advanced Micro Devices, Inc., Motorola, Inc., Computational Logic, Inc., the Defense Advanced Research Projects Agency, the Office of Naval Research, or the U.S. Government.

tools that are part of the ACL2 system. The ACL2 system is essentially a re-implemented extension, for applicative Common Lisp, of the so-called "Boyer-Moore theorem prover" Nqthm [2, 3].

1.1 The Logic

The ACL2 logic is a first-order, essentially quantifier-free logic of total recursive functions providing mathematical induction and two extension principles: one for recursive definition and one for "encapsulation."

The syntax of ACL2 is a subset of that of Common Lisp. Formally, an ACL2 term is either a variable symbol, a quoted constant, or the application of an n-ary function symbol or lambda expression, f, to n terms, written $(f \ t_1 \dots t_n)$. This formal syntax is extended by Common Lisp's facility for defining constant symbols and macros.

The rules of inference are those of Nqthm, namely propositional calculus with equality together with instantiation and mathematical induction on the ordinals up to $\epsilon_0 = \omega^{\omega^{\omega^{\cdot^{\cdot^{\cdot}}}}}$.

The axioms of ACL2 describe five primitive data types: the complex rationals, characters, strings, symbols, and ordered pairs or lists. The complex rationals are complex numbers with rational components and hence include the rationals, the integers and the naturals. Symbols are logical constants denoting words, such as DIV and STEP. Symbols are in "packages" which provide a convenient way to have disjoint name spaces. SMITH::DIV is a different symbol than JONES::DIV; but if the user has selected "SMITH" as the "current package" then the former symbol can be written more succinctly as DIV.

Essentially all of the Common Lisp functions on the above data types are axiomatized or defined as functions or macros in ACL2. By "Common Lisp functions" here we mean the programs specified in [35] or [36] that are (i) applicative, (ii) not dependent on state, implicit parameters, or data types other than those in ACL2, and (iii) completely specified, unambiguously, in a host-independent manner. Approximately 170 such functions are axiomatized or defined. In addition, we add definitions of new function symbols to provide fast multiply-valued functions, an explicit notion of state with appropriate applicative input/output primitives, fast applicative arrays, and fast applicative property lists.

Common Lisp functions are partial; they are not defined for all possible inputs. In ACL2 we complete the domains of the Common Lisp functions, making every function total by adding "natural" values on arguments outside the "intended domain" of the Common Lisp function. For example, if ACL2 is used to evaluate (car 7) no exception is raised and the result is nil; if a Common Lisp were used to evaluate that expression the behavior is unpredictable, depends on which Common Lisp implementation is used, and could include memory violations and system crashes.

ACL2 and Common Lisp agree on the result of a computation provided the functions involved are exercised only on their intended domain. We formalize the notion of the intended domain of each function in our notion of *guard*. A guard is a predicate that recognizes arguments in the intended domain. The guards of the Common Lisp primitives are provided by the system; the user may provide guards for defined functions. The system then provides a mechanism, called *guard verification*, that insures that a function is Common Lisp compliant. Guard verification simply generates and attempts to prove a set of conjectures sufficient to imply that the evaluation of the function on its intended domain exercises its subroutines on their intended domains.

Some comments about guards and guard verification are in order. First, guards and guard verification are optional: ACL2 is a syntactically untyped logic of total functions. Theorems may be proved without considering guards; any ground term can be evaluated to a constant (provided no undefined functions are involved) using the axioms. But such theorems and computations may not be consistent with Common Lisp. If you want assurance that an ACL2 function will run in Common Lisp as predicted by an ACL2 theorem, you must verify the guards on the function and check that the arguments satisfy the guard. Second, guard verification is akin to type checking in that users may provide guards that express arbitrarily strong restrictions. Guard verification then insures that functions are "well-typed." Since guards are arbitrary predicates, guard checking is technically undecidable; but for commonly used primitive guards, guard checking is fast and silent. Finally, ACL2's implementation takes advantage of the fact that ACL2's axioms and Common Lisp agree when guards have been verified. If called upon to evaluate a function on constants, ACL2 can "short-circuit" its interpreter and execute compiled Common Lisp provided the function's guards have been verified and the arguments are in the intended domain. The user wishing to improve the execution speed of an ACL2 model is therefore encouraged to do guard verification. We discuss guards in more detail in [24].

Finally, ACL2 has two extension principles: definition and encapsulation. Both preserve the consistency of the extended logic. See [22]. The definitional principle insures consistency by requiring a proof that each defined function terminates. This is done, as in Nqthm, by the identification of some ordinal measure of the formals that decreases in recursion.

The *encapsulation* principle allows the introduction of new function symbols axiomatized to have certain properties. It preserves consistency by requiring the exhibition of witness functions that can be proved to have the alleged properties. To encapsulate a sequence of logical events — definitions and theorems — one embeds the sequence in an `encapsulate` form and marks certain of the events as `local`. Such an encapsulation is *admissible* to the logic if all the events are admissible; hence the definitions all satisfy the definitional principle and the theorems can be proved in the resulting extension of the logic. But the axiomatic *effect* of an encapsulation is to add to the logic the axioms corresponding to the *non-local* events. Thus, for example, to introduce an undefined function f constrained to be rational, it suffices to define f to be the rational 23, say, prove the theorem that f returns a `rationalp`, and then encapsulate those two events marking the definition as `local`. Effectively the definition serves as a *witness* for the satisfiability of the axiom added.

1.2 The System

Like Nqthm, ACL2's theorem prover orchestrates a variety of proof techniques. As suggested by Figure 1, the user puts the formula to be proved into a pool. The simplifier, the most important proof technique, draws a formula from the pool and either "simplifies" it — replacing it in the pool by the several new cases sufficient to prove it — or passes it to the next proof technique. The simplifier employs many different proof techniques, including conditional (back chaining) rewrite rules, congruence-based rewriting, efficient ground term evaluation, forward chaining, type-inference, the OBDD propositional decision procedure, a rational linear arithmetic decision procedure, and user-defined, machine-verified meta-theoretic simplifiers.

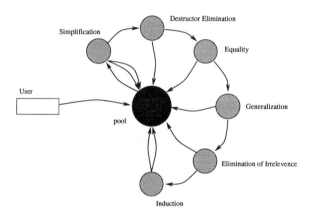

Fig. 1. The Orchestration of Proof Techniques

Roughly speaking, as the formula moves clockwise around the ring in Figure 1 it becomes more general. Eventually, if all else fails, the induction mechanism is applied.

The proof techniques are extensions of those used by Nqthm; see [2]. Most of the techniques are rule-driven. The rules are derived from previously proved theorems. For example, if the user has instructed ACL2 to prove that append is associative

```
(equal (append (append x y) z)
       (append x (append y z)))
```

and to use that fact as a :rewrite rule (from left to right), then — after the associative law is proved — the simplifier will right-associate all append-nests.

Because of the conservative nature of the extensions created by encapsulation, that logical mechanism is also very useful as a proof management tool. A complicated theorem can be derived in an encapsulation in which the uninteresting but necessary details are developed locally. See [22] for a detailed logical justification of encapsulation.

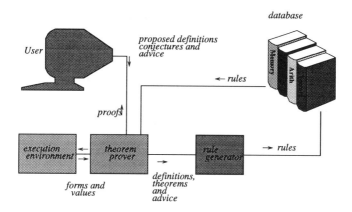

Fig. 2. System Architecture

By the proper choice of theorems to prove, the user can inform the system that a certain type-inference is possible, that a given relation is an equivalence relation to be used in congruence-based rewriting, that a certain function is a correct meta-theoretic simplifier, etc. Via this mechanism the informed user of ACL2 can essentially program it so that it constructs proofs following a certain strategy. By only using rules derived from previously proved theorems we insure that the user's advice cannot lead to logical unsoundness. See Figure 2.

With each proposed definition and theorem the user can supply hints to guide the theorem prover during the admission of the definition or proof of the theorem. A very common hint specifies which rules in the database are to be considered available during a proof or proof step. Other hints allow the user to suggest the use of a particular instance of a given theorem or to skip some step in the ring illustrated in Figure 1. Hints can be computed by ACL2 expressions so the user can codify strategies such as "if the goal contains these function symbols then the following rules should be used."

It is possible to collect together a body of definitions, theorems and advice into a file, here called a *book*. If ACL2 *certifies* the book then it can be loaded into any compatible ACL2 session to extend the logic and rule database appropriately. Books are incremental; multiple books can be loaded provided they compatibly define shared names. Packages allow the authors of books to have disjoint name spaces. Encapsulation can be used to hide unwanted aspects of a book or to extract desirable theorems. Events used in development of books may be marked `local` in order to prevent them from being exported.

The ACL2 system is written in ACL2, except for a relatively small amount of "boot-strapping" code, and is in essence a collection of ACL2 books. Coding ACL2 in its own logic forced many features into the language and insured that the resulting language is efficient and powerful enough for large projects demanding great computational resources. At the moment only one of the ACL2 source books, representing about 15% of the system, has been certified by ACL2; however, all the source books have been checked syntactically by ACL2.

2 Motorola CAP

The CAP is a single-chip, DSP co-processor optimized for communications signal processing, currently under development by Motorola Government and Systems Technology Group, Scottsdale, Arizona [16]. During the first seven months of the project one of us (Brock) relocated from Austin, Texas, to Scottsdale to work in close collaboration with the CAP design team. For the succeeding 18 months a member of the design team (Calvin Harrison) relocated to Austin to work at CLI. The CAP design was evolving throughout this period.

2.1 CAP Architecture

The CAP design follows the 'Harvard architecture', i.e., there are separate program and data memories. The design includes 252 programmer-visible data and control registers. There are six independently addressable data and parameter memories. The data memories are logically partitioned into 'source' and 'destination' memories; the sense of the memories may be switched under program control. The arithmetic unit includes four multiplier-accumulators and a 6-adder array. The CAP executes a 64-bit instruction word, which in the arithmetic units is further decoded into a 317-bit, low-level control word. The instruction set includes no-overhead looping constructs and automatic data scaling. As many as 10 different registers are involved in the determination of the next program counter. A single instruction can simultaneously modify well over 100 registers. In practice, instructions found in typical applications simultaneously modify several dozen registers. Finally, the CAP has a three-stage instruction pipeline which contains many programmer-visible pipeline hazards.

The motivation behind this complexity and unusual design is to allow the Motorola engineers to code DSP applications programs on the CAP and have those programs execute with stunning efficiency. The CAP was designed to execute a 1024-point complex FFT in 131 microseconds. There is no compiler targeting the CAP, perhaps because only a relatively few applications programs are required. For the time being at least, all CAP programs are coded by hand in CAP assembly language (CASM) and then assembled into binary and loaded into ROM. But programming the CAP efficiently and effectively requires a tremendous amount of knowledge and skill. One motivation for our involvement was to demonstrate that it was possible to create and then use a formal model of a state-of-the-art processor to verify applications programs.

2.2 The Model

We began by creating a formal, executable, ACL2 specification of the CAP [5]. This specification closely followed the style of earlier Nqthm work on modeling microprocessors, e.g., [19, 20, 28, 4]. Readers unfamiliar with that style need merely imagine defining, as a Lisp function, an interpreter for the intended machine language. We owe a special debt of gratitude to Yuan Yu, whose techniques in [38] we followed closely.

Our behavioral-level specification describes every well-defined behavior of the CAP including all legal instructions, I/O [21], traps, and interrupts. Only a few hardware and software initiated reset sequences are not modeled by our specification; these sequences were unnecessary to our intended verification work. In an effort to validate the specification we compared its execution with the results from executing Motorola's SPW engineering model of the processor [1].[3] For example, we compared the results of executing an end-to-end application (a QPSK modem) on both the SPW model and the ACL2 model; we found the final states bit-exact for all programmer visible registers. Other tests exposed discrepancies between the two models, most — but not all — of which were specification errors; some bugs were found in the CAP design. The ACL2 model runs several times faster than the compiled SPW model and hence is a potentially valuable debugging tool in its own right. A small part of the hardware implementation of the processor, the XY memory address generation unit, has been formally verified to agree with with the corresponding part of the ACL2 model [18]; this involved the hand-translation into ACL2 (in a very mechanical fashion) of the SPW description of the hardware.

We believe that CAP is the most complex processor for which a complete formal specification has been produced. The MC68020 modeled in [38], the first commercial processor for which a substantially complete formal model was produced, has only sixteen general purpose registers and a simple instruction set, albeit one with 18 addressing modes. Until the CAP work, the most complicated commercial processor subjected to formal modeling at the functional level was probably the Rockwell-Collins AAMP5, a special purpose avionics processor. The functionality of the AAMP5 was partially specified with PVS [15], as described in [26]: 108 of 209 instructions were specified and the two-stage pipelined microcode for eleven was verified. The state of the AAMP5 is much simpler than that of the CAP, involving only two memories, two flags, and six registers; nevertheless, the AAMP5 is far from simple: its 209 variable length CISC-like instructions closely resemble the intermediate output of compilers, and include some real-time executive capabilities such as interrupt handling, task state saving and context switching.

Before turning to the verification of CAP applications programs we undertook the logical elimination of the CAP pipeline. The CAP implementation includes a three-stage instruction pipeline with many visible pipeline hazards. Using ACL2 we demonstrated that under "normal" conditions, which we rigorously defined,

---

[3] As of this writing no CAP chips have been fabricated.

the behavior of CAP programs can be understood without reference to the instruction pipeline. We did this by proving an appropriate correspondence between the pipelined CAP model and a simpler pipeline-free model of the CAP [6]. Aside from making subsequent code proofs easier, this work had the important benefit of identifying (with both precision and assurance) an *equivalence condition* sufficient to avoid hazards yet weak enough to admit CAP application programs. Most of the requirements, in some suitably informal sense, were already part of the evolving "folklore" in the small community of CAP programmers, who generally take great pains to avoid pipeline hazards in their DSP application code. But the equivalence condition is sufficiently complicated that, short of a rigorous proof such as the one we constructed, confidence in its sufficiency is difficult to obtain. Our condition is weak enough to accept every ROM-resident DSP application on the CAP.

Our equivalence proof follows the approach suggested by Burch and Dill [11]. We found their method for stating equivalence, involving the idea of flushing the pipeline, intuitive for writing specifications. Applying it to the full CAP specification was somewhat challenging. The proof requires the symbolic expansion of the both the pipelined and non-pipelined models for three clock-cycles and then the comparison of the resulting symbolic states. Because of the control complexity of the CAP specification and the (necessary) weakness in the equivalence condition, a naive expansion took ACL2 about 10 minutes and produced a term whose printed representation was about 300,000 lines long (about 20 megabytes). After a couple of days' work by the user developing the proper theories, ACL2 can do a three step expansion on arbitrary (hazard-free) code in a few seconds and produce a term of about 3000 lines. Once the libraries needed to show the equivalence of these two models are in place (non-trivial), the proof takes about 4 minutes. ACL2's ability to handle large terms is one of its primary advantages over our earlier theorem prover, Nqthm, when dealing with industrial-sized problems.

The CAP instruction pipeline is certainly not the most complex instruction pipeline that has been formally specified and analyzed. However, although others have considered a few instructions on a pipelined commercial processor design (such as the AAMP5 work, [26]), or all instructions on a simple academic example [33, 34], our results cover *every* instruction sequence in a complete, bit-accurate model of a commercially designed processor. This work is also the only case that we are aware of where the problem of the equivalence of pipelined and non-pipelined models is so dependent on the program being executed, rather than predominately an intrinsic property of the hardware implementation.

Finally, we used the pipeline-free model as a basis for the mechanized verification of CAP application code. We briefly describe two CAP assembly language programs for which we constructed mechanically checked correctness proofs.

The first is a finite impulse response (FIR) filter. The FIR filter is an archetypical DSP application, a discrete-time convolution of an input signal with a set of filter coefficients. This FIR code is one that we developed specifically as an example, although it is based on a FIR algorithm for the CAP originally coded by a Motorola engineer. We prove that the code computes an appropriate fixed-precision result, but do not address any of the numerical analysis or DSP content of the algorithm. The second program is a high-speed searching application. This application uses specialized data paths in the CAP adder array to locate the 5 maxima of an input data vector in just over one clock cycle per input vector element. The 5-peak search is a fascinating example of an application with a straightforward specification, implemented by a clever combination of hardware and software, whose correctness is far from obvious. In this case the code we verified was obtained directly from the CAP program ROM, exactly as written by a Motorola engineer.

2.3 FIR

The finite impulse response (FIR) filter is a commonly used DSP algorithm; for details on its development and utility see for example [30]. Boiled down to its essentials, the FIR filter maps a discrete input signal $x[i]$ to a discrete output signal $y[i]$ by a discrete-time convolution of $x[i]$ with a set of filter coefficients h_k. The equation defining a non-adaptive FIR filter is often shown as something similar to

$$y[n] = \sum_{k=0}^{n-1} h_k \cdot x[n-k].$$

Given that the above equation defines a FIR filter, our goal is simply to prove that certain CAP microcode computes something suitably equivalent. The implementation uses the fixed-precision arithmetic of the CAP; the sense in which it implements the above equation involves the assumption that no overflows, for example, occur. There is no guarantee that every set of coefficients will produce a meaningful result, since the choice of coefficients together with input scaling parameters may lead to catastrophic overflows in the CAP accumulators that compute the sums of products.

The FIR code we verified consists of 30 64-bit microcode instructions. The program simultaneously filters X and Y source memories with a single set of coefficients, writing the results to X and Y destination memories. This can either be interpreted as two real FIR filters on two real signals, or a single real FIR filter on a complex signal. The small size of the FIR program belies its complexity because the CAP instruction set is so complicated. The effects of the single instruction in the inner loop depends upon 9 memories, 3 stacks and 31 registers; 17 registers are simultaneously modified by the instruction. The reader is reminded that we are not considering some idealized algorithm but an actual program written in a new and arcane programming language designed by industrial DSP experts for DSP applications.

The code has been mechanically proved to implement the specification equation. Because the CAP assembly code has no formal semantics, we actually addressed ourselves to the binary object code produced by Motorola's CAP assembler. Our theorem establishes that when (our formal model of) the CAP is run a certain number of steps on a suitably configured initial state the resulting destination memories are configured as required by the equation.

2.4 5PEAK

It is often necessary to perform statistical filtering and peak location in digital spectra for communications signal processing. The so-called "5PEAK" program uses the CAP's array of adder/subtracters with their dedicated registers as a systolic comparator array as shown in Figure 3. The program streams data through the comparator array and finds the five largest data points and the five corresponding memory addresses. In this discussion we largely ignore the fact that

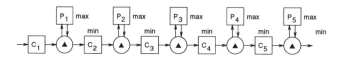

Fig. 3. Abstract View of Comparator Array

the comparator array also maintains the memory address of each data point.

Data points enter the array by way of register C_1 and move through the array, towards the right in the diagram. Maximum values remain in the array in the P_n registers, and the minima are eventually discarded when they pass out of the last comparator. On each cycle the comparator array updates the registers as follows:

$$C_1 = \text{next data point},$$
$$C_n = \min(C_{n-1}, P_{n-1}), \quad n > 1,$$
$$P_n = \max(C_n, P_n).$$

Informally, the peak registers, P_i, maintain the maximum value that has passed by that point in the comparator array.

Suppose we initialize the array with the smallest possible data values (called "negative infinity" but actually just the most negative CAP integer). Then, as we stream an arbitrary amount of data through the array, the peaks accumulate in the peak registers. Immediately after the last data point has entered the array, we know that P_1 is the maximal data point. But P_2 may not be the second highest peak because it will not have been compared to C_2. To cause the necessary comparisons to occur, we must stream more data in. Note that by feeding one "positive infinity" (the most positive CAP integer) into the array we can cause the comparison of C_2 with P_2, along with other comparisons. This also displaces the actual maximal value from P_1 into C_2 and flushes one minimal value from the right end of the array. Thus by feeding five positive infinities in we can accumulate at the right end of the array the five maximal values.

We verified the 5PEAK object code obtained from the CAP program ROM. We proved that when the abstract CAP machine (as defined in ACL2) executes that binary code, on an appropriate initial state and for the appropriate number of cycles, the five highest peaks and their addresses are deposited into certain locations. We defined the "highest peaks and their addresses" by defining, for specification purposes only, a sort function in ACL2 which sorts such address/data pairs into descending order; in our 5PEAK specification we refer to the first five pairs in the ordering.

The argument that 5PEAK is correct is quite subtle, in part because an arbitrary amount of data is streamed through and in part because the positive and negative infinities involved in the algorithm can be legitimate data values but are accompanied by bogus addresses; correctness depends on a certain "anti-stability" property of the comparator array. A wonderfully subtle generalization of a key lemma was necessary in order to produce a theorem that could be proved by mathematical induction, [7]. In addition, as in the FIR example, correctness also depends on the invariance of many low-level properties of the initial CAP state.

2.5 Manpower Breakdown

Our involvement with the CAP project lasted 31 months. Only one formal methods expert (Brock) worked on the project continuously during that time. He was responsible for modeling the CAP in ACL2, producing the non-pipelined abstraction of the CAP, inventing the equivalence condition, mechanically proving the conditional equivalence of the two models, and mechanically proving the two microcode programs reported here. He also developed a library of ACL2 books for use in this work. Except where noted below, he worked on these tasks alone, although in the early months of the project he talked frequently with the CAP design team. Several other researchers and engineers contributed to the formal methods part of the project (e.g., the validation testing of the ACL2 against the SPW, a mechanically checked correctness proof for the address generation unit, an ACL2 macro to help formulate lemmas for expanding function definitions, a CAP assembler, a pin-level specification of the CAP IO interface) but their work has not been the focus of this paper and is not further discussed here. Brock's efforts can be broken down as follows:

- The CAP Specification: 15 months. Brock produced the first executable version of the specification in about 6 months, while resident in Scottsdale and interacting with the design team. During that time he was also learning ACL2[4]. The first model was simple and incomplete. As the project progressed, a 4-valued logic was introduced, and the pipeline, full ALU and IO modules were included. In addition, the CAP design evolved more or less continuously during this interval and the formal model tracked the design. Finally, ACL2 evolved also and Brock had to convert his work from Version 1.7 to Version 1.8, which dramatically changed the treatment of guards, making proofs much easier.

---

[4] He was already an accomplished Nqthm user.

- Reusable Books: 6 months. To do proofs about the CAP specification Brock had to develop many ACL2 books about modular arithmetic, logical operations on integers, hardware arithmetic and bit vectors, arrays, record structures, and list processing. These books are not CAP specific and are hence reusable.
- Equivalence to Non-Pipelined Model: 5 months. Not surprisingly, the non-pipelined model of the CAP shares perhaps 90% of its definition with the (pipelined) CAP specification. It was relatively easy to produce; however it required the reorganization of the CAP specification so that such sharing was possible. The first conditional equivalence proof was carried out on a simplified model of the CAP and then repeated when the IO module was added. A serendipitous visit by David Dill occurred just as Brock was beginning the first proof; Dill explained the methods in [11], which proved very useful. The details of the equivalence condition were derived from failed proofs. The condition was later weakened to so that it would accept the ROM-resident DSP application codes and the (now largely automatic) proof was repeated a third time to confirm the new condition. Brock then incorporated the predicate into an (unverified) automated tool that analyzes CAP programs for pipeline visibility conditions. The tool requires no formal methods expertise to use or to interpret its output.
- FIR: 1 month. Some of the work on reusable books was done in response to difficulties encountered during this task. In addition, CAP-specific books were developed for controlling the unwinding of the CAP model during such proofs. Both the FIR and 5PEAK work benefit from the fact that the non-pipelined model is simpler to reason about.
- 5PEAK: 1 month. About half of the time here was devoted to the development of the specification of the 5PEAK code. After several false starts, it eventually involved the idea of sorting an arbitrary amount of data with a generalization of the 5PEAK algorithm and then collecting the first 5 values. Moore spent one week proving the fundamental properties of the generalized sort algorithm.
- Reporting: 3 months. This includes time to write regular progress reports, travel to meetings, and write the final reports documenting the effort.

The times above are essentially the time it took the user to create the material, discover the proofs and lead the theorem prover to them the first time. The theorem prover can actually carry out the proofs relatively quickly. All of the CAP proofs can be reproduced in 2:30 (2 hours and 30 minutes) on a Sun Microsystems Sparcstation-20/712 with dual 75 MHz SuperSparc-II CPU's (each with 1 megabyte of cache), 256 megabytes of memory and 2 gigabytes of local disk. Many books are shared between the three main proofs (equivalence, FIR, and 5PEAK). A book need be certified only once and thereafter can simply be referenced without proof. The equivalence proof takes 1:45 (1:29 of which is spent certifying books that are re-used in the two applications proofs). FIR can then be done in 33 minutes (17 minutes of which is spent certifying books used also in 5PEAK). 5PEAK then takes 13 minutes.

3 AMD5$_K$86

In the Spring of 1995, the Intel Pentium floating-point division bug was in the headlines. At that time, Advanced Micro Devices, Inc., was working on their first Pentium-class microprocessor, the AMD5$_K$86. Because the AMD5$_K$86 was not yet fabricated, there was time to reassess its division algorithm for bugs and perhaps fix any that might be found. But the algorithm was proprietary so it could not be reviewed by the numerical analysis community; nor was there time for the "social process" of mathematics to review any proffered proof of correctness.

In May, 1995 AMD hired CLI to construct and mechanically check a proof of the correctness of the kernel of the AMD5$_K$86's floating-point division algorithm. The principals of this collaboration were Tom Lynch of AMD, the designer of the division algorithm, and authors Kaufmann and Moore. All three were located in Austin, Texas, which facilitated collaboration. Work commenced in June, 1996 and the proof was completed by mid-August, 1996. Moore was the only person working full-time on the project. No bugs were found; the AMD algorithm was correct. We sketch the algorithm and theorem proved below. See [29] for details.

The AMD5$_K$86 algorithm is supposed to divide p by d, where both are floating-point numbers and $d \neq 0$, and round the result according to a specified rounding mode, m.

There has been much interest in the mechanical analysis of division algorithms since the Pentium bug. Almost all of the work focuses on SRT division [8, 32, 12]. The easiest way to contrast that work with ours is to point out that the results established in the above-mentioned papers do not formalize or discuss the notions of "floating-point number" or "rounding modes." Those concepts have been formalized elsewhere, for example [27], but no significant mechanically checked theorems about them are reported. Our work, on the other hand, focuses almost entirely on the concepts of floating-point number and directed rounding and the properties of the elementary floating-point operations. It is the first substantial body of mechanically checked formal mathematics about floating-point arithmetic.

By an n, , m *floating-point number* we mean a rational that can be represented in the form $\sigma \times s \times 2^e$ where $\sigma \in \{+1, -1\}$, s is a rational, either $s = 0$ or $1 \leq s < 2$ and the binary representation of s fits in n bits, and e is an integer whose (biased) binary representation fits in m bits. A *rounding mode* specifies a rounding procedure and a precision n by which one is to round a rational to an n, , 17 floating-point number. Common procedures include truncation of excessive bits (rounding toward 0), rounding away from 0, and "sticky" rounding in which the least significant bit of the rounded significand is 1 if precision is lost.

The AMD5$_K$86 algorithm uses a lookup table to obtain an 8, , 17 floating-point approximation of $1/d$, twice applies a floating-point implementation of a variant of Newton-Raphson iteration to refine this into a 28, , 17 approximation of $1/d$, uses the approximation to compute four 24, , 17 quotient digits in an algorithm similar to long-division, and then sums the digits using sticky rounding for the first two floating-point additions and the user-specified mode m for the last (and most significant).

Using ACL2 we proved that if p and d are $64,,15$ (possibly denormal) floating-point numbers, $d \neq 0$, and m is a rounding mode specifying a positive precision $n \leq 64$, then the answer described above is the $n,,17$ floating-point number obtained by rounding the infinitely precise p/d according to m. In addition, and of particular interest to the design team, we proved that all intermediate results computed by the algorithm fit in the floating-point registers allocated to them.

Over 1600 definitions and lemmas were involved in this proof. While ACL2 (unlike Nqthm) has built-in support for the rationals, this was the first time that floating-point numbers and directed rounding had been formalized in ACL2. A substantial body of numerical analysis had to be formalized and mechanically checked in order to follow a fairly subtle mathematical argument constructed by the three collaborators. The proof complexity management tools, in particular books, encapsulation and macros, were crucial to the timely completion of this project. Note that only 9 weeks elapsed from the time CLI first saw the microcode to the time the final "Q.E.D." was printed. The final proof script can be replayed by ACL2 in two hours on the Sparcstation-20/712 described above.

4 Conclusion

We have described several theorems proved about commercial microprocessors. Make no mistake, these were "heavy duty proofs" requiring many skills, including great familiarity and insight into the applications areas, engineering issues, mathematics, formal logic, and the workings of the ACL2 proof tool. Furthermore, a fair amount of dedication and persistence were also required. See for example [25] for a case study describing practical problems in the use of a general purpose theorem prover. Simply put, it is hard work producing proofs of conjectures like these, a fact which stands in stark contrast to the impressive results obtained by "lightweight" analysis tools requiring so much less of the user [10, 14, 17, 37].

So why should people use general purpose theorem provers? If correctness is important and the problem cannot be solved by special-purpose "lightweight" tools, then a "heavyweight" tool is not only appropriate but is the only alternative. But is microcode verification within reach of today's model checking tools? If one assumes one has a bit-accurate model of a microcode engine, that the relevant microcode is in ROM (and so does not change during execution) and that all loops are controlled by counting down registers (and so are bounded by the word size), then microcode verification is a finite problem that can, in principle, be done by model checking. The question is simply one of practicality.

It is, of course, difficult to say with certainty that an untried problem is beyond the capabilities of a given model checker. The reason is partly that such tools can often handle designs with surprisingly large state spaces (e.g., several hundred boolean variables). Many techniques exist for reducing the "naive" state-space (e.g., "scaling", the removal of all bits of state that do not support any property being checked); model checkers at higher levels of abstraction are also being developed [13, 9]. But the reason an unequivocal rejection of model checking is so difficult here is that with creative insight (i.e., "heavyweight thinking?") the user of a "lightweight" tool can often abstract the problem into one that is manageable.

An easier question to answer, then, is whether today's model checkers have been used to do microcode verification for machines of industrial interest. The answer seems to be no. While today's model checkers can often succeed at this kind of verification for small values of the input parameters, they rarely do for the full range of values. We believe that the CAP and AMD programs discussed here are beyond the range of today's model checkers. We return to this point shortly.

Assuming that these problems "cannot" be done by "lightweight" tools, theorem proving is the only alternative.[5] We have demonstrated that such problems, in their full complexity, are not beyond the range of ACL2. The manpower requirements of these ACL2 projects are not extreme, considering the amount of manpower industry currently throws at testing.

The manpower requirements in projects focused on code proofs for a given machine (the FM9001 assembly language [28], the MC68020 machine code proofs [4], the CAP work here, and the AAMP5 work [26]) all reflect a common theme: we are repeatedly measuring the startup costs. Building the formal model of the "new" machine or language and getting the appropriate library of rules in place dominates the costs. Yu's work, [4], demonstrated that if one simply focuses on proving one program after another, it begins to get routine. Until automated theorem provers surpass humans in their creative ability to find proofs — today's theorem provers must be regarded primarily as "proof checkers" that fill in the thousands of missing steps in what is commonly called a "proof" by students of mathematics — the mechanical verification of a program will remain essentially a two step process: the user discovers the "intuitive proof" and then gives the machine enough advice to check it. Once the startup costs have been paid for a given machine, the step of intuiting the informal proof becomes much more dominant.

Our CAP and AMD5$_K$86 projects, as well as the AAMP5 work [26], also support the conclusion that theorem-prover based formal methods can "keep up" with an evolving hardware design and make important contributions. Why then is industry apparently so reluctant to adopt these techniques? Part of the answer is that the skills required to use our tools are different than those ordinarily found in a hardware design team. Oddly enough, people with these skills, namely people trained in discrete mathematics, are readily available but their skills are not appreciated by industry. However, another major reason our tools are not more

---

[5] Testing the CAP code, while useful, is probably less likely to find all errors than testing would on a more conventional processor simply because of the sensitivity of the instruction sequencing to large portions of the state. Testing is also unconvincing in the case of sophisticated floating-point algorithms.

widely used is that they are not integrated into the *design process*. The ACL2 specification of the CAP, while efficiently executable and uniquely enabling of mechanically checked proofs, cannot at present be processed by conventional CAD tools. We had intended to build a translator from SPW to ACL2 but, for non-technical reasons beyond our control, could not pursue the task.

Such discouragement not withstanding, the fact remains that general purpose tools, such as ACL2, are truly general purpose. A powerful specification language is provided. As a user one rarely feels *unable* to express the key ideas. Arbitrary complexity can be modeled. The user carries the responsibility of managing this complexity but the system provides a wealth of ways to help, ranging from such traditional mathematical devices as functional composition and lemmas, to system features such as macros, encapsulation, books, symbol packages, and computed hints. The development of these features is the biggest improvement of ACL2 over Nqthm and is the reason that the CAP project could be carried out with ACL2 much more expeditiously, we believe, than with Nqthm. As with a good programming language or other universal tool, one rarely feels that a problem is too big or too complicated to address. Rather, the question is simply how to proceed. In that sense, the system feels open-ended and empowering, especially given that the system carries the burden of logical correctness and leaves the user entirely focused on the larger issues.

Finally, and most importantly, we believe there is value in demonstrating that certain things are merely *possible*. At least one respected numerical analyst told us several years ago that floating-point error analysis was too complicated to imagine checking mechanically. We now know otherwise. After our checking of the division microcode, David Russinoff, a former CLI employee now working at AMD, used ACL2 to analyze the AMD5$_K$86 floating-point square root code; his proof attempt exposed a bug in the original code which had escaped testing. In collaboration with Tom Lynch, the code's designer, the code was changed and proved correct with ACL2. This work would not have been done had we persisted in believing that it was impossible. Similarly, perhaps microcode for such complex processors as the CAP will someday be verified automatically by "lightweight" tools; after all, we know that with general methods it is not only possible but practical and we are now merely talking about the cost.

In summary, general purpose theorem proving tools such as ACL2 are up to the demands of state-of-the-art industrial microprocessor design. In the short run, the contribution of such "heavy duty" tools is merely that they provide answers that can be obtained no other way. Mechanically checked proofs are essentially the only technology allowing the reliable exploration of arbitrarily deep problems. On the frontier of what is thought possible, there will always be a need for such tools and they will likely be in hands of people skilled in both the application area and mathematics. But in the long run, perhaps the main contribution of such "heavy duty" tools is that they allow us to enlarge the realm of what is thought possible.

References

1. K. Albin. *Validating the ACL2 CAP Model.* CAP Technical Report 9, Computational Logic, Inc., 1717 W. 6th, Austin, TX 78703 March, 1995.

2. R. S. Boyer and J S. Moore. *A Computational Logic.* Academic Press: New York, 1979.

3. R. S. Boyer and J S. Moore. *A Computational Logic Handbook*, Academic Press: New York, 1988.

4. R. S. Boyer and Y. Yu. Automated Proofs of Object Code for a Widely Used Microprocessor, *JACM*, **43**(1) January, 1996, pp. 166–192.

5. B. Brock. *The CAP 94 Specification*, CAP Technical Report 8, Computational Logic, Inc., 1717 W. 6th, Austin, TX 78703, July, 1995.

6. B. Brock. *Formal Analysis of the CAP Instruction Pipeline*, CAP Technical Report 10, Computational Logic, Inc., 1717 W. 6th, Austin, TX 78703, June, 1996.

7. B. Brock. *Formal Verification of CAP Applications*, CAP Technical Report 15, Computational Logic, Inc., 1717 W. 6th, Austin, TX 78703, June, 1996.

8. R. E. Bryant. Bit-Level Analysis of an SRT Divider Circuit, CMU-CS-95-140, School of Computer Science, Carnegie Mellon University, Pittsburg, PA 15213.

9. R. E. Bryant and Y. A. Chen. Verification of arithmetic functions with binary moment diagrams. In *Proceedings of the 32nd ACM/IEEE Design Automation Conference* IEEE Computer Society Press, June 1995.

10. J. R. Burch, E. M. Clarke, D. E. Long, K. L. McMillan and D. L. Dill. Symbolic Model Checking for Sequential Circuit Verification, *IEEE Trans. on Computer-Aided Design of Integrated Circuits and Systems* **13**(4) April, 1994, pp. 401–424.

11. J. R. Burch and D. L. Dill. Automatic verification of pipelined microprocessor control. in David Dill, editor, *Computer-Aided Verification, CAV '94,* Stanford, CA, Springer-Verlag *Lecture Notes in Computer Science* Volume 818, June, 1994, pp. 68–80.

12. E. M. Clarke, S. M. German and X. Zhao. Verifying the SRT Division Algorithm using Theorem Proving Techniques, Proceedings of Conference on Computer-Aided Verification, CAV '96, July, 1996.

13. E. M. Clarke, M. Fujita, and X. Zhao. Hybrid Decision Diagrams, ICCAD95, 1995, pp. 159-163.

14. E. M. Clarke, O. Grumberg, H. Hiraishi, S. Jha, D. E. Long, K. L. McMillan and L. A. Ness. Verification of the Futurebus+ Cache Coherence Protocol, *Proc. CHDL*, 1993.

15. J. Crow, S. Owre, J. Rushby, N. Shankar, and M. Srivas. A Tutorial Introduction to PVS, presented at *Workshop on Industrial-Strength Formal Specification Techniques*, Boca Raton, FL, April 1995 (see http://www.csl.sri.com/pvs.html).

16. S. Gilfeather, J. Gehman, and C. Harrison. Architecture of a Complex Arithmetic Processor for Communication Signal Processsing in *SPIE Proceedings, International Symposium on Optics, Imaging, and Instrumentation,* **2296** *Advanced Signal Processing: Algorithms, Architectures, and Implementations V,* July, 1994, pp. 624–625.

17. Z. Har'El and R. P. Kurshan. Software for Analytical Development of Communications Protocols, *AT&T Bell Laboratories Technical Journal,* **69**(1) Jan-Feb, 1990, pp. 45–59.

18. C. Harrison. *Hardware Verification of the Complex Arithmetic Processor XY Address Generator.* CAP Technical Report 16, Computational Logic, Inc., 1717 W. 6th, Austin, TX 78703, August, 1995.

19. W. A. Hunt, Jr. Microprocessor Design Verification. *Journal of Automated Reasoning*, **5**(4), pp. 429–460, 1989.

20. W. A. Hunt, Jr. and B. Brock. A Formal HDL and its use in the FM9001 Verification. *Proceedings of the Royal Society*, 1992.

21. W. A. Hunt, Jr. *CAP Pin-level Specifications*, CAP Technical Report 12, Computational Logic, Inc., 1717 W. 6th, Austin, TX 78703, April, 1996.

22. M. Kaufmann and J S. Moore. High-Level Correctness of ACL2: A Story, URL ftp://ftp.cli.com/pub/acl2/v1-8/acl2-sources/reports/story.txt, September, 1995.

23. M. Kaufmann and J S. Moore. *ACL2 Version 1.8*, URL ftp://ftp.cli.-com/pub/acl2/v1-8/acl2-sources/doc/HTML/acl2-doc.html, 1995.

24. M. Kaufmann and J S. Moore. ACL2: An Industrial Strength Version of Nqthm. In *Proceedings of the Eleventh Annual Conference on Computer Assurance* (COMPASS-96), IEEE Computer Society Press, June, 1996, pp. 23–34.

25. M. Kaufmann and P. Pecchiari. Interaction with the Boyer-Moore and Theorem Prover: A Tutorial Study Using the Arithmetic-Geometric Mean Theorem. *Journal of Automated Reasoning* 16(1–2) March, 1996, pp. 181–222.

26. S. P. Miller and M. Srivas. Formal Verification of the AAMP5 Microprocessor: A Case Study in the Industrial Use of Formal Methods, in *Proceedings of WIFT '95: Workshop on Industrial-Strength Formal Specification Techniques*, IEEECS, April, 1995, pp. 2–16.

27. P. M. Miner. Defining the IEEE-854 Floating-Point Standard in PVS, NASA Technical Memorandum 110167, NASA Langely Research Center, Hampton, VA 23681, 1995.

28. J S. Moore. *Piton: A Mechanically Verified Assembly-Level Language*, Automated Reasoning Series, Kluwer Academic Publishers, 1996.

29. J S. Moore, T. Lynch, and M. Kaufmann. A Mechanically Checked Proof of the Correctness of the Kernel of the $AMD5_K86$ Floating-Point Division Algorithm, March, 1996, URL http://devil.ece.utexas.edu:80/~lynch/divide/-divide.html.

30. A. V. Oppenheim and R. W. Schafer. *Discrete-Time Signal Processing.* Prentice Hall, Englewood Cliffs, New Jersey, 1989.

31. K. M. Pitman *et al. draft proposed American National Standard for Information Systems — Programming Language — Common Lisp; X3J13/93-102.* Global Engineering Documents, Inc., 1994.

32. H. Ruess, M. K. Srivas, and N. Shankar. Modular Verification of SRT Division, Computer Science Laboratory, SRI International, Menlo Park, CA 49025, 1996.

33. M. Srivas and M. Bickford. Formal Verification of a Pipelined Microprocessor, *IEEE Software*, September, 1990, pp. 52–64.

34. V. Stavridou. Gordon's Computer: A Hardware Verification Case Study in OBJ3, *Formal Methods in System Design*, **4**(3), 1994, pp. 265–310.

35. G. L. Steele, Jr. *Common LISP: The Language*, Digital Press: Bedford, MA, 1984.

36. G. L. Steele, Jr. *Common Lisp The Language, Second Edition.* Digital Press, 30 North Avenue, Burlington, MA 01803, 1990.

37. U. Stern and D. L. Dill. Automatic Verification of the SCI Cache Coherence Protocol, in *Proceedings of IFIP WG 10.5 Advanced Research Working Conference on Correct Hardware Design and Verification Methods*, 1995, pp. 21–34.

38. Y. Yu. *Automated Proofs of Object Code for a Widely used Microprocessor*, Technical Report 92, Computational Logic, Inc., 1717 W. 6th, Austin, TX 78703, May, 1993. URL http://www.cli.com/reports/files/92.ps.

Formal Synthesis in Circuit Design – A Classification and Survey

Ramayya Kumar[1], Christian Blumenröhr[2], Dirk Eisenbiegler[2] and Detlef Schmid[1,2]

[1] Forschungszentrum Informatik,
Department of Automation in Circuit Design, Karlsruhe, Germany
e-mail: kumar@fzi.de
[2]Institute of Computer Design and Fault Tolerance
University of Karlsruhe, Germany
e-mail: {blumen,eisen,schmid}@ira.uka.de

Abstract. This article gives a survey on different methods of formal synthesis. We define what we mean by the term formal synthesis and delimit it from the other formal methods that can also be used to guarantee the correctness of an implementation. A possible classification scheme for formal synthesis methods is then introduced, based on which some significant research activities are classified and summarized. We also briefly introduce our own approach towards the formal synthesis of hardware. Finally, we compare these approaches from different points of view.

1 Introduction

In everyday use, synthesis means putting together of parts or elements so as to make up a complex whole. However in the circuit design domain, synthesis stands for a stepwise refinement of circuit descriptions from higher levels of abstraction (specifications) to lower ones (implementations), including optimizations within one abstraction level.

Synthesis can be performed by hand for small circuits. Nowadays more and more automated synthesis tools are in use to harness the complexity of the circuits[McPC90]. Both hand-made and automatically computed hardware implementations may be incorrect with respect to the given input specification. By correctness we mean that the implementation satisfies the specification, in a formal mathematical sense. However, it is possible that the specification itself may be incomplete or erroneous. These anomalies in the specification will have to dealt with separately, either via simulation or by checking certain properties such as safety and liveness. In the rest of the paper, it is assumed that the specification is the reference with which the implementation is checked and the correctness aspects dealt with assume correct specifications.

Examining the correctness of the implementation with respect to a given correct specification, the errors in the hand-made circuits are created by the designer, whereas the errors in synthesized implementations depend on the correctness of the synthesis tools involved. The use of automated synthesis tools significantly increases the reliability as compared to hand designs. During such synthesis, the designer interacts with the system only in a controlled manner. The implementations cannot be created manually and hence the amount of errors are reduced. Furthermore, such systems offer the capability of design space exploration which has a great impact on the quality of designs, especially in the context of designing complex systems.

The "correctness by construction" paradigm that has been propagated in the synthesis domain is nevertheless questionable. Synthesis programs have become more and more complex especially at higher levels of abstraction (algorithmic level, system level). Sophisticated algorithms and data structures are used and they usually programmed as hundreds of thousands of lines of code, in imperative programming languages like C. Due to their complexity, such programs can only be partially tested. Added to that, these programs are implemented by a team of programmers. This scenario enforces careful thought about the complex interfaces between the modules which manipulate the complex data structures representing the hardware. Yet another source of bugs in synthesis programs are wrong interpretations of HDLs (hardware description languages) that are used for specifications. To embed HDLs, one has to implement procedures for converting hardware descriptions given in the HDL into the internal representation and vice versa. To guarantee the overall correctness of a synthesis tool, one also has to consider the correctness of these transformations with respect to the semantics of the HDLs and the semantics of the internal hardware representation. All these reasons motivate us to closely examine the approaches for the correct synthesis of hardware.

We will first give a definition of formal synthesis and delimit it from related approaches towards correct synthesis. In section 3, we will classify formal synthesis approaches. Section 4 briefly summarizes some formal synthesis approaches. We then look at some criteria for comparing approaches for formal synthesis and conclude.

2 Definition and Delimitation of Formal Synthesis

On examining the synthesis process we observe that we can ascertain its correctness *before*, *during* or *after* the synthesis process. Hence we classify the approaches towards formal correctness of synthesis into *pre-synthesis verification*, *formal synthesis* and *post-synthesis verification* (figure 1). All approaches have the same overall aim: proving the correctness of the synthesis output (implementation) with respect to the synthesis input (specification). They all use some kind of logic for deriving proofs within some calculus. However, the time when the logical argumentation takes place differs — in pre-synthesis verification, the proof is derived even *before* synthesis happens, in formal synthesis the proof is formally derived *during* synthesis, and in post-synthesis, the proof is derived after *each* synthesis run.

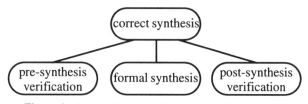

Figure 1. Approaches towards correct synthesis results

Pre-synthesis verification means proving the correctness of the synthesis procedure, i.e. it has to be proven, that the synthesis *program* always produces correct synthesis results with respect to the synthesis input. This is to be done by means of software verification. In pre-synthesis verification the correctness is proven once and for all, whereas in post-synthesis verification, the correctness proof has to be derived for each synthesis run separately. However, software verification is extremely tedious especially for large sized programs such as synthesis tools. Therefore, there is only little work done in this area with respect to hardware.

Formal synthesis means formally deriving the synthesis result within some logical calculus. In conventional synthesis hardware is represented by arbitrary data structures and there are no restrictions as to the transformations on these data structures. In formal synthesis, hardware is represented by means of terms and formulae, and only correctness preserving logical transformations are allowed. Restricting synthesis to only correctness preserving logical transformations guarantees the correctness of the synthesis procedure in an implicit manner. In contrast to conventional synthesis, the result is not only some hardware implementation but also a proof of its correctness, with respect to the specification.

Post-synthesis verification is the most frequently used approach today. In post-synthesis verification the synthesis step is first performed in a conventional manner and then the correctness of the synthesis output with respect to the synthesis input is proved. It is independent of the synthesis procedure.The only information available is the synthesis input and the synthesis output. There is no information on *how* the output was derived from the input. However, general hardware verification techniques are pretty time consuming or even undecidable. This makes verification for large sized circuits tedious or even impossible. [Gupt92] gives an excellent survey of the major trends in post-synthesis verification and we do not discuss them within this paper.

When comparing the above-mentioned approaches towards correct synthesis, one can distinguish between the tool developer's and the circuit designer's points of view.

2.1 Circuit Designer's Viewpoint

For the circuit designer, the characteristics that are relevant are — time taken for synthesis, automation and the requirement of skills in formal methods.

- The *synthesis time* has an additional dependency on the size of the circuit under design and the number of synthesis steps that are performed.
  - With respect to the *size of the design*, the synthesis time is lowest for pre-synthesis verification. Since formal synthesis is restricted to transformations within some calculus, this results in slower synthesis tools. The post-synthesis technique on the other hand suffers from the problem of NP-completeness or undecidability, depending upon the levels of abstraction.
  - The synthesis time with respect to the *number of synthesis steps* is smaller for the pre-synthesis techniques as compared to the formal synthesis approaches, due to the reasons mentioned above. Post-synthesis approaches however, do not depend on the synthesis process and therefore the number of synthesis steps does not have any impact.
- In examining the characteristic of *automation* — it is possible to achieve it at all levels of abstraction in the pre-synthesis approach, since it is possible to write such programs. As far as formal synthesis is concerned, it depends upon the system as seen in section 4. Automation can be achieved only at lower levels of abstraction in the post-synthesis approach via model-checkers and tautology checkers. At higher levels of abstraction, general automation cannot be achieved due to the complexity of the task.
- From the point of view of the circuit designer, the *skills required in formal methods*

is an important factor in its acceptance as a tool. Pre-synthesis verification is the most convenient since the entire logical argumentation is of no concern to the designer. An argumentation similar to the automation holds for the other two approaches.

2.2 Tool Developer's Viewpoint

The criteria to be examined from the tool developer's viewpoint are different — the complexity of tool development, the maintenance of the tool when changes are made, and the skills in formal methods required for tool development.

- The *complexity of the development* of pre-synthesis verification is stupendous and hence it is practically impossible to develop such a tool for realistic synthesis programs, since software verification can be performed only for very small programs (≈ 1000 lines of C code). As far as formal synthesis is concerned, the complexity will increase with the size of the synthesis tool. Although post-synthesis verification tools do not depend upon the size of the synthesis tool, they can only be developed without problems at lower levels of abstraction. The development of post-synthesis verification tools at higher levels of abstraction is beyond decidability.

- As regards the *maintenance of the tool with respect to small changes* are concerned, pre-synthesis verification entails a complete reverification in general. The changes in the formal synthesis tools will be minimal and post-synthesis verification needs none, since changes in the synthesis process does not affect it.

- The *logical skills* required in developing such tools are software verification skills for pre-synthesis approach, and hardware verification skills for formal synthesis and post-synthesis approaches.

2.3 Summary

Summarizing the discussion above, pre-synthesis verification is the most acceptable one for circuit designers and least acceptable for tool developers. The situation is the opposite for post-synthesis verification tools. Formal synthesis is a good compromise between these two extremes (Fig. 2).

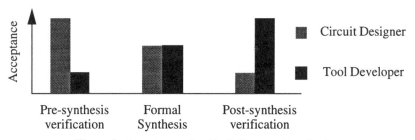

Figure 2. Acceptance of methods for correct synthesis

3 Classification of Formal Synthesis

Formal synthesis can be generally classified into two methods — synthesis step specific verification techniques and transformational derivation (Fig. 3). The former is really a specialized verification technique that has been tuned towards a specific synthesis step. The latter corresponds to a forward derivation of the implementation from a specification within some calculus.

Elaborating further, *transformational derivation* means that the specification will be transformed into an implementation according to a given *core* of elementary transformation rules. This can be further divided into two — transform the specification based on a *hardware-specific calculus* or on a *general purpose calculus*. In a transformational derivation based on hardware-specific calculus, the *core* transformations are circuit transformations whereas in a general purpose calculus they are logical transformations.

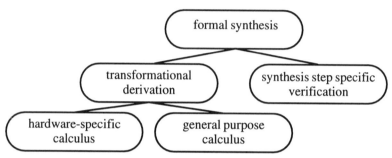

Figure 3. Formal synthesis and related approaches

3.1 Hardware-specific Calculus vs. General-purpose Calculus

In comparing these two approaches for transformational derivation, there are two main criteria — efficiency and security. The efficiency of synthesis is directly determined by the efficiency of circuit transformations, whereas its security is determined by the number and sizes of the core transformations.

The size of the core transformations in the hardware-specific calculus approach is large since the core transformations are the circuit transformations themselves. In contrast, the general purpose calculus approach has a small core of logical transformations. Hence the implementational efforts are high for hardware-specific calculus and low for a general purpose calculus. Additionally, the core transformations are strongly dependent on the application domain and/or the levels of abstraction handled when using a hardware-specific calculus, whereas the core is static for a general purpose calculus. For example, the core transformations for a high-level hardware-specific calculus is completely different from those required for logic synthesis.

Before proceeding further we shall shortly ruminate on the statements in the previous paragraph. It should be noted, that formal derivations within some calculus can only be sound if the transformations in the calculus — a piece of software — are "correct". Verifying the correctness of the calculus itself means the comparison of the derivation within the calculus with the semantics of its logic. This argumentation itself is to be performed within some "meta logic". This exercise is one way to increase the reliability of the calculus. An objective correctness criteria cannot be achieved due to the fact, that the meta logical argumentation itself may be error prone. Therefore, the correctness of the calculus is considered to be more of a theoretical activity which has no practical impact on its usage. Nevertheless, there is at least one major requirement — the calculus should be consistent. It should, for example, be guaranteed, that it is not possible to derive a proof for the equivalence of two circuits as well as the proof for their nonequivalence.

For small sized calculi, such errors may easily be figured out, but for calculi with a big set of complex transformations guaranteeing safety becomes a tedious goal. Due to the smaller core, general purpose calculi can be considered to be more reliable. Furthermore, logical (multi-purpose) calculi are based on some well understood mathematical theories, which also increases their reliability as compared to hardware specific calculi. Usually, multi-purpose calculi are also better tested since they have been applied to different domains.

The major drawback of general purpose calculi is their efficiency. For every hardware transformation step, a series of basic mathematical transformations has to be performed. Figuring out the right series of basic mathematical transformations for implementing some specific synthesis step may be demanding. Hardware specific calculi are much better adapted to the synthesis domain. Hardware transformations are directly implemented, which also results in better runtimes.

3.2 Transformational Derivation vs. Synthesis-Step Specific Verification

Besides transformational derivation, one can also perform *synthesis step specific verification* to guarantee correct synthesis. The difference to post-synthesis verification (see section 2) is that the knowledge about *which* synthesis step has been performed, is exploited during verification. However, in contrast to formal synthesis, the knowledge about *how* the synthesis was performed is unknown. Such verification techniques are optimized with respect to a specific synthesis step thus avoiding some disadvantages of post-synthesis verification. Generally, synthesis step specific verification is independent of the heuristic that is applied during a synthesis step. According to its nature, i.e. (a verification technique) only specification and implementation are considered together with the knowledge about the performed synthesis step. A very important point is that synthesis step specific verification is not applicable to all synthesis steps, which can be easily seen for e.g. state encoding step, where the knowledge that the states were encoded cannot prevent one from having to guess the code. Due to this fact, very few research activities have been started in this verification area ([EiJe96], [HuCC96]).

4 Examples

Based on the classification described in section 3, we shall give a brief overview of some systems that have been developed for formal synthesis. We shall first start with an example of pre-synthesis verification – although it does not fall into the category of systems based on formal synthesis, we deem that it deserves a mention here, due to its uniqueness. Later we describe three systems based on HW-specific calculus and general purpose calculus, respectively. The list of references have been organized into general references and specific references for the described systems. Additionally, a set of miscellaneous references have been listed for the two kinds of formal synthesis approaches. However, these have not been referenced in the text and have been included for the sake of completeness only.

4.1 Proven Boolean Simplifier – PBS

The proven boolean simplifier (PBS) is a verified logic synthesis system developed at the Cornell University [AaLe91, AaLe94, AaLe95]. PBS is a part of the high-level synthesis system called BEDROC, which translates a behavioural description given in the HDL - HardwarePal into an implementation suitable for FPGAs [LeAL91, LCAL+93]. Although some abstract formalization have been performed for the scheduling problem within BEDROC, PBS is the only piece of code that has been totally verified.

PBS has been implemented in a purely functional subset of Standard ML and has been verified using the Nuprl proof assistant [Lees92]. PBS consists of about 1000 lines of SML code and implements the weak-division algorithm for logic minimization [BrMc82]. Weak-division performs logic minimization by removing redundant logic. This is achieved by replacing common sub-expressions among the divisors of functions by new intermediate values, thus realizing them only once within the whole circuit.

In proving the correctness of PBS, the subset of SML used in PBS was embedded in Nuprl and the implementation was then interactively verified by proving over 500 theorems. These theorems involve proofs ranging from the correctness of simple functions, such as membership of an element in a list, up to complex theorems that the input combinational function and the minimized function are equivalent. Theorems were also proven regarding the minimality of the weak-division algorithm. The overall effort required was about 8 man months.

Remarks:
Since PBS falls into the category of pre-synthesis verification, we do not compare it with the other approaches as described in section 5. Hence we summarize some important characteristics before moving on to the other examples.

- Since PBS has been verified, this piece of software can be used with an increased degree of confidence.
- PBS is restricted to the use of weak-division algorithm for logic minimization.
- PBS was conceived with the thought of achieving verified software.
- Verification of PBS was eased due to the use of a functional subset of SML for implementation.

To conclude – although PBS is a first step towards achieving correct synthesis, verifying large synthesis programs written in imperative languages is practically impossible.

4.2 RLEXT – Register Level EXploration Tool

RLEXT has been developed at the University of Illinois for the formal synthesis of register-transfer level datapaths [KnWi89, KnWi92]. This tool falls into the category of hardware-specific calculus for formal synthesis.

The hardware specificity arises from the fact that the design, at the register-transfer level, is looked at from different points of view (dataflow, timing and structure). A large set of types and subtypes defines the objects that make up a design and plausibility constraints have been formulated for defining the well-formedness of a design. The overall design is represented by a 4-tuple – dataflow, timing, structure and binding. The dataflow model captures the behaviour of the design and is conceptually viewed as a tri-partite graph which represents the connection of values to parameters and parameters to operations. The timing model gives the abstract states and the state transitions of the design. The structural model gives the schematic of the design. The binding defines the relationships between the behaviour, timing and structure. For each model a set of plausibility constraints is given to determine the well-formedness of the model. Given that the three models (dataflow, timing and structure) are well-formed, another set of constraints is given for the binding relationship to ensure correct relationships between these models.

The entire program which implements these types and plausibility constraints has been implemented in about 20,000 lines of Lisp code. The major drawback of this approach is the large set of axiomatized plausibility constraints. A positive feature of this approach is that RLEXT automatically fixes small bugs in the design.

4.3 RUBY

Ruby is a language based on relations and functions for specifying VLSI circuits. The language and the synthesis tools around Ruby have been developed at the Programming Research Group - Oxford, University of Glasgow and Technical University of Denmark, Lyngby [Josh90, JoSh91a, JoSh91b, ShRa93, ShRa95]. Although Ruby is only a language, we choose to call the formal synthesis approach as Ruby too. This approach is based on the paradigm of hardware-specific calculus for formal synthesis.

The specification is given as a relation between the terminals of the circuit. Simple relations can be combined into more complex ones by a variety of higher-order functions. The typical derivation of an implementation in Ruby involves the use of term-rewriting using equivalences or conditional equivalences. Since constructs in Ruby have a a graphical interpretation (figure 4), abstract floorplans can also be generated.

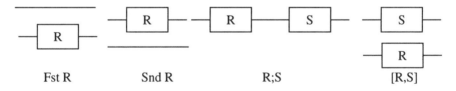

| Fst R | Snd R | R;S | [R,S] |

Figure 4. Graphical interpretation of a few constructs in Ruby

"Fst R", "Snd R" and correspond to the circuit diagrams in figure 4. "R;S" corresponds to a serial connection and [R,S] to a parallel one. Since all these constructs also have a formal basis, it is possible to derive equations such as "Fst R; Snd S = [R,S] = Snd S; Fst R".

Starting from specifications, it is possible to derive equivalent circuits using the transformations that are available in the system. Examples for some transformations are: parallel compositions, retiming, and conversions of regular structures into other forms, e.g. rows to columns. Additionally, it is possible to derive bit-serial and/or parallel circuits from specifications. During such conversions, additional constraints are generated which have to be satisfied.

4.4 DDD — Digital Design Derivation

The DDD system has been developed at the University of Indiana [John84, JoBB88, JoBo91, BoJo93]. The system is based on functional algebra and falls into the category of hardware-specific calculus for formal synthesis.

Starting from a specification in a lisp-like syntax (a derivative of the functional language *Scheme*), a series of transformations are applied to refine them into an implementation which is also represented in a dialect of the same language. DDD can be thought of as an S-expression editor for hardware purposes, with the objective of finding an implementation that satisfies certain constraints. The behaviour is specified as an iterative system using tail recursive functions. This is translated into a sequential description which can be regarded as a network of simultaneous signal definitions comprising of variables, constants, delays and expressions involving operations. Then a concept called *factorization* is used to abstract away multiple instances of common subexpressions which are allocated to data path operators such as ALUs, adders, etc. Complex factorizations ranging from the abstraction of a register and an incrementer into a counter up to the introduction of co-processors are also possible. Failures in factorizations lead to refinements in control, which are accomplished by serializing transformations on the behavioural descriptions. Although all these transformations are based on functional algebra, none of them have been formally verified.

4.5 LAMBDA/DIALOG

LAMBDA/DIALOG has been developed at Abstract Hardware Limited, Uxbridge [FFFH89, FoMa89, FiFM91, MaFo91, HFFM93]. This is the only commercial tool for performing formal synthesis and belongs to the category of general purpose calculus. This tool has also been used in an industrial environment to perform synthesis, as reported in [BoCZ93, BBCC+95].

The core of LAMBDA/DIALOG is its theorem prover based on higher order logic. The theorem prover is called LAMBDA and the GUI (Graphical User Interface), which is a schematic editor, is called DIALOG. The design state is always represented as a formal logical rule. Starting from the trivial rule — *"the overall specification fulfils the overall specification"*, transformations are applied which lead to the rule defining the intermediate stage of design — *"some components and a modified, reduced specification fulfils the overall specification"*. The final design corresponds to the rule — *"some components and some timing conditions fulfil the overall specification"*. The entire process is facilitated by the use of the GUI - DIALOG. Given a specification in higher-order logic, the input and output ports appear on the schematic window. The user can then either interactively apply basic instantiations of modules in the DIALOG library, or apply compound tactics for automatically performing several such steps. For example, each **if** statement in the specification can be individually refined into some kind of a multiplexer, or all of them can be removed in one go by applying a tactic. During this process, the specification is simplified into a set of subgoals which have to be realized and the design continues until all subgoals have been eliminated. A drawback of this tool is that under certain complex situations, direct interactions at the theorem prover level may be necessary in order to continue with synthesis.

4.6 VERITAS

VERITAS is a theorem prover based on an extension of typed higher order logic and has been developed at the University of Kent [HaLD89, HaDa92]. Due to its core, it is a member of the general purpose calculus category.

In VERITAS, synthesis is a recursive interactive design process. Starting from the specification, transformations, called techniques, are interactively applied for converting the specification into a implementation. A technique consists of a pair of functions called the subgoaling function and the validation function. The subgoaling function generates a (possibly empty) set of design subgoals, theorem subgoals and term subgoals. Each of these subgoals must be discharged during the design process. The validation function, which is a derived inference rule in VERITAS logic, constructs the theorem that the implementation satisfies the specification. There are general purpose techniques for splitting conjunctions, rewriting, stripping an existential quantifier, strengthening a specification, etc. An oft used technique is called *claim* — one can claim that an arbitrary specification is an implementation of the original specification. This leads to two subgoals; the first is to implement the new specification (design subgoal) and the second is to prove that the new specification satisfies the original specification (theorem subgoal).

4.7 HASH (Higher order logic Applied to Synthesis of Hardware)

HASH is under development at the University of Karlsruhe [EiKu95a,EiKu95b, EiBK96] and is based on the theorem prover HOL [GoMe93]. This tool falls into the category of general purpose calculus for formal synthesis.

HASH is a toolbox for implementing formal synthesis programs comprising of circuit transformations that correspond to those of conventional synthesis programs. For example, *scheduling* is a transformation in HASH. This transformation gets the dataflow graph and the schedule (calculated outside the logic using ASAP, ALAP, Force-directed, etc.) and transforms the original dataflow graph into a scheduled dataflow graph [EiKu95b, EiBK96]. In a similar manner there are various transformations provided for each synthesis step at the algorithmic and the RT-level.

The main concept of HASH is to split the synthesis process along two dimensions — *design space exploration* and *design transformation*. The design space exploration aspect determines the quality of the synthesized design with respect to the constraints that have been specified, while the design transformation determines the correctness of the design. Due to this split, the heuristics for exploring the design space can be performed outside the logic and the results imported into the design transformation. This implies that for each synthesis step, e.g. scheduling, state minimization, retiming, etc., there exists a unique transformation which performs that synthesis step. It is to be noted here, that these transformations are specific to a synthesis step but independent of the heuristics. This basic concept is shown in figure 5.

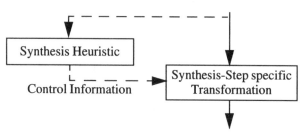

Figure 5. The basic concept of HASH

In this way, the two most important points of circuit design are met independently: the quality and the correctness of the implementation. The quality only depends on the synthesis heuristics involved. The implementations derived within HASH are the same as those achieved with conventional tools. On the other hand, correctness is guaranteed since the circuit transformations are performed inside HOL. From circuit designer's point of view, however, there is no difference between formal synthesis tools based on HASH and conventional synthesis tools. Since all the logical argumentation is performed fully automatically and above all no formal skills are demanded from the user.

5 Comparison

Having presented a brief summary of some systems in the area of formal synthesis, we shall now compare them. We classify the characteristics for comparison into three main criteria — Hardware Representation, Hardware Transformation and Synthesis Control.

5.1 Hardware Representation

This criterion relates to the representation of the specification, the intermediate stages of design and the final implementation. In the systems described in section 4, RLEXT, RUBY and DDD fall into the class of hardware-specific calculus and LAMBDA, VERITAS and HASH are based on general purpose calculus. Additionally, RLEXT is restricted only to RT-level datapaths. The other systems are capable of handling designs at all levels of abstraction, namely from the system level down to the gate level. Due to the strong reliance on the concept of iterative structures, RUBY is mostly suited towards signal processing and module generation applications. All other systems do not have any particular bias towards their application domains. Except RLEXT, all other systems are capable of exploiting hierarchy during the design process.

5.2 Hardware Transformation

This measure compares the conversions used in going from specifications to implementations. Using a theorem prover as a basis has two basic advantages namely, increased reliability and the small number of core transformations. The systems RLEXT, RUBY and DDD are not based on theorem provers. However, due to the use of an algebraic foundation, RUBY and DDD do not have a large number of basic transformations. Before we elaborate this statement, we shall look at the realization of the transformations.

Hardware transformations can be axiomatized or derived from some small set of basic transformations. In the RLEXT system, all the transformations are axiomatized. In RUBY and DDD, some transformations are axiomatized and some of them are derived from these axiomatized transformations based on general algebraic rules. Hence the statement, that these systems do not have a large number of basic transformations. In LAMBDA, VERITAS and HASH all the hardware transformations are derived from the logical transformations provided within the respective theorem prover and hence the size of the core (basic transformations) is small. The user's of the hardware transformations are circuit designers. Hence, we can compare the systems based on the nature of transformations, i.e. are they hardware oriented or logic oriented. RLEXT, LAMBDA and HASH offer transformations which are similar to those used in the conventional synthesis process. The transformations in DDD and VERITAS are more from the logician's point of view. Due to the floorplanning aspect of RUBY, it partially offers a circuit designer's perspective.

5.3 Synthesis Control

This yardstick relates to the use and the practicality of the systems. RLEXT, LAMBDA and HASH provide automation for the circuit designer. Except for RLEXT and HASH, the use of all other systems requires a good knowledge in mathematical notation. Although LAMBDA offers a GUI, there are situations where one has to interact with its theorem prover core. A unique feature of the HASH system is the use of existing heuristics in the synthesis domain and a clear split between the design space exploration task and the actual application of a design transformation.

6 Conclusions

The acceptance of formal methods within the circuit designer community is inversely proportional to the amount of formal knowledge required! The circuit designers would ideally love to have verified implementations for free. Fully automated post-synthesis verification techniques are therefore used wherever possible. However, they are too weak and too time consuming for real sized circuits. For the circuit designer, verified synthesis programs (pre-synthesis verification) would be the best. However, the gap between what can be verified using software verification techniques and the complexity of existing synthesis tools, cannot be bridged today.

In this paper we have presented a summary and classification of formal synthesis techniques. Formal synthesis is constructive in the sense that the proof is derived along with the synthesis process rather than guessed afterwards. Therefore also very expressive logics can be used for formal synthesis. In contrast to post-synthesis verification, the undecidability of a logic is not a handicap. If good tools based on formal synthesis can be developed, hardware verification can be restricted to purely the verification of safety and liveness properties.

Acknowledgements

The authors are grateful to the anonymous referees whose constructive comments have improved the quality of the paper.

References

General

[BrMc82] R.K. Brayton, C. McMullen, "Decomposition and factorization of boolean expressions", Proc. International Symposium on Circuits and Systems, IEEE Computer Society, 1982

[Evek87] H. Eveking, "Verification, Synthesis and Correctness-Preserving Transformations — Cooperative Approaches to Correct Hardware Design", From HDL Descriptions to Guaranteed Correct Circuit Designs, D. Borrione (Ed.), Elsevier Science Publishers, North-Holland, 1987

[GoMe93] M. Gordon, T. Melham, "Introduction to HOL: A Theorem Proving Environment for Higher Order Logic", Cambridge University Press, 1993

[Gupt92] A. Gupta, "Formal Hardware Verification Methods: A Survey", Formal Methods in System Design, Kluwer Academic Publishers, Vol. 1, 1992, pp.151-238

[Lees92] M. Leeser, "Using Nuprl for the verification and synthesis of hardware", Phil. Trans. R. Soc. Lond., Vol. 339, 1992, pp. 49-68

[LeAL91] M. Leeser, M. Aagaard, M. Linderman, "The BEDROC High Level Synthesis System", Proc. ASIC'91, IEEE, 1991

[LCAL+93]M. Leeser, R. Chapman, M. Aagaard, M. Linderman, S. Meier, "High level Synthesis and Generating FPGAs with the BEDROC System", Journal of VLSI Signal Processing, Kluwer Academic Publishers, Vol. 6, No.2, Aug. 1993, pp.191-214

[McPC90] M. C. McFarland, A.C. Parker, R. Camposano, "The high-level Synthesis of Digital Systems", Proc. of IEEE, Vol. 78, 1990, pp. 301-318

Pre-Synthesis Verification

[AaLe91] M. Aagaard, M. Leeser, "A Formally verified system for logic synthesis", Proc. ICCD-91, IEEE, Oct. 1991

[AaLe94] M. Aagaard, M. Leeser, "PBS: Proven Boolean Simplification", IEEE Trans. on Computer Aided Design, Vol. 13, No.4, Apr. 1994

[AaLe95] M. Aagaard, M. Leeser, "Verifying a logic synthesis algorithm and implementation: A case study in software verification", IEEE Trans. on Software Engineering, Oct. 1995

Synthesis Step Specific Verification

[EiJe96] C.A.J. van Eijk, J.A.G. Jess, "Exploiting Functional Dependencies in Finite State Machine Verification", Proc. European Design & Test Conference 1996, Paris, pp. 9-14, IEEE Computer Society Press

[HuCC96] Huang, Cheng, Chen, "On Verifying the Correctness of Retimed Circuits", Great Lakes Symposium on VLSI 1996, Ames, USA

hardware-specific calculus

DDD

[John84] S.D. Johnson, "Synthesis of Digital Designs from Recursion Equations", MIT Press, Cambridge, 1984

[JoBB88] S.D. Johnson, B. Bose, C.D. Boyer, "A Tactical Framework for Digital Design", VLSI Specification, Verification and Synthesis, G. Birtwistle and P. Subrahmanyam, Eds., Kluwer Academic Publishers, Boston, 1988, pp.349-383

[John89] S.D. Johnson, "Manipulation logical organization with system factorizations", Hardware Specification, Verification and Synthesis: Mathematical Aspects, Leeser and Brown (Eds.), LNCS, Vol. 408, Springer, 1989

[JoWB89] S.D. Johnson, R. Wehrmeister, B. Bose, "On the Interplay of Synthesis and Verification", Proc. IMEC-IFIP Intl. Workshop on Applied Formal Methods in Correct VLSI Design, L. Claesen (Ed.), North Holland, 1989, pp. 385-404

[JoBo91] S.D. Johnson, B. Bose, "DDD — A System for Mechanized Digital Design Derivation", ACM/SIGDA Workshop on Formal Methods in VLSI Design, Miami, Florida, Jan. 1991. (Unfortunately, the proceedings of the workshop have not been officially published)

[BoJo93]   B. Bose, S.D. Johnson, "DDD-FM9001: Derivation of a verified microprocessor. An exercise in integrating verification with formal derivation", Proc. IFIP Conf. on Correct Hardware Design and Verification Methods, LNCS, Vol. , Springer, 1993

[RaBJ93]   K.Rath, B. Bose, S.D. Johnson, "Derivation of a DRAM Memory Interface by Sequential Decomposition, Proc. Intl. Conf. on Computer Design - ICCD, IEEE, Oct. 1993, pp.438-441

[RaTJ93]   K. Rath, M.E. Tuna, S.D. Johnson, "An Introduction to Behaviour Tables", Tech. Rep. No. 392, Indiana University, Computer Science Department, December, 1993

[ZhJo93]   Z. Zhu, S.D. Johnson, "An Example of Interactive Hardware Transformation", Tech. Rep. No. 383, Indiana University, Computer Science Department, May, 1993

RLEXT

[KnWi89]   D.W. Knapp, M. Winslett, "A Formalization of Correctness for Linked Representations of Datapath Hardware", Proc. IMEC-IFIP Intl. Workshop on Applied Formal Methods in Correct VLSI Design, L. Claesen (Ed.), North Holland, 1989, pp. 1-20

[KnWi92]   D.W. Knapp, M. Winslett, "A Prescriptive Model for Data-path Hardware", IEEE Trans. on Computer Aided Design, Vol. 11, No.2, Feb. 1992

RUBY

[Shee88]   M. Sheeran, "Retiming and slowdown in Ruby", The fusion of hardware design and verification, G.J. Milne (Ed.), North-Holland, 1988, pp. 245-259

[JoSh90]   G. Jones, M. Sheeran, "Circuit design in Ruby", Formal Methods for VLSI design, J. Staunstrup (Ed.), North-Holland, 1990, pp.13-70

[Ross90a] L. Rossen, "Formal Ruby", Formal Methods for VLSI design, J. Staunstrup (Ed.), North-Holland, 1990, pp. 179-190

[Ross90b] L. Rossen, "Ruby Algebra", Designing Correct Circuits, Oxford, G. Jones and M. Sheeran (Eds.), Springer, 1990, pp. 297-312

[JoSh91a] G. Jones, M. Sheeran, "Deriving bit-serial circuits in Ruby", Proc. VLSI-91, Edinburgh, A. Halaas and P.B. Denyer (Eds.), Aug. 1991

[JoSh91b] G. Jones, M. Sheeran, "Relations and refinements in circuit design", proc. 3rd Refinement Workshop, C.C. Morgan and J.C.P. Woodcock (Eds.), Springer, 1991, pp. 133-152

[ShRa93]   R. Sharp, O. Rasmussen, "Transformational Rewriting with Ruby", Proc. CHDL-93, D. Agnew, L. Claesen and R. Camposano (Eds.), Elsevier Science Publishers, 1993, pp.243-260

[ShRa95]   R. Sharp, O. Rasmussen, "The T-Ruby Design System", Computer Hardware Description Languages and their Applications (CHDL'95), pp. 587-596, 1995

MISCELLANEOUS

[KrSK95]   T. Kropf, K. Schneider, R. Kumar, "A Formal Framework for High Level Synthesis", in TPCD-94, Bad Herrenalb, Germany, R. Kumar and T. Kropf (Eds.), LNCS, Springer, Vol. 901, 1995, pp. 223-238

[GoSS87]   G.C. Gopalakrishnan, M.K. Srivas, D.R. Smith, "From Algebraic Specifications to Correct VLSI Circuits", From HDL Descriptions to guaranteed Correct Design, D. Borrione (Ed.), Elsevier Science Publishers, North-Holland, 1987

[GrRa87]   W.Grass, R. Rauscher, "CAMILOD - A Program System for Designing Digital hardware with Proven Correctness", From HDL Descriptions to guaranteed Correct Design, D. Borrione (Ed.), Elsevier Science Publishers, North-Holland, 1987

[Ueha87]   T. Uehara, "Proofs and Synthesis are Cooperative Approaches for Correct Circuit Designs", From HDL Descriptions to guaranteed Correct Design, D. Borrione (Ed.), Elsevier Science Publishers, North-Holland, 1987

[Camp89]   R. Camposano, "Behaviour-preserving Transformations for High-level Synthesis", Proc. Workshop on hardware Specification, Verification and Synthesis: Mathematical Aspects, Leeser and Brown (Eds.), LNCS, Vol. 408, Springer, 1989

[DrHa89]   D. Druinsky, D. Harel, "Using State-Charts for hardware Description and Synthesis, IEEE Trans. on CAD, Vol. 8, No.7, July, 1989, pp.798-807

[GrMT94] W. Grass, M. Mutz, W.D. Tiedemann, "High-level-Synthesis Based on Formal Methods", Proc. EUROMICRO '94, Liverpool, 1994, 83-91

General Purpose calculus

LAMBDA

[FFFH89] S. Finn, M.P. Fourman, M. Francis, R. Harris, "Formal System Design — Interactive Synthesis based on computer-assisted formal reasoning", Proc. IMEC-IFIP Intl. Workshop on Applied Formal Methods in Correct VLSI Design, L. Claesen (Ed.), North Holland, 1989, pp.97-110

[FoMa89] M.P. Fourman, E.M. Mayger, "Formally based systems design — Interactive hardware scheduling", VLSI-89, G. Musgrave and U. Lauther (Eds.), Munich, North Holland, 1989

[FiFM91] S. Finn, M.P. Fourman, G. Musgrave, "Interactive Synthesis in higher order logic", Proc. Intl. Workshop on the HOL Theorem Prover and its applications, Davis, Calif., IEEE Press, 1991

[MaFo91] E.M. Mayger, M.P. Fourman, "Integration of Formal Methods with System Design", Proc. VLSI-91, Edinburgh, A. Halaas and P.B. Denyer (Eds.), Aug. 1991

[BoCZ93] M. Bombana, P. Cavalloro, G. Zaza, "Specification and Formal Synthesis of Digital Circuits", Proc. Higher Order logic Theorem Proving and its Applications, L.J.M. Claesen and M.J.C. Gordon (Eds.), Elsevier Publishers, Vol. A-20, 1993, pp.475-484

[HFFM93] R.B. Hughes, M.D. Francis, S.P. Finn, G. Musgrave, "Formal Tools for Tri-State Design in Busses", Proc. Higher Order logic Theorem Proving and its Applications, L.J.M. Claesen and M.J.C. Gordon (Eds.), Elsevier Publishers, Vol. A-20, 1993, pp.459-474

[BBCC+95] G. Bezzi, M. Bombana, P. Cavalloro, S. Conigliaro, G. Zaza, "Quantitative evaluation of formal based synthesis in ASIC Design", Proc. TPCD'94, R. Kumar and T. Kropf (Eds.), Bad Herrenalb, Germany, LNCS, Vol. 901, Springer, 1995, pp. 286-291

VERITAS

[HaLD89] F.K. Hanna, M. Longley, N. Daeche, "Formal Synthesis of Digital Systems", Proc. IMEC-IFIP Intl. Workshop on Applied Formal Methods in Correct VLSI Design, L. Claesen (Ed.), North Holland, 1989, pp.532-548

[HaDa92] F.K. Hanna, N. Daeche, "Dependent types and formal synthesis", Phil. Trans. R. Soc. Lond., Vol. 339, 1992, pp. 121-135

HASH

[EiKu95a] D. Eisenbiegler, R. Kumar "An Automata Theory Dedicated towards Formal Circuit Synthesis", Higher Order Logic Theorem Proving and Its Applications HUG'95, Aspen Grove, Utah, USA, September 11-14, 1995

[EiKu95b] D. Eisenbiegler, R. Kumar, "Formally embedding existing high level synthesis algorithms", Proc. CHARME'95, Springer, LNCS, Vol. 987, pp. 71-83

[EiBK96] D. Eisenbiegler, C. Blumenröhr, R. Kumar, "Implementation Issues about the Embedding of Existing High Level Synthesis Algorithms in HOL", International Conference on Theorem Proving in Higher Order Logics, TPHOL'96, Turku, Finland, August 27 - 30, 1996

MISCELLANEOUS

[BaBL89] D.A. Basin, G.M. Brown, M. Leeser, "Formally Verified Synthesis of Combinational CMOS Circuits", "Formal Synthesis of Digital Systems", Proc. IMEC-IFIP Intl. Workshop on Applied Formal Methods in Correct VLSI Design, L. Claesen (Ed.), North Holland, 1989, pp. 251-260

[Basi91] D. Basin, "Extracting circuits from Constructive Proofs", ACM/SIGDA Workshop on Formal Methods in VLSI Design, Miami, Florida, Jan. 1991. (Unfortunately, the proceedings of the workshop have not been officially published)

[Busc92] H. Busch, "Transformational Design in a Theorem Prover", Theorem Provers in Circuit Design, V. Stravidou, T.F. Melham and R.T. Boute (Eds.), Elsevier Science Publishers, North-Holland, 1992, pp. 175-196

[Lars94] M. Larsson, "An Engineering Approach to Formal System Design", in Higher Order Logic Theorem Proving and Its Applications, T.F. Melham and J. Camilleri (Eds.), pp. 300-315, Valetta, Malta, September 1994, Springer

[Lars95] M. Larsson, "An Engineering Approach to Formal Digital System Design", The Computer Journal, Oxford University Press, Vol. 38, No.2, 1995, pp. 101-110

[SGGH92] J.B. Saxe, S.J. Garland, J.V. Guttag, J.J. Horning, "Using Transformations and Verification in Circuit Design", Designing Correct Circuits, J. Staunstrup and R. Sharp (Eds.), Elsevier Science Publishers, North-Holland, 1992

[Wang93] L. Wang, "Deriving a Correct Computer", Proc. Higher Order logic Theorem Proving and its Applications, L.J.M. Claesen and M.J.C. Gordon (Eds.), Elsevier Publishers, Vol. A-20, 1992, pp. 449-458

Formal Specification and Verification of VHDL

Mark Bickford and Damir Jamsek

Odyssey Research Associates

Abstract. We give an overview of our system for the verification of VHDL designs[1], and discuss its rationale. We present a complete example of a simple processor and discuss general methods for specification of state machines and timed and untimed combinational circuits.

1   Introduction

We have developed a system for formal specification and verification of hardware designs written in VHDL. The user writes specifications in the *Larch* [3] style with purely mathematical definitions written in the *Larch Shared Language* (LSL) and the connection between the mathematics and the VHDL entities specified by means of special *interface specifications*. From the specification of an entity in this Larch/VHDL interface language and the architecture of the entity, written in VHDL, the system generates *verification conditions* (VC's) in the form of a Larch theory (called a *trait*). The VC-trait can then be loaded into our (interactive) Larch theorem prover (the Penelope theorem prover, the standalone theorem proving component of the Penelope Ada verification tool [2]) and if all the VC's are proved then the given architecture has been verified w.r.t. the formal specification.

In this paper we describe our methods for formal specification of hardware designs. We also compare our axiomatic formalization of VHDL semantics with alternate formalizations that are defined operationally. We also work through the specification and verification of an example design that shows how we specify and verify "processor-like" circuits consisting of both combinational and state-holding devices.

1.1   Why do HDL Verification?

We begin by addressing the questions, "Why should a verification system be based on a particular hardware description language? Shouldn't we be verifying designs at a purely mathematical, abstract level?".

Our response is that we certainly can and do work at the level of pure mathematics by working in LSL to define and prove properties of the purely mathematical concepts that we need. However, in order to verify a large design, we will want to decompose the design into manageable pieces (even if we don't want to, we will be forced to) that we can specify and verify separately. These pieces

---

[1] This work was supported by Rome Labs contract USAF F30602-94-C-0136.

will need to be combined to form the full design and the way the pieces interact will have to be rigidly defined to enable us to reason about the combined system. To be useful, this decomposition process should be fully hierarchical so that the pieces can be further decomposed. To do this kind of decomposition in a disciplined way we will therefore need all the machinery that is provided by the entities, ports, component instantiations, port maps, etc. provided by a hardware description language like VHDL. To reason about how the state of the design evolves in time, we will want to reason about when individual pieces of data change and how long they have been stable, so we need the equivalents of the signals, events, stable, delayed, etc. provided by VHDL. Typical mathematical structuring constructs such as parameterized Larch traits or parameterized theories in PVS are not directly applicable to such needs.

In short, since all the machinery of an HDL is needed to verify a large design, it makes sense to use an existing well designed language like VHDL to organize the verification effort rather than reinvent such machinery inside a mathematical language.

Of course, we can use verified designs written in VHDL to to communicate the design to other tools for simulation, synthesis, and layout. This is a beneficial feature of an HDL based verification system but it is not the main reason for choosing this approach. Our main goal is specification and verification of designs independent of how they are described, but we claim that the structure provided by the HDL is essential to achieving this goal, so we might as well make use of the hardware description language for these other tasks as well.

1.2 Axiomatic vs. Operational Semantics

Having decided to use a VHDL based verification system we must create a mathematical semantics for VHDL. Since the definition of VHDL as described in the *Language Reference Manual* (LRM) is given in terms of the behavior of a VHDL simulator, a natural way to provide a formal semantics for VHDL is to write a formal description of this simulator. This can be done in many ways (for some examples consult the book [4]) but, for the purpose of this discussion, let us take an abstract (and simplified) view of this formal description of the simulator.

A VHDL design ultimately *elaborates* to a set of *processes*, $P = \{P_1, P_2, \ldots\}$, and a set of *signals*, $S = \{S_1, S_2, \ldots\}$ (with their *drivers*). Each process P_i and signal S_i can be in various states, and the *state* of the simulation is determined by the states of P and S. Given the result of the elaboration, $E = (P, S)$, the simulation kernel K determines the initial state $K(E) = (P_0, S_0)$ and a *next state function*, $(P, S) \mapsto K(P, S)$ (it is a function since VHDL is deterministic). Now we can define the *run*,

$$Run(E) = \langle K(E), K(K(E)), \ldots, K^n(E), \ldots \rangle = \langle Run(E)_0, \ldots, Run(E)_n, \ldots \rangle$$

Now, how might we use such an operational semantics? If we have a design, D, that contains signals x and y, we might want to verify that the values of x and y are always in some specified relation $R(x, y)$ at all states of the simulation.

So we might use the formal translation of entity D to carry out the *formal elaboration* of D to get processes and signals $E_D = (P_D, S_D)$ and x and y would be in S_D. Then we could consider the run $Run(E_D)$ and prove that for all n, in state $Run(E_D)_n$ the relation $R(x, y)$ evaluates to true. Let's write this condition

$$Run(E_D) \models \textbf{always } R(x, y)$$

If D is the entire design, this method would allow us to verify the specification "**always** R(x,y)". Depending on how we represent the processes and signals, this method might allow us to automatically verify such specifications by model checking methods.

But we claim that we must also be able to verify entities that are only pieces of the whole design. In this case, if D is only part of the whole, then to say that D *guarantees* the specification "**always** R(x,y)" we would have to verify that for all elaborations E containing E_D, $Run(E) \models \textbf{always } R(x, y)$. Equivalently, by thinking of E as $E_D \cup E_{ENV}$ the elaboration of D and its environment ENV, we might say D **guarantees always** $R(x, y)$ if

$$(\forall ENV) \; Run(E_D \cup E_{ENV}) \models \textbf{always } R(x, y)$$

Before proceeding with this discussion, let us introduce some notation. In any run, $R(E)$, we have states $K(E), K(K(E)), \ldots$. Let's begin to define a formal specification language L by introducing names for these states. The idea is that the run $R(E)$ will provide a structure (in the model theoretic sense) for the language L. So we introduce a *sort symbol*, *State* and null-ary and unary function symbols,

$$init :\longrightarrow State$$

$$nxt : State \longrightarrow State$$

The run $Run(E)$ provides an interpretation by the recursive definition:

$$init^{Run(E)} = K(E) = Run(E)_0$$

$$nxt(\alpha)^{Run(E)} = K(\alpha^{Run(E)})$$

This only means that the terms $init, nxt(init), nxt(nxt(init)), \ldots$ [2] in the language L are interpreted as the states of the run $Run(E)$. From now on, we will continue to use lower case Greek letters, $\alpha, \beta, \ldots$ to stand for terms in L of sort *State*. For each VHDL type T the language includes a corresponding sort symbol and also a sort for signals of type T that we write as $Signal[T]$. Then if x and y are signals of type T in E, we add constants

$$x, y :\longrightarrow Signal[T]$$

to L. In any state in the run $Run(E)$, we can determine the effective value of the signal x, so we introduce an infix operator:

$$\_@\_ : Signal[T] \longrightarrow T$$

---

[2] We could use the natural numbers $0, 1, \ldots$ but to avoid confusion we prefer not to overload these symbols.

and define its interpretation so that $x@\alpha$ is interpreted as the effective value of signal x in the state named by α.

With this language, our previous definition of $Run(E) \models \textbf{always } R(x, y)$ is equivalent to

$$Run(E) \models (\forall \alpha : State)\ R(x@\alpha, y@\alpha)$$

So we say that the "specification term", $\textbf{always } R(x, y)$ is really the term $(\forall \alpha : State)\ R(x@\alpha, y@\alpha)$ in our language L. In general, if ψ is any Boolean term in our specification language L and D is a VHDL design unit, we say that

$$D \textbf{ guarantees } \psi \Leftrightarrow (\forall ENV)\ Run(E_D \cup E_{ENV}) \models \psi$$

Note that the guarantees relation is monotone in the sense that

$$(D_1 \textbf{ guarantees } \psi_1) \wedge (D_2 \textbf{ guarantees } \psi_2) \rightarrow (D_1 \cup D_2 \textbf{ guarantees } \psi_1 \wedge \psi_2)$$

This means that if we want to show that $D \textbf{ guarantees } \psi$ and we decompose D into components A and B such that $A \textbf{ guarantees } \psi_1$ and $B \textbf{ guarantees } \psi_2$ then it is enough to show that $\psi_1 \wedge \psi_2 \rightarrow \psi$.

It is this kind of *compositional* reasoning that is essential and thus it is the *guarantees relation* between VHDL units and specification terms that is of interest. It can be defined in terms of the operational semantics K of the simulation kernel and an operator E that translates VHDL unit D to its formal elaboration E_D, but the relation can also be defined axiomatically. If we can work directly with the guarantees relation then we do not need the complete operational semantics K, E. The user of the verification tool does not want to reason about operational semantics either, he wants to reason about the properties guaranteed by his design.

The role of the operational semantics should be to prove the necessary properties of the guarantees relation and while it might be desirable to do this with an automated proof checking system using a formal description of K, E, it is not necessary. The properties of the guarantees relation can be proved by standard mathematical analysis of the definition of VHDL in the LRM and then made available to the user of the verification system either as axioms or lemmas in the available theory or by being supplied automatically as hypotheses in the verification conditions.

Now in our experience, based on previous work where we wrote a description of the VHDL simulator in a functional language and proved properties using the Clio theorem prover [1], it is very difficult to prove properties of the guarantees relation directly from the formal definitions of the operational semantics. It is possible, of course, but many of the proofs are "meta-level" arguments where we must prove things by induction on the length of the run quantifying over all possible environments ENV. To do all of this within the formal system requires that a lot of mathematical machinery be developed. None of this machinery would ever be used by the final user of the verification system; only the lemmas proved about the guarantees relation would be used by the final user. For this reason we think it is better to do this kind of analysis outside the formal system of the VHDL verification tool (but perhaps in some other formal proving tool).

Consider, for example, the single concurrent signal assignment statement D : X <= Y; where X is a signal of an unresolved type. This statement elaborates to a single process that updates a driver for X whenever Y has an event. Now one might hope that D **guarantees always** $X = Y$ but because there is a delta-delay between the event on Y and the corresponding event on X and because X and Y need not have the same initial value we can only guarantee

$$D \textbf{ guarantees } (\forall \alpha : State) \ \alpha \neq init \rightarrow \ X@\alpha = Y@prev(\alpha)$$

Even this simple fact would be rather difficult to prove directly from the operational semantics. We need to know that since X is unresolved then in the elaboration of $D \cup ENV$ the only driver for X will be the driver from the elaboration of D, so the value of X is the current value of that driver. We can then prove by induction on α that the driver always contains the previous value of Y and it never gets cancelled.

In a system with an embedded operational semantics and formal elaboration, this simple (or not so simple) fact could be proved as a lemma and automatically instantiated by a proof tactic during the formal elaboration. But given enough such lemmas and tactics the end result of the formal elaboration would be a VC-generator like the one in our verification system. Deriving a VC-generator in this way from an embedded formal operational semantics would give higher assurance that the VC-generator was correct, but it would be a difficult task and the resulting "formal VC-generator" would probably have poor performance. We think that it is more efficient to create the VC-generator directly.[3] We give an overview of our system in the next section.

2 Verification System Overview

Our VHDL verification system consists of a collection of tools integrated under the control of a graphic user interface (gui). The canvas of the gui depicts objects and relations. The objects are such things as libraries, VHDL packages, entities and architectures, Larch theories, specifications of entities, process annotations, and proofs of VC's. For each kind of object there is a menu of operations that can be invoked.

In a typical sequence of interactions, the user creates an entity, writes its VHDL entity specification, and then writes its Larch/VHDL specification in the form of a list of guarantees, ψ_i. He then invokes a command to install the Larch/VHDL specification in the library. At this point the system reports any syntax and sort-checking errors in the specification. The user then creates a VHDL architecture for the entity and invokes the VC-generator. The VC-generator creates a Larch theory that declares constants for all the ports, signals, and generics and asserts the constraints implied by the subtypes of these VHDL

---

[3] We use an attribute-grammar to assign semantic attributes to parsed VHDL and generate VC's as functions of these semantic attributes.

objects. The VC-theory contains an obligation to be proved of the form:

$$\bigwedge \phi_i \rightarrow \bigwedge \psi_i$$

where the ϕ_i are the guarantees of the processes in the architecture body and the ψ_i are the guarantees from the specification of the entity being verified.

The system automatically supplies guarantees, ϕ_i, for simple concurrent statements in the architecture. For component instantiation statements, it instantiates the specification of the entity bound to the component (by substituting the actuals given in the port map for the formals in the entity specification). In the case of VHDL process statements, which have sequential code in their bodies, the guarantee cannot be generated automatically, so for these processes the user must provide the guarantee. A separate VC is then generated to show that the sequential code in the process body implies the user-supplied guarantee.

Once the system has generated the VC, the user creates a proof object and invokes the prover. If all the proof obligations can be proved, then the given architecture satisfies the specification, provided that the components used in the architecture satisfy their respective specifications. This allows us to do a top-down verification. Once all the components are verified the verification task is complete.

The guarantees for simple concurrent statements, which the system generates automatically, do not require any proof from the user. They are instances of facts such as the ones mentioned in the last section that we have proved to be true in general. Most of the statements in an architecture have this form, so this is where our method saves the user a lot of work.

There is usually one automatically generated guarantee for each statement. While proving the VC's the user can easily identify from the signal names which line of the VHDL architecture each hypothesis comes from, so that if he is unable to complete the proof because a particular hypothesis is not strong enough then he knows which line of the VHDL is suspect.

For each VHDL package declaration, the system automatically generates a Larch theory containing mathematical models of the types declared in the package. We currently handle integer and enumerated types and their subtypes, as well as one-dimensional arrays of such types. Constants declared in packages become constants in the theory. Functions declared in the package can be assigned a meaning in the Larch theory by a separate annotation. The system generates a VC to show that the sequential code in each function body implies the function annotation.[4] When a package is used by a VHDL architecture then the corresponding package theory is automatically included in the theory that is available to prove the VC's for the architecture.

3   An Example

We will now develop a complete example that illustrates our specification language, how we specify combinational circuits with and without delays, and how

---

[4] Not implemented yet.

we specify and reason about state-holding devices. Along the way we will discuss our "philosophy" of what makes a good specification.

3.1 Top-level Description

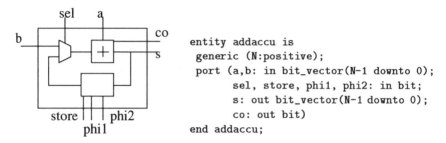

```
entity addaccu is
  generic (N:positive);
  port (a,b: in bit_vector(N-1 downto 0);
        sel, store, phi1, phi2: in bit;
        s: out bit_vector(N-1 downto 0);
        co: out bit)
end addaccu;
```

Fig. 1. Adder/Accumulator Circuit

Figure 1 shows a simple example circuit. The VHDL entity specification shows the types and modes of the ports. The circuit is intended to treat bit vectors **a**, **b**, and **s** as unsigned integers. Input **a** is added to either input **b** or the contents of the internal register, depending on the value of input **sel**. The result of the addition is the output **s** and the carry-out **co**. If input **store** is a '1' then the result **s** of the addition is stored in the register; otherwise the register is unchanged. Inputs **phi1** and **phi2** are two non-overlapping clocks that we use to control the register.

Note that the output **s** may or may not be equal to the internal register. Thus in order to specify the behavior of this device we need to talk about its internal state. The specification of a state machine should specify

1. When we observe the inputs and outputs.
2. What is the next output as a function of the current state and inputs.
3. What is the next state as a function of the current state and inputs.
4. Assumptions about the inputs.

Consider a generic state machine with inputs X,CK, internal state S, and output Y, and suppose that we want to observe the input and output ports at the falling edges of the clock CK. If the relation between the next state and the current input and state is represented by NS, and the relation between the next output and the current input and state is represented by NO, then at consecutive states α and β at which CK is falling, we want the machine to guarantee that $NS(S@\beta, X@\alpha, S@\alpha)$ and $NO(Y@\beta, X@\alpha, S@\alpha)$.

How do we say that α and β are consecutive falling edges of CK? We define, in our mathematical model, a Boolean signal, $falling(CK)$, such that $falling(CK)@\alpha$ is true when CK is falling at state α [5] and we define :

---

[5] That means CK@α = '1' and CK@nxt(α) = '0'

$consecutive(falling(CK), \alpha, \beta) \equiv$
$falling(CK)@\alpha \wedge falling(CK)@\beta \wedge \alpha = last\_state(prev(\beta), falling(CK))$

Here the function $last\_state(\gamma, P)$ is defined to be the last state before or equal to γ at which $P@\gamma$ was true, if there is such a state; otherwise it is *init*, the initial state.

Now we can write the state machine specification as follows:

$(\forall \alpha, \beta : State) \ consecutive(falling(CK), \alpha, \beta) \rightarrow$
$NS(S@\beta, X@\alpha, S@\alpha) \wedge NO(Y@\beta, X@\alpha, S@\alpha)$

This is a common type of specification, so we have added a shorthand for it to our Larch/VHDL interface specification language. The specification above could be written as simply:

```
guarantees always NS(S'next,X,S) and  NO(Y'next,X,S)
           at consecutive falling(CK)
```

In such a specification, the clause "at consecutive P" adds the conditions that α and β are consecutive states at which P is true. In an "always" term a signal X refers to $X@\alpha$ and while X'next refers to $X@\beta$ and the whole term is universally quantified over α, β. If we need to refer to α and β themselves, we use the key words "begin" and "end" (which stand for the beginning and ending of the interval defined by the consecutive $P = falling(CK)$).

For our adder/accumulator circuit, we will observe the ports at the falling edges of clock **phi1**. We will also need to assume that on the interval from α to β clocks **phi1** and **phi2** are really non-overlapping (we will make this precise later). It also seems likely (and is indeed the case) that if any of the inputs **a**, **b**, **sel**, or **store** were to change just at the falling edge of **phi1**, then because there is some delay in the circuit the result that is stored in the register won't be correct. For this example we will not keep track of the actual delays, but even when we make all the internal delays zero, there are still delta-delays, so the circuit won't work unless the the inputs are held stable during all the states that have the same time as the falling edge of clock **phi1**.

How do we specify the assumption of stability? In the VHDL language there is an expression $X'last\_event$ whose value is the time elapsed since X changed. If $X'last\_event > 0$ then X has been stable for some non-zero time and hence has not changed during any of the preceding delta-cycles. Using this formula to specify the stability of X has several advantages. Note that $X'last\_event$ is invariant under changes in the exact number of delta-cycles that have occurred. A reasonable specification should not depend on details that are specific to the VHDL simulation model such as how many delta-cycles have occurred. To add exact timing information to our specification we need only change the assumption $X'last\_event > 0$ to $X'last\_event > del$ where *del* is the actual delay. A major advantage of this stability specification is that reasoning about the stability of signals reduces to reasoning about inequalities, which the theorem prover can handle using a decision procedure.

Figure 2 contains the specification of the entity addaccu. It begins by including two Larch traits, one that defines the conversion from bit vectors to unsigned integers, and another that defines the predicate "nonoverlapping" that defines the notion of nonoverlapping clocks. In order to talk about the internal state we give it a name and a sort, in this case the name **accu** and sort *Vector[Bit]* which is the Larch sort corresponding to the bit_vector type. The effect of this state declaration is to existentially quantify the guarantee with $\exists accu : Signal[Vector[Bit]]$. In the proof that an architecture guarantees the specification we instantiate the existential quantifier with the name of the internal state variable (or a term built from all internal state variables). We specify the *next* value of the internal state as outlined above, but since the adder responds immediately to changes in the inputs, we have to specify the relation between *current* values of the outputs **s** and **co** and the current values of the inputs. Everything else in the specification has already been explained.[6]

To implement the addaccu circuit we need an adder, a register, and a mux. The mux is a single VHDL concurrent statement, but we create components, with their own specifications, for the adder and the register.

```
entity addaccu
includes  (Bit_vector_to_int,NonOverlapping)
state accu:Vector[Bit]
guarantees
    always int(s)+int(co)*2**n
        = int(a)+int(if sel='0' then b else accu)
        and accu'next = (if (store = '1') then s else accu)
    at consecutive falling(phi1)
    whenever a'last_event > 0 and b'last_event > 0
        and store'last_event > 0 and sel'last_event > 0
        and nonoverlapping(phi1,phi2,begin)
        and nonoverlapping(phi1,phi2,end)
end
```

Fig. 2. Adder/Accumulator Specification

3.2 Carry-Lookahead Adder

In order not to bore the reader with another verification of a ripple-carry adder, we decided to use a carry-lookahead adder for this example. The specifications and proofs are hardly more complicated and the example shows how a nice recursive decomposition can be carried out.

---

[6] We haven't explained why the specification uses **begin** and **end**, referring to α and β, as parameters of **nonoverlapping**. For the explanation, see section 3.3.

The left side of figure 3 shows a VHDL entity specification for a carry looka-head adder. Inputs **a** and **b** are N-bit vectors to be added as unsigned integers with carry-in **c**. The output **s** is the resulting N-bit sum, but instead of the usual single bit carry-out there are two carry outputs called **p** and **g** (for "propagate" and "generate"). The idea is that **p** equals '1' if the carry-out will be equal to the carry-in, while **g** equals '1' if the carry-out will be '1' for any value of the carry-in. This means that the carry-out is "**g or (p and c)**".

The Larch/VHDL specification on the right side of figure 3 says that the sum of the integer values of a,b, and c equals the integer value of the sum **s** plus 2^N times the integer value of the carry-out. There is also a clause **with s,g,p delayed by 0 from a,b,c**. To understand what that says, we must digress and discuss the specification of combinational circuits with and without delays.

```
entity carrylookahead is               entity carrylookahead
generic (N: positive);                 includes  (Bit_vector_to_int)
port (c: in bit;                       guarantees always
   a,b: in bit_vector(N-1 downto 0);     int(a)+int(b)+int(c) =
   s: out bit_vector(N-1 downto 0);        int(s)+int(g or (p and c))*2**n
   p,g: out Bit);                       with s,g,p delayed by 0 from a,b,c
end carrylookahead;                    end
```

Fig. 3. Carry-Lookahead Adder (VHDL and Larch/VHDL Spec)

Consider an entity with input X and output Y that is supposed to guarantee **always** $Y = f(X)$. If we implement this in VHDL with all zero-delay signal assignment statements, then we know that whenever X changes, after some number of delta-delays Y will be equal to $f(X)$. The simulator always advances the global clock to the next time that a transaction is scheduled. If that time is the same as the current time then time does not advance and that is a delta-delay. If there are no transactions scheduled for the current time then time will advance by some non-zero amount. In our mathematical model, we represent the state of the global clock by a function $time : State \rightarrow Int$ so that $time(\alpha)$ is the time at state α. The amount the the global clock advances at state α is give by:

$$advance(\alpha) = time(nxt(\alpha)) - time(\alpha)$$

So, when $advance(\alpha) > 0$ then all delta-delays are done and the global clock moves forward, so this might be a good candidate for when to observe the ports of a zero-delay device. We might use the guarantee: $(\forall \alpha : State)\ advance(\alpha) > 0 \rightarrow Y@\alpha = f(X@\alpha)$.

This kind of specification for zero delay devices makes the proofs about them fairly easy. However, when these devices are combined with state-holding devices like the register in our example, this kind of specification has a serious flaw. The specification says that when *all* delta delays are done, then the output is correct. This specification is **not local** in that the value of the condition $advance(\alpha) > 0$

depends on all the signals in the whole design and not just the signals local to the specified entity. When we want to latch the output of this device on the falling edge of clock phi1, there is no way to guarantee that *all* delta delays in *any* environment are done, so we can't guarantee that the output is correct at the state when we need it. The following principle is essential for modularity:

- A specification must be *local*: everything in the specification depends only on the values of signals local to the entity.

To make a specification local, we only need to assume that the input ports (X in this case) are stable, not that *all* signals are stable. But since we aren't observing the output at a state when *all* signals are stable, we should also add to the guarantee the fact that if the inputs are stable then so is the output. In fact we can say more; we can say that the output has been stable for as long as the inputs have been stable. We have already discussed the fact that $X'last\_event > 0$ is a way to say that X is stable. So our new, local, specification for the zero-delay device is the guarantee: $\forall \alpha : State$

$$X'last\_event@(\alpha) > 0 \rightarrow Y@\alpha = f(X@\alpha) \land Y'last\_event@(\alpha) \geq X'last\_event@(\alpha)$$

Since this is our standard form for specifying the assumptions on the stability of the inputs and the guarantee of the stability of the outputs, we have added a shorthand for it to our Larch/VHDL specification language. In the context of an "always term" in which a signal z refers to its value $z@\alpha$ and the term is universally quantified over α, the clause **with z delayed by d from a,b** adds the precondition $min(a'last\_event@\alpha, b'last\_event@\alpha) > d$ and adds the postcondition $z'last\_event@\alpha \geq min(a'last\_event@\alpha, b'last\_event@\alpha) - d$. Note that when the delay d is not zero this clause says that if the inputs have been stable for some time M greater than d then the outputs are stable for at least $M - d$.

So, ending the digression, in the carry lookahead adder specification we have **with s,p,g delayed by 0 from a,b,c** that says that the outputs are stable when the inputs are stable. We are now ready to write a VHDL architecture for the carry-lookahead adder.

Figure 4 shows the VHDL code for a recursive architecture of the carry lookahead adder. When $N = 1$, we generate a simple cell with two **xor** gates and one **and** gate, which the reader will be able to see satisfies the specification. When N is greater than one, we divide the inputs into two slices of length $n1 = N/2$ and $n2 = N - N/2$ and pass them to two smaller instances A1 and A2, of a carrylookahead component. The propagate and generate outputs, p1,g1,p2,g2, from A1 and A2 are combined to give the p and g outputs and the carry in to A2 is set to the carry out "g1 or (p1 and c)" from A1 (where the carry in of A1 is c, the carry in of the whole). The whole architecture consists of generate statements, component instantiation statements and concurrent signal assignment statements. Slice names and indexed names and **and**, **or**, and **xor** expressions are also used. All of these VHDL constructs are fully supported by the current VC-generator, so we generate the VC and invoke the prover.

```
architecture recursive of carrylookahead is
begin
  node: if N > 1 generate
    signal p1,g1, p2,g2,c2: bit;
    begin
        A1: entity carrylookahead generic map (N/2)
            port map (c, a(N/2 -1 downto 0), b(N/2 -1 downto 0),
                      s(N/2 -1 downto 0), p1, g1);
        A2: entity carrylookahead generic map (N-N/2)
            port map (c2, a(N-1 downto N/2), b(N-1 downto N/2),
                      s(N-1 downto N/2), p2, g2);
        p  <= p1 and p2;
        g  <= g2  or (p2 and g1);
        c2  <= g1  or (p1 and c);
  end generate;

  leaf: if N = 1 generate
    begin
        s(0)  <= a(0) xor b(0) xor c ;
        p  <= a(0) xor b(0);
        g  <= a(0) and b(0) ;
  end generate;
end recursive;
```

Fig. 4. Recursive architecture for Carry-lookahead adder.

The proof is not difficult. We have space here only to give a brief summary of the steps that are needed. The proof follows the structure of the VHDL code and considers the cases $N = 1$ and $N > 1$ separately. In the case $N > 1$ we have as hypotheses the guarantees of the components and the signal assignments. Both hypotheses and conclusion have the form $(\forall \alpha : State)$ ... so we strip the quantifiers from all of them and we are now considering a fixed state α. The default guarantee for a zero-delay assignment like p <= p1 and p2; is the one we would have written as:

```
guarantees always p = p1 and p2
            with p delayed by 0 from p1, p2
```

So, in order to deduce that $p@\alpha = p1@\alpha$ and $p2@\alpha$ we have to show that $p1'last\_event@\alpha > 0$ and also for $p2$. That follows from the guarantees for components A1 and A2 provided we can show that their inputs are stable. The slice names in the VDHL are translated into Larch terms like $slice(a, N/2-1, 0)$ and these are the inputs to the components A1 and A2. The specification we are proving has $a'last\_event@\alpha > 0$ as a hypothesis and in our standard theory, which is available in the proofs of all VCs, we have a lemma:

```
last_event_slice: forall x,i,j,alpha::
  slice(x, i, j)'last_event @ alpha >= x'last_event @ alpha
```

Instantiating "last_event_slice" for all the slices in the code allows us to easily discharge all the stability assumptions.

Next we must relate the integer values of a,b, and s to the integer values of their slices. In the theory "Bit_vector_to_int" that defines the integer value of a bit-vector we have a lemma:

```
split: forall b,n:: (b'right<=n and n<b'left) and not b'ascending ->
  int(b) = int(slice(b, n, b'right))
           +2**((n+1)-b'right)*int(slice(b,b'left,n+1))
```

We instantiate "split" appropriately and then after introducing abbreviations a1,a2,b1,b2,s1,s2 for the integer values of the lower and upper slices of a,b, and s, and thinning out irrelevant hypotheses we reach the following sequent. We leave it to the reader to see how a decision procedure that understands integer arithmetic and that operations on bits are isomorphic to operations on Booleans will solve this one.

1. p @ α = (p1 @ α and p2 @ α)
2. g @ α = (g2 @ α or (p2 @ α and g1 @ α))
3. c2 @ α = (g1 @ α or (p1 @ α and c @ α))
4. s2 + int(g1 @ α or (p1 @ α and c @ α)) * 2 ** n2 = (a2 + b2) + int(c @ α)
5. s1 + int(g2 @ α or (p2 @ α and c2 @ α)) * 2 ** n1 = (a1 + b1) + int(c2 @ α)
>> (s2 + 2 ** n2 * s1) + int(g @ α or (p @ α and c @ α)) * (2 ** n1 * 2 ** n2)
   = ((a2 + 2 ** n2 * a1) + (b2 + 2 ** n2 * b1)) + int(c @ α)

3.3 The Register Component

The specification for the adder has the form **always** ψ; it guarantees that something is true in all states of the simulation. To build and reason about state machines we need some kind of *temporal abstraction* so that we need only reason about a subsequence of the simulation states. The register component and its specification will provide the needed temporal abstraction.

The register shown in figure 5, consists of transparent latches and an **and** gate. The transparent latches are shown as boxes with one input labeled with a dot. When that input is '1' then the latch is open and its output is connected to the other input, and when the "dot" input is '0' then the output remains stable. When the register is used in our adder/accumulator example and, in general, in any circuit, its output will feed back to its input after passing through some combinational circuitry, so we need to make sure that at no time is there a closed loop of open transparent latches and combinational circuitry. We accomplish that by using two non-overlapping clocks, **phi1** and **phi2**. We first describe our realization of the transparent latch and then give our formalization of the notion of non-overlapping clocks.

The left side of the text below shows the VHDL entity specification and architecture for the transparent latch. The latch is implemented using a single guarded signal assignment statement. The right side shows the Larch/VHDL specification of the latch.

```
entity BV_Latch is                          entity bv_latch
generic (N:integer);                        guarantees always y = x
port (ck: in Bit;                             with y delayed by 0 from x,ck
  x: in bit_vector(N-1 downto 0);             whenever ck = '1'
  y: out bit_vector(N-1 downto 0));         guarantees always
end BV_Latch;                                 y'last_event >= ck'last_event
                                              whenever ck = '0'
                                            guarantees always
architecture bhvr of BV_Latch is              y = y@last_state(ck='1')
begin                                         whenever ck = '0' and
  B:block (ck='1')                          ck'last_event @ last_state(ck='1')>0
  begin                                       and
   y <= guarded x ;                         x'last_event @ last_state(ck='1')>0
  end block;                                end
end bhvr;
```

```
entity bv_register
includes (NonOverlapping)
guarantees always dout'next =
   (if set = '1' then din else dout)
  at consecutive falling(phi1)
  whenever set'last_event > 0
     and din'last_event > 0 and
     nonoverlapping(phi1,phi2,end)
guarantees always
     dout'last_event'next > 0
  at consecutive falling(phi1)
whenever nonoverlapping(phi1,phi2,end)
end
```

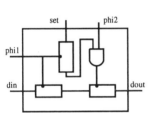

Fig. 5. Register and its Larch/VHDL Spec

In the Larch/VHDL specification there are three guarantees. The first says that when `ck='1'` the latch is like the "wire" `y <= x;` (except that the stability of `y` also depends on `ck`). The second guarantee says that when `ck='0'` then the output has been stable as long as `ck` has been stable, hence `y` has not changed since the last clock edge. This is a useful guarantee but it is not enough to let us deduce that `y` has the same value it had just prior to the last clock edge. The state just prior to the last clock edge is named by the term `last_state(ck='1')`, and the third guarantee say that if both inputs were stable at that state then `y` has the same value it had at that state.

Note that the specification for the transparent latch still specifies its behavior at all simulation states – it is not introducing any temporal abstraction. In contrast, look at the right side of figure 5 which shows the specification for the register. This specification has the form **always** ψ *at consecutive P* and that is our way of abstracting to the subsequence of simulation states at which P is true. The specification of the register specifies it as a simple state machine, giving the next output as a function of the current state and input, where "current" and "next" are defined to be consecutive falling edges of phi1. Since for this simple state machine the state and output are the same, we don't need a name for the internal state and a separate specification of the next-state function.

In addition, the specification of the register has some assumptions on the inputs given in the "whenever" clause. It says that the clocks must be non-overlapping and that the other inputs must be stable at the falling edge of phi1. The second guarantee in the specification says that assuming only that phi1 and phi2 are non-overlapping, the output is stable at falling edges of phi1.

To understand the rest of the specification, we need to describe what the predicate "nonoverlapping" says, and to explain why the word "end" is an argument to "nonoverlapping". Let us address the use of the word "end" first.

In figure 6 is a timing diagram that shows what we mean by non-overlapping clocks. We might specify that clocks ϕ_1 and ϕ_2 are non-overlapping if at *all* states β at which $\phi_1@\beta = 1$ the past values of ϕ_1 and ϕ_2 are as pictured. We could then write a predicate $nonoverlapping(\phi_1, \phi_2)$ that says that. However, using such a predicate in the specification would violate another general specification principle:

– Specifications should be *local* in *time*. They should not depend on the future, and should only depend on the relevant part of the past.

If, for example, a specification depends on the entire past history of the clock, when the specified component is used in a design its assumptions may not be met. This could be merely because they did not hold for some small period of time during the initialization phase.

We therefore write our predicate $nonoverlapping(\phi_1, \phi_2, \beta)$ to take a third argument of sort State, and to say only that *near* state β the clocks are as in the diagram. Recall then that we said earlier that a specification with a clause "at consecutive P" defines an interval $[\alpha, \beta]$ and in the specification we can refer to α and β themselves using the words "begin" and 'end". Thus the assumption nonoverlapping(phi1,phi2,end) in the register specification is assuming that with β being the end of the interval of consecutive falling edges of phi1 the clocks phi1 and phi2 are as pictured in the timing diagram.

On the right side of figure 6 are formulae that say that clocks ϕ_1 and ϕ_2 are as in the diagram. State β is a parameter, and then in terms of β the other edges are defined by the formulae. We need only that the edges are properly interleaved and that the time elapsed between the end of each edge and the beginning of the next edge is strictly positive. If we were keeping track of actual delays, we would need to extend this definition to say that the time between edges was greater than some delay term *del*.

Having the right specification for the register is the most important step in verifying it, but even with the right specification the proof is fairly tricky. However, this component and a small number of similar state-holding components

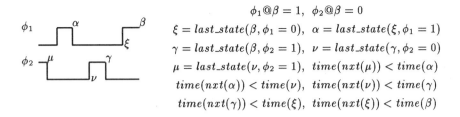

$$\phi_1@\beta = 1, \quad \phi_2@\beta = 0$$
$$\xi = last\_state(\beta, \phi_1 = 0), \quad \alpha = last\_state(\xi, \phi_1 = 1)$$
$$\gamma = last\_state(\beta, \phi_2 = 1), \quad \nu = last\_state(\gamma, \phi_2 = 0)$$
$$\mu = last\_state(\nu, \phi_2 = 1), \quad time(nxt(\mu)) < time(\alpha)$$
$$time(nxt(\alpha)) < time(\nu), \quad time(nxt(\nu)) < time(\gamma)$$
$$time(nxt(\gamma)) < time(\xi), \quad time(nxt(\xi)) < time(\beta)$$

Fig. 6. Non-overlapping clocks

will form the core of a library of reusable components and users of the system will usually be able to choose one of these components to accomplish the needed temporal abstraction. This means that the users will not often have to do the tricky temporal reasoning needed in the proof of the register.

3.4 Verification of Adder/Accumulator

The architecture for the addaccu example is given below. The labeled statements in the body correspond to the three components in the block diagram in figure 1 and the internal signals mx and accu correspond to the two internal wires. Signal p is the propagate carry from the adder and since the unlabeled statement sets the carry-in c to '0' the generate carry from the adder is the carry-out co.

```
architecture structure of addaccu is
   signal mx,accu: bit_vector(N-1 downto 0);
   signal p,c: bit;
begin
   mux: mx <= b when sel='0' else accu;
   c <= '0';
   adder : entity CarryLookahead port map (a,mx,c,s,p,co);
   reg: entity BV_Register port map(s,store,phi1,phi2,accu);
end structure;
```

The system generates the VC for this architecture, and the proof is completely straightforward. All the stability assumptions needed for the component guarantees are either hypotheses of the specification we are verifying or else are guaranteed by the register component given that we have non-overlapping clocks. Once the stability assumptions are discharged simple equational reasoning completes the proof.

3.5 Adding Timing to the Model

To add timing information to the specification for the carry lookahead adder, we do the following. We add a generic parameter **del** to the entity and in the architecture we make all the signal assignment statements that correspond to single **and**, **or**, or **xor** gates have delay **del**. For the new timed carrylookahead entity we write the following specification.

```
guarantees
  always true
  with p,g delayed by (2*log(2,n)+1)*del from a,b
guarantees
  always int(s) + int((g or (p and c))) *2**n = int(a)+int(b)+int(c)
  with s delayed by (4*log(2,n)+2)*del from a,b
        delayed by (2*log(2,n)+1)*del from c
```

We don't have space here to go through the details, but the specification shows that the sum s can be computed in time proportional to log(n). The proof is nearly the same as the proof of the zero-delay version.

4 Conclusions

After years of working on hardware verification using a variety of methods and several different theorem provers, we think that the structure and discipline provided by using a hardware description language and an associated specification language has resulted in the most practical, usable, and most easy to learn verification system we have yet built.

Here are some of the lessons learned:

- A verification system based on VHDL provides structure for complex verification tasks.
- A good interface specification language must provide a means to express complex temporal properties in a natural way.
- Good specifications (local in *space* and in *time*) make the verification easier and make the components easily reusable.
- A well structured HDL and specification language support powerful design and verification methods such as recursive decomposition.
- Timed circuits specified using the methods described here are easily verified.

References

1. Mark Bickford. A Formal Semantics for VHDL and its Use Towards Verifying a Large Design. Technical Report TM-92-0045, ORA Corporation, December 1992.
2. David Guaspari, Carla Marceau, and Wolfgang Polak. Formal verification of Ada programs. *IEEE Transactions on Software Engineering*, 16:1058–1075, September 1990.
3. John V. Guttag and James J. Horning. *Larch: Languages and Tools for Formal Specification*. Springer-Verlag, 1993.
4. Carlos Delgado Kloos and Peter T. Breuer (Editors). *Formal Semantics for VHDL*. Kluwer Academic Publishers, 1995.

Specification of Control Flow Properties for Verification of Synthesized VHDL Designs *

Naren Narasimhan and Ranga Vemuri

Laboratory for Digital Design Environments
Department of ECECS, University of Cincinnati
Cincinnati, OH 45221–210030

Abstract

Behavioral specifications in VHDL contain multiple communicating processes. Register level designs synthesized from these specifications contain a data path represented as a netlist and a controller consisting of multiple communicating synchronous finite state machines. These finite state machines together implement the control flow specified in and implied by the behavioral specification in VHDL. This paper describes a systematic approach to identifying the control flow properties critical to the proper functioning of designs synthesized from VHDL. These properties are then formulated as specifications in Computational Tree Logic (CTL) while presenting a controller model for high-level synthesis. These specifications form a necessary set that must be satisfied by any correct synthesized design. A high-level synthesis system, as a byproduct of creating RTL designs, can automatically generate these CTL specifications.

1 Motivation

VHDL, the VHSIC hardware description language admits behavior level specifications of systems containing multiple communicating processes. A high-level synthesis system accepts such behavioral specifications and produces register transfer level (RTL) designs. An RTL design usually comprises a data path and a finite state controller. The controller can be conveniently organized as a set of hierarchical, communicating synchronous state machines [1]. In VHDL, process communication and synchronization is achieved by *signal* assignment statements and *wait* statements. When a process encounters a wait statement it suspends until the wait predicate is satisfied. When all processes suspend, *signals* can be updated using their projected values stored on transaction lists [2].

The process of verifying a synthesized RTL design can be accomplished by two basic approaches: simulation and formal verification. In the simulation based approach, the behavioral specification is simulated using a set of carefully selected test vectors. The corresponding RTL design is then simulated with the same set of test vectors and the results compared with those from the behavioral level.

---

* This work was partially supported by ARPA and monitored by the FBI under contract number J-FBI-93-116.

This methodology has been addressed in [3]. Since exhaustive simulation of most real-world designs is essentially infeasible, a set of test vectors should be identified that test all possible paths in the given design. This adversely impacts on the effectiveness of simulation centered verification and consequently one's confidence in the design.

Formal verification attempts to establish that the synthesized design is mathematically *correct* [4]. Various notions of correctness as well as various formal verification techniques to support these notions exist. Theorem proving and model checking are two popular formal verification approaches. In theorem proving, the proof that the design realizes the stated behavior is mechanically checked by a theorem prover. Theorem proving based verification efforts are known to be highly interactive requiring many person months of effort even for moderately sized designs [5, 6, 7]. In addition, theorem provers seem to require considerable time to learn and use. Model checking on the other hand, can be fully automated and require little time to learn. In model checking, a set of desired properties of a model of the design are stated in some form of logic and verified using a model checker for that logic. However model checkers suffer from two problems: (1) the design is only as good as the properties stated and verified – it is fairly hard to capture a complete set of properties for all but the most simplistic designs; and, (2) if the number of states is too large, the model checker requires unreasonable amount of time and memory to complete verification – this is known as the state-space explosion problem.

Any design with a reasonable number of registers in the data path cannot be verified by model checking due to state-space explosion. Bradley and Vemuri [8] reported verification efforts for entire designs (data path and controller) using model checking. Table 1 shows excerpts from their results using the Symbolic Model Verifier [9]. These results show that, while the verification efforts for the two behavior level models were successful, verification of the register level models did not terminate even after 15 days on a dedicated Sun Sparcstation 2.

Test Case	Model	State Variables	System States		CPU Time	Memory (in Mb)
			Reachable	Total		
Design 1	Behavioral	29	5897	$3.02(10)^8$	29.15 s	0.999
	RTL	222	*unknown*	$\approx 6.74(10)^{66}$	> 15 d	44.852
Design 2	Behavioral	11	544	2048	0.13 s	0.684
	RTL	81	*unknown*	$\approx 2.42(10)^{24}$	> 15 d	36.272

Table 1. Verification Results

Verification of designs with data path seems generally infeasible using model checking. Fortunately, the register level modules in the data path are well-tested or formally verified [10] before installing them in the module library used in high-level synthesis. Due to this, and the fact that the data path consists entirely of these well-tested modules, one usually has a high degree of confidence in the data path portion of a synthesized design. It is the controller that is generally suspect, especially in the case of synthesis from behavioral VHDL specifications wherein the control flow semantics is rendered complicated in the presence of multiple processes, signals and wait statements. Hence, efforts towards the verification of a synthesized design should focus on these control flow properties. These properties should be carefully specified and formulated. A verification model of the synthesized design could then be tested for the truth of these formulations thus giving enough confidence to the user in the correctness of the design. In addition, various forms of abstractions have been proposed to reduce the state-space of a model. Abstractions make model checking feasible for designs with moderately sized data path. Use of proper abstractions in conjunction with systematic development and specification of control flow properties to be verified will largely alleviate both the disadvantages of model checking mentioned earlier.

In this paper, our focus is on the problem of specifying and verifying control flow properties for a complicated controller model used by the DSS high-level synthesis system [11]. Since this controller model is largely concerned with proper realization of the process synchronization stipulated by VHDL semantics, the properties specified should be useful while verifying designs synthesized by other VHDL-driven synthesis systems as well. More significantly, this paper demonstrates that these control flow properties can be automatically generated as a part of a high-level synthesis process. Readers interested in the abstraction techniques employed to reduce state space of the design should refer to Bradley and Vemuri's work in [8].

2 Background

Hardware Description Languages (HDLs), most notably VHDL, have gained considerable popularity in the specification of hardware designs. VHDL supports process level parallelism. It employs constructs with complicated semantics to achieve concurrency, communication and synchronization among the processes. VHDL constructs such as signal assignment statements and wait statements facilitate deterministic interprocess communication and coordination. One can exploit these features of VHDL to write succinct behavioral descriptions.

Formal techniques for verifying VHDL designs have been well researched and documented. Some of the work done in this area can be found in [12, 13, 14, 15, 16, 17, 18]. for more information on these techniques. In this section, we will present some background material to help us identify the critical control flow properties in a synthesized design. In the next section on Controller Specification we will formulate the specifications in formal logic and discuss their significance.

2.1 Behavioral Representation

The behavior of any design in VHDL can be conveniently represented as a set of modules. Each architecture, process, loop, subprogram and wait statement in the VHDL specification is represented by a corresponding module. We use a Module Call Graph (MCG) shown in Figure 1 to represent the calling hierarchy among these modules. An MCG is a directed acyclic graph, where each node represents a module in the specification and an edge represents the calling relationship between two modules. Each module represents a function that transforms input values into output values. The function of a module is represented as a control/data flow graph (CDFG). The nodes of the CDFG denote the operations within the corresponding module and the edges represent value transfers among the operations. The following control flow semantics are associ-

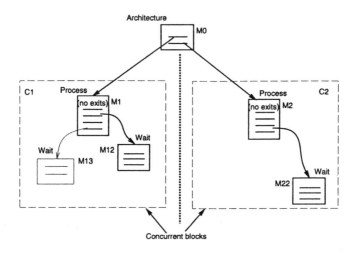

Fig. 1. An Example Module Call Graph

ated with the module call graphs and the CDFG representations of each module: (1) There is a unique top level module called the *architecture module*. The CDFG associated with this module performs all signal initializations and concurrently invokes several process modules, one corresponding to each process within the entity. (2) Process modules execute concurrently and indefinitely as they have no exit statements. (3) A Loop module represents the body of an iterative statement. When an exit condition is encountered the loop returns control back to its caller module. (4) A Subprogram module represents the body of a function or a procedure. At the end of the module or when a return statement is encountered, control is returned to the calling module. (5) Each wait statement is represented as a wait module which contains the CDFG corresponding to the wait predicate. Upon unsuccessful evaluation of the wait condition, control will reach a return node in the CDFG; else control loops back to the top of the CDFG of the Wait module.

We have identified a subset of behavioral VHDL for high-level synthesis and

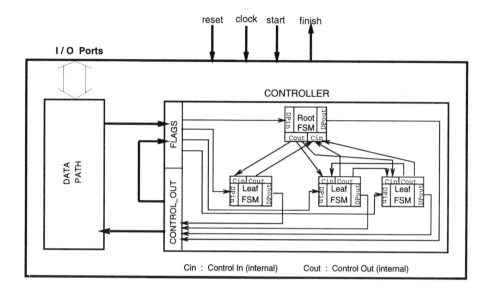

Fig. 2. The Controller Model

will restrict the discussion of the controller models and the specification in the following sections to this subset. This subset is similar to the one discussed in [19]. We will assume that explicit use of time is not allowed in the behavioral specifications. As a result, signal assignments cannot have explicit delay clauses and are therefore realized after a delta cycle delay. Similarly, wait statements cannot have time clauses. They are thus necessarily predicated upon signals and signal attributes.

2.2 Register Level Design

A high-level synthesis system generates register level designs to implement the given behavioral specification subject to constraints on area, throughput rate, clock speed and other performance attributes. High-level synthesis techniques and concerning design optimization techniques have been presented in [20, 21, 22]. Designs synthesized by high-level synthesis systems are usually synchronous and as shown in Figure 2, comprise two interacting components: a data path and a controller. The data path consists of module instances selected from the register level module library, interconnected to realize all the value transfers among the nodes in the CDFG representations. The controller generates control signals to effect register transfers in the data path. The controller is organized as a collection of synchronous FSMs, one for each module in the Module Call Graph. The FSM corresponding to the architecture module is called the *Root* FSM. The

FSMs corresponding to all other modules are broadly known as *leaf* FSMs. They are further classified as *Process, Loop, Subprogram* and *Wait* FSMs. As shown in Figure 2, the controller sends control signals to the data path through the *control_out* registers and the data path communicates with the controller through the *flags* registers.

The controller execution semantics closely resemble the VHDL execution cycle. (1) The *start* signal from the environment triggers the *Root* FSM to invoke all the *Process* FSMs concurrently. (2) After all concurrent threads reach their privileged wait states, the *Root* FSM begins a new execution cycle. (3) All signals are updated at the beginning of the new execution cycle. (4) If all wait evaluations fail, the finish signal is raised and new inputs are admitted from the environment.

Some notable deviations from VHDL simulation semantics, however, are worth mentioning. The test bench participates in the design execution cycle. The primary inputs are internally generated in the test bench and are semantically identical to signals in VHDL. Projected values of a primary input signal are thus placed in a transaction list. The sequence of these transactions is then used to update the projected output waveform of the input signal as the simulation time advances. Clearly, in such a scenario, primary inputs from the test bench are always admitted into the VHDL design whenever wait statements predicated on primary input signals are evaluated. Therefore all wait evaluations do not have to evaluate to false for admitting fresh inputs from the test bench.

On the other hand, a hardware implementation of a VHDL design expects inputs from the environment. As a result there is a need for establishing a handshake protocol to allow the design to communicate with the external world. The external inputs are not placed in any transaction list and therefore all wait statements predicated upon internal signals and the primary inputs fail eventually. The controller raises the finish signal when this happens, indicating to the environment that the no further transactions on the signals in the design are possible given the current primary inputs. The design is now ready to accept the next set of primary inputs from the environment.

VHDL processes execute asynchronously and are synchronized by wait statements. On a single processor machine, "ready" processes are selected and executed in an arbitrary order. In hardware, this would translate to the parallel and asynchronous execution of FSMs in the controller. This further implies that the data path resources cannot be shared and hence have to be duplicated for the hardware realization of each VHDL process. We will assume that the scheduler phase in the HLS system assigns control steps to each operation node in the CD-FGs subject to the constraint that all *Process* modules share data path resource. This ensures that all the FSMs in the register level design execute synchronously to avoid resource contention in the data path.

2.3 Verification Method

In the following section we will present the controller model used in the high-level synthesis system DSS and formulate several specifications that capture its control flow properties. These specifications can then be verified against the verification model of the controller by any model checker. We will now briefly introduce the CTL (Computation Tree Logic) formalism we have used to state our specifications. CTL is a branching time temporal logic developed by Clarke et al. [23]. In addition to the usual boolean connectives ¬, ∧, ∨, ⇒ and ≡, the logic has four operators for expressing temporal relationships [9]:

- **X**, the *next time* operator indicates a condition that holds in the next state.
- **G**, the *global* operator denotes a property that holds globally in all states of all the computation path.
- **F**, the *eventual* operator denotes a property that holds in some future state in the computation path.
- **U**, the *until* operator in fUg holds if formula g is satisfied in some state in the computation path and formula f is satisfied in all states preceding s.

The CTL logic also makes use of the Universal Path Quantifier **A** and the Existential Path Quantifier **E** to check for conditions in multiple computation paths. For example, **AG**f indicates that the formula f should hold over all possible computation paths beginning in the current state. **EX**f says that there exists atleast one computation path where f holds for the immediate successor of the first state in that computation path.

3 Controller Specification

The controller is composed of a hierarchical collection of communicating synchronous finite state machines. Control corresponding to each module in the behavioral description is realized by a finite state machine. Thus, there are as many FSMs in the controller as there are modules in the behavioral description. The structure of each FSM depends on the type of the module whose control flow it is realizing and the structure of the scheduled and bound CDFG attached to the module. The FSMs issue control to the data path based on the flags received from the data path and communicate with each other through signals to coordinate overall sequencing in order to realize the execution semantics of the module call graph. We assume that all FSMs are Moore-type machines.

3.1 Scheduling Assumptions

In a high level synthesis system, the scheduler phase precedes the controller generation phase. We make the following assumptions about the scheduler while describing the controller model. (1) CDFGs associated with the process modules are scheduled together so as to share data path resources. (2) The scheduler schedules the CDFGs associated with loop, subprogram and wait modules individually. While scheduling these modules, the scheduler reserves all the resources in the data path for the execution of these modules. Keeping this in mind, clearly, the scheduler can never schedules more than one call operation at any given control step (a call operation corresponds to the invocation of a subprogram, loop or wait module) [1].

3.2   Design Assumptions

We would like our design models to be well-formed in the sense that they should not deadlock. That is, we expect that each active process can eventually suspend (ie. it has a reachable wait statement); that each suspended process can eventually restart (ie. its *true* exit of the wait predicate in the wait module is reachable) and that each loop and subprogram invocation can eventually exit (ie. the exit node is reachable). We will make these well-formedness assumptions when developing our CTL specifications for the controller model. Our intent is that if a model is not well-formed then one or more of these specifications should evaluate to false.

In the rest of the paper, we will describe a multi-FSM controller model that when incorporated in a high-level synthesis system can help generate RTL designs from complex behavioral VHDL specifications. Each FSM model is expected to preserve the process execution semantics of VHDL and at the same time preserve the order among the CDFG operations dictated by the scheduler. As we discuss each FSM model, we will formulate the CTL specifications to verify its correctness. These specifications can then be checked against a verification model of the controller (in a format recognizable to the model checker). We will show only some of the interesting CTL specifications that capture the control flow properties of our model; a complete list is beyond the space limitations of this paper. We have used universal and existential quantifiers, $\forall, \exists$ in our CTL specifications for succinctness. These quantifiers do not belong to the syntax of CTL but have been bound to range over a finite set of FSMs. They simply denote a set of one or more CTL formulae that should be generated and verified for each synthesized design.

3.3   Root FSM

The *Root* FSM controls the basic execution cycle. It ensures the proper triggering of all the *Process* FSMs and their synchronization at the beginning of every execution cycle. Furthermore, in conjunction with the *Process* FSMs it handles the semantics of signal assignments and wait statements. Figure 3 shows the state diagram template of the *Root* FSM. The machine does not begin executing until the environment raises the *start* signal. Once the *Start (S)* signal goes high, the *Root* FSM, by raising the *Read Inputs (RI)* signal, issues control to read the input stimulus. Subsequently, the *Root* FSM triggers all *Process* FSMs concurrently by raising the *Invoke Processes (IP)* signal. All *Process* FSMs now, start executing and in turn may invoke other leaf FSMs. The Leaf FSMs eventually suspend when they encounter their respective wait states and signal their suspension by raising their respective *SPD* (suspended) flags. The conjunction of the *SPD* signals of all the leaf FSMs gives rise to the *All Processes Suspended (APS)* signal, an input to the *Root* FSM. A high *APS* therefore, indicates the end of the current execution cycle. The *Root* FSM marks the beginning of a new execution cycle by issuing control to update all signals that have been modified by the Leaf FSMs.

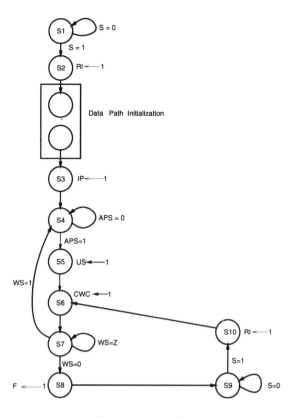

Fig. 3. Root FSM

With the description so far, we can make the following assertions about a correct *Root* FSM. We will back each assertion with a set of one or more CTL formulae that formally capture the essence of the assertion.

- If the *Start* signal is never asserted, the *Root* FSM is always in its *Start* state which in turn means that none of the leaf FSMs leave their *Start* states.

$$\forall i, \ (\neg Start) \Rightarrow (S_i = S_{0_i}) \tag{1}$$

Here i is an FSM in the model, S_i is the state variable of machine i and S_{0_i} is the start state of machine i.

- Once *Start* goes high, eventually the *Root* FSM reads in the inputs and all signals are updated.

$$Start \Rightarrow \mathbf{AF}(Root.RI) \wedge \mathbf{AF}(Root.US) \tag{2}$$

The following specification makes explicit, the order in which the tasks are carried out.

$$((Start \Rightarrow \mathbf{AF}(Root.RI) \Rightarrow \mathbf{AF}(Root.US)) \tag{3}$$

- If the *Start* is never raised, no inputs are ever read and signals are never updated. This stronger relationship between the *Start* signal and the rest of the signals is captured in Formula 4.

$$\mathbf{AG}(\neg Start) \Rightarrow$$
$$\mathbf{AG}\,(\neg Root.RI) \wedge \mathbf{AG}\,(\neg Root.US) \qquad (4)$$

- The *Root* FSM raises *US* to update all signals only after all *Process* FSMs have suspended. The specifications in formulae 5 and 6 captures this notion of order.

$$\mathbf{AG}\,(\neg Root.APS) \Rightarrow \mathbf{AG}\,(\neg Root.US) \qquad (5)$$

$$\mathbf{AG}\,(Root.APS \Rightarrow \mathbf{AF}\,(Root.US) \qquad (6)$$

Process FSMs can resume execution only after their wait conditions are satisfied. The *Check Wait Condition (CWC)* signal shown in Figure 3, is an output signal from the *Root* FSM to all the *Wait* FSMs. It ensures that all wait conditions are evaluated only after all *Process* FSMs have suspended. In other words, the *CWC* signal helps to preserve the synchronization among *Process* FSMs. The outcomes of the wait condition evaluations of all *Wait* FSMs are resolved to form the *Wait Status (WS)* signal, an input to the *Root* FSM. An outcome can be a 1, 0 or 'Z' indicating success, failure or 'yet to be evaluated' respectively. If even one of the outcomes is a 'Z', *WS* is assigned a 'Z'. If all conditions have evaluated and if atleast one of them is a success, *WS* becomes 1. When all wait condition evaluations fail, *WS* is assigned a 0. Subsequently, the *Root* FSM raises the *Finish* signal and waits for the next set of inputs from the environment. With the fresh set of inputs, the wait conditions are re-evaluated and upon a success, the leaf FSMs resume execution. We will conclude the description of the *Root* FSM by making the following additional assertions about its correctness.

- Once the *Root* is triggered it eventually completes execution for the current set of input vectors. In other words, state S_8 is reachable. This is of course subject to our earlier assumption discussed in Section 3.2 that the input specification is well-formed.

$$Start \Rightarrow \mathbf{AF}\,Root.Finish \qquad (7)$$

- If the *start* signal is never raised, the *finish* signal never goes high.

$$\mathbf{AG}\,(\neg Start) \Rightarrow \mathbf{AG}\,(\neg Root.Finish) \qquad (8)$$

- The *Root* FSM accomplishes proper sequencing. The *Start* signal from the environment triggers the *Root* FSM. The machine will eventually exhaust the current set of inputs. All the wait conditions will evaluate to failures thereby suspending the *Process* FSMs indefinitely. As a result, the *Root* FSM raises the *Finish* signal. This in turn prompts the environment to furnish the next set of inputs and re-trigger the *Root* FSM. The *WS* signal finally evaluates to a success indicating that atleast one of the *Leaf* FSMs is free to resume execution. Thereafter, the resumed *Leaf* FSM eventually suspend resulting in a high *APS*.

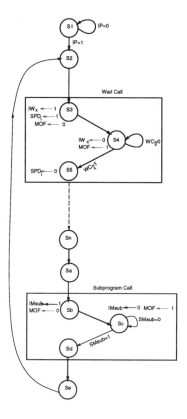

Fig. 4. Process FSM

This prompts the *Root* FSM to update all signals. The following CTL formula captures the above assertion.

$$(((((Start$$
$$\Rightarrow \mathbf{AF}(Root.WS = 0))$$
$$\Rightarrow \mathbf{AF}(Root.Finish))$$
$$\Rightarrow \mathbf{AF}(Root.RI))$$
$$\Rightarrow \mathbf{AF}(Root.APS))$$
$$\Rightarrow \mathbf{AF}(Root.US)) \tag{9}$$

The CTL formulae developed so far capture the process execution semantics and assert that the controller is faithful to it.

3.4 Process FSM

The *Process* FSM embodies the process body in the VHDL description. Since the *Process* FSM would be specific to the design being synthesized, it is not possible to present a template as we did in the case of the *Root* FSM. We will therefore concentrate on defining the design independent control flow properties of the *Process* FSM in this section. As mentioned in Section 2, a *Process* FSM executes a finite number of states before it suspends by invoking a *Wait* FSM either directly or through Subprogram/Loop FSMs. Figure 4 shows a sample *Process* FSM body that, in addition to executing regular operations in the corresponding CDFG also invokes a *Wait* FSM and a *Subprogram* FSM. We will assume that all states not involved in calling the two other FSMs are involved in data path computation that is well behaved.

The *Process* FSM begins execution only after it receives a high *IP* signal from the *Root* FSM. Once triggered it executes for ever. Each state in the FSM is concerned with executing scheduled data path operations. The behavior of the *Process* FSM is of particular significance when it executes call operations. Observe Figure 4 where the *Process* FSM makes a call to a *Wait* FSM followed by a call to a *Subprogram* FSM. The *Invoke Wait* signal, IW_x, is raised to trigger *Wait* FSM x. The *Process* FSM simultaneously raises its *SPD* flag to let the *Root* FSM know that it is now suspended and lowers the *Macro Operation Finish (MOF)* signal for exactly one state. The *MOF* signal plays a key role in data path resource sharing and will be discussed a little later. *Wait* FSM x completes execution and raises WC_x to indicate wait completion. Subsequently, the *Process* FSM lowers *SPD* and resumes execution. The *Subprogram* FSM is invoked in a fashion similar to the triggering of the *Wait* FSM with the exception that the *SPD* signal is never raised.

We will make the following assertions accompanied by the CTL specifications about the control flow properties of the *Process* FSM depicted in Figure 4.

• Once triggered by the *Root* FSM, a *Process* FSM eventually invokes a *Wait* FSM and suspends, or invokes a *Loop/Subprogram* FSM. Note that this is true of any *Process* FSM. The following two specifications unambiguously capture this property of the *Process* FSM.

$$\forall i, \exists x, y : Root.IP \Rightarrow$$
$$\mathbf{AF} ((P_i.IW_x \wedge P_i.SPD_i) \vee P_i.IM_y) \tag{10}$$

where i is a *Process* FSM x is a *Wait* FSM and y is a *Subprogram/Loop* FSM.

$$\forall i, \exists x, y : \mathbf{AG} \neg(Root.IP) \Rightarrow$$
$$\mathbf{AG} \neg((P_i.IW_x \wedge P_i.SPD_i) \vee P_i.IM_y) \tag{11}$$

This establishes the converse of the specification stated in Formula 10. Due to limited space in this paper, we have concentrated only on the design independent specifications of the FSMs. We also generate a set of CTL formulae specific to the *Process* FSM in Figure 4.

3.5 Subprogram and Loop FSMs

Control flow of a subprogram or a loop module is similar to that of a *Process* module with the exception that the process module executes forever while

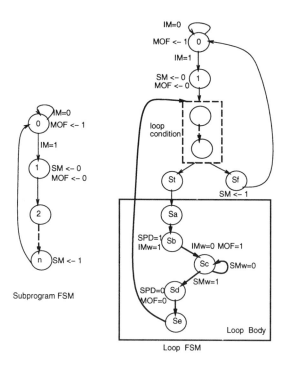

Fig. 5. Subprogram and Loop FSM

the subprogram/loop modules execute until their exit conditions are satisfied. Figure 5 shows the FSMs for a loop module and a subprogram module.

Before we discuss the behavior of these FSMs we need to introduce the concept of macro operations and macro states. All call operations take arbitrary number of control steps and are known as macro operations. The states in the *Process* FSM shown in Figure 4 that execute non-call operations (these include no-ops) are called macro states. From the scheduling assumptions in Section 3.1, recall that the scheduler schedules the loop CDFG, the subprogram CDFG, and the CDFG concerned with the wait evaluations individually. Therefore, when control reaches a *Loop/Subprogram* FSM or when the states within the *Wait* FSM concerned with the wait condition is being executed, all the process FSMs should be frozen in their current states until the *Loop/Subprogram* FSM exits or the wait condition evaluation is completed. This is ensured by the macro state. A *Process* FSM, when in a macro state, makes a state transition only in the presence of an enabling signal, *ENMO*, which is the conjunction of the *MOF* signals of all the *Leaf* FSMs. When a *Subprogram/Loop* FSM is called, it lowers its *MOF* signal. As a result, the *ENMO* signal is disabled and none of the *Process* FSMs or *Wait* FSMs can neither execute any operation nor make any state transitions. After the call operation is completed, the *MOF* signal is raised again and all disabled machines can resume execution.

Both *Loop* and *Subprogram* FSMs execute the operations of their respective module CDFGs and exit. The *Subprogram* FSM raises its *Status Module (SM)* signal indicating completion after the states of the machine have executed all the scheduled operations. The calling FSM can now resume execution. The *Loop* FSM checks the exit condition before executing the semantics of the loop body. If the exit condition is not satisfied it executes the loop body and then returns back to test the loop condition. If the loop condition fails, it raises its *SM* signal and exits enabling its caller to resume execution.

All the specifications that we discussed for the *Process* FSM could be applied to the *Subprogram* and *Loop* FSM. Since they have an exit condition we can make the following statement.

• We assert that the exit state of the FSMs are indeed reachable. This would be valid under the assumption that the design model be well-formed.

$$\forall x, \; x.IM_x \Rightarrow \mathbf{AF}x.SM_x \tag{12}$$

Here x is either a *Subprogram* or a *Loop* FSM.

3.6 Wait FSM

The organization of the *Wait* FSM serves to preserve the synchronization among all *Process* FSMs. In addition to implementing the wait predicate, the *Wait* FSM shown in Figure 6 incorporates the VHDL semantics of the wait statement. Once invoked, the *Wait* FSM eventually tests the wait condition and returns control at the appropriate time to the FSM that invoked it.

Recall from Section 3.1 that wait condition evaluations like other call operations are scheduled individually and hence do not share data path resources with any other operations. To ensure this, the *Wait* FSM lowers its *MOF* signal at the time of wait evaluation (state S_3 in Figure 6) to disable all other operations. The *Wait* FSM begins evaluating the wait condition in the same control step it was invoked. Thus, upon successful evaluation, it returns control to the caller FSM at exactly the right control step thereby maintaining the synchronization among all leaf FSMs. The no-op states (macro states) denoted by the states labeled D in Figure 6 ensure that the *Wait* FSM does indeed evaluate its wait condition in the right control step. The number of control steps to the start of the wait condition evaluation should be equal to the number of control steps in the longest *Process* FSM. This ensures that at the time of wait condition evaluation, the *Wait* FSM is in the control step it was invoked. The role of the *CWC* signal has been discussed while describing the *Root* FSM. If *CWC* is low, the machine steps through a set of no-op states such that it advances to state S_3 on a high *CWC* at exactly the right control step. The wait condition evaluation stage is quite similar to the execution of the *Loop* or *Subprogram* FSM. The *Wait* FSM at this point lowers its *MOF* thereby disabling the other *Leaf* FSMs. Observe that barring the states that are involved in the wait condition evaluation, all other states in the *Wait* FSM are macro states which need a high *ENMO* to make a state transition. If the result of the wait condition evaluation is a success, the

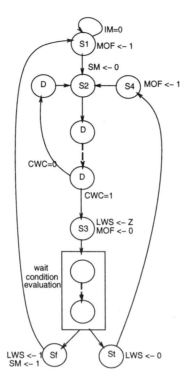

Fig. 6. Wait FSM

Wait FSM sets its *Leaf Wait Status (LWS)* to 1 and signals to its parent FSM that it has completed execution. If the result is a failure, it sets *LWS* to 0 and loops back to the set of no-op states and continues as before. The *LWS* of all the *Wait* FSMs are resolved to form the *WS* signal to the *Root* FSM.

From the description of the *Wait* FSM we can make the following assertions with the corresponding CTL specifications.

• A *Wait* FSM upon invocation, eventually evaluates its wait condition.

$$\forall i, j : \ \mathbf{AG}\,(i.IW_j \ \Rightarrow \ \mathbf{AF}\,\neg(j.LWS = Z)) \tag{13}$$

where i is a *Leaf* FSM which invokes a *Wait* FSMs and j is a *Wait* FSM.

• A high *CWC* from the *Root* FSM implies that eventually all *Wait* FSMs must evaluate their wait condition.

$$\forall i : \ Root.CWC \ \Rightarrow \ \mathbf{AF}\,\neg(i.LWS = Z) \tag{14}$$

• If the *Root* never issues a high *CWC* signal, none of the *Wait* FSMs ever evaluate their wait conditions. Formulae 14 and 15 establish the strong relationship between the *CWC* and the *LWS* signals of each *Wait* FSM.

$$\forall i : \ \mathbf{AG}\,(\neg\,(Root.CWC) \ \Rightarrow \ \mathbf{AG}(i.LWS = Z)) \tag{15}$$

• Once the *Root* FSM issues a *start* signal, the *Wait* FSMs eventually test their wait conditions and atleast one wait predicate evaluates to a success (denoted by $LWS = 1$). As a result, control is transferred back to its parent FSM. The parent then resumes execution and eventually suspends prompting the *Root* FSM to update all the signals. This captures one aspect of the VHDL execution cycle.

$$\forall i, j : (((Start \Rightarrow \mathbf{AF}\,(j.LWS = 0)) \Rightarrow$$
$$\mathbf{AF}\,(\neg i.SPD)) \Rightarrow \mathbf{AF}\,(Root.US)) \qquad (16)$$

where, i is a *Leaf* FSM and j is its child *Wait* FSM.

4 Implementation and Results

Test Case	State Variables	System States		CPU Time	Memory (in Mb)
		Reachable	Total		
Design 1	98	2038	$3.40(10)^{28}$	41.53 s	3.453
Design 2	48	832	$8.16(10)^{13}$	11.58 m	5.276

Table 2. Verification Results Using SMV

We constructed verification models for several synthesized RTL designs and used the Symbolic Model Verifier [9] to model check them. The CTL formulae stated in this paper proved extremely critical in verifying the correctness (with a high degree of confidence) of the control flow properties of the synthesized designs.

Table 2 shows the result of verifying the SMV models of two designs referred in Table 1, against all the CTL formulae developed in the previous section plus a few other problem specific formulae. To make this verification feasible, the two models have been transformed using a temporal abstraction reported in [8]. This temporal abstraction significantly reduces the problem state-space by reducing the number of states and transitions in the data paths.

The CTL specifications formulated for the controller models in this paper played a major role in systematically verifying the correctness of synthesized RTL designs. As a part of this research we have developed and verified a suite of controller models that handle behavioral designs specified by various VHDL subsets. As the subsets get larger the controller models increase in complexity. Table 3 shows statistics of these controller models related to verification time. The models are arranged in increasing order of complexity. Specifications for all these models were formulated in CTL and several designs were subsequently verified. The controller model described in this paper can be classified as type E in the table. It allows multiple processes with no restrictions imposed on the number or positions of wait statements within the processes. Processes share data path resources while loops, subprograms and wait evaluations do not share resources. Controller model A handles single process descriptions with only a single wait statement at the end of the process; Model B handles multiple process descriptions with each process having just one wait statement at the end. Controller model C improves on Model B and allows wait statements anywhere in the description and wait condition evaluations do not share data path resources. Model D is similar to Model E but constrains wait evaluations to share the data path resources with other operations. The controller models get more sophisticated as they accept control constructs of increasing complexity. It is therefore clear, that to gain confidence in the controller models more CTL specifications need to be formulated and verified as the models increase in complexity which has a direct bearing on the time spent on verification.

	System States		CPU
Model	Reachable	Total	Time
A	5	8	0.5 s
B	16	6804	0.68 s
C	50	$1.6(10)^{14}$	0.85 s
D	48	$8.18(10)^{16}$	1.11 s
E	1698	$3.23(10)^{37}$	5.55 s

Table 3. Results for Controller Models

5 Discussion

High Level synthesis systems are being increasingly used to synthesize real world designs. As these systems increase in complexity it becomes necessary to be able to prove them correct. In spite of tremendous progress in formal verification techniques, verification of large software systems, such as a high level synthesis system remains quite impractical. The next best alternative is to automatically verify each synthesized design to be correct. There have been attempts to formally verify entire RTL designs using theorem proving but they continue to be extremely expensive in terms of person effort.

We have concentrated on the controller model verification aspect of synthesized designs using symbolic model checking. We outlined an approach that brings us closer to the problem of formally verifying synthesized designs. By closely examining the execution semantics of a controller model it is possible to identify the crucial control flow properties which can then be formulated as CTL specifications generating a set of rules that any synthesized design must satisfy. However, It is extremely hard to ascertain the sufficiency of these control flow properties in capturing the intended semantics of the controller. In this paper, we streamlined the verification problem by identifying the crucial control flow aspects in the controller model and forming a *necessary* set of rules that any synthesized design should satisfy. This in turn gives the user added confidence in the system and renders it more reliable. It is possible for a high-level synthesis system to automatically generate these CTL formulae at the same time the RTL design is generated. This when combined with the automated generation of SMV-compatible verification models of the RTL designs and the automated application of the abstractions to reduce state-space will essentially, fully automate the verification of control flow properties of synthesized designs.

Acknowledgments

We thank Bill Bradley for his verification efforts using the SMV model checker.

References

1. Naren Narasimhan and Ranga Vemuri. "Synchronous Controller Models for Synthesis from Communicating VHDL Processes". In *Ninth International Conference on VLSI Design*, pages 198–204, Bangalore, India, January 1996.
2. *IEEE Standard VHDL Language Reference Manual, IEEE Std 1076-1993*.
3. Ranga Vemuri et al. "Experiences in Functional Validation of a High Level Synthesis System". In *30th ACM/IEEE Design Automation Conference*, pages 194–201, 1993.
4. Aarti Gupta. "Formal Hardware Verification Methods: A Survey". *Formal Methods in System Design*, 1:151–238, 1992.
5. Warren A. Hunt, Jr., Bishop C. Brock. "The DUAL-EVAL Hardware Description Language and its use in the Formal Specification and Verification of the FM9001 Microprocessor. In *CHDL*, pages 637–642, Makuhari, Japan, September 1995.
6. Mandayam K. Srivas and Steven P. Miller. "Applying Formal Verification to a Commercial Microprocessor". In *CHDL'95*, pages 493–502, Makuhari,Japan, 1995.
7. Tareq Altakrouri. "Verification of a Synthesized Design using the Boyer-Moore Theorem Prover". Master's thesis, ECECS Dept., University of Cincinnati, January 1995.
8. William Bradley and Ranga Vemuri. "Transformations to improve Functional Verification of Synthesized Systems". In *Eighth International Conference on VLSI Design*, New Delhi, India, January 1995.
9. K.L. McMillan. *"The SMV system DRAFT"*, February 1992.
10. D. Verkest, L. Claesen, H.De Man. *"On the use of the Boyer-Moore theorem prover for correctness proofs of parameterized hardware modules"*, pages 99–115. Formal VLSI Specfication and Synthesis. Elsevier Science Publishers B.V. (North-Holland), 1990.
11. Jay Roy, Nand Kumar, Rajiv Dutta and Ranga Vemuri. "DSS: A Distributed High-Level Synthesis System". In *IEEE Design and Test of Computers*, June 1992.

12. Saurin B. Shroff. "Introduction to Formal Verification of VHDL Designs". In *VIUF*, Spring 1996.
13. D. Hemmendinger and J. Van Tassel. "Toward formal Verification of VHDL Specifications". In L. Claesen [24], pages 261–270.
14. P. Narendran and J. Stillman. "Hardware Verification in the Interactive VHDL Workstation". In G. Birtwistle and P. A. Subrahmanyam [18], pages 235–256.
15. D. Borrione, L. Pierre, and A. Salem. "Prevail: A Proof Environment for VHDL Descriptions". In *Workshop on Correct Hardware Design Methodologies*, Italy, June 1991.
16. D. Borrione, L. Pierre, and A. Salem. "Formal Verification of VHDL Descriptions in Boyer-Moore: First Results". In *First European Conf. on VHDL*, France, September 1990.
17. D. Jamsek and M. Bickford. "Formal Verification of VHDL Models". Technical Report RL-TR-94-3, Rome Laboratory, March 1994.
18. G. Birtwistle and P. A. Subrahmanyam, editor. *"VLSI Specification, Verification, and Synthesis"*. Kluwer Academic Publishers, 1988.
19. D. Gajski. "A VHDL Subset for Synthesis". In *VHDL Users Group Meeting*, October 1988.
20. D.D. Gajski, N.D. Dutt, A.C. Wu and S.Y. Lin. *"High-Level Synthesis,Introduction to Chip and System Design"*. Kluwer Academic Publishers, 1992.
21. G. De Micheli. *"Synthesis and Optimization of Digital Circuits"*. McGraw-Hill, 1994.
22. R. Camposano and W. Wolf. *"High-Level VLSI Synthesis"*. Kluwer Academic Publishers, 1991.
23. E.M. Clarke, E.A. Emerson and A.P. Sistla. "Automatic Verification of Finite-State Concurrent Systems using Temporal Logic Specifications". In *ACM Trans. Prog. Lang. Syst.*, volume 8(2), pages 244–263, 1986.
24. L. Claesen, editor. *"Applied Formal Methods for Correct VLSI Design"*. Proceedings of the IMEC-IFIP International Workshop, November 1989.

An Algebraic Model of Correctness for Superscalar Microprocessors

Anthony C. J. Fox* and Neal A. Harman

Department of Computer Science, University of Wales Swansea
Singleton Park, Swansea SA2 8PP, United Kingdom.

Abstract. A set of algebraic tools are presented to model *superscalar* processors, where instructions may be executed in parallel, or out of program order. This has implications for the representation of timing abstraction, the relationship between time at different levels of abstraction, and the concept of the correctness of one representation with respect to another. We illustrate our tools with a simple, superscalar example, and present a *one-step theorem* for simplifying the formal verification of superscalar microprocessors.

1 Introduction

This paper extends a set of algebraic tools for microprocessors ([15] and [7]) to model *superscalar* microprocessor implementations. The tools are modular, and support equational specification and verification techniques, not dedicated to specific software systems. In superscalar microprocessors, the timing of events in an implementation is increasingly different from that of the architecture. We develop the *correctness models* of [15] to accommodate the more advanced timing relationships of superscalar processors. We illustrate our tools with a simple example architecture, and implementation.

We are particularly interested in models of time and temporal abstraction. *Clocks* divide time into (not necessarily equal) segments, defined by the natural timing of the computational process of a device: for example, the execution of machine instructions, or some system clock. We model a computer by the iteration of a map $f : A \to A$ in discrete time $T = \{0, 1, 2, \dots\}$, defined by

$$F(0, a) = a,$$
$$F(t + 1, a) = f(F(t, a)),$$

for $t \in T$, $a \in A$. The set A models the state of the computer and f is the next-state map; thus $F(t, a)$ is the state at time $t \in T$, given initial state $a \in A$. The nature of A and T is determined by the level of abstraction of the computer. In F above, and in the examples that follow, we omit inputs and outputs: however, these may be simply added ([14, 15]). We formally relate clocks by surjective, monotonic maps.

Interesting recent work on microprocessors with complex timing relationships is that of Windley ([34]) on UINTA (a processor of moderate complexity, and its verification in HOL: [10]), and SRI ([26] and [25]) on AAMP5 (a more complex processor,

---
* Supported by EPSRC grant number 94007861.

and its verification in PVS: [27]). In both cases, the development and consideration of timing abstraction (see also [33] and [6]) is similar to our own ([15]). Also of interest is [30] which again has a similar model of time (see also [23] and [24]).

Other work on microprocessors includes: *Gordon's Computer* [9, 19, 29] and [16]; *Viper* [4, 5] and [1]; Landin's SECD machine [20, 11, 12] and [2]; a PDP-11-based processor, the FM8501, and its more advanced successor the FM9001 [18, 17] and [3].

Work at Swansea on modelling hardware and parallel systems is part of the work of Esprit Working Group NADA (No 00 85 33).

2 Preliminaries

We first introduce the algebraic tools and notations used to model time, and micropro-cessors at different levels of abstraction. We assume the reader is familiar with universal algebra; for detailed accounts see [22], and [32].

2.1 Clocks and Retimings

A *clock* is an algebra $(T \mid 0, t+1)$ where: (i) $T = \{0, 1, \dots\}$ is a set of *clock cycles*; (ii) 0 is the initial clock cycle and (iii) $t+1$ is the next cycle function. A *clock* provides a way of identifying discrete time intervals.

Let S and T be clocks. A *retiming* λ, from S to T, is a monotonic, surjective map with domain S, codomain T: see Fig. 1. (This should not be confused with the retimings of [21].) The set of all retimings from S to T is denoted by $Ret(S, T)$. Let $[T \to S]$ be the set of all maps from T to S. The *immersion* $\bar{\lambda} \in [T \to S]$ of a retiming $\lambda \in Ret(S, T)$ is defined by $\bar{\lambda}(t) = (\mu s \in S)[\lambda(s) = t]$; that is, the least cycle $s \in S$ such that $\lambda(s) = t$.

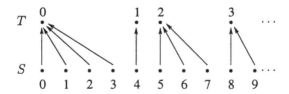

Fig. 1. A retiming from T to S.

Given a non-empty *state space* A, a *state-dependent retiming* $\lambda : A \to Ret(S, T)$ is a retiming parameterised by a state $a \in A$, often representing an initial state.

The function $start : Ret(S, T) \to [S \to S]$ returns the first cycle $s' \in S$ such that, $\lambda(s') = \lambda(s)$:

$$start(\lambda)(s) = \bar{\lambda}(\lambda(s)).$$

Further discussion about retimings can be found in [15, 16] and [13].

2.2 Iterated Maps and Primitive Recursion

Definition 1. Given a clock T, non-empty set A, and two primitive recursive functions $h : A \to A$ and $f : A \to A$ an *iterated map with initialisation function* $F : T \times A \to A$ is a primitive recursive function defined as follows. For all $t \in T$ and $a \in A$

$$F(0, a) = h(a) \quad \text{and} \quad F(t+1, a) = f(F(t, a)).$$

The above equations, denoted E_F, iterate f, given initial value $h(a)$. The solution to E_F is $F(t, a) = f^t \circ h(a)$. In the context of computers the *state function* F returns the state of the machine at time $t \in T$, from initial state $h(a) \in A$, where A is the *state space*. The function f is the *next-state* function and h is the *initialisation function*.

2.3 Decomposition

We modularise a specification by decomposing the state and next-state function into *coordinate* functions: see also [15]. Let T be a clock and $A = A_1 \times \cdots \times A_k$ be a state space , where $A_1, \ldots, A_k$ are non-empty. The next-state function $f : A \to A$ is decomposed as follows:

$$f(a_1, \ldots, a_k) = (f_1(a_1, \ldots, a_k), \ldots, f_k(a_1, \ldots, a_k)),$$

where $a_i \in A_i$ and $f_i : A \to A_i, 1 \le i \le k$. Each function f_i computes the ith component of the next state function f. The state function F may be decomposed accordingly.

To compute the ith component of the state space at $t + 1 \in T$ only a subset of the state at $t \in T$ may be required. We generally modify the domain of f_i such that only the parts of the state required are included. This simplifies the definitions of the next-state functions, and allows the reader to infer the physical connections and dependencies.

3 An Algebraic Model of Correctness

In this paper we are concerned with two levels of microprocessor abstraction: the *programmer's model PM*, which is the *functional specification* of a microprocessor; and the *abstract circuit design AC*, which is the first stage of implementation. The *AC* level describes the structure and operation of the main components within the microprocessor. In Definition 2 we consider a correctness definition for *microprogrammed* and *pipelined* microprocessors, with *in-order execution*. In Sect. 3.1 a definition of correctness is formulated for superscalar processors with *out-of-order* and *parallel execution*.

Definition 2. An iterated map $G : S \times B \to B$ is a *correct implementation* of an iterated map $F : T \times A \to A$ if, given a state-dependent retiming $\lambda : B \to Ret(S, T)$ and a surjective map $\psi : B \to A$, then for all $s = start(\lambda(b))(s)$ and $b \in B$, the following diagram commutes and equation holds.

$$
\begin{array}{ccc}
T \times A & \xrightarrow{\;F\;} & A \\
\big\uparrow{\scriptstyle(\lambda, \psi)} & & \big\uparrow{\scriptstyle\psi} \\
S \times B & \xrightarrow{\;G\;} & B
\end{array}
$$

$F(\lambda(b)(s), \psi(b)) = \psi(G(s, b)).$ ☐

Map ψ must be surjective to ensure correctness for all initial states of F. A *clock condition* $s = start(\lambda(b))(s)$ ensures that an appropriate number of clock cycles $s \in S$ is chosen: for microprocessors we are concerned with correctness at times $s \in S$ corresponding to the start and end of machine instructions. Hence, in the case of simple and pipelined implementations, it is sufficient to be able to identify a unique time $s \in S$ corresponding to the *end* of an instruction. It is immaterial that other instructions will be ongoing at time $s \in S$.

3.1 Superscalar Processors and Correctness

In a *superscalar* implementation of a microprocessor is one in which the execution of more than one instruction may be attempted simultaneously, modulo dependencies, and resource limitations. This means that instructions may be *completing* simultaneously.

Definition 2 is inadequate when considering superscalar implementations of a *PM* specification. There is no surjective map from *AC* clock S to *PM* clock T, because more than one instruction may complete at a given time $s \in S$. Groups of instructions must be considered and this is achieved by using an additional clock.

Let R be an *event clock*, where an event is the completion of one or more instructions. Clock R is used to relate clocks S and T: see Fig. 2, where instructions 0 and 1 complete simultaneously taking four cycles of clock S, then instruction 2 completes after a further cycle of clock S, and so on.

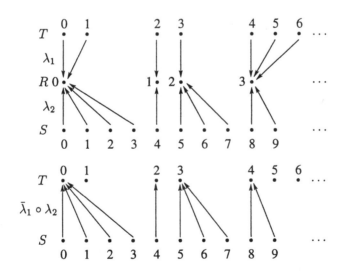

Fig. 2. Retimings λ_1 from T to R and λ_2 from S to R, and the function $\bar{\lambda}_1 \circ \lambda_2$ from S to T.

Definition 3. An iterated map $G : S \times B \to B$ is a *correct implementation (over event clock R)* of an iterated map $F : T \times A \to A$ if, given retimings $\lambda_1 : B \to Ret(T, R)$, $\lambda_2 : B \to Ret(S, R)$ (see Fig. 2), and surjective map $\psi : B \to A$, then for all $s =$

$start(\lambda_2(b))(s)$ and $b \in B$ the following diagram commutes and equation holds with $\rho(b) = \bar{\lambda}_1(b) \circ \lambda_2(b)$.

$$
\begin{array}{ccc}
T \times A & \xrightarrow{F} & A \\
\uparrow{\scriptstyle (\rho,\psi)} & & \uparrow{\scriptstyle \psi} \\
S \times B & \xrightarrow{G} & B
\end{array}
$$

$F(\rho(b)(s), \psi(b)) = \psi(G(s, b))$. $\qquad\qquad\qquad\qquad\qquad\qquad\qquad\qquad$ $\square$

3.2 One Step Correctness Theorems

In this section, we show how the *one step correctness theorems* ([7]) for simple correctness models (Definition 2) can be extended to superscalar correctness. The one step theorems enable us to significantly simplify the verification process for microprocessors. We first introduce some preliminary concepts, then we state the one step event correctness theorem for superscalar processors. Proofs can be found in [8].

Definition 4. A map $F : T \times A \rightarrow A$ is *time-consistent* if, and only if, for all $t \in T$, $a \in A$, $F(t_1 + t_2, a) = F(t_1, F(t_2, a))$. $\qquad\qquad\qquad\qquad$ $\square$

Lemma 5 Time Consistency and Iterated Maps. *An iterated map $F : T \times A \rightarrow A$ with initialisation function (Sect. 2.2) $h : A \rightarrow A$ is time-consistent if, and only if, for all $t \in T$, $a \in A$, $F(t, a) = h(F(t, a))$.* $\qquad\qquad\qquad$ $\square$

Definition 6. Let T and S be clocks, $F : T \times A \rightarrow A$ be a time consistent function, and $dur : A \rightarrow S^+$ be a *duration function*. A state-dependent retiming $\lambda : A \rightarrow Ret(S, T)$ with immersion $\bar{\lambda} : A \rightarrow [T \rightarrow S]$ is *uniform* if, and only if,

$$
\bar{\lambda}(a)(0) = 0,
$$
$$
\bar{\lambda}(a)(t + 1) = dur(F(t, a)) + \bar{\lambda}(a)(t).
$$
$\qquad\qquad\qquad\qquad\qquad\qquad\qquad\qquad\qquad\qquad\qquad\qquad$ $\square$

That is, the number of cycles of S corresponding with any $t \in T$ is only a function of the state of F.

Lemma 7 Uniform Retimings. *Let $F : T \times A \rightarrow A$ be any time consistent map, and $\lambda : A \rightarrow Ret(S, T)$ be a state-dependent uniform retiming defined in terms of F, as in Definition 6. For all $t_1, t_2 \in T$, $a \in A$*

$$
\bar{\lambda}(a)(t_1 + t_2) = \bar{\lambda}(F(t_2, a))(t_1) + \bar{\lambda}(a)(t_2).
$$
$\qquad\qquad\qquad\qquad\qquad\qquad\qquad\qquad\qquad\qquad$ $\square$

Theorem 8 One Step Event Correctness. *Given iterated maps $F : T \times A \rightarrow A$ and $G : S \times B \rightarrow B$, retimings $\lambda_1 : B \rightarrow Ret(T, R)$, $\lambda_2 : B \rightarrow Ret(S, R)$ and a surjective map $\psi : B \rightarrow A$ if (i) λ_1 and λ_2 are uniform retimings defined with maps $dur_1 : B \rightarrow T^+$ and $dur_2 : B \rightarrow S^+$; (ii) $\rho(b) = \bar{\lambda}_1(b) \circ \lambda_2(b)$ for all $b \in B$; (iii) $F(\bar{\lambda}_1(b)(0), \psi(b)) = \psi(G(\bar{\lambda}_2(b)(0), b))$; and (iv) F and G are time-consistent, then for all $s = start(\lambda_2(b))(s)$ and $b \in B$*

$$
F(\rho(b)(s), \psi(b)) = \psi(G(s, b)) \Leftrightarrow F(dur_1(b), \psi(b)) = \psi(G(dur_2(b), b)). \quad \square
$$

To prove the function G implements the function F for all $s = start(\lambda_2(b))(s)$, it is sufficient to show that the function G implements the function F at time $s = dur_2(b)$, when $t = dur_1(b)$ and $r = 1$ for all $b \in B$.

4 Example: Informal Specification and Programmer's Model

The architecture we consider has sixteen registers r_0–r_{15}, a program counter pc and program and data memories mp and md. The disjoint memories avoid *resource conflicts* (see Sect. 5) when simultaneously fetching data and instructions: we may consider mp and md to be abstractions of *instruction* and *data cache* respectively. Instruction words are twenty bits long: bits one and two represent the op-code; the remaining bits represent the operands. The JMP $addr$ instruction performs a conditional pc-relative jump, with offset $addr$, when $r_0 = 0$. The ADD ra, rb, rc ($rc \neq ra$ and $rc \neq rb$) instruction adds ra and rb, and stores the result in rc. The STR ra, $addr$ instruction stores the contents of ra at memory location $addr$. The LDR ra, $addr$ instruction reads location $addr$ and stores the contents in ra. In the above $0 \leq ra, rb, rc \leq 15$ and $1 \leq addr \leq 2^{12} - 1$. A full treatment of instruction formats may be found in [8].

4.1 Programmer's Model

Let T be a program instruction clock. One cycle of the clock T corresponds with the execution of one machine instruction. The specification includes 20-bit, 12-bit and 5-bit words representing machine words, addresses and register indices, abbreviated to R, PC and RC respectively. There are two memory sets $Mp = [PC \rightarrow R]$ and $Md = [PC \rightarrow R]$, and register set $Reg = [RC \rightarrow R]$. The state space C of the machine is:

$$C = Reg \times PC \times Mp \times Md.$$

The state function $COMP : T \times C \rightarrow C$ is defined below:

$$COMP(0, c) = c,$$
$$COMP(t + 1, c) = comp(COMP(t, c)).$$

The next-state function $comp : C \rightarrow C$ is defined as follows:

$$comp(reg, pc, mp, md) =$$
$$\begin{cases} (reg, jump(reg(0), pc, mp(pc)), mp, md), & \text{if } Op(mp(pc)) = 0; \\ (add(reg, mp(pc)), pc + 1, mp, md), & \text{if } Op(mp(pc)) = 1; \\ (reg, pc + 1, mp, store(reg, md, mp(pc))), & \text{if } Op(mp(pc)) = 2; \\ (load(reg, md, mp(pc)), pc + 1, mp, md), & \text{if } Op(mp(pc)) = 3. \end{cases}$$

Each of the cases above corresponds with the execution of one of the four instructions. The definitions of $Op : R \rightarrow \{0, \ldots, 3\}$, which decodes the instruction's op-codes, and $jump$, add, $store$ and $load$ which define instruction execution, are here left as an exercise for the reader: definitions can be found in [8].

5   A Superscalar Implementation

The microprocessor is split into nine units, see Fig. 3. The implementation presen-

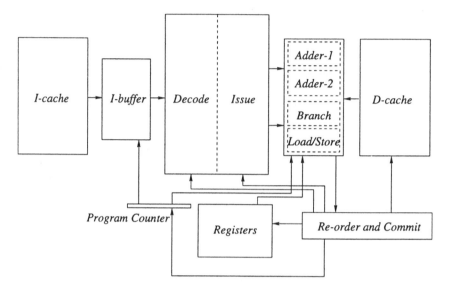

Fig. 3. A simple organisation.

ted here omits optimisation features, present in some superscalar machines, to prevent clouding the example with unhelpful details. Some of the latest techniques are discussed in [28] and may be added at a later date. The important distinction of superscalar processors addressed here is the ability to *commit* to storage more than one instruction in the same clock cycle. It is this feature that requires modification of the definition of implementation correctness used for pipelined and other machines (see [7]). The implementation presented here can commit up to four instructions in a clock cycle.

The Instruction Cache This unit performs the rôle of the *PM* program memory.

The Instruction Buffer This fetches a group of four instructions from the instruction cache. The instructions are read sequentially from the current program counter value. For simplicity, a JMP instruction is *not* treated differently: branches are 'predicted' as not taken.

The Decode Unit This unit decodes the instructions from the instruction buffer and stores this data prior to be being issued. An instruction is decoded into five *fields*: an op-code, three register indices and an address.

The Issue Unit This unit is the most complicated and handles the issue of instructions to one of the four functional units. The unit contains details of register and memory activity/usage to manage dependencies. The unit also contains buffer registers for each functional units. These store the details of the instructions to be executed in the next clock cycle. The instruction issue is based on Tomasulo's algorithm (see [31]) without out-of-order issue. Out-of-order issue does not affect the correctness aspects of this example and may be added by adopting a *register renaming* scheme. Instructions may be issued and executed speculatively, that is, the instruction is issued together with *one* preceding JMP instruction and its results are committed depending on the branch outcome. The issue unit is discussed in Sect. 5.2.

The Functional Units The implementation contains two twenty bit integer adders, a load/store unit and a branch unit. These units are responsible for executing ADD, LDR/STR and JMP instructions respectively. Two adds may be issued simultaneously.

The Reorder and Commit Unit This unit prepares the results, stored in the functional units, prior to writing to the data cache, registers and program counter. A table of committal words is constructed, where each entry specifies the storage unit, destination address and result to be committed. Speculative instructions are added to this table with a bit indicating whether the instruction results should be committed. This enables the issue unit to issue instructions waiting for resources to be freed.

The Storage Units The register unit and data cache are equivalent to their *PM* counterparts with the data cache performing the rôle of the data memory. The program counter unit contains two program counter values: one is equivalent to the *PM* program counter and the other point to the head of the current block of four instructions in the instruction buffer. The storage units are discussed in Sect. 5.3.

5.1 A Formal Description

A full account of a superscalar implementation may be found in [8]. Within this section, the state and next-state functions are defined in terms of coordinate next-state functions for each unit. The state space and next-state functions for the issue unit and storage units are then given in Sections 5.2 and 5.3.

A few typographical conventions have been adopted. State types of the implementation are given in 'sans serif', functions and base state elements are in *italic Roman* and vectors of state elements are in **bold Roman**.

Let S be the *AC* clock. The machine state space is

$$\text{State} = \text{Icache} \times \text{Ibuffer} \times \text{Decode} \times \text{Issue} \times \text{Execute} \times$$
$$\text{Reorder} \times \text{Dcache} \times \text{Reg} \times \text{Counter},$$

where Icache, Dcache and Reg represent non-vectorised state components, and Ibuffer, Decode, Issue, Execute, Reorder and Counter represent vectors of state components. The components of the state space are

$$\text{state} = \textit{icache}, \textbf{ibuffer}, \textbf{decode}, \textbf{issue}, \textbf{execute},$$
$$\textbf{reorder}, \textit{dcache}, \textit{reg}, \textbf{counter}$$

The base state types are Bit, R (register), PC (address), RC (register index) and Ord (instruction order).

The state function of the superscalar implementation $\sigma Comp : S \times \text{State} \to \text{State}$ is defined as follows.

$$\sigma Comp(0, \textbf{state}) = init(\textbf{state}),$$
$$\sigma Comp(s + 1, \textbf{state}) = \sigma comp(\sigma Comp(s, \textbf{state})).$$

The function $init : \text{State} \to \text{State}$ is defined as follows.

$$init(\textbf{state}) = (icache, 0, ibuf_1, \dots, ibuf_4, 0, dec_1, \dots, dec_4, \textbf{issue}, \textbf{execute},$$
$$radd_1, radd_2, rlsr, rbr, 1, com_0, reg, dcache, pc, pc, 1).$$

The function $init$ modifies **state** to be a valid initial state; we omit explanations of some of the state elements above.

The next-state function $\sigma comp : \text{State} \to \text{State}$ is decomposed as follows:

$$\sigma comp = (Icache \circ \pi_{icache}, Ibuffer \circ \pi_{ibuffer}, Decode \circ \pi_{decode}, Issue \circ \pi_{issue},$$
$$Execute \circ \pi_{execute}, Reorder \circ \pi_{reorder}, Storage \circ \pi_{storage}),$$

where $Icache, \dots, Storage$ are the next-state functions of each unit. The family of projection functions π determine the coarse grain connections between the machine components and are defined as follows.

$$\pi_{icache}(\textbf{state}) = (icache),$$
$$\pi_{ibuffer}(\textbf{state}) = (icache, \textbf{ibuffer}, \textbf{counter}),$$
$$\pi_{decode}(\textbf{state}) = (\textbf{ibuffer}, \textbf{decode}, \textbf{reorder}),$$
$$\pi_{issue}(\textbf{state}) = (\textbf{decode}, \textbf{issue}, \textbf{reorder}),$$
$$\pi_{execute}(\textbf{state}) = (\textbf{issue}, \textbf{execute}, dcache, reg, \textbf{counter}),$$
$$\pi_{reorder}(\textbf{state}) = (\textbf{execute}, \textbf{reorder}),$$
$$\pi_{storage}(\textbf{state}) = (\textbf{reorder}, dcache, reg, \textbf{counter}).$$

For example, the instruction buffer shown in Fig. 3 has arrows from the instruction cache and the program counter unit; these state components are projected, together with state of the instruction buffer itself, by the function $\pi_{ibuffer}$.

5.2 The Issue Unit

We now discuss one of the more complex components: the issue unit. We omit definitions of some functions: see [8]. This unit's state space is

$$\text{Issue} = \text{Bit}^4 \times \text{Add}^2 \times \text{Lsr} \times \text{Jmp} \times \text{RegFree} \times \text{Bit} \times \text{DcFree where}$$
$$\text{Add} = \text{Bit} \times \text{Ord} \times \text{Bit} \times \text{RC}^3$$
$$\text{Lsr} = \text{Bit} \times \text{Ord} \times \text{Bit} \times \text{Bit} \times \text{RC} \times \text{PC}$$
$$\text{Jmp} = \text{Bit} \times \text{Ord} \times \text{Bit} \times \text{Bit} \times \text{PC}$$
$$\text{RegFree} = [\text{RC} \to \text{Bit}]$$
$$\text{DcFree} = \text{Bit} \times \text{PC}$$

The components of the state space are as follows

$$\textbf{issue} = issd_1, \ldots, issd_4, \textbf{add}_1, \textbf{add}_2, \textbf{lsr}, \textbf{jmp}, regfree, pcfree, \textbf{dcfree}$$

$$\textbf{add}_1 = toex_{add1}, order_{add1}, cond_{add1}, ra_{add1}, rb_{add1}, rc_{add1}$$

$$\textbf{add}_2 = toex_{add2}, order_{add2}, cond_{add2}, ra_{add2}, rb_{add2}, rc_{add2}$$

$$\textbf{lsr} = toex_{lsr}, order_{lsr}, cond_{lsr}, load, ra_{lsr}, addr_{lsr}$$

$$\textbf{jmp} = toex_{jmp}, order_{jmp}, another, addr_{jmp}$$

$$\textbf{dcfree} = dcfree, dcactive$$

The $issd$ bits keep track of which instructions have been issued; for example, the bit pattern 0, 1, 0, 1 occurs when instructions one and three have been issued. The bits $toex_{add1}$, $toex_{add2}$, $toex_{lsr}$ and $toex_{jmp}$ indicate whether an instruction has been issued to each of the functional units. The $order$ bits encode the relative position of each of the instructions with its block when fetched (00 for the first to 11 for the last). The $cond$ bits indicate whether an instruction is to be conditionally executed after a JMP: that is, results will only be committed if the branch is *not* taken. The registers ra, rb, rc and $addr$ contain the instruction fields required by the functional units to execute the instruction. The $another$ bit enables the unit to prevent the issue of multiple JMP instructions. The $regfree$ table keeps track of register usage and controls instruction dependencies. The $pcfree$ bit indicates whether the program counter may be updated by a branch instruction. The $dcfree$ bit indicates that the cache memory location, whose address is in $dcactive$, may be updated.

The next-state function $Issue :$ Decode $\times$ Issue $\times$ Reorder $\rightarrow$ Issue is defined as follows and is essentially an implementation of Tomasulo's algorithm.

$$Issue(\textbf{decode}, \textbf{issue}, \textbf{reorder}) =$$
$$\begin{cases} (0, \ldots, 0), & \text{if } pccomt = 1; \\ \pi_{issue'}(DoIssue^4(1, \textbf{decode}, SetIssd(issue, decoded), 0, \\ ldots, 0, \pi_{free}(ReleaseRes^4(1, regfree, pcfree, \textbf{dcfree}, com)))). & \text{otherwise.} \end{cases}$$

If $pccomt = 1$ (a flag stored in the reorder unit) then a branch has been taken and the issue unit is cleared. The following applies in all other cases. The function $SetIssd :$ Issue $\times$ Bit $\rightarrow$ Bit^4 updates the $issd$ bits, that is, if the decode unit flags the presence of a new block of instructions all of the $issd$ bits are set to one. The register and program counter usage tables are updated by the function $ReleaseRes :$ $\mathbb{Z}_4 \times$ RegFree $\times$ Bit $\times$ DcFree $\times$ Ct $\rightarrow$ $\mathbb{Z}_4 \times$ RegFree $\times$ Bit $\times$ DcFree $\times$ Ct, using information from the commit unit. That is, $ReleaseRes$ releases registers in use by instructions that have just been committed. The function $ReleaseRes$ is iterated four times because four instructions may be committed per cycle. Instructions are issued ($DoIssue$ defined below) by considering the four instructions in the decode unit. We iterate $DoIssue$ four times, once for each instruction. The projection functions $\pi_{issue'} :$ $\mathbb{Z}_4 \times$ Decode $\times$ Issue $\times$ Reorder $\rightarrow$ Issue and $\pi_{free} :$ $\mathbb{Z}_4 \times$ RegFree $\times$ Bit $\times$ DcFree $\times$ Ct $\rightarrow$ RegFree $\times$ Bit $\times$ DcFree discard unwanted state information from iterations of $DoIssue$ and $ReleaseRes$.

The function $DoIssue : \mathbb{Z}_4 \times$ Decode $\times$ Issue $\to \mathbb{Z}_4 \times$ Decode $\times$ Issue is defined below.

$DoIssue(n, \mathbf{decode}, \mathbf{issue}) =$

$$\begin{cases} (n+1, \mathbf{decode}, AddIssue1(n, \mathbf{decode}, \mathbf{issue}), \\ \qquad\qquad \text{if } CanDo_{add1}(n, \mathbf{decode}, \mathbf{issue}); \\ (n+1, \mathbf{decode}, AddIssue2(n, \mathbf{decode}, \mathbf{issue}), \\ \qquad\qquad \text{if } CanDo_{add2}(n, \mathbf{decode}, \mathbf{issue}); \\ (n+1, \mathbf{decode}, LdrIssue(n, \mathbf{decode}, \mathbf{issue}), \\ \qquad\qquad \text{if } CanDo_{ldr}(n, \mathbf{decode}, \mathbf{issue}); \\ (n+1, \mathbf{decode}, StrIssue(n, \mathbf{decode}, \mathbf{issue}), \\ \qquad\qquad \text{if } CanDo_{str}(n, \mathbf{decode}, \mathbf{issue}); \\ (n+1, \mathbf{decode}, JmpIssue(n, \mathbf{decode}, \mathbf{issue}), \\ \qquad\qquad \text{if } CanDo_{jmp}(n, \mathbf{decode}, \mathbf{issue}); \\ (n+1, \mathbf{decode}, SetAnother(n, \mathbf{decode}, \mathbf{issue}), \\ \qquad\qquad \text{if } op_n = \text{jump and } issd_n = 1 \\ \qquad\qquad \text{and } (toex_{jmp} = 1 \text{ or } pcfree = 1); \\ (n+1, \mathbf{decode}, \mathbf{issue}), \text{otherwise.} \end{cases}$$

The state of the issue unit is updated if the nth instruction in the decode unit can be issued. There are cases for each functional unit (two for load/store which may execute either and LDR or STR instruction). There is one case when a second JMP is encountered; the bit $another$ is set to prevent the issuing of subsequent instructions conditional on more than one branch. Finally there is an otherwise case when no instruction may be issued; for example, when the pipeline is being filled after a branch. The state components ra_n, rc_n and $addr_n$ contain decoded instruction fields and are contained in the decode unit.

The functions $AddIssue1$, $AddIssue2$, $LdrIssue$, $StrIssue$, $JmpIssue : \mathbb{Z}_4 \times$ Decode$\times$Issue $\to$ Issue fill the issue buffers of the appropriate functional units with data from the decode unit. The function $SetAnother : \mathbb{Z}_4 \times$ Decode $\times$ Issue $\to$ Issue sets the $another$ bit to one. The family of functions $CanDo : \mathbb{Z}_4 \times$ Decode$\times$ Issue $\times$ Reorder $\to$ $\mathbb{B}$ determine if an instruction may be issued to a functional unit, for example:

$CanDo_{add2}(n, \mathbf{decode}, \mathbf{issue}) =$

$$\begin{cases} tt, & \text{if } toex_{add1} = 1 \text{ and } toex_{add2} = 0 \text{ and } op_n = \text{add} \\ & \text{and } (pcfree = 0 \text{ or } toex_{jmp} = 1) \\ & \text{and } another = 0 \text{ and } issd_n = 1 \\ & \text{and } (n = 1 \text{ or } issd_{n-1} = 0) \\ & \text{and } regfree(ra_n) = 0 \text{ and } regfree(rb_n) = 0; \\ ff, & \text{otherwise.} \end{cases}$$

An ADD instruction may be issued to the second adder if the instruction has not previously been issued ($issd_n = 1$), an ADD instruction has been issued to the first adder ($toex_{add1} = 1$), an ADD has not been issued to the second adder ($toex_{add2} = 0$), its parameters are free ($regfree(ra_n) = 0$ and $regfree(rb_n) = 0$), the instruction is next in-order ($n = 1$ or $issd_{n-1} = 0$) and may be speculatively executed only when issued at the same time as *one* JMP instruction ($another = 0$ and ($pcfree = 0$ or $toex_{jmp} = 1$)).

5.3 Committing to Storage

We now consider the *storage* units. First, however, we must briefly discuss the *counter* and *reorder* units, whose state information is used by the storage units.

The counter unit's state space is $\mathsf{Counter} = \mathsf{PC}^2 \times \mathsf{Bit}$, and the components are **counter** = $oldpc, pc, isnewpc$. The address $oldpc$ specifies the address of the first instruction in the instruction buffer, and pc is equivalent to the *PM* program counter. The flag $isnewpc$ marks the completion of an instruction block: either by the committal of all the instructions in the current instruction block or by the committal of a taken branch.

The reorder unit contains a table of committal words $ct \in \mathsf{Ct}$ where $\mathsf{Ct} = [\mathbb{Z}_4 \to \mathsf{Com}]$ and $\mathsf{Com} = \mathsf{Bit} \times W_2 \times \mathsf{R} \times \mathsf{PC}$. The committal words are decomposed as follows: **com** = $done, type, word, dest$. The $done$ bit indicates whether a speculatively executed instruction should be committed, with one indicating committal. The $type$ bits specify the storage unit: 00 – none, 01 – register, 10 – data cache and 11 – program counter. The $word$ field specifies the result to be stored and the $dest$ field specifies the precise destination (register number or data cache address).

The function $Storage$: $\mathsf{Reorder} \times \mathsf{Dcache} \times \mathsf{Reg} \times \mathsf{Counter} \to \mathsf{Dcache} \times \mathsf{Reg} \times \mathsf{Counter}$ is defined as follows.

$$Storage(\mathbf{reorder}, dcache, reg, \mathbf{counter}) =$$
$$\pi_{storage'}(DoStorage^4(1, ct, dcache, reg, \mathbf{counter})),$$

where $\pi_{storage'}$: $\mathsf{Com} \times \mathsf{Dcache} \times \mathsf{Reg} \times \mathsf{Counter} \to \mathsf{Dcache} \times \mathsf{Reg} \times \mathsf{Counter}$ discards unwanted state information from the iteration of $DoStorage$.

The function $DoStorage$: $\mathbb{Z}_4 \times \mathsf{Com} \times \mathsf{Dcache} \times \mathsf{Reg} \times \mathsf{Counter} \to \mathbb{Z}_4 \times \mathsf{Com} \times \mathsf{Dcache} \times \mathsf{Reg} \times \mathsf{Counter}$ is iterated four times, for a possible four instructions, and is defined below.

$$DoStorage(n, ct, dcache, reg, \mathbf{counter}) =$$

$$\begin{cases}
\begin{array}{l}(n+1, ct, dcache, \\ reg[\pi_{word}(ct(n))/trim_{PC}^{RC}(\pi_{dest}(ct(n)))], \\ IncPC(\mathbf{counter})), \end{array} & \begin{array}{l}\text{if } \pi_{type}(ct(n)) = 01 \\ \text{and } \pi_{done}(ct(n)) = 1; \end{array} \\[2em]
\begin{array}{l}(n+1, ct, dcache[\pi_{word}(ct(n))/\pi_{dest}(ct(n))], \\ reg, IncPC(\mathbf{counter})), \end{array} & \begin{array}{l}\text{if } \pi_{type}(ct(n)) = 10 \\ \text{and } \pi_{done}(ct(n)) = 1; \end{array} \\[2em]
(n+1, ct, dcache, reg, \pi_{dest}(ct(n)), \pi_{dest}(ct(n)), 1), & \begin{array}{l}\text{if } \pi_{type}(ct(n)) = 11 \\ \text{and } \pi_{done}(ct(n)) = 1; \end{array} \\[1.5em]
(n, ct, dcache, reg, \mathbf{counter}), & \text{otherwise.}
\end{cases}$$

The first three cases correspond with writing to a register, the data cache and the program counter respectively. The final case leaves the state intact when no further instructions are to be committed. The register set $reg[a/b]$ is identical to reg but has $reg[a/b](b) = a$. The projection functions $\pi_{done} :$ Com $\rightarrow$ Bit, $\pi_{type} :$ Com $\rightarrow W_2$, $\pi_{word} :$ Com $\rightarrow$ R and $\pi_{dest} :$ Com $\rightarrow$ PC have obvious definitions.

Consider the case when an instruction is to be committed ($\pi_{done}(ct(n)) = 1$) that writes to a register ($\pi_{type}(ct(n)) = 01$). We update the register set reg by replacing the destination register contents ($trim_{PC}^{RC}(\pi_{dest}(ct(n)))$) by the instruction result ($\pi_{word}(ct(n))$). The function $trim_{PC}^{RC}$ cuts down the 12-bit $dest$ word, which must be large enough to hold a memory address, to a 5-bit register address word. In addition the program counter is updated by $IncPC$.

The function $IncPC :$ Counter $\rightarrow$ Counter is defined as follows.

$$IncPC(\textbf{counter}) = \begin{cases} (pc+1, pc+1, 1), & \text{if } oldpc + 3 = pc; \\ (oldpc, pc+1, 0), & \text{otherwise.} \end{cases}$$

The program counter pc is incremented by one until the end of the instruction block is reached when the value of $oldpc$ becomes the incremented value of pc and the $isnewpc$ flag is set to one.

5.4 A Correctness Definition for the Superscalar Implementation

Recall the correctness definition for superscalar processors in Sect. 3.1. The superscalar implementation in Sect. 5 is correct if (given maps $\rho :$ State $\rightarrow [S \rightarrow T]$ and $\psi :$ State $\rightarrow C$ defined below) the following diagram commutes for all clock cycles $s = start(\lambda(\textbf{state}))(s)$ and $\textbf{state} \in$ State

$$
\begin{array}{ccc}
T \times C & \xrightarrow{\ COMP\ } & A \\
\uparrow {\scriptstyle (\rho, \psi)} & & \uparrow {\scriptstyle \psi} \\
S \times \text{State} & \xrightarrow{\ \sigma\, Comp\ } & \text{State}
\end{array}
$$

The map $\psi :$ State $\rightarrow C$ is defined as follows.

$$\psi(\textbf{state}) = (reg, pc, icache, dcache),$$

where $reg, pc, icache$ and $dcache$ are those parts of $\textbf{state}$ also present in the PM state. Let R be an event clock; each cycle of clock R corresponds with the completion of one instruction block. The map $\rho :$ State $\rightarrow [S \rightarrow T]$ is defined by $\rho(b) = \bar{\lambda}_1(b) \circ \lambda_2(b)$, with the uniform retimings $\lambda_1 :$ State $\rightarrow Ret(T, R)$ and $\lambda_2 :$ State $\rightarrow Ret(S, R)$ defined by their respective immersions below.

$$\bar{\lambda}_1(\textbf{state})(0) = 0,$$
$$\bar{\lambda}_1(\textbf{state})(r+1) = dur_1(\sigma\, Comp(\bar{\lambda}_2(\textbf{state})(r), \textbf{state})) + \bar{\lambda}_1(\textbf{state})(r),$$
$$\bar{\lambda}_2(\textbf{state})(0) = 0,$$
$$\bar{\lambda}_2(\textbf{state})(r+1) = dur_2(\sigma\, Comp(\bar{\lambda}_2(\textbf{state})(r), \textbf{state})) + \bar{\lambda}_2(\textbf{state})(r).$$

The duration functions $dur_1 : \text{State} \to T^+$ and $dur_2 : \text{State} \to S^+$ are defined below.

$$dur_1(\textbf{state}) = NumCmtd(\sigma\,comp(\textbf{state})) +$$
$$\cdots + NumCmtd(\sigma\,comp^{dur_2(\textbf{state})}(\textbf{state})),$$
$$dur_2(\textbf{state}) = \mu s \in S \quad [\pi_{isnewpc}(\sigma\,comp^s(\textbf{state})) = 1].$$

Retiming λ_1 counts the number of instructions committed for each cycle of clock R, which will be in the range of 1–4 per cycle. Retiming λ_2 counts the number of system clock cycles for each cycle of event clock R, by searching for the start of a new block of instructions indicated by $isnewpc$ (see Sect. 5.3). The projection $\pi_{isnewpc} : \text{State} \to$ Bit is defined by $\pi_{isnewpc}(\textbf{state}) = isnewpc$. The function $NumCmtd : \text{State} \to T$ returns the number of instructions to be committed, on each cycle of system clock S, from the commit table ct.

$$NumCmtd(\textbf{state}) = \pi_{committed}(ToCommit^4(1, 0, ct)).$$

The function $ToCommit : \mathbb{Z}_4 \times \mathbb{Z}_4 \times \text{Ct} \to \mathbb{Z}_4 \times \mathbb{Z}_4 \times \text{Ct}$, which accumulates the number of instruction to be committed from ct, is defined by

$$ToCommit(n, committed, ct) =$$
$$\begin{cases} (n + 1, committed + 1, ct), & \text{if } \pi_{type}(ct(n)) \neq 00 \text{ and } \pi_{done}(ct(n)) = 1; \\ (n + 1, committed, ct), & \text{otherwise.} \end{cases}$$

An instruction will be committed, provided it is not conditional ($\pi_{done}(ct(n)) = 1$) and it updates one of pc, reg or $dcache$ ($\pi_{type}(ct(n)) \neq 00$). The projection $\pi_{committed} : \mathbb{Z}_4 \times \mathbb{Z}_4 \times \text{Ct} \to \mathbb{Z}_4$ is defined by $\pi_{committed}(n, committed, ct) = committed$.

6 Conclusions

A set of algebraic tools developed for microprogrammed microprocessors have been applied to a superscalar implementation. The more complex timing relationship between levels of abstraction required a more sophisticated correctness model. We have shown that the one step theorems (Sect. 3.1 and [7]) still apply to this model.

The model of correctness used deserves comment. We establish correctness at times corresponding to the end of a 'block' of instructions. It is possible to identify time-points *within* each block, when one or more instructions are committed. This would give a stronger definition of correctness. If we consider *exceptions*, and the need to identify precise machine states after an error arising from an instruction within a block, a stronger definition of correctness is essential. Unfortunately, a naïve definition of the initialisation function (Sect. 5.1) leads to an iterated map that is not time consistent before the completion of instruction blocks (see Sect. 3.2 and [7]). This means that we cannot use the one step theorems, making formal verification more difficult. In principal however, it should be possible to define an appropriate initialisation function that maintains time consistency in such cases: we are currently investigating this.

Our correctness model says nothing about the number of cycles of clock S required to execute a block of instructions: it simply searches for the time when the current block has finished. We may wish to strengthen our correctness statement to make precise the number of cycles of S required for any block. To do this, we must change the definition of dur_2 to state the number of system cycles required for each possible combination of instructions. Such a definition would be large, but not conceptually complex. Such a correctness condition for a microprogrammed processors can be found in [16].

The one step theorems were developed to simplify formal verification, by reducing the process to a straightforward. The one step theorem for superscalar correctness (Sect. 3.1) is useful, but leaves substantial work to be undertaken. Naïvely, an n-way superscalar processor, with m instructions would have m^n combinations of instructions. In practice, the number would be much smaller, since many combinations would be disallowed. However, considering that each instruction will probably have a number of sub-cases (for example, multi-level cache memories and the complications arising from cache misses) the number of cases could be large. We are exploring techniques to further reduce complexity: for example, by considering further levels of abstraction in the case of cache memories.

Finally, we have made no attempt at formal verification. The focus of current and past work has been on mathematical tools, rather than machine-assisted verification. However, we are currently investigating the application of a number of existing automated reasoning tools to microprocessor verification, based on our algebraic tools.

References

1. T Arora, T Leung, K Levitt, T Schubert, and P Windley. Report on the UCD microcoded viper verification project. In *Higher-Order Logic Theorem Proving and its Applications*, pages 239 – 252. Lecture Notes in Computer Science 780, Springer-Verlag, 1993.
2. G Birtwistle and B Graham. Verifying SECD in HOL. In J Staunstrup, editor, *Formal Methods for VLSI Design*, pages 129 – 177. North-Holland, 1990.
3. B Bose and S D Johnson. DDD-FM9001: Derivation of a verified microprocessor. In L Pierre G Milne, editor, *Correct Hardware Design and Verification Methods*, pages 191 – 202. Lecture Notes in Computer Science 683, Springer-Verlag, 1993.
4. A Cohn. A proof of correctness of the VIPER microprocessor: the first levels. In G Birtwistle and P A Subrahmanyam, editors, *VLSI Specification, Verification and Synthesis*, pages 27 – 72. Kluwer Academic Publishers, 1987.
5. W J Cullyer. Implementing safety critical systems: the viper microprocessor. In G Birtwistle and P A Subrahmanyam, editors, *VLSI Specification, Verification, and Synthesis*, pages 1 – 26. Kluwer Academic Publishers, 1987.
6. D Cyluk. Microprocessor verification in PVS. Technical report, SRI International Computer Science Laboratory Technical Report CSL-93-12, 1993.
7. A C J Fox and N A Harman. Algebraic models of correctness for microprocessors. Technical Report CSR 6-96, University of Wales Swansea, 1996.
8. A C J Fox and N A Harman. Algebraic models of microprocessors: Representation of advanced structures. Technical report, University of Wales Swansea, 1996.
9. M Gordon. Proving a computer correct with the LCF-LSM hardware verification system. Technical report, Technical Report No. 42, Computer Laboratory, University of Cambridge, 1983.
10. M J C Gordon and T Melham. *Introduction to HOL*. Cambridge University Press, 1993.
11. B Graham. *The SECD Microprocessor: a Verification Case Study*. Kluwer, 1992.
12. B Graham and G Birtwistle. Formalising the design of an SECD chip. In M Leeser and G Brown, editors, *Hardware Specification, Verification and Synthesis: Mathematical Aspects*, pages 40 – 66. Lecture Notes in Computer Science 408, Springer Verlag, 1990.

13. N A Harman and J V Tucker. Clocks, retimings, and the formal specification of a UART. In G J Milne, editor, *The Fusion of Hardware Design and Verification*, pages 375 – 396. North-Holland, 1988.

14. N A Harman and J V Tucker. Algebraic models and the correctness of microprocessors. In L Pierre G Milne, editor, *Correct Hardware Design and Verification Methods*. Lecture Notes in Computer Science 683, Springer-Verlag, 1993.

15. N A Harman and J V Tucker. Algebraic models of microprocessors: Architecture and organisation. Technical report, Acta Informatica vol. 33, in press (University of Wales, Swansea, Computer Science Report CSR 9-94), 1995.

16. N A Harman and J V Tucker. Algebraic models of microprocessors: the verification of a simple computer. Proceedings of the 2nd IMA Conference on Mathematics for Dependable Systems, to appear, 1995.

17. W Hunt. *FM8501: A Verified Microprocessor*. Lecture Notes on Artificial Intelligence 795, Springer Verlag, 1994.

18. W A Hunt. Microprocessor design verification. *Journal of Automated Reasoning*, 5(4):429 – 460, 1989.

19. J Joyce. Formal verification and implementation of a microprocessor. In G Birtwistle and P A Subrahmanyam, editors, *VLSI Specification, Verification and Synthesis*, pages 129 – 159. Kluwer Academic Publishers, 1987.

20. P Landin. On the mechanical evaluation of expressions. *Computer Journal*, 6:308 – 320, 1963.

21. C E Leiserson, F M Rose, and J B Saxe. Optimizing synchronous circuitry by retiming. In R Bryant, editor, *Third Caltech Conference on VLSI*, volume 1983, pages 87–116. Computer Science Press, 1803 Research Boulevard, Rockville MD 20850, 1983.

22. K Meinke and J V Tucker. Universal algebra. In T S E Maibaum S Abramsky, D Gabbay, editor, *Handbook of Logic in Computer Science*, pages 189 – 411. Oxford University Press, 1992.

23. T Melham. Using recursive types to reason about hardware in higher order logic. In G J Milne, editor, *The Fusion of Hardware Design and Verification*, pages 27 – 50. North-Holland, 1988.

24. T F Melham. *Higher Order Logic and Hardware Verification*. Cambridge University Press Tracts in Theoretical Computer Science 31, 1993.

25. S Miller and M Srivas. Formal verification of an avionics microprocessor. Technical report, SRI International Computer Science Laboratory Technical Report CSL-95-04, 1995.

26. S Miller and M Srivas. Formal verification of the AAMP5 microprocessor: a case study in the industrial use of formal methods. In *Proceedings of WIFT 95, Boca Raton*, 1995.

27. S Owre, J Rushby, N Shankar, and M Srivas. A tutorial on using PVS. In *Proceedings of TPCD 94*, pages 258–279. Lecture Notes in Computer Science 901, Springer-Verlag, 1994.

28. J E Smith and G S Sohi. The microarchitecture of superscalar processors. In *Proceedings of the IEEE*, volume 83, pages 1609–1624, December 1995.

29. V Stavridou. *Formal Specification of Digital Systems*. Cambridge University Press Tracts in Theoretical Computer Science 37, 1993.

30. S Tahar and R Kumar. Implementing a methodology for formally verifying RISC processors in HOL. In *Higher-Order Logic Theorem Proving and its Applications*, pages 281 – 294. Lecture Notes in Computer Science 780, Springer-Verlag, 1993.

31. R M Tomasulo. An efficient algorithm for exploiting multiple arithmetic units. *IBM J. Res. Develop.*, pages 176–188, January 1967.

32. W Wechler. *Universal Algebra for Computer Scientists*. EATCS Monograph, Springer-Verlag, 1991.

33. P Windley. A theory of generic interpreters. In L Pierre G Milne, editor, *Correct Hardware Design and Verification Methods*, pages 122 – 134. Lecture Notes in Computer Science 683, Springer-Verlag, 1993.

34. P Windley and M Coe. A correctness model for pipelined microprocessors. In *Proceedings of the 2nd Conference on Theorem Provers in Circuit Design*, 1994.

Mechanically Checking a Lemma Used in an Automatic Verification Tool

Phillip J. Windley[1] and Jerry R. Burch[2]

[1] Laboratory for Applied Logic, Brigham Young University
[2] Cadence Berkeley Laboratories, Cadence Design Systems, Inc.

Abstract. Automatic formal verification methods sometimes depend on lemmas for decomposing proofs into parts. The decomposition simplifies the verification task for automatic tools, such as model checkers. Typically the lemmas are proven by hand, and apply to all instances where the automatic tool is applied. Mechanically verifying these lemmas using a theorem prover provides greater assurance that the decomposition is correct and can provide insight into possible simplifications and generalizations. This paper gives an example of such an exercise, proving a theorem used by Burch [Bur96] in the verification of a super-scalar processor model.

1 Introduction

Burch [Bur96] discusses a method for decomposing the verification of a superscalar microprocessor into lemmas that are more easily checked. The decomposition was one of several techniques given for making superscalar microprocessor verification practicable. Other techniques used to verify pipelined microprocessors [BD94, Cyr96, CRSS94, JDB95, SGGH91, SB90, SM95, SM96] and nonpipelined microprocessors [Bea93, Coh88, Hun85, JBG86] have not been demonstrated on superscalar models. The tool described in the paper performs the decomposition according to the theory and then proves the pieces using a decision procedure for a first order hardware description language.

The decomposition provides a simplification of the proof obligations similar to the generic interpreter theory described in [Win93]. The generic interpreter theory has been successfully used to verify several non-pipelined microprocessors [Win95a].

The decomposition described here, like the generic interpreter theory, does several things:

1. The decomposition provides a step-by-step approach to microprocessor specification by enumerating the necessary definitions.
2. Using the decomposition, the verification tool can derive the lemmas that need to be checked.
3. After these lemmas have been proven, the verification tool can use the decomposition to automatically derive the final result from the lemmas.

The generic interpreter theory is a decomposition that frees the user from worrying about those aspects of the verification that need only be done once in the generic

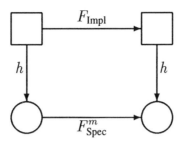

Fig. 1. Commutative diagram for the correctness criteria. Squares designate implementation states; circles designate specification states.

component. But, the generic interpreter theory is not useful in verifying pipelined microprocessors since it requires that the data and temporal abstractions be orthogonal. In a pipelined processor, the abstractions are not orthogonal and thus cannot be separated [Win95b, WC94].

Like the generic interpreter theory, the decomposition described by the pipeline decomposition theory structures the proof and provides standardized formats for key theorems and lemmas, thereby easing the verification burden. Unlike the generic interpreter theory, the pipeline decomposition theory does not require that the data and temporal abstractions be orthogonal.

In this paper, we provide a formal verification of Burch's pipeline decomposition theory in the HOL theorem proving system. The proof not only provides a formal check of the theory, but is done as an abstract theory, making the result useful for structuring proofs about superscalar microprocessors in the theorem proving system.

The next two sections describe the correctness criterion used by Burch *et. al.* [BD94, JDB95] and how that criterion can be decomposed for superscalar processor verification [Bur96]. Section 4 informally outlines the correctness proof for the decomposition technique, while section 5 describes how the decomposition theory is encoded and mechanically checked in the HOL theorem proving system.

2 Correctness Criteria

The primary correctness criterion used by Burch *et. al.* [BD94, JDB95] is a variation of the standard commutative diagram that relates an implementation to a specification through an abstraction function (see Figure 1). The functions F_{Spec} and F_{Impl} are state transition functions for the specification and implementation, respectively, of the processor. The abstraction function h maps an implementation state to the corresponding specification state.

The precise meaning of the commutative diagram is

$$\forall Q_{\text{Impl}}. \exists m. \ h(F_{\text{Impl}}(Q_{\text{Impl}})) = F_{\text{Spec}}^m(h(Q_{\text{Impl}})), \tag{1}$$

where F_{Spec}^m denotes m applications of the function F_{Spec}. The integer m is needed to keep the implementation and the specification "in sync". For example, if the processor does not fetch an instruction in state Q_{Impl} (due to a load interlock, say), then m would be zero. For a superscalar processor, m can be greater than one. The diagram in Figure 1 is said to *commute* if and only if Proposition 1 holds.

By using appropriate control inputs, almost all pipelined processors can be made to continue execution of instructions already in the pipeline while not fetching any new instructions. This is typically referred to as *stalling* the processor. We define F_{Stall} to be a function from implementation states to implementation states that represents the effect of stalling the implementation for one cycle.

All instructions currently in the pipeline can be completed by stalling for a sufficient number of cycles. This operation is called *flushing* the pipeline. We use the function F_{Flush} to model the act of flushing the pipeline. For some sufficiently large integer k,

$$F_{\text{Flush}} = F_{\text{Stall}}^k. \tag{2}$$

Burch and Dill [BD94] showed how the abstraction function h can be constructed automatically using the following equation:

$$h(Q_{\text{Impl}}) = proj(F_{\text{Flush}}(Q_{\text{Impl}})),$$

where the function $proj$ converts an implementation state to a specification state by simply stripping off all but the programmer-visible parts of the implementation state.

There are processors that cannot be stalled: if no new instruction is fetched in a given cycle, then the instructions already in the pipeline do not continue execution. The processor PP used in the Stanford University FLASH project [K+94] is an example of this. Thus, when verifying PP [JDB95], it is not possible to directly construct the abstraction function using F_{Stall} and F_{Flush} as described above. Instead, the processor is modified so that it can be stalled when a special control input is asserted. The modified model is used to construct F_{Stall}, but the original model is used for F_{Impl} in the commutative diagram. Although an error in modifying the processor can lead to a false negative verification result, it can never lead to a false positive.

It is possible for a processor to satisfy Proposition 1 for all reachable implementation states, yet not for all possible implementation states. In this case, the processor is correct but does not satisfy the correctness criterion. This problem can be solved by having the user provide an *invariant*, a predicate which gives an upper bound on the set of reachable states. It must be checked that the invariant contains all initial states and is closed under the implementation transition function. This can be done automatically. Then Proposition 1 is modified to quantify only over implementation states that satisfy the invariant. For the remainder of the paper, universal quantification over the implementation states is understood to be restricted to those states that satisfy the invariant.

3 Decomposition

The theory presented by Burch [Bur96] decomposes the commutative diagram given by Figure 1 into the three new commutative diagrams shown in Figure 2. As described below, these three propositions are significantly easier to verify than the original.

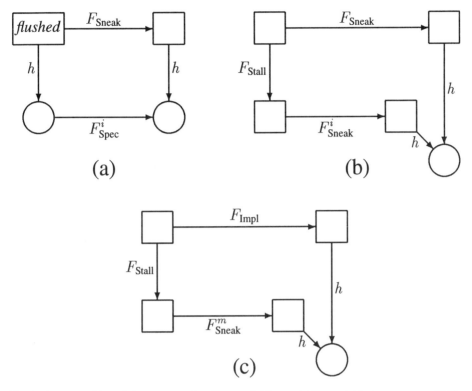

Fig. 2. Decomposition of the commutative diagram in Figure 1. The variable i ranges over $\{0,1\}$ and m ranges from zero to the maximum number of instructions the implementation can fetch in one cycle.

In addition to the propositions shown in Figure 2, the theory requires that a stall have no effect on the abstracted state:

$$\forall Q_{Impl}.\ h(F_{Stall}(Q_{Impl})) = h(Q_{Impl}). \tag{3}$$

The method further requires that the user define a state transition function, F_{Sneak}, which models the effect of a *sneak fetch*: the fetching of one instruction into the processor without having any previously fetched instructions flow down the pipeline. Typically, the only implementation variables changed by a sneak fetch are the instruction queue and the fetch program counter. For implementation states where there is not room in the fetch queue to fetch an instruction, F_{Sneak} should leave the implementation state unchanged.

The user must also provide a predicate P_{Flush} to characterize when an implementation state is flushed (contains no partially executed instructions). The verifier automatically checks that

$$\forall Q_{Impl}.\ P_{Flush}(F_{Flush}(Q_{Impl})). \tag{4}$$

Figure 2(a) checks that the implementation satisfies the specification when executing one instruction at a time. If F_{Impl} were used in the diagram instead of F_{Sneak}, then it would be necessary to consider up to m instructions rather than just one, where m is the maximum number of instructions that can be fetched by the processor in one cycle. In general, the superscript i in Figure 2(a) is either 0 or 1 depending on the implementation state. In practice i is always 1 for the types of processors with which we are familiar. Computationally, the check of the diagram in Figure 2(a) is easy because the initial state is flushed and the fetching and execution of at most one instruction is involved.

Figure 2(b) checks that inserting a stall cycle has no effect on the specification state produced by fetching one instruction and then applying the abstraction function. This can catch a wide variety of errors in interlocks, forwarding logic, etc. The superscript i is a user-provided function from implementation states to 0 or 1. There are potentially many functions that cause the diagram to commute for a correct processor. One possibility is to set i to 0 for implementation states in which F_{Sneak} does not fetch an instruction or the fetched instruction is squashed in the following cycle; otherwise i is set to 1.

Figure 2(c) checks that the implementation fetches instructions properly. The superscript m is a user-provided function from implementation states to the non-negative integers. As with the diagram of Figure 2(b), there are potentially many functions that cause the diagram to commute for a correct processor. One possibility is to set m to the number of instructions that are fetched in the current cycle and not squashed in the following cycle.

The diagrams of Figure 2(b) and Figure 2(c) are computationally more difficult to check than the diagram of Figure 2(a). Even so, they can be much easier to check than Figure 1, for the following reason: typically, the two implementation states on the right side of Figure 2(c) have the same values for most of their state variables. Thus, after the abstraction function h is applied, the expressions that need to be checked for equality have similar structure. This structure can be exploited by the validity checker.

Similar intuition applies to Figure 2(b). However, in that diagram many of state variables of the lower right implementation state differ from the upper right implementation state because of the application of F_{Stall}. The stall cycle causes instructions to have flowed down the pipeline more in one state than in the other. Nonetheless because the abstraction function h is constructed using F_{Stall}, this difference is largely removed after h is applied. So, as in Figure 2(c), the expressions that are checked for equality have similar structure.

In summary, the user provides the following inputs to the verification task:

- the transition functions F_{Spec} and F_{Impl};
- the transition function F_{Stall}, which is usually derived from F_{Impl} by constraining the values of certain control inputs;
- the number k of iterations of F_{Stall} needed to construct F_{Flush};
- the predicate P_{Flush} that characterizes the set of flushed implementation states;
- the transition function F_{Sneak};
- functions from implementation states to non-negative integers that keep the diagrams in Figure 2 "in sync" by providing values for i and m;
- an invariant, if necessary.

The automatic verification tool then checks the following:

- the diagrams in Figure 2 commute;

- Proposition (4);
- the proposition

$$\forall Q_{\text{Impl}}.\ P_{\text{Flush}}(Q_{\text{Impl}}) \implies proj(F_{\text{Stall}}(Q_{\text{Impl}})) = proj(Q_{\text{Impl}}),$$

which is sufficient to show Proposition (3) given Proposition (4) and the definition of h;
- the invariant is inductive (*i.e.,* is closed under F_{Impl}, F_{Stall} and F_{Sneak}) and contains the initial states of the processor.

The original correctness criterion of Figure 1 is not applicable to processors with imprecise interrupts because even a correct processor may not execute the same sequence of instructions as the specification when an imprecise exception occurs. The decomposition theory shares this limitation. Similarly, both theories need to be extended to handle input/output as well as restrictions on the instruction stream (such as load a delay slot). Other features, such as interrupts, precise exceptions and out-of-order execution, can be handled by the decomposition technique.

Notice that among the propositions which the tool checks, only the diagram in Figure 2(a) mentions the specification F_{Spec}. The other propositions provide a self-consistency check for the implementation, similar to the self-consistency checks described by Jones, Seger and Dill [JSD96]. These propositions can be used for partial verification if the specification is not available or is difficult to formalize. If F_{Spec} is available, then all of the propositions (including Figure 2(a)) can be combined to provide full verification.

4   Proof Outline

The proof that the commuting diagram in Figure 1 follows from the commuting diagrams of Figure 2 depends on the two intermediate lemmas shown in Figure 3.

Figure 3(a) is similar to Figure 2(b) except that the F_{Stall} has been replaced by F_{Flush}. Consequently, the lower left corner of Figure 3(a) is flushed. Figure 3(b) is similar to Figure 2(a) except that the starting implementation state (upper left corner) is not flushed.

The proof of Figure 3(a) follows from Figure 2(b) using induction, the fact that flushing is defined in terms of repeated stalls, and the side condition that stalling has no effect on the abstracted state.

The proof of Figure 3(b) depends on the following lemma:

$$\forall s.\ h(F_{\text{Flush}}(s)) = h(s),$$

which follows using induction and the side condition that stalling has no effect on the abstracted state. Using this lemma, along with Figure 2(a) and Figure 3(a), proving Figure 3(b) is straightforward.

The final result, the commuting diagram in Figure 1, is proven using induction, Figure 2(c), and Figure 3(b). The essential idea is to substitute Figure 3(b) into Figure 2(c) m times.

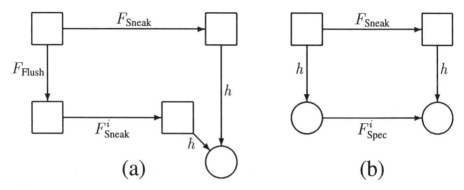

Fig. 3. Representation of intermediate lemmas used in the proof.

5 Mechanically Checked Proof

In this section we describe the formal specification of the pipeline decomposition theory and its mechanical verification using the HOL theorem proving system. This entails describing the theory and its supporting definitions in the HOL specification language and then proving the required properties using the HOL theorem prover.

We begin with a brief overview of the HOL theorem proving system that should suffice for understanding the definitions and theorems presented later. Then we describe our application of HOL to the decomposition theory.

5.1 Brief Introduction to the HOL Theorem Proving System

To ensure the accuracy of our specifications and proofs, we developed them using a mechanical verification system. The mechanical system performs syntax and type checking of the specifications and prevents the proofs from containing logical mistakes. The HOL system was selected for this project because it has higher-order logic, generic specifications, and polymorphic type constructs. These features directly affect the expressibility of the specification language. Furthermore, HOL is widely available and robust, and has a growing world-wide user base. However, there is nothing about our work which specifically requires the HOL theorem proving system. Other theorem proving systems which support abstract theories, such as PVS, would work as well.

HOL is a general theorem proving system developed at the University of Cambridge [CGM87, Gor88] that is based on Church's theory of simple types, or higher-order logic [Chu40]. Higher-order logic is similar to predicate logic, but allows quantification over predicates and functions, thereby permitting more general systems to be described.

For the most part, the notation of HOL is that of standard logic; $\forall$, $\exists$, $\land$, $\lor$, etc. have their usual meanings. There are a few constructs that deserve special attention due to their use in this paper. HOL types are identified by a prefixed colon. Built-in types include :bool and :num. Function types are constructed using $\longrightarrow$. HOL is polymorphic; type variables are indicated by type names that begin with an asterisk.

5.2 Some Auxiliary Definitions

In formalizing the proof in HOL, we found it necessary to define two combinators that perform functions corresponding to the superscripts in the informal presentation.

The R combinator formalizes the superscript m used in the informal presentation. R repeats a function n times.

$$\vdash_{def} \ (\text{R } 0 \text{ f } = \text{ I}) \ \wedge \ (\text{R } (\text{SUC } n) \text{ f } = \text{ f } o \ (\text{R } n \text{ f}))$$

Note that o is standard function composition and I is the identity function.

The Ch combinator is used to formalize the superscript i used in the informal presentation. The combinator either applies a function or not, according to the value of its boolean parameter.

$$\vdash_{def} \ (\text{Ch } \text{T } f = f) \ \wedge \ (\text{Ch } \text{F } f = \text{I})$$

5.3 The Abstract Representation

The abstract representation defines a number of uninterpreted constants. The idea behind an abstract representation is that while the constants are uninterpreted inside the current theory (i.e. the whole of what is presented in this paper), someone who uses the theory can provide an interpretation, as long as it meets the type specification and the theory obligations given in the next section. Thus, the abstract representation becomes a parameterization of the theory.

The abstract representation for the pipeline decomposition theory is shown in Figure 4 and gives the names and types of the important functions from the informal presentation. Here and elsewhere in this section, the names correspond as closely as possible to the names in the informal presentation.

- **F_spec** denotes the function that specifies the abstract behavior of the pipeline. The specification function maps a specification state, *state, into a specification state.
- **F_impl** denotes the function that specifies the implementation of the pipeline. This function maps an implementation state, *state′, into another implementation state.
- **proj** denotes the function that projects an implementation state into a specification state. In practice this usually entails simply stripping off all but the programmer visible portions of the state; although, since the function is uninterpreted, there is nothing here to require that particular implementation.
- **F_sneak** denotes the function that models the effect of a *sneak fetch*. A sneak fetch fetches an instruction into the processor without allowing any of the instructions in the pipeline to proceed. This function maps an implementation state into another implementation state.
- **F_stall** denotes a function which models the effect of a pipeline stall. This function maps an implementation state into another implementation state. In practice, this function may be derived in some manner from F_impl, but our proof does not require that it is.

```
new_abstract_representation 'commute'
    [
      ('F_spec',":*state->*state")
      ;
      ('F_impl',":*state'->*state'")
      ;
      ('proj',":*state'->*state")
      ;
      ('F_sneak',":*state'->*state'")
      ;
      ('F_stall',":*state'->*state'")
      ;
      ('F_flush', ":*state'->*state'")
    ];;
```

Fig. 4. Abstract Representation for the Pipeline Decomposition Theory

- **F_flush** denotes a function that flushes the pipeline. This function maps an implementation state into another implementation state.

The Abstraction Function The abstraction function maps implementation states into specification states. The function is not properly part of the abstract representation. Rather it is defined in terms of two members of the abstract representation, F_flush and proj:

$$\vdash_{def} \forall (Q:*state') \ . \ (h\ Q) \ = \ (proj\ (F\_flush\ Q))$$

5.4 The Theory Obligations

The theory obligations can be looked at in several ways. In one view, they represent a partial semantics of the uninterpreted constants defined in the abstract representation. In this view, the partial semantics is used to prove desired results about the constants of the abstract representation.

In an alternate view, the theory obligations represent the requirements, over and above the type requirements, that are part of the abstract representation. These obligations must be met by any interpretation of the abstract representation. That is, if one looks upon the abstract representation as the parameters of the theory, the theory obligations represent additional restrictions upon any instantiation of those parameters.

The first view is most useful when one is developing the abstract theory, and the second is most useful when one is using it. The two views must be balanced so that we can prove the desired results without imposing an unnecessary burden on users of the theory. Thus, ideally, the partial semantics say just enough of what is required to prove the desired properties, and no more.

In the pipeline decomposition theory, the theory obligation can be broken into two kinds: side conditions which must be verified to provide the desired result but are relatively straightforward to prove, and major propositions which form the heart of the decomposition.

Side Conditions The side conditions, which we assume are true about the abstract representation, take the form of two simple relationships between its members. The first states that the flush function, F_flush, can be implemented using some number of consecutive stalls. The second states that a stall has no effect on the abstracted state.

```
∀s . ∃k . (F_flush  s) = ((R k F_stall) s)

∀s . (h (F_stall s)) = (h s)
```

The first side condition corresponds to Proposition 2 in Section 2 and the second to Proposition 3 in Section 2.

When the pipeline decomposition theory is used, the side conditions will have to be met by whatever functions are used to instantiate the abstract representation. In practice, this is relatively straightforward.

The Major Propositions The major propositions represent the semantics we assume for the abstract representation. These propositions correspond to the diagrams presented in Figure 2 with one exception that will be noted below.

The first proposition corresponds to (a) in Figure 2. The proposition states that sneak fetching on a flushed implementation state s' and then abstracting (recall that abstracting involves flushing again) is equivalent to abstracting s' and then running the specification zero or one times.

```
∀s . ∃i . (h (F_sneak (F_flush s))) =
              ((Ch i F_spec) (h (F_flush s)))
```

There is a difference between this proposition and the one given in Section 4. In part (a) of Figure 2, the state in the upper left corner of the diagram is "flushed," whereas the proposition merely requires that the state be the result of applying the F_flush function. This is a subtle distinction since nowhere have we required that F_flush actually flush the pipeline. Indeed, the formal theory has no notion of what it means for the pipeline to be flushed. In practice, this proposition is easier to check with an automatic verifier when starting from a flushed state. Our formalization does not exclude this since it is a subset of all the possibilities afforded by F_flush. On the other hand, our formalization does not require it either since the user is free to supply any instantiation of F_flush provided it meets the first side condition discussed above.

The second proposition corresponds to (b) in Figure 2. It essentially says that stalling before a sneak fetch has no effect on the resulting abstracted state.

```
∀s . ∃i . (h (F_sneak s)) =
             (h ((Ch i F_sneak) (F_stall s)))
```

The third proposition corresponds to (c) in Figure 2. This proposition requires a correspondence between the implementation transition function and some number of sneak fetches on a stalled implementation state.

```
∀s . ∃m . (h (F_impl s)) =
             (h ((R m F_sneak) (F_stall s)))
```

5.5 The Verification

The main result, that is the commuting diagram shown in Figure 1, follows from the two side conditions and three propositions of the previous section. The proof depends on three lemmas which are discussed next. [3]

Lemma 1: Sneaking Hides a Flush The first lemma corresponds to Figure 3(a). The lemma is similar to the proposition shown in Figure 2(b) except that the stall is replaced by a flush. Thus, the lemma generalizes the first proposition to prove that any number of stalls are acceptable, not just one (recall that first side condition requires that a flush be defined in terms of some number of stalls).

```
∀s . ∃i . (h (F_sneak s)) =
             (h ((Ch i F_sneak) (F_flush s)))
```

The proof of this lemma follows from induction on the number of stalls in F_flush. The base case, where the number of stalls is zero, is trivially solved by choosing i to be true. The inductive step makes use of the inductive hypothesis to create an equality between a path that stalls k times and a path that stalls $k + 1$ times. Because the second side condition discussed above allows us to drop stalls that are subsequently abstracted with h, the two sides can be shown to be equal.

Lemma 2: Abstraction Hides a Flush The second lemma is a statement about how extra stalls affect the abstracted state. We know, from the second side condition that a single stall does not affect the state. This lemma generalizes that to show that any number of stalls (and thus a flush) do not affect the specification state.

```
∀s . (h (F_flush s)) = (h s)
```

---

[3] For clarity, in the presentation of the HOL theorems that follow we have removed an argument on the abstract components which is used to make the abstract theory work in HOL, but which has no meaning otherwise.

The proof of this lemma is relatively easy using induction on the number of stalls in the flush and the second side condition. Both the base and inductive cases can be proven by rewriting with the definitions of combinators R, I, and o.

Lemma 3: No Flush Necessary The third lemma used in the proof corresponds to Figure 3(b). This lemma is a generalization of the proposition shown in Figure 2(a). The form is essentially the same except that there is no requirement that the beginning state be flushed.

```
∀s . ∃i . (h (F_sneak s)) = ((Ch i F_spec) (h s))
```

Rewriting with Lemma 1 replaces the left hand side of this lemma with the right hand side of Lemma 1:

```
∀s . ∃i . (h (F_sneak (F_flush s))) = ((Ch i F_spec) (h s))
```

Rewriting with the first proposition (Figure 2(a)) and Lemma 2 easily solves this.

The Main Result The main result corresponds to the commuting diagram shown in Figure 1. The commuting diagram shows that running the implementation on an implementation state and then abstracting the result produces the same result as abstracting the implementation state and then running the specification on it some number of times (the number can range from 0, in the case of a pipeline stall, to the maximum number of instructions that can be fetched in a single cycle):

```
∀s . ∃m . (h (F_impl s)) = ((R m F_spec) (h s))
```

The proof of the main result begins by rewriting with the third proposition (Figure 2(c)) to produce the following equation:

```
∀s . ∀n . ∃m . (h ((R n F_sneak) (F_stall s))) =
                ((R m F_spec) (h s))
```

The proof proceeds by inducting on n, the number of times a sneak fetch is performed. The base case is quite easily solved. The inductive step makes use of the inductive hypothesis and Lemma 3 from above to replace one sneak fetch with a possible application of F_spec yielding the following intermediate result:

```
∀s . ∀i . ∀m . ∃k . ((Ch i F_spec) ((R m F_spec) (h s))) =
                     ((R k F_spec) (h s))
```

This goal is solved using case analysis on i. Recall that Ch applies a function or not depending on the boolean value of i. In the case where i is true and F_spec is applied, we pick k to be (SUC m). In the case where i is false and F_spec is *not* applied, we pick k to be m.

Discussion of the Proof The proofs of the three lemmas and the subsequent final result are not difficult to formalize in HOL. Indeed, the formal specification and verification of the pipeline decomposition theory follow the informal representation and proof quite closely with only a few minor exceptions.

We found that using a mechanical theorem prover, with its built–in support for induction, made some parts of the proof clearer. Mechanical proof also provided a much needed bookkeeping system for keeping track of case splits, existential witness selection, and so on.

The contribution of the proof is twofold. First the proof provides assurance that the theory used in the automated tool is correct and clarifies some aspects of the proof. Second, the proof can be used inside HOL, through instantiation, to perform pipeline verifications in the theorem proving environment.

6 Conclusions

The preceding sections have presented the proof of the pipeline decomposition theory and its formalization in the HOL theorem proving system. The formalization provides added assurance that the operation of the automatic tool on which it is based is correct. Because automatic tools frequently depend on simple logics for decidability, verifying extra-logical decompositions and simplifications in a theorem prover provides added assurance.

The formalization also provides a clear characterization of the decomposition and all of the conditions on which it depends. This characterization takes the form of an abstract theory in the HOL theorem proving system and thus represents a tool in that system for further work in pipeline verification. Our previous experience with the generic interpreter theory indicates that such formalizations are useful for structuring proofs and providing direction on the kinds of definitions and lemmas required in the specification and verification. The proof is useful outside the HOL theorem proving system since its formalization ensures that the characterization is complete and correct.

The formalization of the pipeline decomposition theory did not find any large omissions in the informal proof or bring to light any inconsistencies in its construction. The informal proof was correct. Still, the formalization gives added assurance that nothing was overlooked in the informal proof and provides a clarity not available in the informal arguments. In addition, our experience in the formalization provides some useful insights:

- The formal result is more general than the informal presentation in some minor, but perhaps useful, ways. For example, the proposition corresponding to Figure 2(a) in the formal presentation does not require that the starting state be flushed, but merely be the result of applying F_{Flush} to some starting state.
- The formal result spells out exactly which side conditions need to be proven for any instantiation of the abstract representation.

There are several areas for future research:

- The generic interpreter theory provides for decomposing a microprocessor proof hierarchically as well as decomposing each step in the hierarchy. The pipeline decomposition theory provides no explicit information about hierarchical decomposition, although such decomposition seems reasonable.
- In addition to the decomposition technique described here, Burch [Bur96] presented a new method for automatically constructing an abstraction function. The method involves stalling the processor in several different modes, as opposed to the one mode represented by F_{Stall}. The verification tool combines this new abstraction function with the decomposition technique which results in a more complicated lemma than using the decomposition technique alone. We hope to use HOL to mechanically check this more complicated lemma.

The exercise presented in this paper is useful, we believe, for anyone contemplating the use of theorem provers and automatic tools. Automatic tools are necessary and useful for ensuring the task can be completed with reasonable effort. Theorem provers provide a means of ensuring the tools themselves are correct and give insight into the processes that the tools use.

Acknowledgments

Dr. Windley is supported in part by NSA contract MDA904-94-C-6115.

References

[BD94]   Jerry R. Burch and David L. Dill. Automatic verification of pipelined microprocessor control. In David L. Dill, editor, *Conference on Computer-Aided Verification*, volume 818 of *Lecture Notes in Computer Science*. Springer-Verlag, June 1994.

[Bea93]   Derek L. Beatty. *A Methodology for Formal Hardware Verification, with Application to Microprocessors*. PhD thesis, School of Computer Science, Carnegie Mellon University, August 1993.

[Bur96]   Jerry R. Burch. Techniques for verifying superscalar microprocessors. In *Design Automation Conference*, June 1996.

[CGM87]   Albert Camilleri, Mike Gordon, and Tom Melham. Hardware verification using higher order logic. In D. Borrione, editor, *From HDL Descriptions to Guaranteed Correct Circuit Designs*. Elsevier Scientific Publishers, 1987.

[Chu40]   Alonzo Church. A formulation of the simple theory of types. *Journal of Symbolic Logic*, 5, 1940.

[Coh88]   A. J. Cohn. A proof of correctness of the Viper microprocessors: The first level. In G. Birtwistle and P. A. Subrahmanyam, editors, *VLSI Specification, Verification and Synthesis*, pages 27–72. Kluwer, 1988.

[CRSS94]   D. Cyrluk, S. Rajan, N. Shankar, and M. K. Srivas. Effective theorem proving for hardware verification. In *Conference on Theorem Provers in Circuit Design*, September 1994.

[Cyr96]   David Cyrluk. Inverting the abstraction mapping: A methodology for hardware verification. In *Conference on Formal Methods in Computer-Aided Design*, November 1996. (This volume).

[Gor88]   Michael J. C. Gordon. HOL: A proof generating system for higher-order logic. In G. Birtwhistle and P. A. Subrahmanyam, editors, *VLSI Specification, Verification, and Synthesis*. Kluwer Academic Press, 1988.

[Hun85]   W. A. Hunt, Jr. FM8501: A verified microprocessor. Technical Report 47, University of Texas at Austin, Institute for Computing Science, December 1985.

[JBG86]   J. Joyce, G. Birtwistle, and M. Gordon. Proving a computer correct in higher order logic. Technical Report 100, Computer Lab., University of Cambridge, 1986.

[JDB95]   Robert B. Jones, David L. Dill, and Jerry R. Burch. Efficient validity checking for processor verification. In *Intl. Conf. on Comp. Aided Design*, 1995.

[JSD96]   Robert B. Jones, Carl-Johan H. Seger, and David L. Dill. Self-consistency checking. In *Conference on Formal Methods in Computer-Aided Design*, November 1996. (This volume).

[K+94]   Jeffrey Kuskin et al. The Stanford FLASH multiprocessor. In *Intl. Symp. on Comp. Arch.*, 1994.

[SB90]   Mandayam Srivas and Mark Bickford. Formal verification of a pipelined microprocessor. *IEEE Software*, 7(5):52–64, September 1990.

[SGGH91]   James B. Saxe, Stephen J. Garland, John V. Guttag, and James J. Horning. Using transformations and verification in circuit design. Technical Report 78, DEC Systems Research Center, September 1991.

[SM95]   Mandayam Srivas and Steven P. Miller. Applying formal verification to a commercial microprocessor. In *IFIP Conference on Hardware Description Languages (CHDL 95)*, Tokyo, Japan, August 1995.

[SM96]   M. Srivas and S. P. Miller. Applying formal verification to the AAMP5 microprocessor: A case study in the industrial use of formal methods. *Formal Methods in System Design*, 8(2):153–88, March 1996.

[WC94]   Phillip J. Windley and Michael Coe. The formal verification of instruction pipelines. In Ramayya Kumar and Thomas Kropf, editors, *Conference on Theorem Provers in Circuit Design*, Lecture Notes in Computer Science. Springer-Verlag, September 1994.

[Win93]   Phillip J. Windley. A theory of generic interpreters. In George J. Milne and Laurence Pierre, editors, *Correct Hardware Design and Verification Methods*, volume 683 of *Lecture Notes in Computer Science*, pages 122–134. Springer-Verlag, June 1993.

[Win95a]   Phillip J. Windley. Formal modeling and verification of microprocessors. *IEEE Transactions on Computers*, 44(1), January 1995.

[Win95b]   Phillip J. Windley. Verifying pipelined microprocessors. In *IFIP Conference on Hardware Description Languages (CHDL 95)*, Tokyo, Japan, August 1995.

Automatic Generation of Invariants in Processor Verification

Jeffrey X. Su, David L. Dill, and Clark W. Barrett

Computer Science Department
Stanford University

Abstract. A central task in formal verification is the definition of *invariants*, which characterize the reachable states of the system. When a system is finite-state, invariants can be discovered automatically.

Our experience in verifying microprocessors using symbolic logic is that finding adequate invariants is extremely time-consuming. We present three techniques for automating the discovery of some of these invariants. All of them are essentially syntactic transformations on a logical formula derived from the state transition function. The goal is to eliminate quantifiers and extract small clauses implied by the larger formula.

We have implemented the method and exercised it on a description of the FLASH Protocol Processor (PP), a microprocessor designed at Stanford for handling communications in a multiprocessor. We had previously verified the PP by manually deriving invariants.

Although the method is simple, it discovered 6 out of 7 of the invariants needed for verification of the CPU of the processor design, and 28 out of 72 invariants needed for verification of the memory system of the processor. We believe that, in the future, the discovery of invariants can be largely automated by a combination of different methods, including this one.

1 Introduction

As microprocessors become more complex, an increasingly large fraction of the design time is spent on validation. Validation techniques which rely entirely on simulation are limited because of the many possible cases which must be accounted for. Sometimes, significant bugs are found after a processor is commercially released, which is both embarrassing and costly.

Formal verification techniques have been steadily improving over the last decade, and several simple microprocessors have been verified [3, 6, 8]. The methods used in these verification efforts are based on theorem provers, which require a great deal of expert guidance. In addition, some automatic techniques used on simple processors (i.e. [9]) are not applicable to pipelined processors. And even though some pipelined processors have been successfully verified [4, 14, 15, 17], they are either very simple or require a great deal of work to verify.

Burch and Dill [1] proposed a new method for verifying the control of microprocessors. The verification method compares two behavioral descriptions of the processor: a pipelined implementation and a simpler, unpipelined specification. A symbolic simulator is used to generate a collection of formulas in a quantifier-free fragment of first-order logic including propositional connectives, equality, and uninterpreted functions. These formulas are checked for universal truth by a *validity checker*. For an incorrect implementation, this validity checker can produce a specific example where the implementation of the processor contradicts its specification. Significant effort has gone into making this validity checker fast and efficient [7, 18].

For all but the simplest descriptions, the proofs require *invariants*: a logical formula characterizing a superset of the states reachable from the initial state of the processor. Currently, finding appropriate invariants is the single most labor-intensive part of the verification method.

The most complex design to which we have applied the method is the Stanford Protocol Processor (PP) [16], designed by the Stanford FLASH group. The proof was divided into two parts: verification of the CPU and verification of the memory system. Most of the invariants had to be found by trial and error: we attempted verification; when it failed, we analyzed the results to see if the problem was a design error (very rarely), an error from translating the original design to our input language (frequent), an error in the invariant (frequent), or an invariant that was too weak (frequent). This process took a few weeks for the CPU, and a few months for the memory system. It became clear from this experience that even partial automation of invariant discovery would be a major step towards making processor verification more practical.

The invariant discovery methods described here were inspired by inspecting the logical formulas involved in the verification of the PP. The method is to find a formula characterizing the set of states reachable from at least one other state. The disjunction of this formula with a formula characterizing the initial state is an invariant, as is anything implied by the disjunction. However, the formula has existential quantifiers, which is a problem since we are working in a quantifier-free logic. So we employ several simple transformations to extract quantifier-free formulas that are implied by the original formula (and, thus, also invariants).

Perhaps the most surprising result is that, although the method is seemingly weak, it finds a significant number of the invariants that are needed in the proof of the PP. Indeed, it discovers 6 out of the 7 invariants required for the CPU verification, and 28 out of 72 invariants required in the memory system. From this, we can reasonably conclude that the difficulty of verifying the PP would have been greatly reduced if we had used this method to find the invariants instead of doing it manually. However, we believe that the the greatest potential of the method will be realized when it is used in combination with other invariant discovery techniques.

The rest of the paper is organized as follows: related work in invariant generation is discussed in Section 2. The logic and processor model are described in Section 3. Our methods for automatically finding invariants along with proofs of their correctness are presented in Section 4. Experimental results are provided in Section 5, and some concluding remarks are given in Section 6.

2 Generating Invariants

There are existing methods to find invariants automatically. Manna and Pnueli show how to generate invariants for proving safety properties in fair transition systems [11], and these methods have been extended by Bensalem et al [13]. Their methods can be classified as either top-down or bottom-up.

Top-down invariant generation begins with an assertion that we desire to prove for a given system. If this assertion is not valid in general, then various heuristics are applied to strengthen the assertion. However, this method is not guaranteed ever to produce a valid assertion. Bottom-up methods, on the other hand, simply look at the system and use it to deduce assertions which are always valid. Bottom-up invariants tend to be simple properties of the system, and few practically interesting properties are likely to be generated using only bottom-up methods. The two methods can be combined by using invariants generated by the bottom-up method as strengthening assertions for a top-down approach. The Stanford Temporal Prover (STeP) is a system which implements these methods in a tool for program verification [10]. Many properties of simple examples can be proved with very little manual effort by using the automatic methods provided by STeP.

Although there are inherent differences between our verification model and that used by STeP, the ideas of top-down and bottom-up invariant generation are applicable to the approach we take. Ours are bottom-up syntactic methods for extracting simple invariants from the state transition function of an implementation. These invariants are then used to limit the initial state space in our verification model. If the automatic invariants are not sufficient, additional assertions can be added manually in a top-down manner until a valid verification condition is obtained.

3 Logic and Processor Model

A processor is modeled as a Mealy machine $M=(Q, \Sigma, \Delta, \delta, \eta, q_0)$, where Q is a finite set of states, $q_0 \in Q$ is the initial state, Σ and Δ are finite sets of inputs and outputs, δ is the state transition function which maps $Q \times \Sigma$ to Q, and η is the output function which maps $Q \times \Sigma$ to Δ. We require δ to be total (defined for all inputs and states). In the rest of the paper, we will assume that M is given.

The techniques we use to find invariants are based on syntactic manipulation of formulas in first-order logic. We will use the standard Boolean connectives as well as an if-then-else (ite) construct: we write $(ite\ \alpha\ \beta\ \gamma)$ to mean *if α then β else γ*. Within the formulas, we will use the following variable conventions: $\mathbf{x} \in \Sigma$, and $\mathbf{w}, \mathbf{z} \in Q$, where $\mathbf{w}$ is the old or previous state and $\mathbf{z}$ represents the new or current state.

We also assume that δ can be expressed as a formula in quantifier-free first-order logic (we obtain this formula by symbolic simulation). This formula is dependent on a state $\mathbf{w}$ and an input $\mathbf{x}$. We will denote the syntactic representation of the state transition function depending on variables $\mathbf{w}$ and $\mathbf{x}$ by $\tau[\mathbf{w}, \mathbf{x}]$.

We will deal primarily with Boolean logical formulas containing a single free variable which can range over the states of M. Such a formula will be called a *state predicate* and will be written as a symbol followed by the state variable (usually z) in square brackets (i.e. $I[z]$). As a special case, we define the *initial state predicate* as:

$$Q_0[z] \equiv z = q_0.$$

We also define the *predecessor state predicate* as:

$$P[z] \equiv \exists w, x. \, z = \tau[w, x].$$

Intuitively, $P[z]$ holds exactly when z has at least one predecessor.
A state predicate which is true for all reachable states is called an *invariant*. Formally,

Definition 1 (Invariant) *Let $I[z]$ be a state predicate. $I[z]$ is an invariant if:*
   Base Rule: $Q_0[z] \Rightarrow I[z]$, *and*
   Induction Rule: $I[w/z] \Rightarrow I[\tau[w, x]/z]$ [1]
are valid logical formulas (i.e. true for all possible interpretations of free variables). □

By this definition, the predecessor state predicate is an invariant. In practice, however, it is both difficult and computationally expensive to deal with invariants which contain quantifiers (especially in automatic methods). Thus we seek to find equivalent or weaker invariants which are implied by the predecessor state predicate and which do not contain quantifiers. We use the following theorem as the basis for three techniques to automatically generate such invariants.

Theorem 1. *If $I[z]$ is a state predicate and $\forall z. \, P[z] \Rightarrow I[z]$ is valid, then $Q_0[z] \vee I[z]$ is an invariant.*

Base Rule : Clearly, $Q_0[z] \Rightarrow Q_0[z] \vee I[z]$ is valid.
Induction Rule : We must show that $Q_0[w/z] \vee I[w] \Rightarrow Q_0[\tau[w, x]/z] \vee I[\tau[w, x]/z]$
   is valid. Consider just the formula $I[\tau[w, x]/z]$ and let q and σ be arbitrary assignments to w and x respectively. We know that $P[z] \Rightarrow I[z]$ is valid. By inspection, we see that $P[z]$ is true under the assignment $z \leftarrow \delta(q, \sigma)$. It must then be the case that $I[z]$ is also true under the same assignment. But this is equivalent to $I[\tau[w, x]/z]$ since τ simply computes δ. Thus $I[\tau[w, x]/z]$ is valid and it follows that $Q_0[w/z] \vee I[w] \Rightarrow Q_0[\tau[w, x]/z] \vee I[\tau[w, x]/z]$ is valid.

   □

Note that we did not use the hypothesis in the induction step. Thus, as was previously mentioned, these are weak invariants.

---

[1] We use the notation $I[y/z]$ to denote the result of substituting y for z in $I[z]$.

4 Automatic Generation of Invariants

In the previous section, for the sake of simplicity, we assumed that the state of a processor can be represented by a single state variable. In practice, it is much more useful to think of the state as being divided into many individual state variables. In the following, we will assume that states are actually n-tuples, so that any state $\mathbf{z}$ can be written as $(z_1, ..., z_n)$ and $\tau[\mathbf{w}, \mathbf{x}]$ can be written as $(\tau_1[\mathbf{w}, \mathbf{x}], ..., \tau_n[\mathbf{w}, \mathbf{x}])$. We will also assume that $\mathbf{w} = (w_1, ..., w_n)$ and $\mathbf{x} = (x_1, ..., x_m)$. We can then write the predecessor state predicate as follows:

$$P[z_1, ..., z_n] = \exists \mathbf{w}, \mathbf{x}. \bigwedge_{i=1}^{n} (z_i = \tau_i[\mathbf{w}, \mathbf{x}]).$$

We now use this formula to find valid quantifier-free state predicates. This is done by applying some simple syntactic transformations with the goal of eliminating the existentially quantified state and input variables (which we refer to as bound variables in the following discussion). These methods are discussed in the following subsections.

4.1 Method 1

Our first method is to replace subexpressions containing bound variables in $P(\mathbf{z})$ with state variables wherever possible. We accomplish this by discovering common subexpressions which are equivalent to the state transition function for some state variable. Formally, for each clause $z_i = \tau_i[\mathbf{w}, \mathbf{x}]$ in $P[\mathbf{z}]$, whenever $\tau_i[\mathbf{w}, \mathbf{x}]$ appears as a subexpression of $\tau_j[\mathbf{w}, \mathbf{x}]$ where $i \neq j$, we replace the occurrence of $\tau_i[\mathbf{w}, \mathbf{x}]$ in $\tau_j[\mathbf{w}, \mathbf{x}]$ with z_i. This results in a new set of state transition formulas which depend not only on the input and previous state variables, but on the current state variables as well. We will denote these new formulas by $(\tau_1'[\mathbf{w}, \mathbf{x}, \mathbf{z}], ..., \tau_n'[\mathbf{w}, \mathbf{x}, \mathbf{z}])$. We can then write a new predecessor state predicate $P'[\mathbf{z}]$ which is logically equivalent to $P[\mathbf{z}]$:

$$P'[\mathbf{z}] = \exists \mathbf{w}, \mathbf{x}. \bigwedge_{i=1}^{n} (z_i = \tau_i'[\mathbf{w}, \mathbf{x}, \mathbf{z}]).$$

This new predicate $P'[\mathbf{z}]$ has fewer occurrences of bound variables.

Now, if there are any clauses in $P'[\mathbf{z}]$ in which all bound variables have been eliminated as a result of the transformation, they can be moved outside the scope of the existential quantifiers. Call the conjunction of all such clauses $I_1[\mathbf{z}]$ and the remaining term $P_1[\mathbf{z}]$ so that $P'[\mathbf{z}] = I_1[\mathbf{z}] \wedge P_1[\mathbf{z}]$. For simplicity, we assume that the removed clauses are those associated with the highest numbered state variables so that:

$$P_1[\mathbf{z}] = \exists \mathbf{w}, \mathbf{x}. \bigwedge_{i=1}^{n_1} (z_i = \tau_i'[\mathbf{w}, \mathbf{x}, \mathbf{z}]).$$

where $n_1 \leq n$.

Since $I_1[\mathbf{z}]$ is a subexpression of $P'[\mathbf{z}]$, it follows that $P'[\mathbf{z}] \Rightarrow I_1[\mathbf{z}]$ is valid, and thus $Q_0[\mathbf{z}] \vee I_1[\mathbf{z}]$ is an invariant by Theorem 1. We will now apply this method to the circuit in Figure 1.

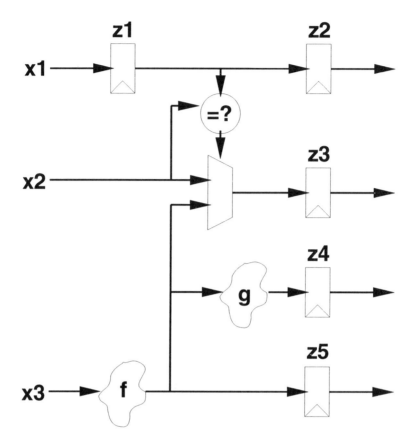

Fig. 1. A simple circuit.

Example 1 *The predecessor state predicate of this circuit is given below (recall that* **w** *represents the previous state and* **z** *represents the current state):*

$$\exists \mathbf{w}, \mathbf{x}. \, [z_1 = x_1 \, \wedge \, z_2 = w_1 \, \wedge \, z_3 = (ite \, (x_2 = w_1) \, x_2 \, f(x_3))$$
$$\wedge \, z_4 = g(f(x_3)) \, \wedge \, z_5 = f(x_3)]$$

where f and g represent functions computed by combinational logic. We now substitute state variables z_2 and z_5 for w_1 and $f(x_3)$:

$$\exists \mathbf{w}, \mathbf{x}. \, [z_1 = x_1 \, \wedge \, z_2 = w_1 \, \wedge \, z_3 = (ite \, (x_2 = z_2) \, x_2 \, z_5)$$
$$\wedge \, z_4 = g(z_5) \, \wedge \, z_5 = f(x_3)].$$

Since the clause $z_4 = g(z_5)$ does not have any bound variables and is the only such clause, we factor it out:

$$\exists \mathbf{w}, \mathbf{x}. \, [z_1 = x_1 \, \wedge \, z_2 = w_1 \, \wedge \, z_3 = (ite \, (x_2 = z_2) \, x_2 \, z_5)$$
$$\wedge \, z_5 = f(x_3)] \, \wedge \, z_4 = g(z_5).$$

Let $\mathbf{z} = \mathbf{q_0}$ *be the initial state predicate for this circuit. Then*

$$\mathbf{z} = \mathbf{q_0} \, \vee \, z_4 = g(z_5)$$

is an invariant for this circuit. □

Usually, this method generates many invariant clauses. However, there are more invariants that can be automatically generated from the predecessor state predicate. These methods are described in the next two subsections.

4.2 Method 2

The second method is to convert existential quantification into disjunction when the quantified variable has a finite domain. For example, if x_k has a domain of $v_1, v_2, ..., v_l$, then the function $f(x_k)$ can be written as $f(v_1) \lor f(v_2) \lor ... \lor f(v_l)$.

Note first that since $P_1[\mathbf{z}]$ may in general contain a conjunction of many clauses, a direct application of this method to $P_1[\mathbf{z}]$ would result in an expression size exponential in the number of bound variables. In order to avoid this, we first move the conjunction outside the scope of the quantifiers. This still results in a size increase which is linear in the size of the finite domain. Though efficient for Boolean variables, it can produce a sizable blow-up for other variables. Thus we apply Method 1 first in order to eliminate as many bound variables as possible.

By moving the conjunction outside the scope of the quantifiers, we get an approximation which is not equivalent to, but is implied by the predecessor state predicate. Call the result $P_1'[\mathbf{z}]$:

$$P_1'[\mathbf{z}] = \bigwedge_{i=1}^{n_1} [\exists \mathbf{w}, \mathbf{x}. \, z_i = \tau_i'[\mathbf{w}, \mathbf{x}, \mathbf{z}]]$$

We apply the finite-domain transformation described above to each of the clauses in the conjunction $P_1'[\mathbf{z}]$, replacing each clause with a disjunction (these disjunctive clauses are simplified to eliminate duplicate terms). Note that we can substitute for functions as well as variables which can be very advantageous: if we have a function $f(x)$ whose range is small, we can perform the transformation on $f(x)$ even if x has a very large domain. As in the first method, after performing these transformations, some of the clauses may no longer have bound variables. These clauses are then combined via conjunction to form $I_2[\mathbf{z}]$. Since $P_1'[\mathbf{z}] \Rightarrow I_2[\mathbf{z}]$ is valid, by transitivity $P[\mathbf{z}] \Rightarrow I_2[\mathbf{z}]$ is valid so that $Q_0[\mathbf{z}] \lor I_2[\mathbf{z}]$ is an invariant by Theorem 1.

In contrast to the first method, which generally produces invariants that are strong constraints on state variables, this method tends to produce weaker constraints which enumerate all the possible values of a state variable or all the possible relationships among several state variables. We illustrate by continuing with the same example.

Example 2 *At the end of the last example, we were left with the following residual expression $P_1[\mathbf{z}]$:*

$$\exists \mathbf{w}, \mathbf{x}. \, [z_1 = x_1 \land z_2 = w_1 \land z_3 = (ite \, (x_2 = z_2) \, x_2 \, z_5) \land z_5 = f(x_3)].$$

We then distribute the quantifiers inside the conjunction to get:

$$[\exists \mathbf{w}, \mathbf{x}. \, z_1 = x_1] \land [\exists \mathbf{w}, \mathbf{x}. \, z_2 = w_1] \land [\exists \mathbf{w}, \mathbf{x}. \, z_3 = (ite \, x_2 \, w_2 \, z_5)]$$
$$\land [\exists \mathbf{w}, \mathbf{x}. \, z_5 = f(x_3)].$$

Now, suppose that the bound variable x_3 has the domain $\{v_1, v_2, v_3\}$ and that it is the only variable with a finite domain. We make use of this fact to get:

$$\cdots \land [z_5 = f(v_1) \lor z_5 = f(v_2) \lor z_5 = f(v_3)].$$

Using the same initial state predicate as before, we have the following new invariant:

$$\mathbf{z} = \mathbf{q_0} \lor z_5 = f(v_1) \lor z_5 = f(v_2) \lor z_5 = f(v_3). \qquad \Box$$

4.3 Method 3

The third and final method consists of two transformations that can be used to eliminate expressions containing bound variables within ite expressions. The first is much like Method 2: in particular, the expression $z_i = (ite \ \alpha \ \beta \ \gamma)$ is transformed to $(z_i = \beta) \lor (z_i = \gamma)$ when α contains a bound variable, and β and γ do not (the difference from Method 2 is that we do not explicitly enumerate the possible values of α). The transformed expression is implied by the original, so it can be used in an invariant.

The other transformation applies when α is of the form $a = b$: a may be substituted for b in β since β can only affect the truth of the expression if $a = b$. This is useful when a and γ contain no quantified variables, and b contains the only bound variables in β. A dual transformation can be performed if α is of the form $a \neq b$. In this case, we can substitute a for b in γ since γ can only affect the truth of the expression if it is not the case that $a \neq b$ (which means $a = b$).

Example 3 *We continue with the residual expression left from the end of Example 2:*
$$[\exists \mathbf{w}, \mathbf{x}. \ z_1 = x_1] \land [\exists \mathbf{w}, \mathbf{x}. \ z_2 = w_1] \land [\exists \mathbf{w}, \mathbf{x}. \ z_3 = (ite \ (x_2 = z_2) \ x_2 \ z_5)].$$
We transform the last clause to get:
$$\cdots \land [z_3 = z_2 \lor z_3 = z_5].$$
Thus we have the following invariant: $\mathbf{z} = \mathbf{q_0} \lor z_3 = z_2 \lor z_3 = z_5.$ □

4.4 Invariant extraction procedure

The three methods we have presented are most effective when used together. As mentioned, Method 1 is applied first, and Methods 2 and 3 are then applied to the remaining predicate. It is important to apply Method 1 first because the bound variables eliminated by Method 1 may significantly reduce the blow-up caused by Method 2. Similarly, it is important to apply Method 3 last because Method 3 requires choices to be made based on which expressions contain bound variables and the other two methods may be able to eliminate some of those bound variables. In our example we were able to produce three invariants. The individual invariants can then be combined. For our example, the strongest invariant we can generate is thus:

$$\mathbf{z} = \mathbf{q_0} \lor [z_4 = g(z_5) \land (z_5 = f(v_1) \lor z_5 = f(v_2) \lor z_5 = f(v_3)) \\ \land (z_3 = z_2 \lor z_3 = z_5)].$$

5 Experimental Results

In this section, we present the results of applying our three methods for finding invariants to the Stanford Protocol Processor (PP). We first briefly describe the PP architecture.

5.1 Protocol Processor

PP is a pipelined microprocessor with a dual pipeline. The two sides (A and B) of the pipeline have similar functionality except for control and memory instructions. Control instructions are always executed on the A-side, while memory instructions are executed on the B-side. The processor has one delay slot for branch and load instructions, separate instruction and data caches which are two-way associative, and five stages in its pipeline which are very similar to the stages in the DLX pipeline [5]. PP does not have hardware interlocks, relying on compiler scheduling to avoid dependency problems. It does allow a load to bypass a store if they are accessing different cache lines, but if they access the same cache line, PP stalls the whole pipeline for two cycles to resolve the conflict. There is also a write back buffer to speed up cache replacement. All these architecture features are very difficult to verify all at once. Thus, we decided to divide the verification into two parts: the pipeline controller and the cache controller.

5.2 PP Pipeline Controller

We first wrote descriptions of the specification and implementation in our behavior description language. The specification was translated from the PP instruction set architecture specification [16], and the implementation was translated from the PP Verilog description. Our PP implementation has nine instructions: ALU, ALUI, B, BR, J, JR, JAL, LD, SD. The ALU and ALUI instructions abstract all of the ALU instructions (non-immediate and immediate respectively). We believe that these nine instructions represent the main features of PP.

Controller	Manual Method	Automatic Method
Pipeline	7	6
Cache	72	28

Table 1. Number of invariants needed for manual method compared to those generated by the automatic methods in the verification of PP controllers.

A manual top-down approach to verifying the controller required the addition of seven strengthening invariants (see table 1). Our automatic methods found six of these invariants. Although the automatic method did not find all of the needed invariants, it only took a few hours, while the manual method took several weeks.

5.3 PP Cache Controller

We next turned our attention to the PP cache controller. The cache controller has three finite state machines: the conflict FSM, the miss FSM, and the replacement FSM. Because of complicated interactions among these three state machines, the verification of the cache controller is much more difficult than the verification of the pipeline controller. Thus, larger and more complicated invariants are required. Using a top-down manual approach, we found that 72 invariants were necessary to verify correctness. This took several months. The automatic method found 28 of these invariants (in fact, the automatic method produced 61 simpler invariants, which implied 28 out of 72 of the manually-discovered invariants). Thus, we still needed to provide 44 invariants by hand. Still, the automatic invariants are generated in just a few hours, so the overall

Manual Method	Automatic Method

(⇔ DEXTREAD-S2
  (= EXTSTATE-S1 @EXT_WAITOK2))
	(⇔ DEXTREAD-S2
	  (= EXTSTATE-S1 @EXT_WAITOK2))

(⇔ DEXTREQ-S2
  (or (= FLUSHSTATE-S2 @F_WAITOK)
    (= EXTSTATE-S2 @EXT_WAITOK2)))
	(⇔ DEXTREQ-S2
	  (or (= FLUSHSTATE-S2 @F_WAITOK)
	  DEXTREAD-S2))

(and
(ite
 (and (or INSTRISLOAD-S2M
   INSTRISSTORE-S2M
   (≠ CONFSTATE-S2 @CONF_IDLE))
  (= EXTSTATE-S2 @EXT_IDLE)
  DTAG1-RW1-S2)
 DTAGREAD-S2M
 true)
...
)
	(or
	(⇔ (or (and (≠ CONFSTATE-S2
	        @CONF_IDLE)
	     DTAG1-RW1-S2)
	  (= EXTSTATE-S2 @EXT_PROBE)
	  DEXTREAD-S2)
	 DTAGREAD-S2M)
	(⇔ (or DTAG1-RW1-S2
	  (= EXTSTATE-S2 @EXT_PROBE)
	  DEXTREAD-S2)
	 DTAGREAD-S2M))

Table 2. A comparison of manually and automatically generated invariants in the PP cache controller verification. ⇔ denotes "if and only if", *ite* denotes "if then else", and @ indicates a symbolic constant. Variable names are in all capital letters while logic operations use lower-case letters. The first pair of invariants show that DEXTREAD-S2 is true if and only if the variable EXTSTATE-S1 is in the @EXT_WAITOK2 state. The automatically and manually generated invariants are identical. The second pair of invariants constrain the variable DEXTREQ-S2. The two invariants are in different forms, but are equivalent. The third pair of invariants are very different. Each of them constrain the variable DTAGREAD-S2M, but the manual method includes this as part of a larger and more complicated constraint.

time required for verification is significantly reduced. Some typical invariants in this verification are listed in table 2.

The design of the memory system is qualitatively different from the design of the CPU, in that the memory system is much more *control-intensive*. Indeed, it has three finite-state machines that are explicitly implemented in the description. Many of the necessary invariants deal with characterizing the different reachable combinations of states from these machines. The method described above is ill-suited for finding these combinations.

There exist several extremely efficient ways of finding the reachable state combinations of a collection of machines already (indeed, this is the central task in finite-state verification). An obvious approach, which we intend to explore, is to separate out the finite state machines from a description, analyze the state combinations separately, then combine the results with invariants extracted by the method described above. We believe this approach could find most of the invariants required for the PP memory system.

6 Conclusions and Future Work

We have presented some simple heuristics for automatic invariant discovery in processor verification. Surprisingly, these simple methods are effective in discovering many of the invariants required to verify a real microprocessor. Although we have focused on the Burch and Dill verification paradigm, these techniques should be useful in any methodology in which a syntactic description of the state transition function can be obtained. Combining the method with more powerful tools and other ways of discovering invariants should produce even better results.

An obvious generalization of our methods is to simulate for more than one cycle and use the multi-cycle next state function to generate invariants. Our experimentation to date has shown that the results are too complex to be useful, but the idea may prove to be useful on future designs.

Acknowledgments

This research was sponsored by ARPA contract ARMY DABT63-95-C-0049-P00002, and the third author is supported by a National Defense Science and Engineering Graduate Fellowship. The authors also wish to thank Robert Jones for his help and advice in preparing the final revision of the paper.

References

1. J. Burch and D. Dill, "Symbolic Verification of Pipelined Microprocessor Control", 6th Computer Aided Verification, 1994.
2. J. Burch, E. Clarke, K. McMillan, D. Dill, and L. Hwang, "Symbolic Model Checking: 10^{20} States and Beyond", 5th Annual IEEE Symposium on Logic In Computer Science, 1990.
3. A. Cohn, "A Proof of Correctness of the VIPER Microprocessors: The First Level", In VLSI Specification, Verification and Synthesis, 1988.
4. D. Cyrluk, "Microprocessor Verification in PVS: A Methodology and Simple Example", Technical Report SRI-CSL-93-12, SRI Computer Science Laboratory, Dec. 1993.
5. J. Hennessy and D. Patterson, "Computer Architecture: A Quantitative Approach", Morgan Kaufmann, 1990.
6. W. Hunt, Jr., "Microprocessor Design Verification", Journal of Automated Reasoning 5: p429-460, 1989.
7. R. Jones, D. Dill and J. Burch, "Efficient Validity Checking for Processor Verification", IEEE/ACM International Conference on Computer Aided Design, 1995.
8. J. Joyce, G. Birtwistle, and M. Gordon, "Proving a Computer Correct in Higher Order Logic", Technical Report 100, Computer Lab., University of Cambridge, 1986.
9. M. Langevin and E. Cerny, "Verification of Processor-like Circuits", In Advanced Research Workshop on Correct Hardware Design Methodologies, June 1991.
10. Z. Manna, et al., "STeP: the Stanford Temporal Prover", Technique Report, STAN-CS-TR-94, Computer Science Department, Stanford, 1994.
11. Z. Manna and R. Waldinger, "The Deductive Foundations of Computer Programming", Addison Wesley, 1993.
12. Z. Manna and A. Pnueli, "Temporal Verification of Reactive Systems: Safety", Springer-Verlag, 1995.

13. S. Bensalem, Y. Lakhnech, and H. Saidi, "Powerful Techniques for the Automatic Generation of Invariants", To appear in CAV96.

14. A. Roscoe, "Occam in the Specification and Verification of Microprocessors", ACM Trans. Prog. Lang. Syst., 1(2):245-257, Oct. 1979.

15. J. Saxe, S. Garland, J. Guttag, and J. Horning, "Using Transformations and Verification in Circuit Design", Technical Report 78, DEC System Research Center, Sept. 1991.

16. R. Simoni, "PP Instruction Set Architecture Specification", version 1.7, Stanford FLASH group, 1995

17. M. Srivas and M. Brickford, "Formal Verification of a Pipelined Microprocessor", IEEE Software, 7(5):52-64, Sept. 1990.

18. C. Barrett, D. Dill, and J. Levitt, "Validity Checking for Combinations of Theories with Equality", To appear in FMCAD, 1996.

A Brief Study of BDD Package Performance

Ellen M. Sentovich

Ecole Nationale Supérieure des Mines de Paris
Centre de Mathématiques Appliquées
B.P. 207, F
06904 Sophia-Antipolis Cedex, FRANCE
ellen@cma.cma.fr

Abstract. The use of BDD techniques has become essential in a variety of CAD problems, and particularly in employing formal methods for design and verification. There are now many BDD packages available with many different features. Given a particular application, it is difficult to choose which package is best-suited for solving the problem, despite extensive published results on the packages. This paper contains a study of a few state-of-the art BDD packages. Results comparing the packages are reported on a variety of examples, for a variety of applications, and on several computing platforms. Great care has been taken to make fair and meaningful comparisons. The study is offered to compare current available BDD technology, and to encourage BDD package authors to use this experimental foundation and to make their results and code available to the research community.

1 Introduction

It has been 10 years since Bryant's publication [3] of BDD (*binary decision diagram*) algorithms launched intense activity in BDD theory and package development. There are now many flavors of BDDs, a large body of published work both theoretical and practical, and a number of BDD packages in use. It is difficult to judge their relative merit based on the results published in the papers. Though there are theoretical results on size and space bounds for the algorithms for various definitions of BDDs, these cannot predict the performance for a given application. The practical results published in the result tables for well-known benchmarks are difficult to assess as well. Experiments are run on a variety of platforms and under a variety of computing conditions. Results are reported on different sets of metrics, usually designed to highlight the features of the new package. Although care has been taken to clearly report, for example, the number of BDD nodes, the size of each node, the size of auxiliary hash tables, etc., even these concrete numbers must be computed and examined carefully.[1] For the BDD package user presented with an application and a set

---

[1] Size has been a particularly troublesome measure for BDD packages, and is nearly impossible to estimate based on the BDD. Packages can allocate memory for several nodes at once even though they are not yet needed, garbage collection can be invoked at any time, memory can be reclaimed in different ways, etc.

of BDD packages, it is difficult to choose the most appropriate package.

The CAD community, and particularly in the formal methods area, is using BDDs ubiquitously – in formal verification, logic synthesis, test generation, and simulation. Further, the research community heavily uses available BDD packages for experimentation with new algorithms. The issue of selecting an appropriate package is critical and will remain so. Often heard in the conference corridors is "and which BDD package did you use?" followed moments later by "but I hear N's is X-times faster." There is no consensus or basis for meaningful comparison. Furthermore, the BDD research community states the new and useful features of their work but not its weaknesses.

This paper contains a brief study and comparison of several state-of-the art, publicly available BDD packages. The goal is to highlight useful features of these packages for several applications, rather than to run a race and declare a winner. The study is oriented toward the BDD user, and as such the packages are applied as is, without tuning parameters per application or per platform. The experiments are performed in a controlled computing environment, with both standard benchmark circuits and new benchmarks generated from higher-level languages. The goals of the study are to provide data on current BDD package technology, to highlight directions for further research, and to encourage further development and reporting to be done under the same, publicly available, experimentation environment.

The remainder of the paper is organized as follows. A brief review of BDDs is given in Section 2. Relevant general BDD package features and brief descriptions of the packages studied here are given in Section 3. The experimental environment is described in Section 4, and the results in Section 5. Conclusions and discussions of other relevant packages to be studied in the future are given in Section 6.

2   Background: BDDs

A description of the original canonical BDD and associated algorithms is found in [3]; the technology is reviewed and applications described in [4]; an update on the most important recent developments is in [5].

Briefly: a BDD is a DAG representation of a function, with each nonterminal node labeled by a variable, each terminal node labeled with a value, and the two children of a node each representing the function with that variable set to one of its two values (0/1). The representation is made canonical by restricting the ordering of the variables and forcing sharing of common subgraphs. This results in a *reduced, ordered binary decision diagram*, or *ROBDD*, herein called a BDD. A canonical BDD admits polynomial time algorithms for many Boolean functions (e.g., verifying the equivalence of two functions); for this reason, BDDs and efficient packages for manipulating them have become essential tools in CAD applications.

There are now many types of BDDs (see the "alphabet soup" list in [5]), which differ in ordering restrictions, functional decomposition, use of edge values, use of multiple terminal values, and reduction rules. The focus here is on the basic *ROBDD*, with *complement edges* and implemented with an *efficient memory management scheme* including caching of intermediate results and storage of unique entries for each variable [2]. These are the essential elements for any state-of-the-art package and are now considered standard. They are the basis for many public and private BDD packages, which are widely used in CAD applications for representation of Boolean circuits and related functions. The packages selected for the study are implemented in this way.

3 BDD Packages

The set of BDD packages has been chosen based on the following requirements.

- They are publically available and free, so the results are immediately useful to the entire community. These packages are excellent and widely used, so this does not restrict the scope to a weak subset of existing packages.
- The packages employ some of the latest algorithms in variable ordering, traversal techniques, memory management, and special applications.
- The source code is available. Thus, features of the packages can be explored and tuned by the general public.

The BDD packages under study are CAL, CMU, CUDD; a description of each follows. A number of other packages were considered to broaden the study, but could not be included at this date for a variety of reasons. They are listed in Section 6.

BDD packages can be distinguished by the following characteristics:

Variable ordering: Since BDD size is dependent on variable ordering, selecting a good ordering is essential. Good heuristics for circuits have been developed [9, 14], but these break down for some examples. It is difficult to determine a good ordering a priori. *Dynamic ordering techniques* (DVO) [21] change the ordering while the BDD is being built to control its size; in particular, *sifting* exploits efficient swapping of adjacent variables to search for better orderings. Many heuristics for employing dynamic ordering have been reported (e.g., [19]).

Memory management: The now-standard memory management scheme is described in [2]. Recent research demonstrates improvements based on localizing memory allocation and access [18, 20]. A report on cache performance of the BDD packages studied here is contained in [15].

Traversal algorithms: The original algorithms traverse the BDD in depth-first order. Breadth-first traversal has been proposed [18, 20] to localize memory access and thus improve performance. In a related development, the addition of temporary new variables, sifting and quantifying them while building BDDs, has been proposed [10] to improve BDD size and run-time during creation.

Specialized algorithms: Some packages have been optimized for particular applications. For example, building partial BDDs on partitions and using these to reorder or obtain a final result without building the complete BDDs has been employed [17, 8, 16]. In another work, the DD data structure and algorithms are optimized to reduce cache misses in MTBDDs [11].

3.1 CAL : UC Berkeley

The CAL package is the latest BDD package from UC Berkeley [20]. Its main goal is to improve performance by using breadth-first traversal and by storing all nodes of a single variable contiguously in memory. This scheme exploits memory hierarchy by reducing accesses *across* the hierarchy (e.g., from cache to main memory). It thus preserves locality of reference in an architecture-independent way. This is particularly important when the size of the BDD exceeds main memory. It also uses pipelining and superscalarity to increase the performance by issuing multiple operations simultaneously. The implementation of the BFS algorithm, which involves a first pass down the BDD to issue computation requests (APPLY) and a second pass up the BDD to process them (REDUCE), implies a complete computed table, or cache of intermediate results. Thus there is never a cache-miss on an intermediate result and this can improve performance. Only limited sifting is available, and hence all experiments were run without it. The same is true for the reachable states computation.

3.2 CMU : David Long

There are many BDD packages from Carnegie Mellon University. The one called CMU here was written by David Long [12]. It was integrated in the SIS logic synthesis system [22] and has been widely distributed stand-alone and through SIS for many years. It has an efficient memory management scheme and contains support for dynamic variable ordering via sifting and window permutations. We use the last publicly available version, released in November 1993.[2]

3.3 CUDD : UC Boulder

The most recent release of the CUDD package (Colorado University Decision Diagram), 1.1.1, became available in June 1996. Its primary distinguishing feature is extensive heuristics for dynamic variable ordering. In particular, support is provided for random variable exchange, sifting, group sifting, window permutations (all variable-order permutations among a group of adjacent variables in the BDD are tested), identification and linking of symmetric variables, simulated annealing, and genetic algorithm-based reordering. It has a very efficient memory management scheme, and finely tuned heuristics for automatically adjusting cache sizes and controlling dynamic variable ordering.

---

[2] The latest version is said to be significantly upgraded: performance ×2 compared to the 1993 version [13]. It was developed at AT&T and is not available publicly.

4 Experimental Environment

We performed these experiments because we rely heavily on BDD-based algorithms to synthesize and verify large industrial designs. As a result, the experimental focus here is on *computations in real applications* rather than specific operations of BDD packages. While some packages could be engineered to perform better for a given application, we prefer to run them "as-is" – in a real synthesis environment and on real examples to test their current performance.

4.1 Computing Environment

Experiments were carried out on three different machines, each named below and listed with its characteristics.

Alpha: DEC Alpha 2100, 512MB RAM, running OSF/1 v3.0
Sparc20: SS20, 128MB RAM, running SunOS 4.1.3
Linux: DEC Pentium 120 MHz, 32MB DRAM, running Linux/GNU

For consistency and comparison across experiments, most of the results are reported solely for the Sparc20. The general trends noted were observed on all machines. Results on other machines are included occasionally for breadth or to demonstrate a significant difference for some experiments.

4.2 Implementation

The packages were all integrated and run within SIS. The SIS program provides a common interface between a network representing a circuit and the BDD package. Given a particular ordering, the BDDs for the network can built by calling functions in the interface. This ensures that *for a given network and a given ordering, precisely the same external calls to the BDD packages are made.* This is extremely important for package comparison as small differences in the starting point or order of processing can have a large effect on the results. Furthermore, with all packages implemented in the same environment, the time and memory usage statistics can be measured at precisely the same points with precisely the same overhead.

There are some inefficiencies in the SIS implementation that will result in performance degradation roughly equal in all packages. First, when building the BDD for a network node, the BDDs for all intermediate network nodes are saved. This implies memory overhead when building output BDDs, but also effectively implements a lossless cache of computed results on the intermediate nodes – which may slightly improve efficiency. Second, SIS makes several calls to basic BDD operations for each sum-of-products function at a node, rather than making a single call to a more complex function when possible.

For the CAL package, the implementation follows precisely that recommended by its authors, and hence is slightly different to take advantage of the specialized memory access and superscalarity (without this, a comparison is meaningless). As part of this implementation, the initial network is decomposed into AND and invert gates, and again the BDDs are saved at the intermediate nodes. The initial ordering is determined based on the network before decomposition to guarantee the same starting point for all packages.

4.3 Examples

Two sets of examples have been selected: standard benchmark circuits (IS-CAS85), and large industrial examples generated from the high-level language Esterel [1] and from the POLIS hardware/software codesign system [7]. These two categories represent a mix of circuits: combinational and sequential, those notoriously difficult to verify, those with very large BDDs compared to the number of latches, those that are highly redundant, and those that are optimized.

The characteristics of the Esterel-based circuits are given in Table 1. The final column gives the number of iterations required to compute the reachable states reported in the previous column. Each of the Esterel circuits was derived from a real design specified in Esterel and synthesized using the Esterel compiler. In the case of the *p1* and *dash* examples, the synthesis was done with POLIS. *tcintnc* is a version of *tcint* without the counters. *p1opt* is *p1* optimized with the standard script in SIS. *dashs* is *dash* with trivially redundant logic removed via the *sweep* command in SIS.

Circuit	pi/po	latches	SOP lit	Rstates	iter
att	47/255	178	3934	226938	14
bm	42/168	112	2192	850609	20
noyau	41/98	179	3743	2524161	11
prosa	46/227	73	1621	7361	27
renault	23/179	66	1838	257	12
sequenceur	55/98	121	1784	?	?
snecma	23/5	70	1689	10241	28
tcint	19/20	96	1232	2826	26
tcintnc	19/20	90	1084	310	10
train	59/81	866	25651	?	?
p1	49/12	63	6886	17620993	123
p1opt	29/8	59	825	17620993	123
dash	49/17	443	33673	?	?
dashs	49/17	356	12133	?	?

Table 1. Characteristics of Esterel-based Circuits

The Esterel examples are control-oriented, and furthermore the Esterel compiler extracts only the control when generating the BLIF files. However, some of the above BLIF specifications were generated using the Esterel compiler in conjunction with a library of arithmetic elements in BLIF format.[3] In partic-

---

[3] Although some examples contain arithmetic parts, we do not focus on arithmetic circuits as these are known to be better candidates for other types of BDDs, e.g. [6].

ular, *train* contains a number of counters, and the POLIS-generated circuits contain many arithmetic subcircuits (in fact, they are mostly arithmetic as they represent the portion of the design destined to hardware).

4.4 Experiments

Three categories of experiments have been run: basic BDD building, reached state computation, and BDD building and reached state computation interleaved with logic optimization. The packages are compared for different initial variable orderings (but the same ordering for each package), different initial circuit configurations, and with and without dynamic variable ordering. Finally, experiments are interleaved with some optimization steps, demonstrating performance as circuit configuration changes.

Results are measured by CPU time in seconds as reported by *time*, memory usage in megabytes as reported by *ps*, and BDD node count. Care was taken to ensure that the BDD nodes counts were the same for each package, where they were expected to be the same.

5 Results

Due to limited space, quite clearly the results tables for all experiments under all packages and platforms cannot be included. Results are described below, with tables included to illustrate important trends and conclusions.

5.1 Initial Variable Ordering

In the first experiment, all packages were compared under two different initial variable orders: the first based on the input order in the BLIF file, and the second on a topological order generated by *order_dfs* in SIS ([14]). The BLIF order almost always produced larger BDDs and took more time for all packages, even when dynamic variable ordering was invoked. This illustrates that care must be taken to select a good initial order. A notable exception is C499 on Sparc, where the number of nodes is 31195 for the dfs order, and 1761 for the blif order. On Alpha, dfs yields 45974 and blif 45962. Note that the dfs ordering algorithm produces different results on the Alpha and Sun machines. For this reason, care was taken to always provide the same initial ordering for each package and experiment on a given platform. After the blif experiments, all were performed with the initial ordering given by *order_dfs*.

5.2 Building Output BDDs

Results for building output BDDs with *order_dfs* and no DVO are reported on Sparc20 for all packages in Table 2 (ISCAS) and Table 3 (Esterel). *train310* is a version of the *train* example described in Section 5.5. **m** indicates that the process ran out of memory. Failure to produce a BDD should be taken as failure subject to a particular *platform, memory constraint, time constraint, and initial ordering*, rather than "failure of package X to produce BDDs for benchmark Y". The goal is to make relative comparisons and not produce absolute numbers.

The node columns give the number of BDD nodes in the intermediate result (**int:** with the BDDs for all intermediate nodes) and the final result (**final:** just the output BDDs) for the original circuit. The former gives an indication of the memory overhead in saving the intermediate BDDs, and will track to some extent the memory usage reported by *ps*.[4] The double columns labeled CMU2 and CUDD2 represent runs on the decomposed circuit, as is done in the CAL package, for comparison on the same initial circuit. The node counts are not given for the intermediate BDDs on the decomposed circuit, though they appear in Tables 4 and 5 in the CAL column. Node counts are the same for the final BDDs. The totals are given at the bottom of the tables, though for Table 3 they are obviously dominated by the large examples.

Circuit	CAL time	CAL mem	CMU time	CMU mem	CUDD time	CUDD mem	Nodes int	Nodes final	CMU2 time	CMU2 mem	CUDD2 time	CUDD2 mem
C432	4.9	6.5	10.6	4.5	7.2	7.2	146384	31195	10.1	5.4	8.2	7.4
C499	3.8	5.8	4.5	1.9	3.4	2.2	43364	33214	6.4	3.6	5.4	4.6
C880	2.1	3.9	1.3	1.6	1.2	2.0	28316	7926	2.5	2.0	2.2	2.4
C1355	4.7	6.4	6.6	4.3	5.0	4.7	135914	33214	8.0	5.0	6.8	5.2
C1908	4.4	6.1	3.8	2.8	3.4	3.0	42929	12734	5.6	3.5	5.0	3.7
C2670		m		m		m				m		m
C3540		m		m		m				m		m
C5315	12.7	12.1	5.5	5.1	5.0	6.5	49095	21236	14.1	7.9	12.2	8.7
C7552		m		m		m				m		m
Total	32.6	40.8	32.3	20.2	25.2	25.6			46.7	27.4	39.8	32.0

Table 2. Output BDDs on Sparc20

For the ISCAS circuits that completed under this fixed variable ordering, the expense in time and memory is not very high. CAL incurs some overhead in both time and memory for these small examples on the original circuit. This is probably due to its allocation of memory in larger blocks at the outset. It is the fastest package on the decomposed circuits. CMU consistently requires more time but less memory than CUDD.

For the Esterel circuits, CAL requires less time on all the larger examples (for the original and decomposed versions), and more memory on all examples. Again CMU usually takes more time and less memory than CUDD on the original circuits, but note its particularly good memory performance for *p1* with respect to CUDD. This CMU memory advantage is lost on the larger decomposed networks. The relative memory increase for CMU with example size is continued on more intensive applications (e.g., reachable states, Section 5.4).

---

[4] *ps* also includes the hash tables, which are sized and resized differently for each package.

Circuit	CAL time	CAL mem	CMU time	CMU mem	CUDD time	CUDD mem	Nodes int	Nodes final	CMU2 time	CMU2 mem	CUDD2 time	CUDD2 mem
att	25.1	19.9	29.1	14.7	21.4	15.1	417307	161531	38.0	17.7	28.4	17.2
bm	13.2	13.1	21.8	9.5	14.8	11.1	249317	74794	23.4	11.6	18.6	12.0
noyau	23.0	21.4	33.1	14.5	21.5	14.8	445180	171005	38.4	19.2	28.4	17.2
prosa	11.2	13.0	15.7	8.9	11.6	10.0	233894	153006	21.7	11.2	15.9	11.5
renault	5.0	6.3	2.3	2.6	2.0	2.9	18269	8017	5.5	3.8	5.1	4.1
seq		m		m		m				m		m
snecma	66.0	50.1	130.5	41.5	91.2	35.3	1793385	1270801	129.7	55.9	90.2	44.0
tcint	3.3	5.1	.7	1.3	.7	1.6	2613	1446	3.0	2.2	2.9	2.5
tcintnc	2.7	4.8	.5	1.2	.5	1.4	2622	1431	2.5	2.0	2.5	2.3
train310	164.9	91.9	255.1	96.4	160.6	82.3	4106721	1122116	291.9	117.8	175.1	87.1
p1	152.2	72.3	699.8	6.7	176.0	13.4	139025	70268	240.3	48.2	211.2	40.4
p1opt	3.1	4.4	1.0	1.0	.9	1.3	4726	2967	3.0	2.1	3.0	2.2
dash		m		m		m				m		m
dashs		m		m		m				m		m
Total	469.7	302.3	1189.6	198.3	501.2	189.2			797.4	291.7	581.3	240.5

Table 3. Output BDDs on Sparc20

The decomposed network is consistently more costly than the original (disregarding the outlying CPU time for CMU on *p1*). This is undoubtedly due to the storage of all intermediate node BDDs, though we have observed in general that an optimized network is a better starting point than a non-optimized one, even when the BDDs are released immediately.[5]

5.3 Dynamic Variable Ordering

Next, each example was run with dynamic variable ordering. The sifting algorithm reported in [21] is implemented in CMU and CUDD. CMU also supports 3-variable window permutations and CUDD many other reordering techniques; only sifting was employed in these experiments. The results are shown in Tables 4 and 5. The previous CAL results are repeated here for reference (without sifting). The number of nodes reported includes BDDs at the intermediate networknodes, since the package was forced to save these during the computation. Totals are given for the examples that complete for CMU and CUDD.

CMU and CUDD complete all ISCAS examples. CMU completes all but two Esterel examples, and CUDD all but one. CUDD reports much better run times than CMU, and roughly the same memory usage. CAL fails to complete six examples, but otherwise reports the best run times (excepting *p1*).

Sifting dramatically increases runtime, but is essential for completing the large examples. The heuristics for choosing how aggressively to sift must be

---

[5] This is not always true, just as starting from a good initial ordering and building BDDs with DVO does not necessarily produce a better result than starting with a bad ordering and using DVO. The general trends hold, but care must be taken to avoid particularly costly behavior on outlying examples.

	CAL			CMU			CUDD		
Circuit	time	mem	nodes	time	mem	nodes	time	mem	nodes
C432	4.9	6.5	192671	5.9	1.0	5171	3.8	.9	4674
C499	3.8	5.8	109741	218.1	2.4	46342	161.4	2.1	38643
C880	2.1	3.9	32506	33.5	1.5	16551	15.8	2.0	15623
C1355	4.7	6.4	140038	525.6	5.8	121334	400.9	4.8	121335
C1908	4.4	6.1	58009	53.0	2.5	20762	34.2	3.0	21155
C2670		m		49.7	2.9	16041	55.8	4.8	20903
C3540		m		396.4	9.2	148456	139.8	5.4	126317
C5315	12.7	12.1	71621	13.0	4.2	11622	14.6	4.4	10954
C7552		m		255.9	7.6	50950	83.5	7.8	47289
Total				1551.1	37.1		909.8	35.2	

Table 4. Output BDDs with sifting on Sparc20

	CAL			CMU			CUDD		
Circuit	time	mem	nodes	time	mem	nodes	time	mem	nodes
att	25.1	19.9	494131	1706.0	13.2	112367	76.4	5.9	55648
bm	13.2	13.1	340941	57.8	2.8	18761	78.9	4.7	41395
noyau	23.0	21.4	526481	443.4	6.2	61675	88.0	5.6	34567
prosa	11.2	13.0	355241	128.7	2.9	28032	13.3	2.8	9941
renault	5.0	6.3	28096	14.8	2.2	8764	8.6	2.8	3649
seq		m		t(1hr)			2840.1	12.4	293187
snecma	66.0	50.1	2241334	1528.3	12.5	183705	t(1hr)		
tcint	3.3	5.1	2815	.7	1.3	2613	.7	1.6	2613
tcintnc	2.7	4.8	2770	.6	1.2	2622	.6	1.4	2622
train310	164.9	91.9	4257859	t(1hr)			920.2	11.9	180225
p1	152.2	72.3	1614809	9.7	1.5	4805	11.0	2.2	2196
p1opt	3.1	4.4	15134	1.1	1.0	4726	2.5	1.2	3467
dash		m		254.0	10.6	34338	116.1	14.6	12531
dashs		m		69.5	4.7	32418	76.7	8.6	10089
Total				2686.3	47.6		472.8	51.4	

Table 5. Output BDDs with dfs and sifting on Sparc20

well-engineered. In these tables and in the remainder of the paper, we note the most consistent behavior from the CUDD package, due to a judicious choice in trading off such parameters. Nonetheless, CUDD with sifting fails on the *snecma* example with a timeout after one CPU hour.

As another illustration of the unpredictability of BDD performance, consider the data in Table 6, containing sifting results on the Alpha and Linux platforms. For *dashs*, CUDD is slower than CMU on Sparc and Linux machines, but faster on the Alpha. This is undoubtedly due in large part to a difference in variable orderings computed by *order_dfs* across different platforms which will subsequently trigger sifting differently. Given that the two packages begin with the same ordering nonetheless, the difference in performance is notable.

	Alpha				Linux			
	CMU		CUDD		CMU		CUDD	
Circuit	time	mem	time	mem	time	mem	time	mem
p1	5.7	6.8	6.4	8.3	5.7	3.4	5.8	4.1
p1opt	.6	5.8	1.7	6.5	2.3	2.9	2.4	3.0
dash	180.1	24.3	66.9	29.4	172.7	11.6	50.6	15.6
dashs	47.7	12.4	38.7	20.4	22.9	6.7	39.3	10.4

Table 6. DFS/Sifting Results

The results in Tables 4 and 5 indicate that DVO is more useful for completing large examples than BFS methodology, but that the BFS algorithm is faster when it does complete. This confirms the findings in [15], where it was conjectured that the performance advantages in the CAL package are due more to the lossless computed table and superscalarity than due to the data locality implied by the BFS approach. For a more in-depth study of the memory effects of these three packages, the reader is referred to [20] and [15].

5.4 Application: Reachable States Computation

For the reachable states computation, the BDDs are computed for the next-state functions and the reached state set. The latter has proven to be difficult and to cause memory explosions. This experiment thus reflects a new dimension in stressing the packages.

The CMU and CUDD experiments were done using the reachable states package in SIS, with sifting enabled.

Also included in this table are the runtimes for computing the reachable states in the TiGeR package with sifting enabled. This package is described in Section 6; limited results are included here only to illustrate its efficient reached-states algorithm. The algorithm builds the BDDs for the next state functions *at each iteration* of the reached-states computation, while simplifying each with respect to the current reached state set. Thus, the complete BDDs for the next state variables are never built. For many of our large examples, this is the only practical method for generating the reached state set.

Results on computing the reachable states are given in Table 7 (run on Alpha). TiGeR completes all examples, though not always in the least amount of time.[6] Nonetheless it is effective in consistently determining the reached state set. This algorithm is separate from core BDD technology, but can and should be implemented on top of any BDD package used in an FSM-type verification system. CMU fails on four examples and CUDD on three in one CPU hour. Where they do complete, no real conclusions can be drawn about their relative efficiency, unlike the previous experiments.

			CMU		CUDD		TiGeR
Circuit	Rstates	iter	time	mem	time	mem	time
att	226938	14	t(1hr)		t(1hr)		1398.7
bm	850609	20	628.5	17.1	1115.6	24.0	1883.2
noyau	2524161	11	t(1hr)		t(1hr)		680.3
prosa	7361	27	528.7	16.8	209.5	17.6	195.9
renault	257	12	21.1	10.6	17.1	1.5	6.6
snecma	10241	28	t(1hr)		t(1hr)		59.2
tcint	2826	26	38.8	10.9	173.6	11.1	140.6
tcintnc	310	10	7.9	9.3	1495.6	10.2	19.5
train310	25	2	t(1hr)		966.2	34.8	9.8
p1	17620993	123	108.5	9.8	105.7	13.1	102.4
p1opt	17620993	123	72.1	8.5	36.9	9.8	80.1

Table 7. Reachable States Computation on Alpha

Note that although the CUDD package usually outperformed the CMU in previous experiments, the results are quite different here.

5.5 An Example

The *train* example is our biggest and most difficult example. In order to complete synthesis and verification, we must alternate traditional logic optimization techniques (e.g. SIS) and optimization using reached state sets.

In one phase of the design process for this example, a bug was introduced. As a result, the underlying state graph went from millions of reachable states to just two. This was not an easy change to detect: the buggy design has millions of *equivalent* reachable states. The BDD packages applied to the buggy design ran for hours and produced no results. Only after simple logic optimization brought the design from 866 latches to 310, could TiGeR determine that there were 25 reachables states (in that version of the design). At this point, it was clear there was a bug, but we continued optimization to reduce the circuit as

---

[6] We found that TiGeR often performed better with sifting disabled; these results are not reported here.

much as possible and test the BDD packages. We used the 25 reachable states computed by TiGeR to compute and further remove redundant latches resulting in a 3-latch design. One final optimization step in SIS brought the design down to its final 1-latch, 2-state version.

The BDD packages were tested on several versions of this circuit. Recall that it is a large complex controller with counters. *train* is the original, correct design. *trains* is the original design after SIS *sweep*, and *traino* is the original after the SIS standard optimization script.

trainb is the original buggy design. *train_310* is *trainb* after *sweep*, with 310 latches. *train_254* is *trainb* after the SIS standard script, with 254 latches. *train_31* is a 31-latch version produced from *train_310* by TiGeR. (TiGeR has the capability of removing latches based on the reached state set.) *train_3* is a 3-latch version produced from *train_31* by our latest latch removal algorithms which are based on the reachable state set.

The results are given in Table 8 for BDD building and reachable states computation with sifting enabled. The examples are listed in increasing estimated difficultly. The experiments were run in this order, stopping when a package could not complete an example, as this uniformly led to failure as well in further testing. The t indicates a timeout after one CPU hour.

	CAL		CMU				CUDD				TiGeR	
	BDDs		BDDs		States		BDDs		States		BDDs	States
Circuit	time	mem	time	mem	time	mem	time	mem	time	mem	time	time
train3	7.8	6.5	3.0	1.5	2.4	1.7	6.9	2.2	3.5	2.5	1.6	1.5
train31	11.6	10.8	36.5	3.9	59.8	5.3	7.0	4.4	7.7	5.2	2.6	3.5
train254	147.4	91.3	t		t		363.9	6.8	2439.1	10.5	1267.1	15.1
train310	175.2	92.0					908.9	12.0	t		1687.0	12.3
trainb		m					2349.7	26.8			t	m
traino							3415.1	13.3			t	t
trains							3008.8	24.1				
train							2931.2	26.5				

Table 8. Train Example on Sparc20

The trends previously observed are repeated in this example: CAL is by far fastest for the examples in which it completes (only BDD building could be run); CMU requires more time but less memory than CMU for the examples in which it completes (albeit only the two smallest); CUDD returns a consistent performance in completing the BDDs for all circuits and returning better run times than TiGeR for BDD building; the TiGeR algorithm returns the best runtimes and the most circuits completed for the reachable states computation.[7]

---

[7] The BDDs are in TiGeR are usually built after the reachable states computation, and this is much more efficient than direct BDD-building as reported in the table.

6 Conclusions

Some interesting features of BDD packages have been demonstrated, and signal areas for further cooperative development. In particular,

- The CAL package is fast, perhaps due to its lossless computed table (note the high memory usage) and superscalarity features. These features should be further explored for integration in other packages.
- The CUDD package exhibits an excellent tradeoff in memory usage and CPU time consumed. It has well-tuned heuristics for controlling memory allocation and sifting.
- The use of simplified BDDs during the reachable states computation, as that employed in TiGeR, is invaluable and should be incorporated in public domain verification systems.

6.1 Other BDD Packages

A number of additional packages were investigated and may be included in subsequent studies, or otherwise of interest to the reader:

- Ken McMillan's BDD package is part of the SMV verification system, and is rumored to perform significantly better than the Long package.
- A BDD package from the Technical University of Eindhoven is used in the PVS verification system. The code is available but not yet ready for integration in SIS.
- Nils Klarlund and Theis Rauhe have a BDD package that is designed to reduce cache misses. The code has very recently become available and now is ready to be compared to other packages.
- The AMORE [10] package is currently under development at Albert-Ludwigs-University, Germany with plans for public availability soon. It uses a novel idea for creating BDDs: temporary auxiliary variables are introduced at the top of the BDD, sifted to the bottom, and quantified out. The authors report improved efficiency with respect to the traditional algorithms, but the package needs to be compared in a common environment.
- The TiGeR package [8] used for part of this study is sold through DEC. As such, it is publicly available but not free. It was included here solely to illustrate the efficacy of the reachability algorithm, which can be implemented on top of other packages.

7 Acknowledgements

Extensive discussions with Fabio Somenzi and Rajeev Ranjan are gratefully acknowledged. Our industrial partners are acknowledged for providing examples: AT&T, DEC, Renault, Snecma, and especially Yves Auffray of Dassault and Reinhard von Hanxleden of Daimler-Benz. This work was supported in part by NSF grant INT-9505943, by NSF/NATO grant GER-9552795, and by the French GENIE MESR INRIA project.

References

1. G. Berry and G. Gonthier. The ESTEREL Synchronous Programming Language: Design, Semantics, Implementation. *Science of Comp. Prog.*, 19(2):87–152, 1992.
2. K. Brace, R. Rudell, and R. Bryant. Efficient Implementation of a BDD Package. In 27th *DAC*, pages 40–45, June 1990.
3. R.E. Bryant. Graph-Based Algorithms for Boolean Function Manipulation. *IEEE Transactions on Computers*, C-35(8):677–691, August 1986.
4. R.E. Bryant. Symbolic Boolean Manipulation with Ordered Binary-Decision Diagrams. *ACM Computing Surveys*, 24(3):293–318, September 1992.
5. R.E. Bryant. Binary Decision Diagrams and Beyond: Enabling Technologies for Formal Verification. In *ICCAD*, pages 236–243, November 1995.
6. R.E. Bryant and Y.-A. Chen. Verification of Arithmetic Circuits with Binary Moment Diagrams. In *Proceedings of the 32nd Design Automation Conference*, pages 535–541, June 1995.
7. M. Chiodo, P. Giusto, H. Hsieh, A. Jurecska, L. Lavagno, and A. Sangiovanni-Vincentelli. A Formal Specification Model for Hardware/Software Codesign, June 1993. UC Berkeley Technical REport UCB/ERL M93/48.
8. O. Coudert, J.-C. Madre, and H. Touati, December 1993. TiGeR Version 1.0 User Guide, Digital Paris Research Lab.
9. M. Fujita, H. Fujisawa, and N. Kawato. Evaluation and Improvements of Boolean Comparison Method Based on Binary Decision Diagrams. In *ICCAD*, pages 2–5, November 1988.
10. A. Hett, R. Drechsler, and B. Becker. MORE: Alternative Implementation of BDD-Packages by Multi-Operand Synthesis. In *Proc of EuroDAC*, 1996. To appear.
11. N. Klarlund and T. Rauhe. BDD algorithms and cache misses, January 1996. Unpublished manuscript.
12. David Long, November 1993. Personal communication.
13. David Long, April 1996. Personal communication.
14. S. Malik, A.R. Wang, R.K. Brayton, and A. Sangiovanni-Vincentelli. Logic Verification using Binary Decision Diagrams in a Logic Synthesis Environment. In *ICCAD*, pages 6–9, November 1988.
15. S. Manne, D. C. Grunwald, and F. Somenzi. Remembrance of Things Past: Locality and Memory in BDDs, April 1996. Unpublished manuscript.
16. P.C. McGeer, K.L. McMillan, A. Saldanha, A.L. Sangiovanni-Vincentelli, and P. Scaglia. Fast Discrete Function Evaluation using Decision Diagrams. In *ICCAD*, pages 402–407, November 1995.
17. M.R. Mercer, R. Kapur, and D.E. Ross. Functional Approaches to Generating Orderings for Efficient Symbolic Rep. In 29th *DAC*, pages 624–627, June 1992.
18. H. Ochi, N. Ishiura, and S. Yajima. Breadth-First Manipulation of SBDD of Boolean Functions for Vector Processing. In 28th *DAC*, pages 413–416, June 1991.
19. S. Panda, F. Somenzi, and B.F. Plessier. Symmetry Detection and Dynamic Variable Ordering of Decision Diagrams. In *ICCAD*, pages 628–631, November 1994.
20. R.K. Ranjan, J.V. Sanghavi, R.K. Brayton, and A. Sangiovanni-Vincentelli. High Performance BDD Package Based on Exploiting Memory Hierarchy, June 1996.
21. R. Rudell. Dynamic Variable Ordering for Ordered Binary Decision Diagrams. In *ICCAD*, pages 42–47, November 1993.
22. E.M. Sentovich, K.J. Singh, L. Lavagno, C. Moon, R. Murgai, A. Saldanha, H. Savoj, P.R. Stephan, R.K. Brayton, and A.L. Sangiovanni-Vincentelli. SIS: A System for Sequential Circuit Synthesis, May 1992. UC Berkeley Technical Report UCB/ERL M92/41.

Local Encoding Transformations for Optimizing OBDD-Representations of Finite State Machines

Christoph Meinel, Thorsten Theobald*

FB IV – Informatik, Universität Trier, D–54286 Trier, Germany
{meinel,theobald}@ti.uni-trier.de

Abstract. Ordered binary decision diagrams are the state-of-the-art representation of switching functions. In order to keep the sizes of the OBDDs tractable, heuristics and dynamic reordering algorithms are applied to optimize the underlying variable order. When finite state machines are represented by OBDDs the state encoding can be used as an additional optimization parameter. In this paper, we analyze local encoding transformations which can be applied dynamically. First, we investigate the potential of re-encoding techniques. We then propose the use of an XOR-transformation and show why this transformation is most suitable among the set of all encoding transformations. Preliminary experimental results illustrate that the proposed method in fact yields a reduction of the OBDD-sizes.

1 Introduction

Ordered binary decision diagrams (OBDDs) which have been introduced by Bryant [Bry86] provide an efficient graph-based data structure for switching functions. The main optimization parameter of OBDDs is the underlying variable order. In order to find a good order two techniques were applied so far: the use of heuristics which try to exploit the structure of a circuit representation (see e.g. [MWBS88]), and dynamic reordering techniques [Rud93]. Unfortunately, there are many applications, in particular in the field of sequential analysis, where these two optimization techniques for OBDDs reach their limits. Hence, one essential problem in logic synthesis and verification is to find new techniques to minimize OBDDs in these applications.

When OBDDs are used to represent finite state machines the OBDD-size does not only depend on the variable order but also on the state encoding. For a fixed state encoding there are many finite state machines whose OBDD-representations are large w.r.t. all variable orders [ATB94]. Therefore the relationship between the OBDD-size and the state encoding becomes of increasing interest, see e.g. [QCC+95, TM96]. The importance of this relationship is underlined by recent ideas to apply heuristic state re-encoding techniques to speed up a verification process between similar-structured finite state machines [QCC+95].

---

* Supported by DFG-Graduiertenkolleg "Mathematische Optimierung".

The underlying general problem of all these efforts is the following: Given the OBDDs for the next-state and output functions of a finite state machine – if one is interested in the input/output behavior of the machine, in how far can the internal state encoding be exploited to minimize OBDD-sizes ? Our approach targets at applying *local* encoding transformations, i.e. transformations which involve only a limited number of encoding bits. These transformations can be interpreted as a re-encoding of the symbolic states. The aim is to minimize OBDD-sizes by the iterated application of local transformations. The advantage of this approach is that the costs for applying these transformations are still manageable.

The paper is structured as follows: We begin with recalling some important definitions and point out the principle potential of state re-encodings w.r.t. OBDD-sizes. Then, in Section 4, we analyze the advantages of local encoding transformations. In Section 5 we propose the application of the *XOR-transformation* and show why this transformation is most promising among the set of all encoding transformations. At the end of the paper we describe an implementation of this transformation and give some first experimental results which illustrate the positive impact of the presented ideas.

2 Preliminaries

2.1 Finite state machines

Let $M = (Q, I, O, \delta, \lambda, Q_0)$ be a finite state machine, where Q is the set of states, I the input alphabet, O the output alphabet, $\delta : Q \times I \to Q$ the next-state function, $\lambda : Q \times I \to O$ the output function and Q_0 the set of initial states. As usual, all components of the state machine are assumed to be binary encoded. Let p be the number of input bits, n be the number of state bits and m be the number of output bits. In particular, with $I\!B = \{0, 1\}$, δ is a function $I\!B^n \times I\!B^p \to I\!B^n$, λ is a function $I\!B^n \times I\!B^p \to I\!B^m$, and Q_0 is a subset of $I\!B^n$.

2.2 Binary decision diagrams

Ordered binary decision diagrams (OBDDs) [Bry86] are rooted directed acyclic graphs representing switching functions. Each OBDD has two sink nodes which are labeled 1 and 0. Each internal (= non-sink) node is labeled by an input variable x_i and has two outgoing edges, labeled 1 and 0 (in the diagrams the 1-edge is indicated by a solid line and the 0-edge by a dotted line). A linear variable order π is placed on the input variables. The variable occurrences on each OBDD-path have to be consistent with this order. An OBDD computes a switching function $f : I\!B^n \to I\!B$ in a natural manner: each assignment to the input variables x_i defines a unique path through the graph from the root to a sink. The label of this sink defines the value of the function on that input.

The OBDD is called *reduced* if it does not contain any vertex v such that the 0-edge and the 1-edge of v leads to the same node, and it does not contain any distinct vertices v and v' such that the subgraphs rooted in v and v' are isomorphic. It is well-known that reduced OBDDs are a unique representation of switching functions $f : \mathbb{B}^n \rightarrow \mathbb{B}$ with respect to a given variable order [Bry86]. The *size* of an OBDD is the number of its nodes. Several functions can be represented by a multi-rooted graph called *shared OBDD*. In the following, all next-state and output functions are represented by a shared OBDD.

2.3 The transition relation

For a finite state machine M, the characteristic function of its *transition relation* is defined by

$$T(x_1, \ldots, x_n, y_1, \ldots, y_n, e_1, \ldots, e_p) = T(x, y, e) = \prod_{1 \leq i \leq n} (y_i \equiv \delta_i(x, e)).$$

Hence, the function T computes the value 1 for a triple (x, y, e) if and only if the state machine in state x and input e enters the state y. The variables $x_1, \ldots, x_n$ are called *current-state variables* and the variables $y_1, \ldots, y_n$ are called *next-state variables*.

3   Motivation: The Potential of Re-encoding

In order to demonstrate how much the size of an OBDD-representation depends on the choice of the state encoding, let us consider an *autonomous counter*, a finite state machine with a very simple structure:

Example 1. An *autonomous counter* (see for example [GDN92]) with 2^n states $q_1, \ldots, q_{2^n}$ is an autonomous (i.e. input-independent) finite state machine with $\delta(q_i) = q_{i+1}$, $1 \leq i \leq 2^n - 1$, and $\delta(q_{2^n}) = q_1$.

Fig. 1. Autonomous counter

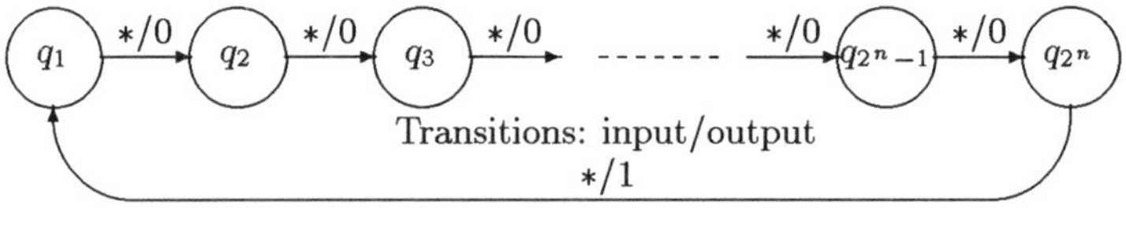

Transitions: input/output
*/1

The following theorem shows that almost all encodings for the autonomous counter lead to exponential-size OBDDs, even for their optimal variable order.

Theorem 1. *Let $e(n)$ denote the number of n-bit encodings for the autonomous counter with 2^n states which lead to a (shared) OBDD of size at most $2^n/n$ w.r.t. their optimal variable order. Let $a(n) = (2^n)!$ denote the number of all possible counter encodings. Then the ratio $e(n)/a(n)$ converges to zero as n tends to infinity.*

The proof of the theorem can be found in the appendix. It is based on ideas of [LL92] and classical counting results of Shannon [Sha49]. An analogous result can be established for the characteristic function of the transition relation and its OBDD-size.

Definition 2. An *encoding transformation*, shortly called a *re-encoding*, is a bijective function $\rho : I\!B^n \rightarrow I\!B^n$ that transforms the given state encoding to a new encoding. (For an example see Figure 2.) If a state s is encoded by a bit-string $c \in I\!B^n$, then its new encoding is $\rho(c)$.

Fig. 2. Encoding transformation $\rho(c_1, c_2) = (c_2, \overline{c_1})$

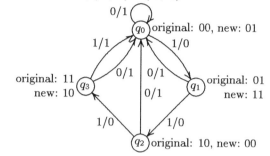

symbolic state	original encoding	new encoding
q_0	00	01
q_1	01	11
q_2	10	00
q_3	11	10

This modification of the internal state encoding does not modify the input/output behavior of the state machine. The machine with the new encoding is denoted by $M' = (Q', I, O, \delta', \lambda', Q_0')$. Its encoded next-state function, output function and set of initial states are computed as follows:

$$\delta'(s, e) = \rho(\delta(\rho^{-1}(s), e)),$$
$$\lambda'(s, e) = \lambda(\rho^{-1}(s), e), \tag{1}$$
$$Q_0' = \rho(Q_0),$$

where $\delta, \delta' : I\!B^n \times I\!B^p \rightarrow I\!B^n$, $\lambda, \lambda' : I\!B^n \times I\!B^p \rightarrow I\!B^m$, and $Q_0, Q_0' \subset I\!B^n$.

The transition relation of the re-encoded machine M' can be obtained from the transition relation of M as follows:

Lemma 3. *Let* $T(x, y, e)$ *be the characteristic function of the transition relation of* M. *Then characteristic function* $T'(x, y, e)$ *of the transition relation of* M' *is*

$$\prod_{i=1}^{n} \left(\rho_i^{-1}(y) \equiv \delta_i(\rho^{-1}(x)) \right).$$

Therefore $T'(x, y, e)$ *can be obtained from* $T(x, y, e)$ *by the substitutions* $y_i \mapsto \rho_i^{-1}(y)$ *and* $x_i \mapsto \rho_i^{-1}(x)$, $1 \leq i \leq n$.

Proof. The lemma is a consequence of the following equivalences:

$$T'(x, y, e) = 1 \iff \forall i \; y_i = \rho_i(\delta(\rho^{-1}(x))) \iff y = \rho(\delta(\rho^{-1}(x)))$$
$$\iff \rho^{-1}(y) = \delta(\rho^{-1}(x)) \iff \prod_{i=1}^{n} \left(\rho_i^{-1}(y) \equiv \delta_i(\rho^{-1}(x)) \right) = 1 \square$$

Example 1 (ctd.). The large potential of re-encoding techniques can now be demonstrated at the example of the autonomous counter: There exists an encoding such that the transition relation of the autonomous counter with 2^n states and n encoding bits has at most $5n - 1$ nodes [TM96] even if the variable order is fixed to $x_1, y_1, \ldots, x_n, y_n$. Hence, for each given encoding of a finite state machine, there exists a re-encoding which leads to OBDDs of linear size. As according to Theorem 1 most encodings lead to OBDDs of exponential size, the gain between the original OBDD and the OBDD after a suitable re-encoding is exponential in most cases. The aim now is to *find* the suited re-encoding that leads to small OBDD-sizes.

4 Local Re-encodings

In the previous section we have shown that re-encodings may have a large impact on the OBDD-size. It is possible that the OBDD becomes much smaller, but in the case of a badly chosen re-encoding the OBDD could even become much larger. This situation is comparable to the problem of finding a good variable order for an OBDD. When changing the variable order of an OBDD, the graph may become much smaller in the best case or much larger in the worst case. This sensitivity is the main reason why it is hard to *find* a good re-encoding or a good variable order.

For the effective *construction* of good variable orders it has turned out that the most efficient strategies are based on local exchanges of variables. The presently best strategies for finding good variable orders dynamically are based on the *sifting algorithm* of Rudell [Rud93, PS95]. The main principle of this algorithm is based on a subroutine which finds the optimum position for one variable, if all other variables remain fixed. This subroutine is repeated for each variable. There are two main reasons why this strategy works so efficiently:

Bounded size alteration: If one variable x_i is moved to another position in the OBDD, the size of the OBDD cannot change arbitrarily much, in particular it cannot explode. [BLW95] have shown the following theorem:

Theorem 4. *Let P be an OBDD. If a variable x_i is moved to a later position in the variable order, then the size of the resulting OBDD P′ satisfies*

$$size(P)/2 \le size(P') \le size(P)^2.$$

If a variable x_i is moved to an earlier position in the variable order, then the size of the resulting OBDD P′ even satisfies the relation

$$size(P)^{1/2} \le size(P') \le 2 \, size(P). \qquad \square$$

Practical studies have shown that in most cases the resulting sizes are even far below the worst-case estimations. Hence, each application of the above mentioned subroutine keeps the size of the OBDD manageable. However, this bounded size alteration for the subroutine does not mean that the optimization potential is limited. The iteration of this subroutine allows to minimize OBDDs very effectively.

Continuity: The procedure for moving a variable x_i to a different position in the order works continuously: During this process only the variables between the original and the new position of x_i are involved, and all nodes labeled by the remaining variables remain untouched. In particular, the time complexity of this operation is very small if x_i is moved to an adjacent position, and it increases with the number of variables between the original and the new position of x_i in the variable order.

In the case of re-encoding the situation is analogous. It seems to be very hard to find the right *global* re-encoding, whereas it is very promising to combine and iterate operations with restricted local effect. Our approach to construct *local* re-encodings $\rho : I\!B^n \rightarrow I\!B^n$ is to keep most of the bits fixed (i.e. $\rho_i(x) = x_i$) and vary only on a small number of bits. In particular, if we vary only 2 bits, we will speak of *two-bit re-encodings*. In this case it follows from the worst-case bounds for the synthesis and the substitution of OBDDs that the OBDDs remain polynomial.

Example 2. The *exchange variables* re-encoding $\rho_{i,j} : I\!B^n \rightarrow I\!B^n$, $1 \le i, j \le n$, is defined as follows (The following definition shows the case $i < j$, the case $i > j$ is defined analogously):

$$(q_1, \ldots, q_n) \mapsto (q_1, \ldots, q_{i-1}, q_j, q_{i+1}, \ldots, q_{j-1}, q_i, q_{j+1}, \ldots, q_n)$$

Obviously, the *exchange variables* re-encoding has the same effect on the next-state functions as exchanging the state variables x_i and x_j in the variable order. From all the $(2^n)!$ possible encoding transformations $n!$ can be generated by the iterated application of this transformation type. The inverse mapping ρ^{-1} from Equation 1 does not affect the size of the resulting OBDDs as this mapping only causes the renaming of the two functions δ_i and δ_j.

Note, that the transformation which exchanges the encodings of two fixed states may not be seen as a local operation, although the transformation seems to be very simple.

5   The XOR-Transformation

We will now propose *XOR-transformations*. This transformation is a local re-encoding which operates on two bits.

Definition 5. An *XOR-transformation* $\rho_{i,j}$, $1 \le i, j \le n$, is defined by

$$(q_1, \ldots, q_n) \mapsto (q_1, \ldots, q_{i-1}, q_i \oplus q_j, q_{i+1}, \ldots, q_n)$$

Short notation: $q_i \mapsto q_i \oplus q_j$. For an example see Figure 3.

Fig. 3. XOR-transform. $q_1 \mapsto q_1 \oplus q_2$

original encoding		new encoding	
q_1	q_2	q_1^{new}	q_2^{new}
0	0	0	0
0	1	1	1
1	0	1	0
1	1	0	1

Indeed, XOR-transformations provide a solid basis for the design of effective re-encodings due to the following facts:

1. The number of possible re-encodings generated by the iterated application of XOR-transformations is much larger than the number of possible variable orders. Thus, XOR-transformations considerably enlarge the optimization space. On the other hand, the number of these re-encodings is much smaller than the number of all re-encodings which makes it possible to keep the search space manageable.
2. The size influence of this transformation is bounded in a reasonable way like in the case of local changes in the variable order.
3. A precise analysis even shows that an XOR-transformation contains the same asymmetry as the movement of one variable in the variable order. Namely, the bounds for the effect of a transformation $x_i \mapsto x_i \oplus x_j$ depends on the position of x_i and x_j in the variable order.
4. The XOR-transformation is in fact the only new possible re-encoding on two variables.
5. The XOR-transformation can be implemented efficiently like an exchange of two variables in the order.

In the following subsection we will prove these statements.

5.1   Enumeration results

The following combinatorial statements characterize the size of the optimization space provided by the use of XOR-transformations.

Lemma 6. *(1) Let $t(n)$ be the number of possible encoding transformations that can be generated by the iterated application of XOR-transformations. It holds $t(n) = \prod_{i=0}^{n-1}(2^n - 2^i)$.*
(2) The quotient of $t(n)$ and the number of all possible encoding transformations converges to zero as n tends to infinity.
(3) Let $v(n) := (2n)!$ denote the number of possible variable orders for the transition relation of an autonomous finite state machine with n state bits. The fraction $v(n)/t(n)$ converges to zero as n tends to infinity.

Statement 3 says that in the case of autonomous state machines, there are much more encoding transformations generated by XOR-transformations than variable orders for the transition relation. This relation also holds when the number of input bits is fixed and the number of state bits becomes large.

Proof. (1) Obviously, each XOR-transformation is a regular linear variable transformation over the field $\mathbb{Z}_2$. Moreover, the XOR-transformations provide a generating system for all regular linear variable transformations. Therefore the state encodings which can be obtained by iterated XOR-transformations are in 1-1-correspondence with the regular $n \times n$-matrices over $\mathbb{Z}_2$.

The number of these matrices can be computed as follows: The first row vector b_1 can be chosen arbitrarily from $\mathbb{Z}_2^n \setminus 0$. The i-th row vector b_i, $2 \leq i \leq n$, can be chosen arbitrarily from $\mathbb{Z}_2^n \setminus \left\{ \sum_{j=1}^{i-1} \lambda_j b_j : \lambda_1, \ldots, \lambda_{i-1} \in \mathbb{Z}_2 \right\}$. These are $2^n - 2^{i-1}$ possibilities for the vector b_i. This proves the claimed number.

(2) This statement follows from the relation

$$\frac{t(n)}{(2^n)!} \leq \frac{2^{n^2}}{(2^n)!} \leq \frac{(n^2)!}{(2^n)!} \to 0 \text{ as } n \to \infty.$$

(3) It holds

$$\frac{(2n)!}{t(n)} = \frac{2n(2n-1)}{2^n - 2^0} \cdot \frac{(2n-2)(2n-3)}{2^n - 2^1} \cdots \frac{2 \cdot 1}{2^n - 2^{n-1}} \to 0 \text{ as } n \to \infty. \qquad \square$$

In particular, the number of possible encoding transformations which can be generated by the iterated application of XOR-transformations is smaller than 2^{n^2} which is exactly the number of all $n \times n$-matrices over $\mathbb{Z}_2$.

It follows from the previous proof that all exchanges of two state variables can be simulated by the iterated application of XOR-transformations.

5.2 Bounded size alteration

Let ρ be the XOR-transformation $q_i \mapsto q_i \oplus q_j$. Then the inverse transformation is defined by

$$\rho^{-1} : q_i \mapsto q_i \oplus q_j,$$

i.e. we have $\rho = \rho^{-1}$. The effect of $\rho(\delta(\rho^{-1}(\cdot)))$ in Equation 1 can be split into two parts:

1. Substitute the current-state variable x_i by $x_i \oplus x_j$.
2. Replace the function δ_i by $\delta_i \oplus \delta_j$.

It does not matter which of these two steps is executed first.

Lemma 7. *Let $P_1, \ldots, P_n$ be the OBDDs for $\delta_1, \ldots, \delta_n$ and $P_1', \ldots, P_n'$ be the OBDDs after the application of the XOR-transformation $q_i \mapsto q_i \oplus q_j$. The following holds:*

$$\Omega\big(size(P_i)^{1/2} / size(P_j)^{1/2}\big) \leq size(P_i') \leq \mathcal{O}\big(size(P_i)^2 \cdot size(P_j)^2\big),$$
$$\Omega\big(size(P_k)^{1/2}\big) \leq size(P_k') \leq \mathcal{O}\big(size(P_k)^2\big), \qquad k \neq i.$$

The upper bound immediately follows from the facts that the substitution of an OBDD P_2 into one variable of an OBDD P_1 leads to an OBDD of size at most $\mathcal{O}(size(P_1)^2 \cdot size(P_2))$, and that the operation $P_1 \oplus P_2$ leads to an OBDD of size at most $\mathcal{O}(size(P_1) \cdot size(P_2))$. For the lower bounds it suffices to observe $\rho = \rho^{-1}$.

In case of the transition relation both the current-state variables and the next-state variables have to be substituted. This leads to the result

$$\Omega(size(P)^{1/4}) \leq size(P') \leq \mathcal{O}(size(P)^4),$$

where P and P' are the original and the re-encoded OBDD for the transition relation, respectively.

5.3 Stronger bounds

For a more refined analysis of the XOR-transformation we use the following theorem from [SW93]. In particular, we will refine the analysis for the substitution of a variable x_i by $x_i \oplus x_j$ in an OBDD.

Theorem 8. *The reduced OBDD representing f with the variable order $x_1, \ldots, x_n$ contains as many x_i-nodes as there are different functions f^S, $S \subset \{1, \ldots, i-1\}$, depending essentially on x_i (i.e. $f^S_{x_i=0} \neq f^S_{x_i=1}$). Here $f^S = f_{x_1=a_1, \ldots, x_{i-1}=a_{i-1}}$ where $a_j = 0$ if $j \notin S$, and $a_j = 1$ otherwise.* □

Let s_k be the number of nodes labeled by x_k in the OBDD P and s_k^* be the number of nodes labeled by x_k in the OBDD P' which is the result of the transformation.

Theorem 9. *The size of an OBDD w.r.t. the variable order $x_1, \ldots, x_n$ after the application of the substitution $x_i \mapsto x_i \oplus x_j$ is bounded from above by*

$$2(s_1 + \ldots + s_{i-1}) + 1 + s_{i+1} + \ldots + s_n + 2 \quad \text{in case } j < i$$

and by

$$s_1 + \ldots + s_i + (s_{i+1} + \ldots + s_{j-1} + s_{j+1} + \ldots + s_n^* + 2)^2 \quad \text{in case } i < j.$$

Due to space limitations the proof is omitted here and can be found in [MT96]. It applies ideas from [BLW95], in which local changes in the variable order are analyzed.

Corollary 10. *Let P be an OBDD and P' the resulting OBDD after the substitution $x_i \mapsto x_i \oplus x_j$. Then*

$$\begin{aligned} size(P)/2 \leq size(P') \leq 2\, size(P) \quad &\text{if } x_j < x_i, \\ size(P)^{1/2} \leq size(P') \leq size(P)^2 \quad &\text{if } x_i < x_j. \end{aligned}$$
□

The analogy between the behavior of the XOR-transformation and the local changes in the variable order recommends to use XOR-transformations for the optimization of OBDD-sizes.

The XOR-transformation $x_i \mapsto x_i \oplus x_j$ for $j < i$ and $j + 1 \leq k \leq i$ can be visualized as shown in Figure 4. Let A and B be the two sub-OBDDs whose roots are the children of an x_i-node. Consider a path from x_j to x_i. If this path contains the 0-edge of x_j, the subgraph rooted in x_i remains unchanged. If instead the path contains the 1-edge of x_j, the 0- and the 1-successor of the x_i-node are exchanged. This modification can prevent subgraph-isomorphisms in the new sub-OBDDs which are rooted in an x_k-node, $j + 1 \leq k \leq i$.

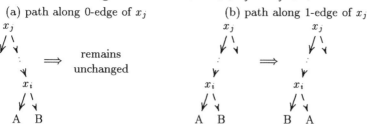

Fig. 4. Mutation $x_i \mapsto x_i \oplus x_j$ for $x_j < x_i$

(a) path along 0-edge of x_j (b) path along 1-edge of x_j

5.4 General two-bit re-encodings

The effect of each two-bit re-encoding can be split into the two-parts "Substitute the two variables x_i and x_j by some functions" and "replace the two functions δ_i and δ_j by some functions". The variable substitution has an impact on all functions which depend essentially on x_i or x_j, whereas the function replacement only affects the functions δ_i and δ_j.

The next table shows that all re-encodings which are induced by the $4! = 24$ bijective functions $f : I\!B^2 \to I\!B^2$ can be obtained by a combination of maximal one XOR-transformation, an exchange variable transformation and the identity. Hence, beside the exchange variable transformation merely XOR-transformations are needed to produce all two-bit re-encodings. We write a two-bit re-encoding which is induced by f as

$$\rho_{i,j}^f : (x_1, ..., x_n) \mapsto (x_1, ..., x_{i-1}, f_1(x_i, x_j), x_{i+1}, ..., x_{j-1}, f_2(x_i, x_j), x_{j+1}, ..., x_n)$$

	f(00)	f(01)	f(10)	f(11)	function		f(00)	f(01)	f(10)	f(11)	function
1	00	01	10	11	$(x_1, x_2) = $ id	13	10	00	01	11	$(x_1 \oplus x_2, x_1)$
2	00	01	11	10	$(x_1, x_1 \oplus x_2)$	14	10	00	11	01	$(\overline{x_2}, x_1)$
3	00	10	01	11	(x_2, x_1)	15	10	01	00	11	$(x_1 \oplus x_2, x_2)$
4	00	10	11	01	$(x_1 \oplus x_2, x_1)$	16	10	01	11	00	$(\overline{x_2}, x_1 \oplus x_2)$
5	00	11	01	10	$(x_2, x_1 \oplus x_2)$	17	10	11	00	01	$(\overline{x_1}, x_2)$
6	00	11	10	01	$(x_1 \oplus x_2, x_2)$	18	10	11	01	00	$(\overline{x_1}, x_1 \oplus x_2)$
7	01	00	10	11	$(x_1, x_1 \oplus x_2)$	19	11	00	01	10	$(x_1 \oplus x_2, \overline{x_2})$
8	01	00	11	10	$(x_1, \overline{x_2})$	20	11	00	10	01	$(\overline{x_2}, x_1 \oplus x_2)$
9	01	10	00	11	$(x_2, x_1 \oplus x_2)$	21	11	01	00	10	$(x_1 \oplus x_2, x_1)$
10	01	10	11	00	$(x_1 \oplus x_2, \overline{x_2})$	22	11	01	10	00	$(\overline{x_2}, \overline{x_1})$
11	01	11	00	10	$(x_2, \overline{x_1})$	23	11	10	00	01	$(\overline{x_1}, x_1 \oplus x_2)$
12	01	11	10	00	$(x_1 \oplus x_2, \overline{x_1})$	24	11	10	01	00	$(\overline{x_1}, \overline{x_2})$

The substitution $x_i \mapsto \overline{x_i}$ does not affect the size of the OBDD. As $\overline{x_i \oplus x_j} = \overline{x_i} \oplus x_j = x_i \oplus \overline{x_j}$, each of the above 24 transformations has the same effect w.r.t. the OBDD-size as a combination of the exchange variables transformation, the XOR-transformation and the identity operation. Moreover, for each of the 24 transformations, a combination of at most *two* of the "basis" transformations suffices.

5.5 Implementation aspects

In this section we will describe how to implement the XOR-substitution $x_i \mapsto x_i \oplus x_j$ efficiently. Our starting point is the consideration of local changes in the variable order. In order to modify the variable order of OBDDs we iterate exchanging variables in adjacent levels. Since an exchange of adjacent variables is a local operation consisting only of the relinking of nodes in these two levels, this can be done efficiently as shown in Figure 5. In order to move a variable x_i behind an arbitrary variable x_j in the order, the exchanges of adjacent variables are iterated.

Fig. 5. Exchanging two neighboring variables

Level exchange
$$\Longrightarrow$$

In case of the XOR-operation and adjacent variables x_i and x_j, we can proceed analogous to the level exchange. Figure 6 shows the case where x_j is the direct successor of x_i in the order. The case where x_i is the direct successor of x_j in the order works analogously. If x_i and x_j are not adjacent, it would of course be helpful if we could simulate the substitution $x_i \mapsto x_i \oplus x_j$ by a sequence like $x_i \mapsto x_i \oplus x_{i+1}$, $x_i \mapsto x_i \oplus x_{i+2}$, ..., $x_i \mapsto x_i \oplus x_j$. However, this straightforward idea does not work, as this would require operations in the intermediate steps which influence more than two adjacent levels.

Fig. 6. Performing $x_i \mapsto x_i \oplus x_j$ for two neighboring variables x_i, x_j

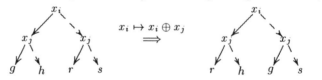

$$x_i \mapsto x_i \oplus x_j$$
$$\Longrightarrow$$

A method that works and is only slightly more expensive than the exchange of two non-adjacent variables is the following: First, shift the variable x_i to a new position in the order which is adjacent to x_j. Then perform the XOR-operation, and then shift the variable x_i back to its old position. This technique retains the locality of the operation, as only nodes with a label x_k are influenced whose position in the order is between x_i and x_j.

6   Do All FSM-Descriptions Profit From XOR-Re-encod. ?

In principle, the applicability of the XOR-transformation is not restricted to the use of OBDDs as underlying data structure. It can also be applied to other data structures for Boolean functions. However, the strong relationship between the XOR-transformation and local changes in the variable order like in the case of OBDDs does not always transfer to other representations. We will demonstrate this effect on OFDDs.

Ordered functional decision diagrams (OFDDs) [KSR92] are a modification of OBDDs which seem to be more compact for arithmetic functions. Each node v with label x_i in an OBDD represents a Shannon decomposition

$$f = x_i \cdot g + \overline{x_i} \cdot h,$$

whereas each node v with label x_i in an OFDD represents a Reed-Muller decomposition

$$f = g \oplus x_i \cdot h.$$

In both decompositions the functions g and h are independent of x_i and are the functions which are represented by the subgraphs rooted in the two successor nodes of v.

It has been shown in [BLW95] that local changes in the variable order have the same effect on OFDDs like on OBDDs. In particular, the exchange of two variables x_i and x_j in the order only affects the nodes of an OBDD resp. OFDD which are labeled by a variable whose position in the order is between x_i and x_j. These observations justify the notions of *local* changes. From the proof of Theorem 9 it follows that the XOR-transformation for OBDDs has also this pleasant local property. However, in spite of the fact that the Reed-Muller decomposition seems to operate well with XOR-transformations, the substitution $x_i \mapsto x_i \oplus x_j$ for OFDDs does not have the local property like in the case of OBDDs.

Fig. 7. OFDD for $f = g \oplus x_2 \cdot h \oplus x_1 \cdot r \oplus x_1 \cdot x_2 \cdot s$

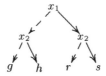

To prove this statement, consider the OFDD in Figure 7 which represents the function

$$f = g \oplus x_2 \cdot h \oplus x_1 \cdot r \oplus x_1 \cdot x_2 \cdot s$$

for some functions g, h, r, s independent of x_1, x_2. The function $f'(x_1, \ldots, x_n) = f(x_1 \oplus x_2, x_2, \ldots, x_n)$ is

$$f' = g \oplus x_2 \cdot h \oplus (x_1 \oplus x_2) \cdot r \oplus (x_1 \oplus x_2) \cdot x_2 \cdot s$$
$$= g \oplus x_2 \cdot (h \oplus r \oplus s) \oplus x_1 \cdot r \oplus x_1 \cdot x_2 \cdot s.$$

Hence, there must exist a node in the OFDD for f' representing the function $h \oplus r \oplus s$. As h, r and s are arbitrary functions, the substitution operation $x_i \mapsto x_i \oplus x_j$ does not have the local property. However, as the $\oplus$-operation is a polynomial operation on OFDDs, the result of the substitution remains polynomial.

A tight relationship between our re-encoding techniques and specific OBDD-variants is the following: In a more general setting, the concept of domain transformations has been proposed for the manipulation of switching functions [BMS95, FKB95]. The corresponding variants of OBDDs are called TBDDs. In the following we show that our encoding transformations can also be interpreted as TBDD-transformations.

Definition 11. Let $f : \mathbb{B}^n \to \mathbb{B}$ be a switching function, and let $\tau : \mathbb{B}^n \to \mathbb{B}^n$ be a bijective mapping. A τ-*TBDD-representation* of f is an OBDD-representation of $f \circ \tau$, where $\circ$ denotes the composition of functions.

It turns out that every re-encoding function ρ defines a transformation within the TBDD-concept. However, the OBDDs for the next-state functions of a re-encoded machine with re-encoding function ρ are not isomorphic to the ρ-TBDDs, as in the OBDDs of the re-encoded machine, the transformation ρ^{-1} is also involved (see Equation (1) in section 3).

7 Experimental Results

In this section we present some very first experimental results on the extended optimization techniques for OBDDs. We built up some routines on top of the OBDD-package of D. Long and used the ISCAS89 benchmark circuits s1423, s5378, s9234 which have a large number of state bits and have also formed the set of examples in [RS95]. Each optimization run consists of three phases: First, we applied Rudell's sifting algorithm [Rud93] for finding a good variable order. Then some minimization based on XOR-transformations is performed. Finally, sifting is applied once more to re-establish a suitable variable order. The table shows the obtained shared OBDD-sizes of the next-state functions in comparison to the sizes that were obtained *without* the minimization by XOR-transformations.

Circuit	# state bits	without XOR	with XOR
s1423	74	3054	2853
s5378	121	1801	1517
s9234	228	4621	4359

The minimization based on XOR-transformation works as follows: In a pre-processing step we compute promising pairs (i, j) for an XOR-transformation. The heuristic criteria for considering a pair (i, j) as promising are:

1. the next-state functions δ_i and δ_j have a nearly equal support, or
2. the variables x_i and x_j appear in nearly the same functions.

Then, as long as improvements are possible, the best XOR-transformation among these pairs are applied. In order to avoid the expensive computations $\delta_i \oplus \delta_j$, we only perform this step if the variable substitution $x_i \mapsto x_i \oplus x_j$ yields a good intermediate result.

It must be admitted that in our experiments, the running times are significantly higher than the running times for pure sifting. This is due to a non-optimal implementation of the XOR-transformation and due to the large number of performed XOR-transformations. However, we think that intensive studies of different strategies (analogous to the intensive and finally quite successful studies of the variable order) should be able to improve these results by far. Hence, we think that these experimental results underline the optimization potential of the XOR-transformation.

An efficient implementation of the XOR-transformation and the construction of effective minimization strategies is in progress.

8 Conclusion and Future Work

We have proposed and analyzed new re-encoding techniques for minimizing OBDDs. In particular, we have proposed the XOR-transformation and shown that this transformation is in fact the only new transformation on two state variables. This transformation can in certain cases significantly enrich the set of basic operations for the optimization of OBDDs. In the future we propose to study heuristics for choosing the right transformation pairs and efficient combinations of variable reordering techniques with the new proposed transformations. Furthermore, the dynamic application of the new re-encoding technique in the traversal of a finite state machine seems promising: It helps to reduce the OBDD-sizes for the set of reached states and also to reduce the efforts of the image computations.

9 Acknowledgements

We wish to thank Stefan Krischer, Jan Roßmann, Anna Slobodová and Fabio Somenzi for interesting discussions and many valuable comments.

References

[ATB94]   A. Aziz, S. Taziran, and R. K. Brayton. BDD variable ordering for interacting finite state machines. In *31st ACM/IEEE Design Automation Conference*, pp. 283–288, 1994.

[BLW95]   B. Bollig, M. Löbbing, and I. Wegener. Simulated annealing to improve variable orderings for OBDDs. In *International Workshop on Logic Synthesis*, 1995.

[BMS95]   J. Bern, Ch. Meinel, and A. Slobodová. OBDD-based Boolean manipulation in CAD beyond current limits. In *32nd ACM/IEEE Design Automation Conference*, 1995.

[Bry86]   R. E. Bryant. Graph-based algorithms for Boolean function manipulation. *IEEE Transactions on Computers*, C–35:677–691, 1986.

[FKB95]   M. Fujita, Y. Kukimoto, and R. K. Brayton. BDD minimization by truth table permutations. In *International Workshop on Logic Synthesis*, 1995.

[GDN92]   A. Ghosh, S. Devadas, and A. R. Newton. *Sequential Logic Testing and Verification*. Kluwer Academic Publishers, 1992.

[KSR92]   U. Kebschull, E. Schubert, and W. Rosenstiel. Multilevel logic synthesis based on functional decision diagrams. In *European Design Automation Conference*, pp. 43–47, 1992.

[LL92]   H.-T. Liaw and C.-S. Lin. On the OBDD-representation of general Boolean functions. *IEEE Transactions on Computers*, 41:661–664, 1992.

[MT96]   Ch. Meinel and T. Theobald. Local encoding transformations for optimizing OBDD-representations of finite state machines. Technical Report 96-23, Universität Trier, 1996.

[MWBS88] S. Malik, A. Wang, R. K. Brayton, and A. Sangiovanni-Vincentelli. Logic verification using binary decision diagrams in a logic synthesis environment. In *IEEE International Conference on Computer-Aided Design*, pp. 6–9, 1988.

[PS95]   S. Panda and F. Somenzi. Who are the variables in your neighborhood. In *IEEE International Conference on Computer-Aided Design*, 1995.

[QCC+95]   S. Quer, G. Cabodi, P. Camurati, L. Lavagno, E. M. Sentovich, and R. K. Brayton. Incremental FSM re-encoding for simplifying verification by symbolic traversal. In *International Workshop on Logic Synthesis*, 1995.

[RS95]   K. Ravi and F. Somenzi. High-density reachability analysis. In *IEEE International Conference on Computer-Aided Design*, 1995.

[Rud93]   R. Rudell. Dynamic variable ordering for ordered binary decision diagrams. In *IEEE International Conference on Computer-Aided Design*, pp. 42–47, 1993.

[Sha49]   C. E. Shannon. The synthesis of two-terminal switching circuits. *Bell System Technical Journal*, 28:59–98, 1949.

[SW93]   D. Sieling and I. Wegener. NC-algorithms for operations on binary decision diagrams. *Parallel Processing Letters*, 3:3–12, 1993.

[TM96]   T. Theobald and Ch. Meinel. State encodings and OBDD-sizes. Technical Report 96-04, Universität Trier, 1996.

Appendix: Proof of Theorem 1

Proof. Let $K = \lfloor 2^n/n \rfloor$. Suppose there are c_i variables with label x_i in the OBDD, and let $c_{n+1} = K - \sum_{i=1}^{n} c_i$. c_{n+1} serves to cover the case $\sum_{i=1}^{n} c_i < K$. Since each c_i, $1 \le i \le n+1$, is a nonnegative integer, the equation $\sum_{i=1}^{n} c_i = K$ has exactly $\binom{n+K}{K}$ solutions. Suppose that all the vertices are ordered by increasing positions in the variable order, then each of the two edges of a vertex can only connect to its successors, including the nonterminal vertices 0 and 1. Hence there are at most $(K+1)^2 K^2 (K-1)^2 \cdot \ldots \cdot 2^2 = ((K+1)!)^2$ ways to place the edges. Since there are $n!$ different variable orders, there are at most

$$n! \binom{n+K}{K} ((K+1)!)^2$$

shared OBDDs with at most K vertices for their optimal variable order. Due to the cyclic symmetry of the autonomous counter it follows

$$e(n) \le 2^n n! (n+K)! ((K+1)!)^2 \le 2^n (2n + 3K + 2)!$$
$$\le 2^n (2n + 3 \cdot 2^n/n + 2)! \le 2^n (4 \cdot 2^n/n)! \quad \text{for sufficiently large } n,$$

and therefore

$$\frac{e(n)}{a(n)} \to 0 \quad \text{as } n \to \infty. \qquad \square$$

Decomposition Techniques for Efficient ROBDD Construction

Jawahar Jain[1]   Amit Narayan[2]   C. Coelho[3]   Sunil P. Khatri[2]

A. Sangiovanni-Vincentelli[2]   R. K. Brayton[2]   M. Fujita[1]

Abstract. In this paper, we address the problem of memory-efficient construction of ROBDDs for a given Boolean network. We show that for a large number of applications, it is more efficient to construct the ROBDD by a suitable combination of top-down and bottom-up approaches than a purely bottom-up approach. We first build a decomposed ROBDD of the target function and then follow it by a symbolic composition to get the final ROBDD. We propose two heuristic algorithms for decomposition. One is based on a topological analysis of the given netlist, while the other is purely functional, making no assumptions about the underlying circuit topology. We demonstrate the utility of our methods on standard benchmark circuits as well as some hard industrial circuits. Our results show that this method requires significantly less memory than the conventional bottom-up construction. In many cases, we are able to build the ROBDDs of outputs for which the conventional method fails. In addition, in most cases this memory reduction is accompanied by a significant speed up in the ROBDD construction process.

1 Introduction

Reduced Ordered Binary Decision Diagrams (ROBDDs) [4] are frequently used as the representation of choice to solve various CAD problems such us synthesis, digital-system verification and testing. However, the construction of ROBDDs is often a time and memory intensive process. Techniques that can help in a more efficient construction of ROBDDs are of great practical significance.

Traditionally, ROBDDs for a given netlist are built in a bottom-up manner. To construct the ROBDD for a given node, ROBDDs of all the nodes that are present in the transitive fan-in of that node are constructed in terms of the primary inputs before the ROBDD of the target node is constructed. In this method, the peak intermediate memory requirement often far exceeds the final representation size of the given function. Although we are usually interested in obtaining the ROBDD of only the output, the large intermediate peak memory requirement often limits our ability to construct it. This places a limit on the complexity of circuits that can be processed using ROBDDs, and also usually dictates the time required for ROBDD construction. In this paper, we

---

[1] Fujitsu Laboratories of America, San Jose, CA 95134
[2] Department of Electrical Engineering and Computer Sciences, University of California, Berkeley, CA 94720
[3] Department of Electrical Engineering, Stanford University, Stanford, CA 94305

present techniques to reduce the intermediate peak memory requirement. This is achieved by a suitable combination of bottom-up and top-down approaches. We show that by using these techniques, ROBDDs for many circuit outputs can be constructed for which the conventional method fails. In addition, we observe that the reduction in peak memory is often accompanied by a significant speed up in the ROBDD construction process. Our approach is fully compatible with other approaches of reducing memory (like variable reordering) and can be seamlessly integrated within any ROBDD package. Therefore, we believe that there is no real trade-off in using it.

The presence of an intermediate peak followed by a subsequent reduction in the memory requirement indicates that Boolean simplification has occurred in the circuit. We attempt to capture this simplification so as to reduce the peak memory requirement. We create a small decomposed representation of the output by introducing decomposition points in the circuit. These points are introduced in such a manner that the decomposed function captures the Boolean simplification occurring in the circuit. Finally, we compose this function to get the canonical ROBDD of the output. By focusing only on the target ROBDD, we avoid the intermediate peak and are able to reduce the overall memory requirement.

We propose two separate methods to combine the top-down and bottom-up approaches. In the first method, we conduct a topological analysis to determine good decomposition points for the circuit. In the second, we do a functional decomposition. Here, a bottom-up construction of the ROBDD is attempted; decomposition points are introduced when the number of nodes in the ROBDD manager grows beyond a predetermined threshold. This approach makes no assumptions about the underlying topology of the circuit and there is no physical correspondence between the decomposition points and the nodes in the circuit. As such, this technique is applicable to any arbitrary sequence of ROBDD operations besides building ROBDDs of the outputs of a given netlist. To emphasize this point, in Section 6 we present a method to do image computation for sequential circuit verification using the decomposed representation of the transition relation.

We present the results on some hard industrial circuits in addition to the standard ISCAS benchmarks. We conduct experiments both with and without dynamic variable ordering and show impressive gains in each case.

Rest of the paper is organized as follows. In Section 2 we review the previous work. In Section 3, we define the terminology used in the sequel. We describe our decomposition/composition approach in Section 4, and discuss the results obtained in Section 5. In Section 6, we discuss the application of decomposition technique in sequential circuit verification and some extensions in Section 7. We present the conclusion of our study in Section 8.

2   Previous Work

Though BDDs have been researched for about four decades [15, 1], they found widespread use only after Bryant [4] showed that such graphs, under some restrictions, are canonical and can be easily manipulated. The two restrictions imposed are that the graph is reduced (i.e. no two nodes have identical subgraphs), and that a total ordering of the variables is enforced. The resulting BDD is called a Reduced Ordered BDD (ROBDD). Two important symbolic manipulation procedures were presented by Bryant to combine two identically ordered ROBDDs. The first operation is *Apply* which allows ROBDDs to be combined under some Boolean operation, and the second is *compose* which allows the substitution of an ROBDD variable by a function.

ROBDDs are typically constructed using some variant of Bryant's *apply* procedure[4]; the ROBDD for a gate g is synthesized by the symbolic manipulation of the ROBDDs of its inputs, based on the functionality of g. The gates of the circuit are processed in a depth first manner until the ROBDD of the desired (or *target*) gate is constructed. The ITE method [3] for constructing ROBDDs is similar; there are equivalent ITE operators for *apply* and *compose*. For details on ROBDDs, and the implementation of a typical ROBDD package, please refer to [4, 5, 3].

The size of a ROBDD is strongly dependent on the variable ordering. Finding the best variable ordering for a given Boolean function is a hard problem. Many different ordering heuristics have been proposed in the literature but the most commonly used methods are those of [9, 21].

While the problem of variable ordering has received a widespread attention, very little attention has been paid towards decomposition/composition techniques to reduce the memory. This is perhaps because the final ROBDD size was perceived to be the main bottleneck in using ROBDD based methods. Since the final ROBDD size is a constant for a given variable ordering, it was believed that the only way to reduce the memory requirement was to have a better variable ordering. While this is true in the sense that the final ROBDD size provides a lower bound on the amount of resources needed to construct the ROBDD, we observe that in a large number of cases it is not the final ROBDD size but the large intermediate ROBDD which limits the ROBDD construction. In such cases we can use our techniques to reduce the overall memory requirement. Although the dynamic variable reordering (DR) techniques [21] do attempt to reduce the intermediate memory also, we show that our techniques give significant additional gains over what is obtained by DR. In this sense our techniques exploit a different aspect of the memory explosion problem and complement the variable ordering research.

Decomposition in the ROBDD context has received some attention in the past. Fujita et. al. briefly discussed a few interesting decomposition experiments in [10] while computing efficient variable orders for a given circuit. However, the results obtained by decomposition were not always encouraging; further no intuitively sound decomposition procedure was outlined.

Berman [2] found a relation between the register allocation problem and efficient circuit decomposition. For a bounded-width circuit of n variables, that is, a circuit whose elements can be linearly ordered such that any cut crosses at most w wires, it was shown that there exists a variable order such that the size of the ROBDD will be bounded by $n * 2^w$. While the decomposition problem was conceptually well analyzed, efficient and well-tested procedures to find good decompositions were not provided. McMillan [17] generalized the results of Berman and proved that if w_f bounds the number of cut wires in the forward direction, and w_r bounds the number of cut wires in the reverse direction, then the size of the resulting ROBDDs is bounded by $n2^{w_f}2^{w_r}$

Jain et. al., in [12], discussed the possibility of obtaining decompositions for efficient "hashing" of Boolean circuits. However no method was described to obtain such decompositions.

The notion of introducing auxiliary (decomposition) variables was also used in [14]. In [14], auxiliary variables were used to create a new data structure called the Extended BDD. The data structure was not canonical and generating a canonical representation from this structure required a storage of $O(|p|2^n)$ intermediate results in the worst case. The authors of [14] observed that it was not practical to generate canonical ROBDDs from this structure. A technique to verify combinational circuit was presented using this data structure. Very limited experiments were performed and results reported were not very encouraging.

Experiments on using auxiliary variables were reported in [7] for creating more efficient symbolic state-space traversal techniques. General decomposition techniques, however, were not presented. The benchmark consisted of sequential circuits with data paths having a very regular structure which was exploited to manually generate the decomposition points.

None of the above studies have focussed on the problem of decomposition to reduce the intermediate peak memory requirement in the ROBDD construction phase. Our main contribution lies in studying the problem in this context. In addition, the decomposition techniques that we propose are fully automated and work on arbitrary Boolean circuits without making any assumptions about their structure.

3 Preliminaries

In this section we establish the terminology for the rest of this paper. Assume that we are given a circuit representing a Boolean function $F \equiv F : B^n \rightarrow B^o$, with n primary inputs (PIs) $\{x_1, \ldots, x_n\}$, and o primary outputs (POs). To simplify the discussion, let us focus on a single output g. Let $G(X)$ represent the ROBDD of g in terms of PIs X (In future, we will not make any distinction between a function g and a ROBDD representing it in terms of the PIs). Let $G_d(\Psi, X)$ represent the ROBDD of g expressed in terms of primary inputs X and a set of auxiliary variables Ψ, where $\Psi = \{\psi_1, \psi_2, \ldots, \psi_m\}$. Here each ψ_i corresponds to a *decomposition point* and $\Psi \equiv \{\psi_1, \ldots, \psi_k\}$ is called a *decomposition set* of the circuit.

For every variable (decomposition point) ψ_i in Ψ, there exists a ROBDD $\psi_{i_{bdd}}$ representing the functionality of ψ_i in terms of PIs as well as (possibly) other $\psi_j \in \Psi$, where $\psi_j \neq \psi_i$. We ensure that the elements of Ψ can be ordered such that $\psi_{j_{bdd}}$ depends on ψ_i only if $i < j$. Let $\Psi_{bdd} \equiv \{\psi_{i_{bdd}}, \ldots, \psi_{k_{bdd}}\}$ represent the array containing the ROBDDs of decomposition points in Ψ.

The variable ψ_i can be substituted by the function $\psi_{i_{bdd}}$ using the composition operation [4]. The composition of ψ_i in $G_d(\Psi, X)$ is denoted by $G_d(\Psi, X).(\psi_i \leftarrow \psi_{i_{bdd}})$ where,

$$G_d(\Psi, X).(\psi_i \leftarrow \psi_{i_{bdd}}) = \overline{\psi_{i_{bdd}}}.G_d(\Psi, X)_{\overline{\psi_i}} + \psi_{i_{bdd}}.G_d(\Psi, X)_{\psi_i} \qquad (1)$$

Here, $G_d(\Psi, X)_{\psi_i}$ represents the *restriction* of $G_d(\Psi, X)$ at $\psi_i = 1$, and is obtained by directing all the incoming edges to the node with variable id ψ_i to its $\psi_i = 1$ branch and reducing the resulting graph. The compose operation can be done in $O(|G_d|^2 |\psi_{i_{bdd}}|)$ by using the ITE operation[4] and caching the intermediate results. Here $|G_d|$ and $|\psi_{i_{bdd}}|$ represent the number of nodes in the ROBDDs G_d and $\psi_{i_{bdd}}$ respectively.

Another way to compose ψ_i in G_d is by smoothing out ψ_i from the product of G_d and the characteristic function representing $\psi_{i_{bdd}}$. This is represented by the following equation:

$$G_d(\Psi, X).(\psi_i \leftarrow \psi_{i_{bdd}}) = \exists_{\psi_i}[G_d(\Psi, X).(\psi_i \overline{\oplus} \psi_{i_{bdd}})] \qquad (2)$$

The vector composition of the Ψ in $G_d(\Psi, X)$ is denoted by $G_d(\Psi, X).(\Psi \xleftarrow{\pi} \Psi_{bdd})$ and represents successive composition of the variables in Ψ in G_d by their corresponding $\psi_{i_{bdd}}$s. Here, π represents the order in which ψ_is are composed in G_d.

The monolithic ROBDD G of the output in terms of PIs can be obtained from the decomposed ROBDD $G_d(\Psi, X)$ and the decomposition set Ψ (along with the corresponding Ψ_{bdd} array) by successive composition of Ψ in G_d i.e.

$$G = G_d(\Psi, X).(\Psi \xleftarrow{\pi} \Psi_{bdd}) \qquad (3)$$

The problem of decomposition is to find a set Ψ with the corresponding Ψ_{bdd} such that the monolithic ROBDD G can be obtained by the successive composition of Ψ in G_d.

4   Our Approach

Let us look at an example where the memory requirement for a bottom-up scheme is exponential while the decomposition/composition approach requires only polynomial resources. Consider the function shown in Figure 1. Here f and g are two internal nodes. Assume that the ROBDD of g is exponential in terms of the primary inputs (PIs) for any variable ordering. Further assume that all the other internal nodes of the function require only polynomial memory resources. If we try to build the ROBDD of the primary output y in a bottom-up fashion, we will need to build the ROBDD of g in terms of the PIs. But since the ROBDD of g is exponential for any given variable ordering, the peak memory required in the bottom up scheme will be exponential. The functionality of y can be expressed in terms of f and g by the equation $y = f \vee (f \wedge g)$. This simplifies to $y = f$. Therefore, to construct the ROBDD of y we do not need to construct the ROBDD of g. We can introduce a new variable representing g and build the ROBDD of y in terms of this variable. Since g is eventually not present in y, we need not ever construct the ROBDD of g. In this manner we can get an exponential reduction in the peak memory requirement and extend the class of circuits that can be efficiently processed using ROBDDs.

The previous example shows that in a typical ROBDD construction procedure there is frequent functional simplification due to Boolean Absorption: $x \vee (x \wedge y) = x$, Boolean Cancellation: $x \wedge \overline{x} \wedge y = 0$. This means that the final output function is often much simpler (i.e. has a smaller ROBDD) than the intermediate functions that are used to implement it. In fact it can be proved that for general Boolean functions, ROBDDs are almost always exponential. It is only a statistical coincidence that ROBDDs for the functions that are generally encountered in real life do not exhibit the worst case behavior. But there do exist useful circuits (like Multiplier) which exhibit the exponential worst case. The traditional bottom-up method will fail when any node in the transitive fan-in of the target node exhibits the worst case behavior whereas our method is limited (at least in principle) only by the worst case behavior of the target node and hence is more likely to succeed.

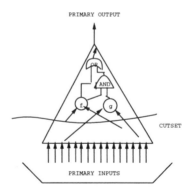

Fig. 1. *Example where decomposition can avoid exponential blowup*

Our strategy to avoid intermediate memory blowup is very simple. We try to avoid building ROBDDs of the intermediate functions having a large ROBDD representation by introducing suitable decomposition points (ψ_is). We build the ROBDD of the output (G_d) in terms of these decomposition points and PIs. The functionality of the decomposition points is expressed as ROBDDs ($\psi_{i_{bdd}}$s) in terms of previously introduced decomposition points and PIs. We then compose $\psi_{i_{bdd}}$s into G_d to obtain canonical ROBDD of the output function. In deference to the intractability of such analysis problems, we note that one can work out examples where this mixed bottom-up/top-down approach will require exponentially more resources than a purely bottom-up (or top-down) technique used in isolation. Thus our aim is to find heuristics which perform well in most cases.

Two issues need to be addressed here:

- Finding a good decomposition set (Ψ)
- Determining a good order of composition (π) of Ψ into G_d so that the intermediate memory explosion during the composition phase is low.

In Section 4.1 and 4.2 we describe our decomposition techniques. In Section 4.1 we present a structural (topology based) approach augmented using functional information. In Section 4.2 we present a purely functional approach. Section 4.3 describes out composition approach.

4.1 A Structural Analysis Based Decomposition Approach

A good decomposition set, Ψ, should have the following properties:

- Small $|G_d(\Psi)|$ as the complexity of composition is quadratic in the size of G_d.
- Small cardinality so that fewer compositions are needed to obtain the canonical ROBDD of the output G in terms of the primary inputs.
- Large average fanout of the decomposition points. When an internal signal fans out to a large number of gates in a circuit, and is used as a decomposition point, it is more likely to be present in many expressions. Given a DNF expression, the greater the number of variables common among its different terms, the higher is the likelihood that the disjunction is simpler due to the application of Boolean absorption.

It is empirically observed that the size of an ROBDD increases as its support set increases. A cutset of a circuit is a set of nodes such that any path from a primary input to a primary output has to cross through one of the nodes in the circuit. A cutset with minimum cardinality is defined as a mincut of the circuit. A mincut of the circuit is a good decomposition set as it minimizes the number of variables on which G_d depends and also the number of compositions needed in the end to derive the final ROBDD. We also look at the *gradient* of the circuit, which is the rate at which the cutset size of the circuit increases as we traverse the circuit from the primary inputs to the primary outputs. A cutset having a large gradient will have a large average fanout, increasing the likelihood of Boolean simplification. We call a cutset having a large gradient and small cardinality a *neck* of the circuit. This forms our initial decomposition set.

If the *neck* is very close to the primary inputs or the primary outputs then almost all ROBDD resources will be spent in the top-down or bottom-up procedures respectively. In this case, any contribution of decomposition is negated. Hence we restrict our search for cutsets to a *candidate region*, which comprises of the entire circuit except the regions at either end of the circuit.

Generally, the resources required in generating the ROBDDs for G_d as well as all $\psi_{i_{bdd}}$s, for $1 \leq i \leq m$, are significantly less than the resources spent in constructing the function G. Thus we can experiment with various decompositions without paying an excessive space or time penalty.

After selecting the best cutset in the candidate region, we perform a local alteration on it. As already mentioned in Section 3, composing an ROBDD $\psi_{i_{bdd}}$ into an ROBDD G_d can require $O(|G_d|^2|\psi_{i_{bdd}}|)$ resources. Thus, if Ψ_a and Ψ_b are two decomposition sets having the same cardinality and the same average fanout, and if the ROBDDs of the decomposition points are approximately the same, then the cutset having a smaller decomposed ROBDD (G_d) is a better choice. We perform local alterations on the initial cutset (obtained using the minimum cardinality and the largest gradient criteria) to obtain a new cutset having a smaller G_d, without significantly affecting the sizes of the $\psi_{i_{bdd}}$s.

After performing local changes to the cutset, we obtain a new cutset, say $\Psi = \{\psi_1, \psi_2 \ldots \psi_k\}$. We build the ROBDDs of each $\psi_i \in \Psi$ in terms of its immediate fanins and successively compose it into G_d. We continue this process, moving the cutset towards the primary inputs. We stop when we detect a disproportionate increase in the size of G_d. The cutset at this point represents the decomposition set. We construct the ROBDDs of these decomposition points in terms of primary inputs (using the conventional bottom-up technique) and then compose them into G_d to get the final canonical representation of the output G. The entire algorithm is referred to as the "XTop" algorithm.

4.2   A Functional Decomposition Approach

The above approach is based towards a netlist based representation of Boolean functions, and may not be suitable for processing other abstract representations of Boolean functions. Also the above scheme will fail if there is a memory explosion while trying to build $\psi_{i_{bdd}}$. In multilevel description of Boolean network, each node represents a complex function of its inputs. It is possible that the memory explosion happens while performing an ROBDD operation *within* a circuit node. Since structural decomposition points correspond to actual nodes in the circuit, the above scheme may not be able to handle this case very effectively.

Our functional approach avoids this explosion by choosing decomposition points based on the increase in the ROBDD sizes during the intermediate stages of ROBDD construction. Starting in a depth-first manner from the primary inputs, the ROBDD for the output G is built using *apply*. If for any Boolean operation during the construction of G, the graph size increases by a *disproportionate measure* we introduce a decomposition point. Such a decomposition point captures instances of difficult functional manipulations. Therefore, any simplification of the graph made on any such decomposition variable is likely to decrease the total resource requirement. Note that no decomposition is done when the function can be processed without any memory explosion.

In our current implementation, we decompose a function when the total ROBDD manager size increases by more than a user-defined factor, due to some Boolean operation. This check is done only if the manager size is larger than a predetermined minimum. Also, decomposition points are added when the individual ROBDD grows beyond another threshold value. Based on the growth of the ROBDDs, this threshold can be dynamically adjusted. Other researchers have explored ideas of introducing new variables based on the ROBDD growth (see Section 2) but our emphasis is different. Most of the previous approaches try to work using the decomposed representation throughout the procedure. What we are showing here is that in many cases, we can compose back all the decomposition points and get the same canonical representation as the one given by an exclusively bottom-up scheme but consume significantly less resources (sometimes even exponentially less) in the processes.

4.3   Order of Composition

The problem of finding good composition orders has been addressed in [11, 6, 23] In [11, 6, 23] the problem is studied in the context of reachability computation for sequential machines. In this problem, the transition relations (relating the present state variables to the next state variables) are built for individual latches. Transition relation for the entire sequential circuit is a conjunction of transition relations of the individual latches. Each step of reachability calculation is done by existentially quantifying out the present state variables from the conjunction of the relations representing the current reached state set and the transition relation. The problem consists of scheduling the conjunctions and quantifications so that the memory utilization is minimized. By using equation 2, it can be shown that the problem of finding optimal composition order is the same as the quantification-conjunction scheduling problem. In [11] it was shown that the quantification-conjunction scheduling problem is NP-complete. A number of heuristics were proposed in [19, 11] and the conclusion was that greedy heuristics which try to minimize the number of new variables in the support set of G_d at each stage generally give the best time-memory efficiency trade-off.

We use the traditional compose algorithm given by equation 1. In our approach, the decomposition points are frequently nested inside other decomposition points. This means that $\psi_{i_{bdd}}$ representing a decomposition point can have ψ_j in its support if ψ_j was introduced earlier to ψ_i. Out of all the decomposition points present in the support set of G_d at any given time, we restrict the *candidate set* for composition to only those variables which are not present in the support of any other $\psi_{i_{bdd}}$. This ensures that every ψ_i is composed only once and our algorithm requires only $O(n)$ compositions where n is the number of decomposition points. At each stage, we compose a variable from the candidate set which results in the minimum increase in the support set of G_d. The dependency information of the decomposition variables is kept as a 0-1 adjacency matrix and is updated after each composition step. This method of composition is now recursively applied till a fully composed graph is obtained. The advantage of this method is that it is very fast. Almost no calculation is needed to decide which decomposition point to compose next. So far we have not done experiments comparing our order of composition with the previous methods. It will be an interesting experiment to determine which method of composition is more efficient in reducing the intermediate memory.

5 Results

In this section we present the results of our experiments on the structural and functional methods. We report results on the standard ISCAS85 benchmark circuits along with some hard industrial circuits. Results are reported for the outputs which are considered 'hard' for building ROBDDs. The implementation of our structural scheme was done in the SLIF tools environment, on a DECstation 5000/240 with 64MB of memory. Our functional scheme was implemented in the SIS [22] environment, on a DECstation 5000/260 with 128MB of memory.

The "size" entry in all the following tables refers to the maximum ROBDD manager size. Tables 1 and 2 report the results for the structural techniques. Tables 3, and 4 report the results for the functional decomposition approach. In each case, we present results with and without the dynamic variable reordering (DR) [21] Tables 1 and 3 use Malik's ordering (MO) [9] without DR while Tables 2 and 4 use MO with DR. Since the structural and functional methods were implemented in different environments, the results of structural methods can not be directly compared with the functional methods. In Table 5, we show the performance of the functional decomposition method on some large industrial circuits.

In all the tables, the first column shows the peak memory utilization and the time taken by the reference algorithm. The reference algorithm builds the monolithic ROBDD of the outputs in terms of the primary inputs in a depth first order. For this purpose, we changed the code in the *ntbdd_node_to_bdd* function of SIS, so that if an ROBDD of a node is not required in subsequent computations, it is freed, to reduce memory utilization.

Tables 1 and 2 report the results for our "XTop" structural algorithms. Without DR, our overall memory gains are about 2.5X and the time gains are about 2.3X averaged over all examples. The reference algorithm is unable to make the ROBDD for the 14th output of C6288 (multiplier) in 1 million nodes while the XTop algorithm finishes it in less than 205 seconds. In the absence of DR (Table 1), XTop almost always gives better results. When DR is invoked, XTop is less effective. This is because a small G_d, selected under one variable ordering, may not be small under a different variable ordering. Despite this, the XTop algorithm performs satisfactorily compared to the reference algorithm with substantial gains in both time and memory (1.95X gain in memory and 1.42 gain in time), especially in the larger examples.

In Tables 3 and 4, the first column represents the reference algorithm while the second column reports the time and memory requirements for the functional decomposition approach. Once again, excellent time and space improvements are observed in almost all cases. Average improvements in time and memory are between 1.5X to 3X. Note that we use the same memory explosion thresholds for all the experiments. We are able to obtain much better results if we

Ckt	Out	Reference		XTop		Ref./XTop	
	#	Time	Size	Time	Size	Time	Size
C880	26	3.34	23818	2.57	21246	1.30	1.12
C1355	32	1.86	25482	1.37	16284	1.36	1.57
C1908	24	5.65	34205	3.12	18102	1.81	1.89
C1908	25	4.53	18288	4.43	17241	1.02	1.06
C2670	140	-	>1000000	-	>1000000	-	-
C3540	22	-	>1000000	-	>1000000	-	-
C5315	123	28.36	300035	30.32	300098	0.94	1.00
C6288	12	41.41	289344	25.32	80007	1.63	3.61
C6288	13	126.98	808482	26.08	108538	4.87	7.45
C6288	14	-	>1000000	203.48	581086	inf	inf
Total		212.13	1499654	93.21	561516	2.28	2.67

Table 1. Structural Methods, Malik Ordering without dyn. reorder, 1 Million node limit

Ckt	Out	Reference		XTop		Ref./XTop	
	#	Time	Size	Time	Size	Time	Size
C880	26	6.84	4598	7.86	4530	0.87	1.02
C1355	32	122.42	32600	127.07	30167	0.96	1.08
C1908	24	29.65	8863	14.85	8876	2.00	1.00
C1908	25	14.30	8870	14.57	8889	0.98	1.00
C2670	140	126.05	12609	413.02	111907	0.31	0.11
C3540	22	2169.08	316824	3650.85	418335	0.59	0.76
C5315	123	25.35	9934	28.29	10000	0.59	0.76
C6288	12	1727.46	286992	625.36	77863	2.76	3.69
C6288	13	3665.51	805793	659.20	99370	5.56	8.11
C6288	14	-	>1000000	15296.35	590869	inf	inf
Total		7886.66	1487083	5541.07	760937	1.42	1.95

Table 2. Structural Methods, Malik Ordering with dyn. reorder, 1 Million node limit

experiment with a few different thresholds. By using the functional decomposition method, we can build the ROBDD for the 14th output of C6288 (Table 3) while the reference algorithm fails (for a 1 million node limit). Note that the total improvement figures do not reflect the outputs for which the reference fails. Therefore the actual gains are even better than what the numbers in the 'Total' row suggest. Interestingly, space and time improvements are more prominent for bigger circuits and circuits which are known to be difficult for ROBDDs (like C6288).

Table 5 presents the performance of our algorithm on some industrial circuits. These circuits have more than 200 inputs and 2000 complex multi-level gates. The first circuit, NINPTB is interesting because the ROBDDs for the outputs are not very big, yet the decomposition procedure gives a speedup of 10X to 60X. For the other circuit, decomposition method takes less memory. Again RCore3 and Ind1 can be completed in about 300K and 100K nodes while the reference method fails in 1Million node limit. This demonstrates that hard intermediate points do occur in circuits whose outputs are relatively simple and our technique can give orders of magnitude gains in dealing with such cases.

Ckt	Out #	Reference Algo.		Functional.		Ref./Functional	
		Time	Size	Time	Size	Time	Size
C880	26	0.20	6467	0.19	6468	1.05	1.00
C1355	32	0.48	9729	0.31	9422	1.55	1.03
C1908	24	0.86	11307	0.63	12276	1.36	0.92
C1908	25	1.01	13904	0.72	12059	1.40	1.15
C2670	140	1.40	29282	1.33	15590	1.05	1.88
C3540	22	–	>1000000	–	>1000000	-	-
C5315	123	1.86	9971	1.74	9971	1.07	1.00
C6288	12	30.59	268076	13.90	160498	2.2	1.67
C6288	13	64.57	633423	37.28	378995	1.73	1.67
C6288	14	–	>1000000	109.66	878639	inf	inf
Total		100.97	982161	56.10	605279	1.80	1.62

Table 3. Functional Methods, Malik Ordering without dyn. reorder, 1 Million node limit

6 Formal Verification of Sequential Circuits

ROBDDs are extensively used in the formal verification of sequential circuits[8, 23, 18]. There are two flavors of the problem. In the first, we have to check whether two sequential circuits produce the same outputs sequence for any given sequence of inputs. The second problem consists of verifying whether a system (specified as a state transition graph) satisfies a given property. In both of these problems a key step consists of finding out all the states that the system can reach. This computation is called the *reachability analyses*. For reachability analysis we need the following two relations:

- Transition Relation $(T(s, s'))$, which gives the set of states $N(s')$ that can be reached from the present set of states in one step.
- Set of Reached States $(R(s))$, representing the present set of reached state.

Ckt	Out #	Reference Algo.		Functional.		Ref./Functional	
		Time	Size	Time	Size	Time	Size
C880	26	0.29	6467	0.44	6468	0.66	1.00
C1355	32	0.46	9729	0.30	9422	1.53	1.03
C1908	24	1.71	11307	1.37	12276	1.25	0.92
C1908	25	11.12	10679	5.64	9795	1.97	1.09
C2670	140	49.76	9566	52.37	9578	0.95	1.00
C3540	22	2856.73	331653	1572.94	178436	1.82	1.86
C5315	123	1.86	9971	1.91	9971	0.97	1.00
C6288	12	1821.21	280845	726.07	77775	2.51	3.61
C6288	13	7237.81	735401	1378.26	133426	5.25	5.51
C6288	14	–	>1000000	–	>1000000	-	-
Total		11980.95	1405618	3737.30	447147	3.21	3.14

Table 4. Functional Methods, Malik Ordering with dyn. reorder, 1 Million node limit

Ckt	Out #	Reference Algo.		Functional.		Ref./Functional	
		Time	Size	Time	Size	Time	Size
NINPTB	1	131.13	17238	5.87	8043	22.34	2.14
NINPTB	2	381.98	13227	6.19	9054	61.71	1.46
NINPTB	3	120.25	14170	23.12	9520	5.20	1.49
NINPTB	4	397.53	14264	5.98	7009	66.48	2.04
NINPTB	5	51.79	12638	3.32	7186	15.60	1.76
RCore3	1	932.15	54287	1167.51	35787	0.80	1.52
RCore3	3	–	>1000000	6911.33	293930	inf	inf
Ind1	1	–	>1000000	2547.77	100852	inf	inf
Ind2	1	–	>1000000	17754.99	993345	inf	inf

Table 5. Industrial Circuits: Functional Methods, Malik Ordering with dyn. reorder, 1 Million node limit

The set of next states, $N(s')$, is calculated using the following equation 4 (also known as *image computation*:

$$N(s') = \exists_s [T(s, s') R(s)] \tag{4}$$

The set of reached state is computed by adding the next state relation to the present state relation and iteratively doing the above image computation step to reach a fixed point.

In many cases, we can not build the transition relation $T(s, s')$ using the bottom-up method. We can use our method to build $T(s, s')$ in some of these cases. For other cases we can use the decomposed representation to directly do the reachability computation. Let us assume that the decomposed transition relation is given by $T_d(\Psi, s, s')$ and each $\psi \in \Psi$ has a corresponding ψ_{bdd}. Using equation 2, the transition relation is given by equation 5:

$$T(s, s') = \exists_\Psi [T_d(\Psi, s, s') \Pi(\psi_i \overline{\oplus} \psi_{bdd})] \tag{5}$$

Substituting equation 5 in equation 4 we get:

$$N(s') = \exists_s [(\exists_\Psi [T_d(\Psi, s, s') \Pi(\psi_i \overline{\oplus} \psi_{i_{bdd}})]) R(s)] \tag{6}$$

Since, $R(s)$ doesn't have any variables from Ψ in its support set we can rewrite the above equation as follows:

$$N(s') = \exists_s \exists_\Psi [T_d(\Psi, s, s') \Pi(\psi_i \overline{\oplus} \psi_{i_{bdd}}) R(s)] \tag{7}$$

From the above equation it is easy to see that the image computation (i.e. the computation of $N(s')$) can be done using the decomposed representation $T_d(\Psi, s, s')$ without ever building the monolithic ROBDD T.

7 Dominators as Decomposition Points

In this section we will discuss a possible extensions that can be made to the decomposition schemes. In the decomposition/composition approach, the composition phase is the most time and memory consuming part. If we can introduce decomposition points which simplify the process of composition then we will be able to further reduce the resources required in ROBDD construction. Let $f = g \odot h$, where $\odot$ represents any ROBDD operation between g and h. Also assume that the support of g and the support of f are disjoint. We make the following two observations:

- The optimal order for f is a concatenation of the optimal orders of g and h [9]
- The composition of h in g ($g(h \leftarrow h_{bdd})$) can be done in $O(|g| + k|h_{bdd}|)$, where k represents the number of occurrences of h in the ROBDD of g, as opposed to $O(|g|^2|h_{bdd}|)$ for the general case [4]. This is achieved by structurally replacing the nodes labeled with variable id h in the ROBDD of g with h_{bdd} and directing the branches going to terminals in h_{bdd} to the appropriate child of the node being replaced.

We can effectively use the notion of dominators to benefit from the above observations. A gate g dominates another gate h, i.e., $g \vdash h$, in a flowgraph, if every path from h to the output has g on it. Note that $\vdash$ gives a partial order on the gates in any circuit, and can be represented by a *dominator tree*. This dominator tree can be constructed in almost linear time by using an algorithm given in [16]. It is easy to see that if we construct the ROBDD of a node g in terms of the decomposition points which are dominated by g, then the ROBDD of g can be composed in the ROBDD of any other function, f by just structural replacement of g in f. Therefore, the nodes in the dominator tree of a given Boolean function can prove to be effective decomposition points. Currently we are in the process of implementing the dominator algorithm.

8 Conclusion

In this paper we discussed a combined bottom-up/top-down approach to reduce the intermediate memory requirement during the ROBDD construction process. Results on ISCAS85 benchmarks as well as some industrial circuits show impressive reductions in both memory and time utilization. In many cases we are able to build the ROBDD where the conventional method fails.

In the past, much attention has been paid towards reducing the memory requirement by changing the ROBDD variable ordering. We find that in many cases similar reductions can be obtained by changing the way in which ROBDDs are constructed. Our methods perform consistently well under different variable ordering schemes and provide additional gains. Hence, they are exploiting a complimentary aspect of the memory explosion problem. Further, the reduction in peak memory is generally accompanied by a reduction in the time to construct the ROBDD. Therefore, there is no apparent trade-off in using this approach. This is in contrast to dynamic variable ordering where a significant time penalty is associated with the memory reduction.

In addition, for cases where the final memory requirement itself is large, the decomposed representation can be used as a starting point to generate partitioned-ROBDDs which provide a compact, canonical and efficiently manipulable representation for Boolean functions [20].

Future research is directed towards identifying better decomposition techniques and studying various computational issues in efficient function composition. Also, we plan to use this approach to reduce the resources required in computing transition relations, reachability, and also for some alternate BDD approaches such as [13].

9 Acknowledgements

The second author was supported by CA State MICRO program grant #94-110 and the SRC 95-DC-324. The fourth author was supported by California State Micro program grant #94-023.

References

1. Sheldon B. Akers. Binary decision diagrams. *IEEE Transactions on Computers*, C-27:509–516, June 1978.
2. C. L. Berman. Circuit width, register allocation, and ordered binary decision diagrams. *IEEE Transactions on Computer-Aided Design*, pages 1059–1066, August 1991.
3. K. S. Brace, R. L. Rudell, and R. E. Bryant. Efficient Implementation of a BDD Package. In *DAC*, pages 40–45, June 1990.
4. R. Bryant. Graph-based Algorithms for Boolean Function Manipulation. *IEEE Transactions on Computers*, C-35:677–691, August 1986.
5. R. E. Bryant. Symbolic boolean manipulation with ordered binary decision diagrams. *ACM Computing Surveys*, 24:293–318, September 1992.

6. G. Cabodi and P. Camurati. Symbolic fsm traversals based on the transition relation. *submitted to Transaction on Computer-Aided Design*, 1994.

7. G. Cabodi, P. Camurati, and Stefano Quer. Auxillary variables for extending symbolic traversal techniques to data paths. *31st DAC*, pages 289–293, June 1994.

8. O. Coudert, C. Berthet, and J. C. Madre. Verification of Sequential Machines Based on Symbolic Execution. In *Proc. of the Workshop on Automatic Verification Methods for Finite State Systems*, Grenoble, France, 1989.

9. S. Malik et. al. Logic Verification using Binary Decision Diagrams in a Logic Synthesis Environment. In *ICCAD*, pages 6–9, November 1988.

10. M. Fujita, H. Fujisawa, and N. Kawato. Evaluation and improvements of Boolean comparison method based on binary decision diagrams. *ICCAD*, pages 2–5, November 1988.

11. R. Hojati, S.C. Krishnan, and R. K. Brayton. Heuristic Algorithms for Early Quantification and Partial Product Minimization. Technical Report UCB/ERL M93/58, Electronics Research Lab, Univ. of California, Berkeley, CA 94720, July 1993.

12. J. Jain, J. Bitner, D. S. Fussell, and J. A. Abraham. Probabilistic verification of Boolean functions. *Formal Methods in System Design*, 1, July 1992.

13. J. Jain, J. Bitner, M. Abadir, D. S. Fussell, and J. A. Abraham. Indexed BDDs: Algorithmic Advances in Techniques to Represent and Verify Boolean functions. *to appear in IEEE Transactions on Computers.*

14. S.-W. Jeong, B. Plessier, G. Hachtel, and F. Somenzi. Extended BDDs: Trading Canonicity for Structure in Verification Algorithms. In *ICCAD*, pages 464–467, November 1991.

15. C. Y. Lee. Representation of switching circuits by binary-decision programs. *Bell Syst. Tech. J.*, 38:985–999, 1959.

16. T. Lengauer and R. Tarjan. A fast algorithm for finding dominators in a flowgraph. *ACM Transactions on Programming Languages*, 1:121–141, July 1979.

17. K. L. McMillan. Symbolic model checking: An approach to the state explosion problem. *Ph.D Thesis, Dept. of Computer Sciences, Carnegie Mellon University*, 1992.

18. K. L. McMillan. *Symbolic Model Checking.* Kluwer Academic Publishers, 1993.

19. A. Narayan, S. P. Khatri, J. Jain, M. Fujita, R. K. Brayton, and A. Sangiovanni-Vincentelli. A Study of Composition Schemes for Mixed Apply/Compose Based Construction of ROBDDs. In *Intl. Conf. on VLSI Design*, January 1996.

20. A. Narayan, J. Jain, M. Fujita, and A. L. Sangiovanni-Vincentelli. Partitioned-ROBDDs - A Compact, Canonical and Efficiently Manipulable Representation for Boolean Functions. *ICCAD*, November 1996. To appear.

21. R. L. Rudell. Dynamic Variable Ordering for Ordered Binary Decision Diagrams . In *ICCAD*, pages 42–47, November 1993.

22. E. M. Sentovich, K. J. Singh, L. Lavagno, C. Moon, R. Murgai, A. Saldanha, H. Savoj, P. R. Stephan, R. K. Brayton, and A. L. Sangiovanni-Vincentelli. SIS: A System for Sequential Circuit Synthesis. Technical Report UCB/ERL M92/41, Electronics Research Lab, Univ. of California, Berkeley, CA 94720, May 1992.

23. H. J. Touati, H. Savoj, B. Lin, R. K. Brayton, and A. L. Sangiovanni-Vincentelli. Implicit State Enumeration of Finite State Machines using BDD's. In *ICCAD*, pages 130–133, November 1990.

BDDs vs. Zero-Suppressed BDDs : for CTL Symbolic Model Checking of Petri Nets

Tomohiro Yoneda[1], Hideyuki Hatori[1],
Atsushi Takahara[2], Shin-ichi Minato[2]

[1] Tokyo Institute of Technology ({yoneda,hatori}@cs.titech.ac.jp)
[2] NTT LSI Laboratories ({taka,minato}@aecl.ntt.jp)

Abstract. This paper proposes using Zero-Suppressed BDDs for the CTL symbolic model checking of Petri nets. Since the state spaces of Petri nets are often very sparse, it is expected that ZBDDs represent such sparse state spaces more efficiently than BDDs. Further, we propose special BDD/ZBDD operations for Petri nets which accelerate the manipulations of Petri nets. The approaches to handling Petri nets based on BDDs and ZBDDs are compared with several example nets, and it is shown that ZBDDs are more suitable for the symbolic manipulation of Petri nets.

1 Introduction

Petri nets have been used to model various concurrent systems such as network protocols, asynchronous circuits, and so on. Trace theoretic verification[Dil88] is one famous application of Petri nets in the area of formal verification.

For the formal verification of practical systems, it is very important to avoid state explosion. It is reported that the symbolic manipulation of states based on Binary Decision Diagrams (BDDs) has succeeded in handling huge state spaces[BCD+92]. Petri nets have also been managed in symbolic manner in some works[HHY92, RCP95]. One important feature of Petri nets is that their state spaces are very sparse. For example, it is not surprising that a Petri net with a hundred places has 10,000 reachable states: its state space (2^{100}) is approximately 10^{26} larger than the reachability set.

In this paper, especially for the CTL (Computation Tree Logic) symbolic model checking of Petri nets, we have tried to use a different representation of state spaces, Zero-Suppressed Binary Decision Diagrams (ZBDDs)[Min93]. ZBDDs are reported to manage sets of binary vectors efficiently even if the space of the vectors is very sparse. Roughly speaking, the ZBDD nodes are related to 1s in the vectors of the set. Thus, if a token in a Petri net corresponds to 1 in the vectors, then the ZBDD size will depend on the number of tokens in the net. Hence, we expect that the sparse state space of Petri nets is represented efficiently by using ZBDDs.

Further, we propose special BDD/ZBDD operations to handle Petri nets efficiently. If we use only original BDD/ZBDD operations, the number of those operations needed for handling one transition is proportional to the number of input and output places of the transition, and each operation needs the traversal of the BDD/ZBDD, which is very inefficient. The proposed special operations achieve every necessary manipulation for one transition during one traversal of the BDD/ZBDD.

Finally, we have compared the approaches based on BDDs and ZBDDs with several examples.

The rest of this paper is organized as follows. In the following section, we briefly review the Petri nets and CTL. In section 3, BDDs and ZBDDs are introduced. Three approaches to the symbolic manipulations of Petri nets based on BDDs and ZBDDs are shown in section 4. In the next section. we propose the special BDD/ZBDD operations for Petri nets. In section 6, the experimental results for various examples are shown and the approaches are compared. Finally, we summarize the discussion.

2   Petri Nets and CTL

A *Petri net* N is a tuple $N = (P, T, F, \mu_0)$, where P is a finite set of *places*, T is a finite set of *transitions* $(P \cap T = \emptyset)$, $F \subseteq (P \times T) \cup (T \times P)$ is the *flow relation*, and $\mu_0 \subseteq P$ is the *initial marking* of the net. For any transition t, $\bullet t = \{p \mid (p,t) \in F\}$ and $t\bullet = \{p \mid (t,p) \in F\}$ denote the *set of the input places* and the *set of the output places* of t, respectively. In particular, we say that $p \in \bullet t - t\bullet$ is a *true-input* place, $p \in t \bullet - \bullet t$ is a *true-output* place, and $p \in \bullet t \cap t\bullet$ is an *in-out* place.

A *marking* μ of N is any subset of P. A transition is *enabled* at a marking μ if $\bullet t \subseteq \mu$ (all its input places have tokens at μ); otherwise it is *disabled*. When a transition t enabled at a marking μ *fires*, a new marking

$$\mu' = (\mu - \bullet t) \cup t \bullet \qquad (1)$$

is obtained. Let $\mu \to \mu'$ denote that μ' is obtained from μ by firing some transition enabled at μ.

The initial marking is *reachable*, and μ' is reachable if $\mu \to \mu'$ for a reachable μ. A Petri net is *one-safe* if at any reachable marking μ and for any t enabled at μ, $\mu \cap t\bullet = \emptyset$. In this paper, we handle only one-safe Petri nets.

The CTL formulas are defined inductively as follows:

- true is a formula.
- Every propositional variable is a formula.
- If ψ is a formula, then $\neg\psi$, $\mathbf{EX}\psi$, and $\mathbf{AX}\psi$ are formulas.
- If ψ_1 and ψ_2 are formulas, then $(\psi_1 \wedge \psi_2)$, $\mathbf{E}(\psi_1 \mathcal{U} \psi_2)$ and $\mathbf{A}(\psi_1 \mathcal{U} \psi_2)$ are formulas.

Additional boolean connectives false, $\vee$, $\to$ are defined as usual, and

$$\begin{aligned}
\mathbf{EF}\ \psi &= \mathbf{E}[true\ \mathcal{U}\ \psi], & \mathbf{AF}\ \psi &= \mathbf{A}[true\ \mathcal{U}\ \psi], \\
\mathbf{AG}\ \psi &= \neg\mathbf{EF}\ \neg\psi, & \mathbf{EG}\ \psi &= \neg\mathbf{AF}\ \neg\psi.
\end{aligned}$$

Since our goal is to reason about the temporal properties of Petri nets, the set of propositional variables is the set P of places. That is, at a marking μ, a propositional variable $p \in P$ is true, iff $p \in \mu$. For a nonnegative integer i, a *path (starting) from* μ_i is $\mu_i \mu_{i+1} \mu_{i+2} \cdots$ such that for $k \geq i, \mu_k \rightarrow \mu_{k+1}$. The validity of CTL formula ψ at a marking μ_i, denoted by $\mu_i \models \psi$, is defined inductively as follows:

- $\mu_i \models$ true.
- $\mu_i \models p$, iff $p \in \mu_i$ for $p \in P$.
- $\mu_i \models \neg\phi$, iff $\mu_i \not\models \phi$.
- $\mu_i \models \phi_1 \wedge \phi_2$, iff $\mu_i \models \phi_1$ and $\mu_i \models \phi_2$.
- $\mu_i \models \mathbf{AX}\phi$, iff for any μ' such that $\mu_i \rightarrow \mu'$, $\mu' \models \phi$.
- $\mu_i \models \mathbf{EX}\phi$, iff for some μ' such that $\mu_i \rightarrow \mu'$, $\mu' \models \phi$.
- $\mu_i \models \mathbf{A}[\phi_1 \mathcal{U} \phi_2]$, iff for any path $\mu_i \mu_{i+1} \cdots$ from μ_i, there exists a nonnegative integer $n \geq i$ such that $\mu_n \models \phi_2$, and for $i \leq j < n$, $\mu_j \models \phi_1$.
- $\mu_i \models \mathbf{E}[\phi_1 \mathcal{U} \phi_2]$, iff for some path $\mu_i \mu_{i+1} \cdots$ from μ_i, there exists a nonnegative integer $n \geq i$ such that $\mu_n \models \phi_2$, and for $i \leq j < n$, $\mu_j \models \phi_1$.

3 BDDs and Zero-Suppressed BDDs

Both BDDs and ZBDDs consist of *decision nodes*, *0-edges* and *1-edges* attached to all decision nodes, and two types of terminal nodes called *0-terminal nodes* and *1-terminal nodes* as shown in Fig.1(a)-(e). A decision node with no upper nodes is called a root node. As shown in the figure, a variable is assigned to every decision node such that every path from the root node to one of the terminal nodes obeys the same variable ordering.

It is well known that BDDs efficiently represent boolean functions based on node elimination and node sharing (Fig.1(a) [3]). Further, boolean operations such as logical AND, logical OR, and so on can be achieved by using BDD manipulations, which have an average time complexity almost propositional to the size of BDDs if some cache mechanism is used to contain the results of recent operations.

On the other hand, ZBDDs represent sets of binary vectors $(x_1 x_2 \cdots)$ efficiently, where for $i \geq 1$, $x_i \in \{0, 1\}$. Any set of n bit binary vectors is represented by a ZBDD with n variables. Each binary vector in the set corresponds to a path from the root node to 1-terminal node, which is called *1-path*. More precisely, for a binary vector and the corresponding 1-path,

- if a variable x_i is 1 in the vector, the node for x_i has 1-edge in the 1-path, and
- if a variable x_i is 0 in the vector, either the node for x_i has 0-edge in the 1-path or such a node does not exist in the 1-path.

For example, consider the ZBDD with 2 variables x_1 and x_2 as shown in Fig.1(b). Note that this ZBDD has three 1-paths, $x_2 \xrightarrow{0} x_1 \xrightarrow{0} 1$, $x_2 \xrightarrow{0} x_1 \xrightarrow{1} 1$, and $x_2 \xrightarrow{1}$

---

[3] Some BDD nodes in this figure can be deleted, if we use attributed edges.

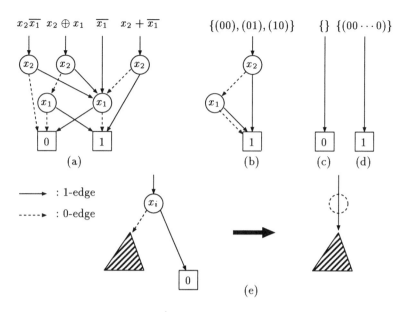

Fig. 1. BDDs and ZBDDs.

1, where $\overset{0}{\to}$ and $\overset{1}{\to}$ denote 0-edge and 1-edge, respectively. Thus, this ZBDD represents the set $\{(00), (01), (10)\}$ of 2 bit binary vectors $(x_2 x_1)$. In particular, Fig.1(c) shows a ZBDD representing an empty set, and Fig.1(d) represents a set $\{(00 \cdots 0)\}$ because (c) has no 1-paths, and (d) has one 1-path where no variable has 1-edge in the path.

If a node for a variable x_i has 1-edge pointing directly to 0-terminal node as shown in Fig.1(e), then this node clearly has no 1-edge in any 1-path through this node. Therefore, we can eliminate this node, as shown in the figure, without changing the set represented by the ZBDD. This is the node elimination rule for ZBDDs.

By using ZBDD manipulations, set operations such as union, intersection, and so on can be achieved efficiently. For example, Fig.2 shows the procedure for computing the union of two sets f and g (represented by ZBDDs). In the figure, $f.top$ and $g.top$ denote the variables assigned to the root nodes of f and g, respectively. $\mathsf{order}(var)$ gives the variable order of variable var : If a node n' is a successor of a node n, and v, v' are assigned to n, n', then $\mathsf{order}(v') < \mathsf{order}(v)$ holds. $f.0$ denotes the ZBDD pointed to by the 0-edge of the root node of f. $f.1$ denotes the ZBDD pointed to by the 1-edge of the root node of f. $g.0$ and $g.1$ are defined similarly. $\mathsf{ZGetNode}(var, f, g)$ returns the ZBDD such that its root node has the variable var, and the 0-edge and 1-edge of the root node point to f and g, respectively. The above node elimination for ZBDDs is implemented in $\mathsf{ZGetNode}(var, f, g)$.

There are three more set operations : $\mathsf{Subset1}(f, var)$, $\mathsf{Subset0}(f, var)$ and $\mathsf{Change}(f, var)$, which are essential for manipulating Petri nets. For a ZBDD f

```
ZbddUnion(f, g) {
    if f is an empty set, then return g;
    if g is an empty set, then return f;
    if f == g, then return f;
    if order(f.top) > order(g.top), then
        return ZGetNode(f.top, ZbddUnion(f.0, g), f.1);
    if order(f.top) < order(g.top), then
        return ZGetNode(g.top, ZbddUnion(f, g.0), g.1);
    if order(f.top) == order(g.top), then
        return ZGetNode(f.top, ZbddUnion(f.0, g.0),
                                ZbddUnion(f.1, g.1));
}
```

Fig. 2. Union operation.

and a variable var, Subset1(f, var) gives the ZBDD obtained by extracting the subset of f such that $var = 1$, and by changing the value of var in the subset to 0. That is, for a set $M = \{(101), (011), (110)\}$ of binary vectors $(x_3x_2x_1)$, Subset1(M, x_1) first extracts $\{(101), (011)\}$ because the values of x_1 in those elements are 1. Then, $\{(100), (010)\}$ will be given by changing the values of x_1 to 0. Subset0(f, var) simply returns the ZBDD representing the subset of f such that $var = 0$. Finally, Change(f, var) inverts the value of var $(0 \rightarrow 1, 1 \rightarrow 0)$ in every element of f. These operations are implemented efficiently by recursive procedures similar to that shown in Fig.2.

In order to compare BDD representations and ZBDD representations, let us consider the net shown in Fig.3(a), and its reachable marking set

$$M = \{(11000), (10010), (01100), (01001), (00110), (00011)\},$$

where each marking is represented by a binary vector $(p_1p_2 \cdots p_5)$: $p_i = 1$, iff the place p_i has a token. This set M is also represented by the logical function F_M such that

$$F_M(p_1, p_2, \cdots, p_5) = 1 \iff (p_1p_2 \cdots p_5) \in M.$$

Such a function is called a *characteristic function* for M. Fig.3(b) shows the BDD[4] for this F_M with respect to the variable order $p_1 > p_2 > \cdots > p_5$. Similarly, M is represented by the ZBDD shown in figure (c). In this representation, each 1-edge in a ZBDD corresponds to a token in at least one reachable marking, and the total number of 1-edges in a ZBDD is equal to the number of its decision nodes. Thus, the size of ZBDDs does not depends on the number of places in the net, but the number of tokens in the reachable markings. On the other hand, the size of BDDs somewhat relates to the number of places. This is because in a BDD representation a node for each place variable is necessary for each reachable marking in order to represent the existence / absence of a token in the place. Since the number of tokens in one-safe nets is usually less than one

---

[4] If we use attributed edges for negation, two nodes in the figure can be deleted.

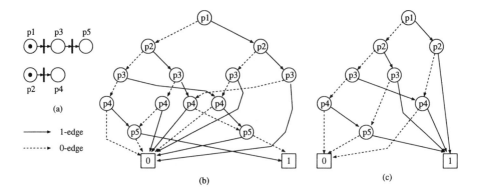

Fig. 3. BDD representation and ZBDD representation.

half of the number of places, we expect that ZBDDs handle such Petri nets more efficiently than BDDs.

4 Symbolic Manipulations of Petri Nets

In this section, we show three approaches to manipulating Petri nets symbolically: the first two use rather standard techniques for BDD based symbolic manipulation of Petri nets and CTL model checking, and the last one is a ZBDD version of the second approach.

For each place $p \in P$, we consider variables p and p'. p' is a variable for the next marking, which is used only in approach 1 below. As shown in the previous example, each variable p (p') has the value 1, iff the corresponding place has a token in the current (next) marking. A binary vector $(p_1 p_2 \cdots p_n)$ represents a marking. In the first two approaches, a BDD represents a characteristic function for a set of markings of this form while in the last approach, a ZBDD represents the set directly.

4.1 Approach 1

Suppose that a marking μ' is obtained from μ by firing a transition t. μ and μ' are represented by $(p_1 p_2 \cdots p_n)$ and $(p'_1 p'_2 \cdots p'_n)$, and the transition relation TR_t by the firing of t is expressed as follows according to the firing rule (1):

$$TR_t(p_1, p_2, \cdots, p_n, p'_1, p'_2, \cdots, p'_n) = \bigwedge_{p \in \bullet t} p \wedge \bigwedge_{p \in t\bullet} p' \wedge \bigwedge_{p \in (\bullet t - t\bullet)} \neg p' \wedge \bigwedge_{p \notin (\bullet t \cup t\bullet)} (p = p').$$

The first factor expresses that all input places of t should have tokens before the firing. The second and third factors are for the markings after the firing: all output places of t have tokens, and the true-input places of t have no tokens. Note that the prime variables p' are used to express the condition after the firing. The last factor says that tokens in other places should not move.

The transition relation TR of the whole net is then obtained by computing the disjunction of the transition relation for each transition:

$$TR(\boldsymbol{p}, \boldsymbol{p}') = \bigvee_{t \in T} TR_t(\boldsymbol{p}, \boldsymbol{p}'),$$

where $\boldsymbol{p} = (p_1, p_2, \cdots, p_n)$ and $\boldsymbol{p}' = (p_1', p_2', \cdots, p_n')$. Once we have the transition relation of the net, computing the characteristic function R of reachable markings with respect to the initial marking μ_0 is easy :

$R(\boldsymbol{p})$: $R(\boldsymbol{p}) = R_{new}(\boldsymbol{p}) = M_0(\boldsymbol{p});$
  loop {
    $M_{tmp}(\boldsymbol{p}) = \exists \boldsymbol{p}'[R_{new}(\boldsymbol{p}') \wedge TR(\boldsymbol{p}', \boldsymbol{p})];$
    $R_{new}(\boldsymbol{p}) = M_{tmp}(\boldsymbol{p}) \wedge \neg R(\boldsymbol{p});$
    if $R_{new}(\boldsymbol{p}) == 0$, then terminate;
    $R(\boldsymbol{p}) = R(\boldsymbol{p}) \vee R_{new}(\boldsymbol{p});$
  }

where M_0 is the characteristic function for the set $\{\mu_0\}$.

The characteristic function M_ψ for the set of markings satisfying CTL formulas ψ is computed as follows :

$M_{\text{true}}(\boldsymbol{p}) = 1,$ $\qquad M_{p_i}(\boldsymbol{p}) = p_i,$
$M_{\neg \phi}(\boldsymbol{p}) = \neg M_\phi(\boldsymbol{p}),$ $\qquad M_{\phi_1 \wedge \phi_2}(\boldsymbol{p}) = M_{\phi_1}(\boldsymbol{p}) \wedge M_{\phi_2}(\boldsymbol{p}),$
$M_{\mathbf{EX}\phi}(\boldsymbol{p}) = F_X(M_\phi, \text{"E"}),$ $\qquad M_{\mathbf{E}[\phi_1 \mathcal{U} \phi_2]}(\boldsymbol{p}) = F_U(M_{\phi_1}, M_{\phi_2}, \text{"E"}),$
$M_{\mathbf{AX}\phi}(\boldsymbol{p}) = F_X(M_\phi, \text{"A"}),$ $\qquad M_{\mathbf{A}[\phi_1 \mathcal{U} \phi_2]}(\boldsymbol{p}) = F_U(M_{\phi_1}, M_{\phi_2}, \text{"A"}),$

where

$F_X(M_\phi, md)$:   if $md == \text{"E"}$, then
        $F_X(M_\phi, md) = \exists \boldsymbol{p}'[TR(\boldsymbol{p}, \boldsymbol{p}') \wedge M_\phi(\boldsymbol{p}')]$
      else if $md == \text{"A"}$, then
        $F_X(M_\phi, md) = \forall \boldsymbol{p}'[TR(\boldsymbol{p}, \boldsymbol{p}') \rightarrow M_\phi(\boldsymbol{p}')],$

$F_U(M_{\phi_1}, M_{\phi_2}, md)$: $M_{old}(\boldsymbol{p}) = M_{\phi_2}(\boldsymbol{p});$
      loop {
        $F_U(M_{\phi_1}, M_{\phi_2}, md) =$
          $M_{\phi_2}(\boldsymbol{p}) \vee \left(M_{\phi_1}(\boldsymbol{p}) \wedge F_X(M_{old}, md) \right);$
        if $F_U(M_{\phi_1}, M_{\phi_2}, md) == M_{old}(\boldsymbol{p})$, then terminate;
        $M_{old}(\boldsymbol{p}) = F_U(M_{\phi_1}, M_{\phi_2}, md);$
      }.

Because the state spaces of Petri nets are very sparse, the inverse image computation (F_X) gives a lot of unreachable markings. Thus, we initially compute the conjunction $TR(\boldsymbol{p}, \boldsymbol{p}') \wedge R(\boldsymbol{p})$, and use it instead of $TR(\boldsymbol{p}, \boldsymbol{p}')$. It often keeps the obtained BDDs smaller.

Practically, the BDD operations for large BDDs are very slow. Thus, the transition relations TR^i for small subsets of transitions satisfying $TR = TR^0 \vee$

$TR^1 \vee TR^2 \vee \cdots$ are computed. We can efficiently obtain R and M_ψ by using such partitioned transition relations without actually using TR.

One disadvantage of this approach is that it needs $2|P|$ variables, which sometimes makes BDDs large, even when the transition relation is partitioned.

4.2 Approach 2

Let $M(\boldsymbol{p})$ be the characteristic functions for the given set of markings. The following $TR(M,t)$ is the characteristic function for the set of the next markings obtained by firing t from markings in M :

$$TR(M,t) = \exists(\bullet t \cup t\bullet)[M \wedge \bigwedge_{p \in \bullet t} p] \wedge \bigwedge_{p \in (\bullet t - t\bullet)} \neg p \wedge \bigwedge_{p \in t\bullet} p.$$

The characteristic function R for the set of the reachable markings of the net is computed by using $TR(M,t)$ as follows:

$R(\boldsymbol{p})$: $R(\boldsymbol{p}) = R_{new}(\boldsymbol{p}) = M_0(\boldsymbol{p})$;
     loop {
          $M_{tmp}(\boldsymbol{p}) = R_{new}(\boldsymbol{p})$;
          for $(t \in T)$ {
               $M_{tmp}(\boldsymbol{p}) = M_{tmp}(\boldsymbol{p}) \vee TR(M_{tmp}(\boldsymbol{p}), t)$;
          }
          $R_{new}(\boldsymbol{p}) = M_{tmp}(\boldsymbol{p}) \wedge \neg R(\boldsymbol{p})$;
          if $R_{new}(\boldsymbol{p}) == 0$, **then terminate**;
          $R(\boldsymbol{p}) = R(\boldsymbol{p}) \vee R_{new}(\boldsymbol{p})$;
     }.

For the CTL model checking, we need the inverse image $TR^{-1}(M,t)$ such that M is obtained by firing t from the markings in $TR^{-1}(M,t)$. This computation is similar to the above.

$$TR^{-1}(M,t) = \exists(\bullet t \cup t\bullet)[M \wedge \bigwedge_{p \in t\bullet} p \wedge \bigwedge_{p \in (\bullet t - t\bullet)} \neg p] \wedge \bigwedge_{p \in (t\bullet - \bullet t)} \neg p \wedge \bigwedge_{p \in \bullet t} p.$$

Note that we have to confirm that the true-input places of t have no tokens. For example, for M including only the marking shown in Fig.4 (a), the marking

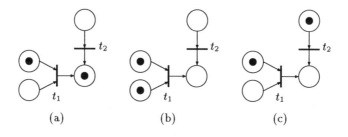

Fig. 4. Inverse image computation.

(b) is not in $TR^{-1}(M, t_1)$, but the marking (c) is in $TR^{-1}(M, t_2)$. This kind of consideration for true-output places in $TR(M, t)$ is not necessary because we only consider one-safe nets.

By using this inverse image, F_X is computed as follows :

$F_X(M_\phi, md)$: if $md ==$ "E", then
$$F_X(M_\phi, md) = \bigvee_{t \in T} \left(R(\boldsymbol{p}) \wedge TR^{-1}(M_\phi(\boldsymbol{p}), t) \right)$$
else if $md ==$ "A", then
$$F_X(M_\phi, md) = R(\boldsymbol{p}) \wedge$$
$$\neg \bigvee_{t \in T} \left(R(\boldsymbol{p}) \wedge TR^{-1}(R(\boldsymbol{p}) \wedge \neg M_\phi(\boldsymbol{p}), t) \right).$$

F_U in approach 1 works also for approach 2, and M_ψ is obtained similarly as in approach 1.

4.3 Approach 3

This approach is very similar to the previous one. The only difference is that sets of markings are directly represented by sets of binary vectors which are manipulated by the set operations of ZBDDs. First, we extend Subset and Change operations of ZBDDs for a set $V = \{v_1, v_2, \cdots, v_m\}$ of variables :

$$\text{Subset1}(f, V) = \text{Subset1}(\text{Subset1}(\cdots(\text{Subset1}(f, v_1), \cdots), v_{n-1}), v_n),$$
$$\text{Subset0}(f, V) = \text{Subset0}(\text{Subset0}(\cdots(\text{Subset0}(f, v_1), \cdots), v_{n-1}), v_n),$$
$$\text{Change}(f, V) = \text{Change}(\text{Change}(\cdots(\text{Change}(f, v_1), \cdots), v_{n-1}), v_n).$$

$TR(M, t)$ and $TR^{-1}(M, t)$ are computed by using these extended set operations.

$$TR(M, t) = \text{Change}(\text{Subset1}(M, \bullet t), t \bullet),$$
$$TR^{-1}(M, t) = \text{Change}\left(\text{Subset1}(M, t\bullet) \cap \text{Subset0}(M, \bullet t - t\bullet), \bullet t\right).$$

These are very straightforward: for example, $TR(M, t)$ is obtained by first applying Subset1 to extract the markings where all input places of t have tokens and to make those places empty (see the definition of Subset1), and then by applying Change to produce tokens in the output places of t. We can compute R and M_ψ in the same way as in approach 2.

5 Special BDD/ZBDD Operations for Petri Nets

In approaches 2 and 3, BDD or ZBDD operations are repeated for each input and output place of t in order to compute $TR(M, t)$ or $TR^{-1}(M, t)$. This is very inefficient because each operation must traverse a BDD/ZBDD and sometimes copy nodes as shown in Fig.2. In this section, we propose special BDD/ZBDD operations to handle all operations needed for one transition within one traversal of a BDD/ZBDD.

When a transition t fires, the variables for the true-input places of t should change from 1 to 0, the variables for the in-out places should remain 1, and the variables for the true-output places should change from 0 to 1. Based on this idea, $TR(M,t)$ for approach 3 can be implemented by the single ZBDD operation which does the followings during the traversal of the given ZBDD :

- For each true-input place variable p of t, eliminate each node n for p, and modify the ZBDD such that the edge pointing to n directly points the sub-ZBDD which was originally pointed to by the 1-edge of n. This modification is expressed by :

$$\langle n, h, \langle m, f, g \rangle \rangle \Rightarrow \langle m, f, g \rangle,$$

where a ZBDD is represented by a tuple $\langle n, f, g \rangle$ with the root node n and two sub-ZBDDs f, g pointed to by the 0-edge and the 1-edge of n. If such a node n does not exists, the 0-terminal node is returned. See Fig.5(a).
- For each in-out place variable p of t, modify the ZBDD such that the 0-edge of each node n for p points to the 0-terminal node. That is,

$$\langle n, f, g \rangle \Rightarrow \langle n, 0, g \rangle.$$

If such a node n does not exist, the 0-terminal node is returned (Fig.5(b)).
- For each true-output place variable p of t, modify the ZBDD such that the 1-edge of each node n for p points to the sub-ZBDD which was originally pointed to by the 0-edge of n and the 0-edge of n points to the 0-terminal node. That is,

$$\langle n, f, g \rangle \Rightarrow \langle n, 0, f \rangle.$$

If such a node does not exist, then a new node n' for p is created, and the lower ZBDD is pointed to by the 1-edge of n'. The 0-edge of n' points to the 0-terminal node. That is,

$$\langle m, f, g \rangle \Rightarrow \langle n', 0, \langle m, f, g \rangle \rangle.$$

See Fig.5(c).

Fig.6 shows the procedure ZbddTR to implement the above ZBDD operation. In this implementation, the information of input or output places of each transition is given as a list $plist$ of the data structures such that each data structure consists of the variable var for the corresponding place and the $type$ which specifies the type of place (i.e., true-input, true-output, or in-out). This list is arranged based on the variable order of the ZBDD. Thus, all ZBDD manipulations for one transition are achieved during one (depth-first) traversal of the ZBDD. $TR^{-1}(M,t)$ can also be implemented in the same way as above.

$TR(M,t)$ and $TR^{-1}(M,t)$ for approach 2 can also be implemented based on the same idea, although we omit those explanations in this paper because of the page limitation.

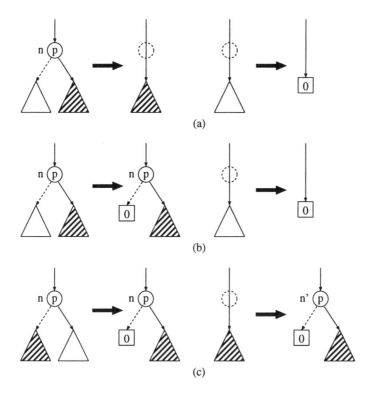

Fig. 5. ZBDD manipulation to handle one transition.

6 Experimental Results

In order to compare the performance of each approach, we have developed four programs. The first one, denoted by A_1, is based on approach 1. It uses the BDD library developed by David Long[Lon93]. Transitions are partitioned into groups such that each group consists of ten transitions. For the computation of the reachability set, we use the technique proposed in [HY93], where the fixed point of the reachability set with respect to one transition group is first computed, and then the obtained reachability set is used for the fixed point computation with respect to the next transition group.

The second and third programs, denoted by A_{2a} and A_{2b}, are based on approach 2. A_{2a} is implemented by using the standard BDD operations, whereas A_{2b} uses the special BDD operation for Petri nets proposed in the previous section. The last program, denoted by A_3, is based on approach 3. A_3 uses the special ZBDD operations proposed in the previous section. A_{2a} uses Long's BDD library. In order to do a precise comparison of A_{2b} and A_3, we have developed a BDD/ZBDD library by ourselves, and implemented the special BDD/ZBDD operations for Petri nets. In particular, A_{2b} and A_3 use the same hash function, the same cache mechanism, the same memory management strategy, the same data

```
ZbddTR(f, plist) {
    if plist.type == "true-input", then return InProc(f, plist);
    else if plist.type == "in-out", then return InOutProc(f, plist);
    else if plist.type == "true-output", then return OutProc(f, plist);
}

InProc(f, plist) {
    if plist is null, then return f;
    if order(f.top) < order(plist.var), then return 0;
    if order(f.top) == order(plist.var), then
        return ZbddTR(f.1, plist.next);
    if order(f.top) > order(plist.var), then
        return ZGetNode(f.top, InProc(f.0, plist), InProc(f.1, plist));
}
InOutProc(f, plist) {
    if plist is null, then return f;
    if order(f.top) < order(plist.var), then return 0;
    if order(f.top) == order(plist.var), then
        return ZGetNode(f.top, 0, ZbddTR(f.1, plist.next));
    if order(f.top) > order(plist.var), then
        return ZGetNode(f.top, InOutProc(f.0, plist), InOutProc(f.1, plist));
}
OutProc(f, plist) {
    if plist is null, then return f;
    if order(f.top) < order(plist.var), then
        return ZGetNode(plist.var, 0, ZbddTR(f, plist.next));
    if order(f.top) == order(plist.var), then
        return ZGetNode(f.top, 0, ZbddTR(f.0, plist.next));
    if order(f.top) > order(plist.var), then
        return ZGetNode(f.top, OutProc(f.0, plist), OutProc(f.1, plist));
}
```

Fig. 6. ZbddTR operation.

structures for BDD/ZBDD nodes, and so on. In A_{2a} and A_{2b}, the CTL model checking algorithm is not yet implemented. Our BDD/ZBDD library does not use any attributed edges.

For the comparison of the above approaches, we have used various nets which mainly model asynchronous circuits. Table 1 shows the number of places, the number of transitions, the number of tokens in the initial markings, and the number of reachable markings of these nets. "c" nets model 2-bit asynchronous registers and their environment found in [HHY92]. "DME-spec" nets are the high level specification of DME cells in [McM95], while "DME-cir" nets are the gate level representation of DME cells in [Dil88]. The digits in their names represent the number of DME cells in the nets. "comb" net also models an asynchronous combinational circuit in [YY96]. "JJreg" nets model the registers designed for Josephson Junction devices. In gate level circuits such as "DME-cir" and "comb" nets, all gates are represented by Petri nets as shown in [YY96].

The examples, except for "c" nets, are obtained by making product of small nets for circuit components which are originally prepared for the trace theoretic verification. For A_1 and A_3, we have model-checked several CTL formulas of the form $\mathbf{AG}[p_x \to \mathbf{A}(p_x \mathcal{U} p_y)]$.

Table 2 shows the sizes of BDDs/ZBDDs representing reachability sets and the CPU time for each step of the programs. Since Long's library uses attributed edges for negation, the BDD sizes obtained by A_1 and A_{2a} are a little smaller than those by A_{2b}. "TR" is the step to construct the transition relation of each transition group in approach 1. In "RS" step, the reachability sets are computed. "MC" is the step for the CTL model checking: for $k = 1, 2, \cdots, f_k$ denotes the model-checked CTL formula. All experiments were done on HP-9000/J210 (120MHz, 650MB). The BDD/ZBDD node limit was set at 10,000,000 for every case. The variable orders were decided according to the given files of the nets. Since the net files were generated from the descriptions of the circuit diagrams, the obtained variable orders somewhat reflect the circuit structures, and such variable orders sometimes give good results. Every program uses the same variable order for each example.

From these results, we can first see that the ZBDD sizes for the reachability sets are approximately one half to one third of the BDD sizes. These ratios coincide with the ratios of the number of initial tokens to the number of places. We know that the number of tokens in these nets is almost constant in every reachable marking. Hence, these results substantiate our claim that the ZBDD sizes relate to the number to tokens in the markings.

Second, program A_3 is much faster than the others, especially for the nets with less tokens. Its performance is achieved by using the ZBDD representation and the special operations for Petri nets. Program A_{2b} behaves as we expected for the "c" and "comb" nets. However, it is slower than A_{2a} for the other nets. This is probably caused by the naive algorithms and codes for the cache mechanism and BDD node management used in A_{2b} (Long's library used in A_{2a} is very sophisticated). But, note that A_3 also uses the same algorithms and codes for the cache mechanism and BDD node management as A_{2b}.

Table 1. Example nets

Nets	#pl	#tr	#token	#reach
c1	45	44	21	37
c2	104	115	52	11568
c3	253	188	125	766178
DME-spec8	137	128	33	786432
DME-spec9	154	144	37	3538944
DME-cir5	491	820	231	859996
DME-cir7	687	1148	323	90446346
comb	169	376	82	31492
JJreg_a	251	249	81	1807781
JJreg_b	251	249	81	119709

Table 2. Experimental results

Nets	Pro-gams	#nodes for RS	TR	RS	MC			
c1	A_1	211	0.18	0.08	f_1: 0.03	f_2: 0.06	f_3: 0.02	f_4: 0.06
	A_{2a}	211	–	0.07	–	–	–	–
	A_{2b}	213	–	0.03	–	–	–	–
	A_3	112	–	0.01	f_1: 0.02	f_2: 0.02	f_3: 0.01	f_4: 0.02
c2	A_1	3431	1.64	12.54	f_1: 0.72	f_2: 13.15	f_3: 1.19	f_4: 5.46
	A_{2a}	3431	–	14.99	–	–	–	–
	A_{2b}	3433	–	2.28	–	–	–	–
	A_3	1928	–	0.99	f_1: 0.31	f_2: 3.75	f_3: 0.63	f_4: 1.85
c3	A_1	57604	8.25	121.59	f_1: 464.98	f_2: 1119.13	f_3: 23.01	f_4: 44.26
	A_{2a}	57604	–	114.38	–	–	–	–
	A_{2b}	57606	–	45.44	–	–	–	–
	A_3	30448	–	12.62	f_1: 47.96	f_2: 428.96	f_3: 12.97	f_4: 0.51
DME-spec8	A_1	108941	1.55	227.38	f_5:162.63			
	A_{2a}	108941	–	225.11	–			
	A_{2b}	108943	–	482.14	–			
	A_3	32178	–	14.34	f_5: 6.69			
DME-spec9	A_1	242701	1.98	636.75	f_5: 410.69			
	A_{2a}	242701	–	623.35	–			
	A_{2b}	242703	–	3423.22	–			
	A_3	71602	–	38.98	f_5: 19.57			
DME-cir5	A_1	184897	38.63	2602.35	f_5: 910.29			
	A_{2a}	184897	–	6741.01	–			
	A_{2b}	184899	–	10291.54	–			
	A_3	92214	–	621.97	f_5: 682.47			
DME-cir7	A_1	N.A.	83.11	O.F.	f_5: N.A.			
	A_{2a}	1012851	–	81223.21	–			
	A_{2b}	N.A.	–	(> 48000)	–			
	A_3	504324	–	10205.21	f_5: 13917.71			
comb	A_1	171263	3.78	243.93	f_6: 269.35			
	A_{2a}	171263	–	265.32	–			
	A_{2b}	171265	–	86.81	–			
	A_3	89287	–	27.44	f_6: 110.95			
JJreg_a	A_1	N.A.	5.02	O.F.	f_7: N.A.			
	A_{2a}	2005654	–	17397.31	–			
	A_{2b}	N.A.	–	(> 36000)	–			
	A_3	952246	–	2325.97	f_7: 1637.38			
JJreg_b	A_1	395846	4.78	1038.25	f_7: 116.50			
	A_{2a}	395846	–	615.87	–			
	A_{2b}	395848	–	1122.87	–			
	A_3	181701	–	42.62	f_7: 30.44			

"–" : model checking algorithm is not implemented.
"O.F." : node overflow (node limit is 10,000,000).
"N.A." : not available.

7 Conclusion

In this paper, we have proposed using ZBDDs for the CTL symbolic model checking of Petri nets. The experimental results show that the sparse state spaces of Petri nets are managed very well by ZBDDs. However, these results are obtained by the fixed variable order for each example. Therefore, for the fair comparison, we need further experiments by using better variable order for each approach, which will be computed for BDD based approaches by the dynamic reordering techniques implemented in the Long's library or by some variable ordering algorithms ([SY95] and others).

We have also proposed special BDD/ZBDD operations which manipulate Petri nets efficiently. Although our BDD/ZBDD library is still naive, it is shown from the experimental results that the ZBDD based approach with the special operations is much faster than the others.

In the future, we would like to apply the proposed ZBDD approach to other Petri net based verification methods such as the trace theoretic verification.

References

[BCD+92] J. R. Burch, E. M. Clarke, D. L. Dill, L. J. Hwang, and K. L. McMillan. Symbolic model checking: 10^{20} states and beyond. *Academic Press*, 98(2):142–170, 1992.

[Dil88] D. L. Dill. *Trace theory for automatic hierarchical verification of speed-independent circuits.* MIT press, 1988.

[HHY92] K. Hamaguchi, H. Hiraishi, and S. Yajima. Design verification of asynchronous sequential circuits using symbolic model checking. *Proc. of International Symposium on Logic Synthesis and Microprocessor Architecture*, pages 84–90, 1992.

[HY93] K. Hamaguchi and S. Yajima. Symbolic model checking for Petri nets using block partitions. *FTC workshop (in Japanese)*, 1993.

[Lon93] D. E. Long. BDD library man page. 1993.

[McM95] K. L. McMillan. Trace theoretic verification of asynchronous circuits using unfoldings. *LNCS 939 Computer aided verification*, pages 180–195, 1995.

[Min93] Shinichi Minato. Zero-suppressed BDDs for set manipulation in combinatorial problems. *Proc. of 30th DAC*, pages 272–277, 1993.

[RCP95] Oriol Roig, Jordi Cortadella, and Enric Pastor. Verification of asynchronous circuits by BDD-based model checking of Petri nets. *LNCS 935 Application and theory of Petri nets 1995*, pages 374–391, 1995.

[SY95] Alexei Semenov and Alexandre Yakovlev. Combining partial orders and symbolic traversal for efficient verification of asynchronous circuits. *Proc. of CHDL'95*, pages 567–573, 1995.

[YY96] T. Yoneda and T. Yoshikawa. Using partial orders for trace theoretic verification of asynchronous circuits. *Proc. of Second International Symposium on Advanced Research in Asynchronous Circuits and Systems*, pages 152–163, 1996.

HDL-Based Integration of Formal Methods and CAD Tools in the PREVAIL Environment

D. Borrione, H. Bouamama, D. Deharbe [1], C. Le Faou, A. Wahba

Université Joseph Fourier, Laboratoire TIMA, Grenoble, France

Abstract. We present an open environment for the integration of formal methods applied to HDL descriptions of circuits. The system currently accepts SMAX[4] and VHDL, and provides equivalence checking, model checking, theorem proving, and automatic diagnosis of simple design errors. After an overview of the system, we discuss the salient features of the common intermediate format, of the diagnosis tools, and of the automatic generation of NQTHM[11] models from VHDL functional descriptions.

1. Introduction

Automatic formal verification methods are today in the state where simulation was in the early seventies: the technology exists, many approaches have been proposed and demonstrated, many tools are developed. Yet, no single algorithm is efficient for all verification tasks, no proof tool applies at all description levels and for all categories of circuits, and algorithms for performance improvement are being actively studied and published continuously. Thus, the current state of the art requires that several proof tools be available, very much like the previous situation in the simulation area, where the designer could perform RT-level, gate-level, switch-level and electrical simulation, by invoking different tools on different models. Far from being outdated, conventional simulation is extensively used for the initial validation of system specifications, and for error-correction purposes. The need for system descriptions that can be both simulated and formally verified has therefore emerged.

Contrarily to the situation twenty years ago, standard hardware description languages (HDL's) are now available [1, 2]. These languages have become an essential integration factor among the various CAD tools, which take as input the same circuit description; thus the designer has greater confidence that what is simulated is also what is synthesized and tested (by ATPG techniques). We chose VHDL due to its wide acceptance as a standard in Europe. In order to integrate formal verification with the other design tools, it is essential that HDL's be given formal semantics compatible with the simulation and synthesis semantics traditionnally attached to them.

These considerations motivated the development of the PREVAIL proof environment, which is the result of a cooperative effort between Université de Provence - Marseille, Université Joseph Fourier - Grenoble and Technische Hochschule Darmstadt. An early version of the system was described in [3]. Most of the proprietary software has been totally rebuilt to create the current version, for portability and efficiency reasons,

---

[1] Currently a visiting researcher at Carnegie-Mellon University, CS Dept.

and new tools have been created and added to the system. We here present the new ideas that have emerged in the past years, and refer the reader to previous publications for older concepts or extended discussions. Non trivial issues dealt with in this paper are critical to the implementation of any similar integrated system, and the association of automatic diagnosis with formal verification is crucial to the acceptance of formal methods in industry.

This paper is organized as follows. Section 2 gives an overview of the system. Section 3 presents the semantic model and the common intermediate format. Section 4 discusses the integration of diagnosis tools. Section 5 introduces the principles of our translator to the Boyer-Moore logic. We finally present our conclusions in relation with other works.

2. Overview of PREVAIL

PREVAIL is a multi-HDL, multi-tool formal verification environment, which presents to the designer a unified interactive graphical user interface (see Fig. 1).

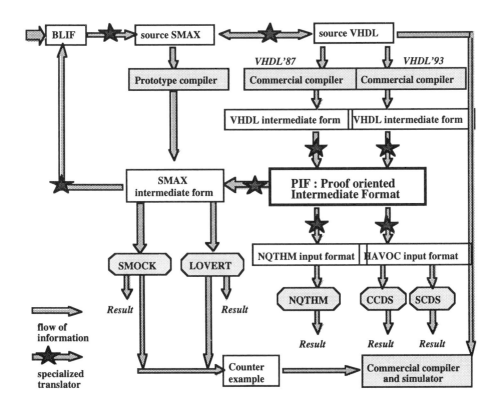

Fig. 1: The overall PREVAIL system

2.1. Source Description Languages

Currently, two HDL's may be used as input language: SMAX and VHDL. SMAX was designed in Darmstadt using the CONLAN definition method (which includes formalized syntax and semantics)[4]; it is specially intended for reasoning at the bit-vector level, with appropriate first order logic semantics, on combinational circuits and synchronized unit delay Finite State Machines (FSM). It is semantically equivalent to a subset of VHDL.

As in most verification systems with a VHDL front-end, only a subset of the standard VHDL is recognized. Basically, the "P-VHDL" subset of PREVAIL corresponds to a synthesizable subset extended with generic parameters. No effort has been made to conform to a particular synthesis tool subset; rather we kept or discarded each construct based on its semantic expressiveness, trying to retain as large and consistent a subset as possible. Thus, contrarily to other systems, we recognize the use of guarded blocks to convey the meaning of clock-synchronized register loads in a data-flow description; conversely, we tend to prefer the construct *not clock'stable*, to which mathematical semantics can easily be given, rather than *clock'event* which is defined in terms of the simulation execution model.

The underlying common semantic framework is the FSM model, with repetition, hierarchy, and genericity. All types are assumed to be discrete (including integers and enumerated symbolic types, arrays and records with arbitrary number of dimensions/fields). Generic models are models where the number of sub-components or the data path width is kept a parameter: typically, in VHDL, they correspond to *unconstrained* arrays and *for...generate* constructs with generic bounds. Hierarchical models must be fully configured: in VHDL terms, all components must be bound to a specific *entity-architecture* pair.

2.2. Compiler Front-Ends

In PREVAIL, a "Proof-oriented Internal Format" (PIF) is the data structure representing the common scheme mentioned above. The SMAX Intermediate Format is a subset of PIF (for instance, PIF includes repetitive statements, which must be formally expanded in the SMAX Intermediate Format). The environment can be extended to other HDL's, either by providing source to source translators, or by implementing source to PIF compilers. As an example, a BLIF to SMAX translator has been written.

Due to the modest size of our group, we decided to use existing compiler front-ends, whenever possible. Thus PREVAIL relies on the SMAX compiler from Darmstadt, and the VTIP[2] software for VHDL'87. LVS[3] is currently being incorporated, for parsing VHDL'93. In the case of VHDL, both VTIP and LVS produce a decorated syntax tree, from which VHDL2PIF, our proprietary second level compiler front-end, constructs the PIF structure.

---

[2]VTIP is distributed by Compass
[3]LVS is distributed by Leda

2.3. Verification Tools

Several proof systems are embedded in the PREVAIL environment. The control of the verification steps and the selection of the appropriate prover are the responsibility of the designer.

LOVERT is both a tautology checker, and a FSM equivalence checker, implemented by the group of H. Eveking [5] on the principles of Coudert et al [6]. SMOCK is a proprietary symbolic model checker inspired by the seminal work of Ed Clarke's group [7], which checks formulas written in CTL-P, i.e. CTL [8] extended with modalities looking back to the past [9]. Both LOVERT and SMOCK are built over an OBDD package developed at Tech. Hochscule Darmstadt [10]. Like all BDD-based tools, they apply to components where generic parameters, if any, have been fixed to constant values. Such tools are now commercially available (Vformal, CheckOff, FormalCheck ...) and will not be discussed further here.

When specifications are given by arithmetic functions, or when the object to verify is a parameterized function or library module that must be proven correct for all admissible values of its parameters in an unbounded set, elaborate reasoning capabilities, including induction, are needed. To this end, the NQTHM general purpose theorem prover of Boyer and Moore [11] distributed by CLI has been integrated in PREVAIL. We chose this system for its powerful proof strategies, and its capabilities to handle large hardware proofs automatically [12]. After modelling the VHDL primitives in the NQTHM logic, we implemented a special purpose translator which transforms the PIF structure into a set of function definitions in the NQTHM input language, and generates accompanying lemmas.

2.4. Automatic Diagnosis

When a design has been found erroneous, the designer faces the difficult and time consuming task of locating and correcting the error. We have developed prototype automatic diagnostic tools for logic level designs, which locate the error and suggest a correction with a 100% hit ratio in the case of a single gate or connection errors: CCDS is the diagnoser for combinational circuits [13], and SCDS for sequential circuits [14] .

2.5. User Command and Control

In designing PREVAIL, the goal was to provide an open and user-friendly environment, integrating new formal tools as they become available. Because some verification tasks may take long to execute, multi-tasking has been implemented: the user may start simultaneous applications; task control is provided to monitor, interrupt, visualize the results of one or more tasks. Online documentation is integrated into the system. All tools are invoked through a unified graphical interface implemented in Tcl/Tk.

The architecture of the system is composed of five specialized main modules: the Starter, the Watcher, the Helper, the Executer and the Lock Manager (Fig. 2). All are Tcl/Tk based, but only the first three are graphically visible to the user.

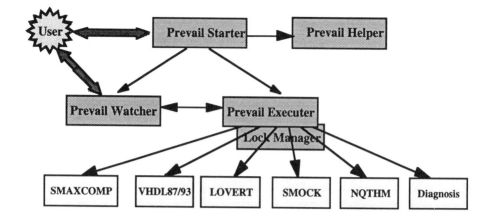

Fig. 2: Architecture of the PREVAIL User Interface

The Prevail Starter is the entry point to Prevail. It coordinates the entire environment. Interactions with the user, global initializations and entries to the other modules are done here. A new module can be added to Prevail by pluging it into the Starter.

The Prevail Watcher manages the display of messages and status of the tasks currently launched. The user can see which tool is running, follow output messages displayed by a given task and even stop any task at any moment. This is useful when such tasks are blocked, slow or no more needed.

The Prevail Helper integrates the documentation, providing online explanations on:
* How to use Prevail from the graphical interface,
* How to execute individual tools in the batch mode using a line interface, which is desirable for time consuming tasks,
* Access to the manuals of the different tools (when proprietary),
* Setting the various environment variables that control the system.

The Prevail Executer is directly linked to the basic tools such as text editors (not shown on Fig. 1), compilers, provers and diagnosers. For each tool selected by the user from the graphical interface, the Executer runs a child process, launching the specified application in the background. This feature allows the user to run several tools concurrently.

The Lock Manager protects data accessed by concurrent tasks, and prevents the destruction or modification of critical data during the execution of a given task. For instance, if a VHDL architecture that uses several components is being translated into PIF, the Lock Manager insures that these components are not recompiled until the translation is finished.

3. The Common PIF Format

From a circuit description in a conventional HDL, a more mathematical model can be extracted, which is the functional semantic model of the description. Indeed, despite the syntactic idiosyncrasis of the individual HDL's, and the presence or absence of particular primitives, a number of concepts have emerged which constitute the common conceptual scheme onto which the individual HDL's can be mapped. Providing translators from one or more HDL's to a mathematical model of this common scheme is the method by which individual proof tools can be applied to circuit designs written in various HDL's. An additional benefit of this method is that it allows comparison of designs, or of model libraries, written in different HDL's; a direct application is the formal verification of the translation from one HDL to another one.

The conceptual model in PREVAIL is the hierarchy of synchronous interconnected FSM's. A VHDL *component* (or SMAX *description*) is translated into a FSM. In the case of hierarchical designs, the interconnected sub-components are translated into the parallel composition of their corresponding FSM's. The timing statements of VHDL (*after* and *until* clauses, *'delay* attribute) are currently not recognized in the model, and time is abstracted in cycles of a master clock, syntactically identified by a special attribute (these restrictions are implicit in SMAX). In *clock-synchronized* designs, state changes occur on the clock edge. Under some syntactic and semantic constraints [3,15], a clock-synchronized VHDL description can be modeled by a deterministic FSM that has one state change per clock cycle (under these constraints, cycle-based simulation is semantically equivalent to the simulation algorithm of [1]). If these constraints are relaxed, the FSM has one state change per simulation cycle [16]. The current version of the VHDL2PIF translator enforces the most constrained clock-synchronized approach, and checks that the expected conditions are fulfilled.

A dataflow style description is in direct correspondance with the semantic model. In VHDL, guarded signal assignments, where the guard expression involves the clock edge, identify the state elements, and other signals are combinational. Combinational signals other than the inputs and outputs are eliminated (if the elimination fails, a zero delay loop is detected and flagged as an error).

A behavioral description requires more elaborate processing. In VHDL, it is an architecture with one or more processes. Combinational processes correspond to combinational circuits which assign one or more output signals in reaction to any change on their input signals. Such processes must be sensitive to all signals referenced in the process' statements, and must have no hidden memory (e.g. all variables must be assigned before being referenced, the same signals must be assigned in all alternatives of conditional *if/case* statements, etc...) [3,15]. All signals assigned in a combinational process are a combinational function of the signals in the process sensitivity list, and all internal variables are eliminated.

Sequential processes have all their *wait* statements of the form:
      wait on CK until ...
      wait on ... until (not CK'stable and ...)
where CK is declared a *clock*, and " ... " may be empty or may involve other signals.

All signals assigned in a sequential process are state elements. Variables require a specific processing, in order to eliminate all those which are not referenced across clock edges, and only keep the minimum amount of independant state elements. The principles of variable elimination and merging of all wait statements into a single final wait statement per process are described in [16]. After this transformation, the resulting sequential process is directly translated into a set of state assignments.

The **Proof-oriented Intermediate Format** (PIF) implements the semantic model. PIF was created to reduce the amount of programming effort needed to add new tools to Prevail, and to assure independance of our tools from commercial VHDL compiler front-ends. The translation from the intermediate structure produced by compiler front-ends to PIF corresponds to a proof-oriented *elaboration* of the description.

The PIF syntax is based on the SMAX intermediate format designed in Darmstadt, of which it is an extension. It is written in Common Lisp syntax, and is fully recognizable by a Lisp interpreter.

A PIF file is associated either to a couple entity-architecture, to a package, or to a couple package-package-body. Each VHDL identifier is represented by a symbol, and given an internal name in PIF. A PIF file is decomposed in two parts:

- The header part defines global data, and the names of all symbols, grouped by object kind (signal, variable, generic, type, for...generate clause, component instantiation, function, procedure, etc...).
- The detailed part gives, for each internal name, the list of properties that characterize its associated symbol [17]. Three properties are most important:
  - CKIND: represents the object kind.
  - ACTUALNAME: represents the full name of the object as defined in the VHDL/SMAX source. This information is specially useful for diagnosis.
  - STATEMENTS: contains the assignment function for a signal or a variable, and the lines of the source statements which assign to it.

Component instantiations, *for/if...generate* and *for...loop* statements are bound to special symbols in PIF, allowing dedicated treatments when translating from PIF to specialized formats: these statements are expanded in translating to the input of BDD-based provers, and are transformed into recursive functions for theorem proving.

The translation preserves the modular structure of the VHDL code. Thus, the hierarchy of components is not flattened, rather each component instance appears as a symbol in PIF. Likewise, the *use* of a *package* is translated as a pointer to that package's PIF file. This permits reusability of already translated source codes.

4. Automatic Diagnosis

Automatic diagnosis is a natural complement to formal verification. When an implementation is found to be erroneous it must be corrected. Most advanced verifiers provide counterexamples, in the form of input patterns that witness the error. The tedious work of finding the error and correcting it is left for designers, who use these counterexamples to simulate their designs.

In the last years automated synthesis tools have received a considerable attention, since they are capable of generating *correct-by-construction* designs. However, fine level optimizations, and manual "patches" due to late changes in the specification, are still common. During this manual phase, errors may be inserted unintentionally. Examples from VLSI industry show that obtaining correct-by-construction products is not always possible [18-20]. The place of the diagnosis in the design life cycle is shown in Fig. 3.

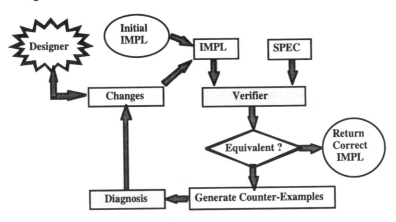

Fig. 3: The place of the diagnosis n the design life cycle

CCDS (Combinational Circuits Diagnostic System) and SCDS (Sequential Circuits Diagnostic System), that we present briefly here, address the correction of bit and bit-vector level designs, when the proof of equivalence between the specification and the implementation has failed. The specification may be described in any style. The implementation is assumed to be a logic network.

Both tools are composed of three cooperating basic modules: a generator of special *diagnosis-oriented test patterns*; a three valued logic simulator; and a diagnosis engine that contains all the diagnostic rules. The information flow between these modules is shown in Fig. 4.

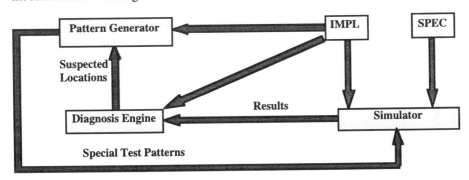

Fig. 4 : The three basic modules of the diagnosis system

The principle of the diagnosis algorithm is as follows. Initially all the circuit gates are suspected (i.e. the error may be situated at any gate). The pattern generator selects one of these gates and generates a special pair of test patterns for it [13]. This pair is sent to the simulator, and the specification and the implementation are simulated. The simulation result is sent to the diagnosis engine, that applies the diagnosis rules to limit the number of suspected locations. The reduced set of suspected locations is then sent back to the pattern generator which, in turn, selects another suspected gate and the whole operation is repeated, until the error is located precisely.

4.1. Error Model and Basic Assumptions

Abadir [21] classified simple design errors into *gate errors* and *connection errors*. Table 1 summarizes these errors. Our diagnosis tool CCDS can diagnose both classes in combinational circuits. SCDS treats only gate errors in sequential circuits; its extension to connection errors is currently being studied. These tools work under the following assumptions:
1. One error, at most, exists in the cone of influence of each primary output.
2. The error is one of the above mentioned ones.
3. The implementation is given as a network of the following gate types: AND, NAND, OR, NOR, XOR, XNOR, BUFFER, and NOT.
4. The error does not introduce extra loops in the implementation.

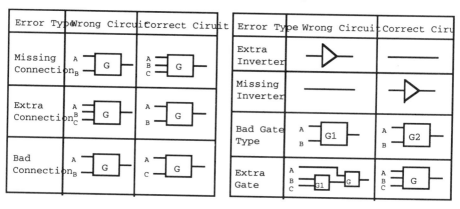

Connection Errors   Gate Errors

Table 1. Simple Design Errors

4.2. Error Hypotheses

The diagnosis problem in its general form is known to be NP-complete [22]. To limit its complexity, we use the principle of the *diagnosis by error hypotheses*. An error type is assumed and the diagnosis is made considering only this type. If the error is not found another type is selected and so on, until the error is found. In our tools, the error types are considered in decreasing order of their frequency, according to the statistical study made by Aas [23]. The following six error hypotheses cover all the error types listed in Table 1, as explained in [13]:

HYP-0: An extra/missing inverter
HYP-1: A gate replacement of type 1 (OR <-> AND, NOR <-> NAND).
HYP-2: A gate replacement of type 2 (OR <-> NAND, NOR <-> AND).
HYP-3: An extra wire error.
HYP-4: A missing wire error.
HYP-5: A bad connection error.

4.3. The Backward Propagation Technique

The backbone of our diagnosis algorithm is the *backward propagation technique*. The implementation is simulated under the application of the specially generated test patterns, and it is then scanned from its primary outputs back towards its primary inputs. Diagnostic rules are applied during this scan to limit the number of gates to examine and the number of suspected locations of the circuit.

4.4. Diagnosis of Sequential Circuits

Diagnosing sequential circuits is much more complex than combinational ones, since the specification and the implementation may have a different number of state variables and different state encodings. Thus , there is no reference to which the value of the memory elements of the implementation can be compared. The specification is considered a black box, from which only the primary inputs and outputs can be observed.

The algorithm used in SCDS is based on the new concept of *possible next states* [14]. They are the states reachable from a given state (or a set of states), under the application of a given input vector, by inserting an error at the different possible locations in the circuit. The sequential circuit is unfolded over the time frames using the *iterative logic array* model, and combinational diagnostic rules are applied in each time frame under the application of each one of the possible next states.

4.5 Practical Considerations

The CCDS and SCDS tools are implemented in PROLOG. They take as input the translation of circuits in a proprietary netlist-type format called HAVOC, that describes circuits as Prolog facts. They were tested on the ISCAS'85 and ISCAS'89 benchmarks [24,25]. Thousands of experiments were made with randomly inserted errors. In all the experiments the error was found after the application of a small number of test patterns (CCDS: about 12 patterns on average for the circuit c6288 (2416 gates). SCDS: about 20 test vectors within 7 test sequences for the circuit s5378 (2779 gates and 179 flip/flops)). The tools were also tested on industrial circuits supplied by Thomson-TCS.

5. Integration of NQTHM

A previous attempt at providing automatic input to NQTHM from a VHDL description was limited to proving that a repetitive *structural* circuit implemented a specified arithmetic function [3]. The new approach now taken is dedicated to *behavioral* descriptions, and was motivated by our contribution to the recently voted VHDL standard synthesis packages, and the formal validation of one package during its development. There, the problem was not only proving mathematical properties of the individual functions, including the fact that the function body correctly implements the intended operation, but also properties of function compositions (e.g. to show that a function is the inverse of the other). Dealing with VHDL functions on bit vectors of arbitrary size, we met the following problems:

- to correctly model the package, where functions returning a vector have the index direction of their result explicitly specified, reducing bit-vectors to the list of their elements, as is usually done since[12], was inadequate: the VHDL array indexing mechanism had to be modeled.

- the user error handling mechanism of VHDL, typically written:
  if error_case then assert false severity error;
  which interrupts further computations, was to be portrayed; in recursive functions, a clear distinction between the "base case" and the "error case" became necessary. Conceptually, all basic types have been extended with an error element, which acts as an absorbing element for all functions. Formally, let F be a n-ary function, if $\exists$ i $(1 \le i \le n)$ ti = *error* then F(t1, t2, ...tn) = *error*

To set up the basic knowledge for further proofs, we created a VHDL-oriented level over the "Ground-Zero-Logic" of NQTHM, that we called "Ground-VHDL-Logic"; its definitions and lemmas are part of the object code of the PIF2NQTHM translator.

5.1. The Ground-VHDL-Logic

Ground-VHDL-Logic is a library of shell definitions, functions and lemmas, built on top of Ground-Zero-Logic, that model the VHDL primitives and their characteristic properties[26]. It borrows largely to [12], and contains:

- A shell for the error
- Basic discrete data types as Bit, Boolean, Naturals, Integers
- Unconstrained mono-dimensional arrays, with the special type Bit_Vector.
- Functions for all operators on these basic types
- Functions for all VHDL type attributes
- Theorems stating the characteristics of these functions, including error and type checking

Because NQTHM is not meant to handle arrays, and the indexing mecanism of VHDL is very general and complex, an important part of the library is dedicated to vectors. The notion of vector is implemented by the shell named Vector, with the following three constructors:

- Vlist : represents the list of values of the vector.
- Vleft : represents the left bound of the vector.
- Vascending is the direction of the vector.

This library is automatically linked with the NQTHM format generated by the translator, or can be used for writing a manual specification for a VHDL package.

5.2. Principles of the PIF to NQTHM Translator

The current translator is applied to behavioral descriptions, and imposes restrictions on procedure and loop statements. It transforms VHDL subprograms and processes to a functional representation within the logic of Boyer and Moore extended with the Ground-VHDL-Logic.

The translation keeps the modularity of the VHDL code. In particular, importing the definitions of other packages (by the VHDL *use* clause) corresponds to generating a load of its NQTHM library, and loading the tables of the translator with the names of all the functions in the library (function identifiers may be overloaded in VHDL, but not in NQTHM). Naming conventions are adopted to generate NQTHM code easy to map to its corresponding part in the VHDL source, and avoid name conflicts. They should be self-explanatory in the examples below.

Type declarations are translated into recognizers, which are boolean NQTHM functions. In particular, enumerated data types defined in extension are translated into a function that checks equality of its argument with one of the enumerated values.

Constant declarations are expanded into the generated code and no special functions are generated for them. A variable or a signal is translated into the function which computes its value, by composition of all the statements which assign to it; intermediate variables are eliminated in that process. If a variable or signal is assigned in a *for...loop*, an intermediate recursive function is created for the assignment in the loop, and the global function for the object contains a call to it.

As proofs may be hard to perform on vectors, which are not an inductive type but a structure as in programming languages, the translation process is done in two steps: intermediate functions are first defined on the *Vlist* (of the shell list) before generating functions on vectors. The choice of this strategy is driven by the difficulties encountered when proving lemmas handling vectors directly: proofs are more efficient, and require fewer interactions from the user, when working on lists first.

The translator automatically generates some lemmas, according to a predefined template that depends on the VHDL statement being translated. Some of the essential lemmas are:

- **Type checking**: states that if the arguments of a function are of the correct type then the function returns an element of its result type.
- **Error checking**: states that if one of the arguments of a function is an error then the returned value is also an error.

As the NQTHM prover is very sensitive to the way the VHDL code is translated, and a proof can fail or succeed depending on how the functions are written, we give great importance to the simplification process included in the translator. Some translation schemes simplify the generated code in such a way that the prover can do the job with minimum user interaction.

We now discuss and illustrate some of the points mentioned above in this section.

5.2.1. Vector Types and Subtypes

In the case of array declarations, two recognizers instead of one are generated: one recognizer for the shell list defining the list of elements of the appropriate type, and another one for the vector with its left bound and index direction which corresponds to the translation of the VHDL type or subtype.

The following example shows VHDL type and subtype declarations, with their translated NQTHM code and the associated automatically generated lemmas (the comments have been added manually).

VHDL source

```
type UNSIGNED is array ( NATURAL range <> ) of BIT;
subtype T16_UNSIGNED is UNSIGNED(0 to 15);
```

NQTHM source

```
;Recognizer of type UNSIGNED on the shell list: the function is defined recursively
(defn p{list_unsigned} (x)
 (if (p{list} x)
    (if (lemptyp x) t
       (and (p{bit} (lcar x)) (p{list_unsigned} (lcdr x)))) f))
```

```
;Recognizer of the type UNSIGNED on vectors
; x must be a vector and its Vlist part must be of type p{list_unsigned}
(defn p{unsigned} (x) (and (p{vector} x) (p{list_unsigned} (vlist x))))
```

```
;Error checking on the type UNSIGNED
 (prove-lemma th_t_error_p{unsigned} (rewrite)
    (implies (p{unsigned} x) (not (errorp x))))
```

```
;Recognizer of T16_UNSIGNED: x is a p{unsigned} and satisfies range constraints
 (defn p{T16_unsigned} (x)
    (and (p{unsigned} x) (equal (av{left} x) 0) (equal (av{right} x) 15)
       (equal (av{ascending} x) t)))
```

```
; Error checking on type T16_UNSIGNED
(prove-lemma th_t_error_p{T16_unsigned} (rewrite)
   (implies (p{T16_unsigned} x) (not (errorp x))))
```

```
;Checking that T16_UNSIGNED is a subtype of UNSIGNED
(prove-lemma th_subtype_p{T16_unsigned} (rewrite)
   (implies (p{T16_unsigned} x) (p{unsigned} x)))
```

5.2.2. Subprograms and For-loops

Special processing is done on subprograms and *for* loops having vectors as arguments or returning vector values. First a function is generated that deals only with lists (the argument vectors and the returned value become lists). Then the main function is generated thar returns a normalized vector, the list part of which being a call to the previous function on lists.

For each function generated, a predefined set of lemmas is added. These lemmas verify the correctness and the coherence of the translated code, and are submitted to the theorem prover. For instance type checking and error checking are done on each function. In the case of procedures, we generate one function (or function pair) per output argument, which computes its value.

For loops are handled like procedures with some differences. First, all signals and variables assigned in the *for* loop statement are assumed to be outputs of the loop. For each output a NQTHM expression is generated, that gives the value of this signal or variable depending on a list of other signals or variables assumed to be inputs. Here again we distinguish between vectors and scalars, in that if the output or one of the inputs is a vector then we generate an expression on lists first. The next step is the simplification and the transformation of this expression into a recursive NQTHM function. The transformation can take one of several predefined translation schemes, depending on whether or not the loop index interval can be related to the dimension of an array that is referenced in the loop. Whenever possible, we eliminate the loop index and the explicit array indexing, and generate recursive functions on the array list of elements, which are easier to prove.

5.2.3. Simplification of the Generated Code

A number of simplifications are done during the translation process, to produce as simple a NQTHM code as possible. The different steps of the simplification can be summarized as follows:

- The use of normalized vectors: presently we recognize two kinds of normalization. One with the range attribute equal to (SIZE-1 downto 0) and one with the range attribute (1 to SIZE). SIZE is the length of the vector.
- The use of low functions from the Ground-VHDL-Logic. This means functions hard-coded in the Ground-VHDL-Logic and using efficiently the prover facilities. Such functions were tested and several lemmas stating their characteristics were proved and integrated in the library.
- Simplification of static expressions on naturals. Such expressions, when badly written, may presently cause the proof to fail.

5.2.4. Example

Function RESIZE is extracted from the Numeric_Bit package. It takes an unsigned ARG and a natural NEW_SIZE, and returns a vector of size NEW_SIZE. If the length of ARG is less than NEW_SIZE, the result is equal to ARG left-extended with zeroes. In the other case the high order bits in excess are truncated.

VHDL source

function RESIZE (ARG: UNSIGNED; NEW_SIZE: NATURAL) return UNSIGNED
is

```
  constant ARG_LEFT: INTEGER := ARG'LENGTH-1;
  alias XARG: UNSIGNED(ARG_LEFT downto 0) is ARG;
  variable RESULT: UNSIGNED(NEW_SIZE-1 downto 0) := (others => '0');
begin
  if (NEW_SIZE < 1) then return NAU; end if;
  if XARG'LENGTH = 0 then return RESULT; end if;
  if (RESULT'LENGTH < ARG'LENGTH) then
    RESULT(RESULT'LEFT downto 0) := XARG(RESULT'LEFT downto 0);
  else
    RESULT(RESULT'LEFT downto XARG'LEFT+1) := (others => '0');
    RESULT(XARG'LEFT downto 0) := XARG;  end if;
  return RESULT;
 end RESIZE;
```

NQTHM translation

```
;Function on shell list
(defn f{list_resize_11} (arg new_size)
 (if (not (P{LIST} arg)) (err)
   (if (lessp new_size 1) (lempty)
     (if (equal (size arg) 0) (lfill '0 new_size)
       (if (lessp new_size (size arg)) (lslice arg 0 new_size)
         (lappend arg (lfill '0 (difference new_size (size arg)))))))))

;Function on vectors
 (defn f{resize_11} (arg new_size)
   (if (and (p{unsigned} arg) (numberp new_size))
     (vector (f{list_resize_11} (vlist arg) new_size) (sub1 new_size) f) (err)))

;This lemma guarantees that the list part of the returned vector of this function
; is a call to the function on lists.
 (prove-lemma th_vector_to_list_f{resize_11} (rewrite)
   (implies (and (p{unsigned} arg) (numberp new_size))
        (equal (vlist (f{resize_11} arg new_size))
                (f{list_resize_11} (vlist arg) new_size))))

;Type checking on function on lists
 (prove-lemma th_signature_f{list_resize_11} (rewrite)
   (implies (and (p{list_unsigned} arg) (numberp new_size))
        (p{list_unsigned} (f{list_resize_11} arg new_size))))

; Type checking on function on vectors
 (prove-lemma th_signature_f{resize_11} (rewrite)
   (implies (and (p{unsigned} arg) (numberp new_size))
            (p{unsigned} (f{resize_11} arg new_size)))
;; Some automatically generated hints to guide the prover
   ((expand (f{resize_11} arg new_size))
    (use (th_signature_f{list_resize_11} (arg (vlist arg)) (new_size new_size)))))

;Error checking lemmas are omitted for space reasons
```

6. Conclusions

Many research groups have proposed formal semantics for VHDL, to apply formal verification techniques on designs described in that language. The FSM model for clock synchronized circuits closest to ours is Bull's [15], but currently only machines with identical state encodings can be compared in the commercial verifier VFORMAL [27]. More recently, another approach close to ours for synchronous circuits has been published [28]. Cycle-level synchronization semantics of VHDL [16, 40] serve as a basis for Siemens' verification environment CVE [41] and a model checking system under development at Carnegie-Mellon University [29]. Models based on Petri Nets [30,31] tend to create an excessively large number of states, and a phase of state space reduction is required before model checking tools are applied. The semantic definition of a VHDL subset in SIGNAL and the construction of an interface between SIGNAL and VHDL is the basis for a research on high level synthesis and verification of hybrid systems [32]. The use of theorem provers has also been explored. A model of a small subset of VHDL was made in HOL [33], and a larger subset in ML for linking to Lambda [34], both describing " delta cycle " state changes. The NQTHM definition of [35] is very different from ours: it defines a symbolic simulator for structural timed models, on which properties can be proven; to our knowledge, no automatic translator from VHDL source has been implemented.

Seen from the viewpoint of tool integration, the definition of an intermediate format common to independently designed tools is crucial. The " Synchrone " working group in France has defined a common format to integrate the " reactive synchronous " languages Lustre, Esterel and Signal. The VIS system translates Verilog descriptions into the Blif-MV format, to allow the cooperation of synthesis and model checking on non-deterministic FSM's [36,37]. Translating from PIF to Blif-MV would give a VHDL input to VIS, and make it possible to model check non deterministic VHDL models (e.g. through the use of *shared* variables), currently rejected in Prevail. A more elaborate common format is currently under study for the specification of refinements and proof obligations in design verification, to be used as input to cycle simulation, model checking and theorem proving [38].

The concept of verification environment for VHDL has also been implemented in the FORMAT project, which integrates tautology checking, model checking and the Lambda interactive theorem prover, and provides automatic translation from the output of LVS to the input of these tools; a graphical interface is provided, in particular for editing the symbolic timing diagrams used as input to the model checker [39]. FORMAT, however, includes no automatic diagnosis tool.

It is our intention to use Prevail in teaching and in formal technology transfer towards industry. New tools will be incorporated as they become available (as a test of the openness of the system, we imported VFORMAL in Prevail in a few weeks). Our future research works will focus on specification languages, automatic diagnosis, links between verification and high level synthesis and the verification of libraries.

Acknowlegements

The development of PREVAIL has been supported by the ESPRIT Basic Research

Action CHARME, and by the EUREKA JESSI-AC3 project. The authors are grateful to their colleagues of CHARME for many helpful discussions and friendly cooperation over many years.

References

[1]   IEEE: " *Standard VHDL Language Reference Manual* ", IEEE Standard 1076-1993, 1993

[2]   Thomas D. E., Moorby P. R.: *" The Verilog Hardware Description Language "*, Second edition, Kluwer, 1995

[3]   Borrione D., Pierre L., Salem A.: " Formal verification of VHDL Descriptions in the Prevail Environment ", *IEEE Design and Test of Computers*, Vol 9, N°2, pp. 42-56, 1992

[4]   Eveking H.: "Axiomatizing hardware description languages". *International Journal of VLSI Design*, Vol.2, N° 3, pp. 263-280, 1990.

[5]   A.Bartsch, H.Eveking, H.J.Faerber, M.Kelelatchew, J.Pinder, U.Schellin : "LOVERT- A Logic Verifier of Register-Transfer Descriptions". In *Formal VLSI Correctness Verification*, L.Claesen Ed., North Holland (1990), ISBN 0444 88689 3.

[6]   O.Coudert, C.Berthet, J.C.Madre : "Verification of synchronous sequential machines based on symbolic execution". In *Automatic Verification Methods for Finite State Systems*, LNCS n°407. Spinger Verlag 1989 (pp 365-373).

[7]   J.R.Burch, E.M.Clarke, K.L.McMillan, D.L.Dill : "Sequential circuit verification using symbolic model ckecking", Proc. 27th Design Automation Conf, 1990.

[8]   Clarke E.M., Emerson E.A., Sistla A.P.: " Automatic Verification of Finite-State Concurrent Systems Using Temporal Logic Specifications ", *ACM Trans. on Programming Languages and Systems,* Vol 8, N° 2, pp. 244-263, 1986.

[9]   Deharbe D., Borrione D.: "Symbolic Model Checking with Past and Future Temporal Modalities: Fundamentals and Algorithms". Current Issues in Electronic Modeling, Vol 1 on *Model Generation in Electronic Design*, Kluwer, March 1995.

[10] Höreth S.: " Improving the performance of a BDD-based tautology checker," Proc. Advanced Research Workshop on *Correct Hardware Design Methodology*, Turin, North-Holland, 1991

[11] Boyer R. S., Moore J. S.: "*A Computational Logic Handbook* ", Academic Press Inc., 1988

[12] Hunt W.A.: " FM8501: A Verified Microprocessor ", Technical Report 47, University of Texas at Austin, 1986

[13] Wahba A., Borrione D.: " A Method for Automatic Design Error Location and Correction in Combinational Logic Circuits", to appear in *Journal of Electronic Testing: Theory and Applications*, Kluwer.

[14] Wahba A., Borrione D.: "Design error diagnosis in sequential circuits." Proc. Advanced Research Working Conference on *Correct Hardware Design and Verification Methods*, Frankfurt, 2-4 October 1995. LNCS, Springer Verlag.

[15] Debreil A., Oddo P.: " Synchronous Designs in VHDL ", Proc. EuroDAC/EuroVHDL, pp.486-491, 1993

[16] Deharbe D., Borrione D.: "Semantics of a verification. Oriented subset of VHDL." Proc. Advanced Research Working Conference on *Correct Hardware Design and Verification Methods,* Frankfurt, 2-4 October 1995. LNCS, Springer Verlag.

[17] Bouamama H, Borrione D.: "VHDL subset for parameterized models: the proof-oriented intermediate form Version 1.1", technical report, JESSI-AC3, 1994

[18] L. Yang, D. Gao, J. Mostoufi, R. Joshi, and P. Loewenstein, ``System Design Methodology of UltraSPARC^{TM}-I,'' *Proceedings of the 32nd Design Automation Conference DAC'95*, pp. 7-12, 1995.

[19] T. W. Albrecht, ``Concurrent Design Methodology and Configuration Management of the SIEMENS EWSD-CCS7E Processor System Simulation,'' *Proceedings of the 32nd Design Automation Conference DAC'95*, pp. 222-227, 1995.

[20] A. Aharon, D. Goodman, M. Levinger, Y. Lichtenstein, Y. Malka, C. Metzger, M. Molcho, and G. Shurek, ``Test Program Generation for Functional Verification of PowerPC Processors in IBM," *Proceedings of the 32nd Design Automation Conference DAC'95*, pp. 279-285, 1995.

[21] M. S. Abadir, J. Ferguson, and T. E. Kirkland, ``Logic Design Verification via Test Generation," *IEEE Transactions on Computer-Aided Design*, Vol. 7, No. 1, pp. 138-148, Jan. 1988.

[22] M. R. Garey, and D. S. Johnson, ``Computers and Interactibility: A Guide to the Theory of NP-Completeness," Freeman, New York, 1979.

[23] E. J. Aas, K. A. Klingsheim, and T. Steen, ``Quantifying Design Quality: A Model and Design Experiments," *Proc. EURO-ASIC'92*, pp. 172-177, 1992.

[24] F. Brglez, and H. Fujiwara, "A neutral netlist of 10 combinatorial benchmark circuits and a target translator in FORTRAN," in *Proc. IEEE Int. Symp. Circuits and Systems*, pp. 663-698, June 1985.

[25] F. Brglez, D. Bryan and K. Kozminski, "Combinational Profiles of Sequential Circuits," *Proc. IEEE International Symposium of Circuits and Systems (ISCAS'89)*, Portland, OR, May 1989.

[26] Borrione D., Bouamama H., Suescun R.; "NQTHM Library for VHDL", JESSI-AC3 Tech. Report, 1995

[27] CLSI-Solutions: " VFORMAL User's Manual ", Version 1.0, 1993

[28] D.Eisenbiegler, R. Kumar, and J. Muller, "A Formal Model for a VHDL Subset of Synchronous Circuits", in Proc. of APCHDL'96, Bengalore, India

[29] D. Déharbe: "The CV system". URL http://www.cs.cmu.edu/~deharbe/project.html

[30] Olcoz S., Colom J.M.: "A Colored Petri Net Model of VHDL", *Formal Methods in System Design*, Vol 7, N° 1-2, pp.101-123, 1995.

[31] E. Encrenaz: "Une methode de verification de propriétés de programmes VHDL basée sur des modèles formels de reseaux de Petri", PhD Thesis, Universite Paris VI, Dec 1995 (in French)

[32] M. Belhadj : "Conception d'architectures en utilisant SIGNAL et VHDL", PhD Thesis, Universite de Rennes 1, Dec. 1994 (in French)

[33] J.P. van Tassel: "Femto-VHDL: The semantics of a subset of VHDL and its embedding in the HOL theorem prover", PhD thesis, University of Cambridge, 1993

[34] G. Humbreit: "Providing a VHDL interface for proof systems", Proc. EDAC, Paris, 1992

[35] D. Russinoff: " A formalization of a subset of VHDL in the Boyer-Moore Logic", *Formal Methods in System Design*, Vol 7, No 1/2, Kluwer, August 1995

[36] R. Brayton et al.: "VIS: A System for Verification and Synthesis", Proc. Computer Aided Verification, CAV'96 (to appear)

[37] VIS Development Group: "Description of BLIF-MV, An Intermediate Format for Verification and Synthesis of Hierarchical Networks of FSMs", Tech. Report, U.C. Berkeley, CAD Group, hhtp://www-cad.eecs.berkeley.edu/Respep/Research/vis

[38] K.L.McMillan: private communication.

[39] W. Damm, G. Dohmen, P. Kelb, H. Pargmann, R. Schlör, R. Herrmann: " Verification Flow ", Seminar 5 on " Specification and verification of VHDL-based System-level Hardware Designs " Proc. of APCHDL'96, Bengalore, India

[40] J. Lohse, J. Bormann, M. Payer, G.Venzl: "VHDL-Translation for BDD-based Formal Verification". Siemens Internal Report, München, Germany, 1994.

[41] J. Bormann, J.Lohse, M. Payer and G.Venzl: "Model Checking in Industrial Hardware Design", Proc of DAC' 95.

List of Authors

Springer
and the
environment

At Springer we firmly believe that an international science publisher has a special obligation to the environment, and our corporate policies consistently reflect this conviction.
We also expect our business partners – paper mills, printers, packaging manufacturers, etc. – to commit themselves to using materials and production processes that do not harm the environment. The paper in this book is made from low- or no-chlorine pulp and is acid free, in conformance with international standards for paper permanency.

 Springer

Lecture Notes in Computer Science

For information about Vols. 1–1083

please contact your bookseller or Springer-Verlag